22nd Offshore Symposium 2017

Redefining Offshore Development: Technologies and Solutions

Houston, Texas, USA
2 February 2017

ISBN: 978-1-5108-4361-5

Printed from e-media with permission by:

Curran Associates, Inc.
57 Morehouse Lane
Red Hook, NY 12571

CURRAN ASSOCIATES INC. proceedings.com

Some format issues inherent in the e-media version may also appear in this print version.

Printed by Curran Associates, Inc. (2017)

For permission requests, please contact Society of Naval Architects and Marine Engineers (SNAME) Texas Section at the address below.

Society of Naval Architects and Marine Engineers (SNAME) Texas Section
99 Canal Center Plaza, Suite 310
Alexandria, VA 22314
USA

Phone: +1-703-997-6701

www.sname.org/texassection

Additional copies of this publication are available from:

Curran Associates, Inc.
57 Morehouse Lane
Red Hook, NY 12571 USA
Phone: 845-758-0400
Fax: 845-758-2633
Email: curran@proceedings.com
Web: www.proceedings.com

2016 OFFSHORE SYMPOSIUM MANUSCRIPT LIST

The Texas Section
The Society of Naval Architects and Marine Engineers

The 22nd Offshore Symposium

Redefining Offshore Development: Technologies and Solutions

Houston, Texas
February 02, 2017

Event Sponsors

SNAME
INTERNATIONAL COMMUNITY
FOR MARITIME AND OCEAN PROFESSIONALS

SILVER SPONSOR

BUREAU VERITAS

EVENT SPONSORS

Evening Reception: SOFEC

Delegate Bags: MARIN

Afternoon Coffee Break: Halton

Lanyards: WOOD GROUP

CERRADO

MARIN

WOOD GROUP
Delivering successful offshore projects for three decades.
www.woodgroup.com

THE SOCIETY OF NAVAL ARCHITECTS AND MARINE ENGINEERS

TEXAS SECTION

SNAME

THE INTERNATIONAL COMMUNITY
FOR MARITIME AND OCEAN PROFESSIONALS

*** Formatting Issue - Best Available Paper/Graphic ***

SNAME

THE INTERNATIONAL COMMUNITY
FOR MARITIME AND OCEAN PROFESSIONALS

A POWER LAW FORMULATION EQUIVALENT TO THE STEP FUNCTION FOR USE IN WIND TUNNEL TESTING

JOHN ANDREW BREUER (M)
PRESIDENT, BREUER CONSULTING

BRIAN CHEATER (M)
DIRECTOR OF NAVAL ARCHITECTURE, GUSTOMSC US INC

CARL CHAPMAN (AM)
OFFSHORE CONSULTANT

TIRTHARAJ BHAUMIK (M)
SENIOR NAVAL ARCHITECT, KBR-GVA

MARCO MASCIOLA (M)
SENIOR ENGINEER, ABS

ANDREW PEAK (M)
PRINCIPAL NAVAL ARCHITECT, AQUALIS

TAO WANG (M)
LEAD NAVAL ARCHITECT, AKER SOLUTIONS

JAMES N. BREKKE (M)
CHAIR OFFSHORE T&R COMMITTEE, BREKKE OFFSHORE CONSULTING LLC

Proceedings of the 22nd Offshore Symposium, February 2017, Houston, Texas
Texas Section of the Society of Naval Architects and Marine Engineers

*** Formatting Issue - Best Available Paper/Graphic ***

Proceedings of the 22nd Offshore Symposium, February 2017, Houston, Texas
Texas Section of the Society of Naval Architects and Marine Engineers

ABSTRACT

This paper presents a historic account of the step function which has become the basis for calculating wind loads in stability analysis. The step function used in conjunction with accepted stability criteria provides a proven level of performance but it is discontinuous and cannot be directly modelled physically. For wind tunnel testing, a continuous power law formulation is required. This paper traces the origin and development of the step function. A power law with a reference height of 50feet and exponent of 0.1 was originally used to develop the profile; this was projected downward at 50-foot intervals causing discrete steps. This paper proposes a continuous power law formulation of the resulting step function with the commonly used 10m reference height. The exponent is based on a least-squares fit to the original height coefficients. The paper also examines the effect of different exponents on the wind over-turning moment and wind forces for a range of vessel types with different area versus height profiles. These are calculated using both a spreadsheet and available industry software. It was found that a power law of 10m-0.0883 provides the best fit to the final step function and is a continuous function that is recommended for wind tunnel testing for stability purposes.

Keywords: SNAME, Offshore, Step Function, Power Law, Wind Tunnel Testing, Stability

THE BOUNDARY LAYER PROFILE

The increase of wind velocity experienced as we rise above the sea level has been established early in history. The relation between wind speed and height is called the wind profile or wind gradient. The phenomenon, in the atmospheric layer immediate to the earth is known as the Atmospheric Boundary Layer. This velocity gradient is mostly caused by friction between the atmosphere and the ground or sea surface. The relationship between wind speed and elevation has been expressed by a number of authors using formulations that are commonly known as the Boundary Layer Profile (BLP).

The first published standard for floating offshore units is the 1966 – US Coast Guard Merchant Marine Technical Note No. 6-66 (Ref 1 - MMTN). This document includes a stability criterion based on a 100-knot wind and a boundary layer profile based on what is known as the "Power Law" expressed as:

$$\frac{}{_0} = \left(\frac{}{_0}\right)^{\propto} \qquad (1)$$

Where:

is the velocity at elevation h;

0

is the nominal velocity at a designated reference elevation h_0;

is the elevation above the sea level;

0 is the reference height where the nominal velocity is established.

Application of the above BLP was constrained by the practical calculation considerations of the time.

To simplify the methodology, the formulation was converted into tabular values calculated at regular intervals and this gave way to what is today known as the "Step Function" (SF).

A number of more sophisticated logarithmic formulations which account for turbulence intensity have since been adopted in codes such as API 2MET. These may be more physically accurate but the SF is used in regulations and will continue to be so. However the SF in discontinuous and cannot be physically modeled. This paper will provide a continuous function that can be physically modeled. This will allow for calibration of wind tunnel tests against the regulatory requirements as stated for example in the US CFR Regulations.

Wind and its Boundary Layer have multiple applications for offshore engineering but the three main ones are Stability, Structural, and Station Keeping. Stability for offshore units is unique in that wind is the only environmental load considered and consequently the stability evaluation is strictly a benchmark. On the other hand, structural analysis and station keeping consider wind as one of several loads contemplated in the design and their analysis is a strict application of first principles of engineering.

The benchmark evaluation of the stability of offshore units has been time tested with great success to the point that no catastrophic events have been determined to be caused by the basic stability concerns addressed by the standards applied in the design.

TRACKING THE EARLY REGULATORY HISTORY

Offshore drilling started near-shore in the early 1900s and by 1947 had moved offshore beyond the sight of land. Growth in the offshore industry prompted ABS to convene a special committee on MODUs and prompted the USCG to publish Merchant Marine Technical Note 6-66.

1966 – US COAST GUARD MERCHANT MARINE TECHNICAL NOTE NO. 6-66 (MMTN)

MMTN 6-66 is the first publication to propose a boundary layer profile (BLP). The document includes, a BLP defined by a power law, and introduces the idea to correct the winds loads by using a Height Correction C_h formulated as follows:

$$= \left(\frac{}{_0}\right)^{2} = \left(\frac{}{30'}\right)^{\frac{2}{7}} \qquad (2)$$

The following items are noted:

- An exponential equation or power law is used to evaluate the Boundary Layer;
- The reference height h_0 used: 30' (9.14m);
- The exponent used: 1/7 = 0.143;
- The MMTN indicates that each ". . .area should be broken down into suitable blocks and the proper height correction factors applied to each block". Design values are tabulated as shown below in Table 1.

Table 1 – Height Coefficients per MMTN 6-66

h (ft)	*Ch*	$(\frac{\ }{30})^{\frac{2}{7}}$
0	0.000	0.000
30	1.000	1.000
60	1.220	1.219
80	1.330	1.324
100	1.415	1.411
120	1.485	1.487
140	1.550	1.554
160	1.610	1.614
180	1.665	1.669
200	1.710	1.720

- Some differences are noted due to rounding between calculated and tabular values; the rounding is believed to be due to limitations on the calculation methods used available at the time (slide rule).
- The sequence of heights in the table is not even and its highest value (200') falls short of the typical height of a modern MODU.

1968 - Classification Rules for MODUs first published by ABS

The development of the ABS Rules (Ref 2) had the benefit of industrywide input through the Special Committee on MODUs. This provides a mechanism for accumulated industry experience to be maintained and for updates to be introduced as necessary. The following items are of note:

- The increase of C_h (and the wind velocity) is a linear function of the elevation h (in feet) that may be expressed as:

$$= \frac{1}{500} \quad + 0.9 \qquad (3)$$

- The concept of a "Step Function" Boundary Layer Profile is introduced.
- The table of C_h is provided for evenly spaced (50') intervals (steps) as shown below in Table 2.

Table 2 – Height Coefficients per ABS 1968

- The Rules have no specific directive but most designers have applied it by breaking every exposed surface into 50-foot bands coinciding with the range of elevations of each "interval" in the table.
- The maximum elevation (250') is higher than the one in the MMTN but it falls short of the MODU under design in the late 60's.
- The C_h between 0 and 50 feet elevation is 1.0. Theoretically, the wind velocity at sea-level is nil; yet, the nominal velocity, established at the top of this range applies to the entire band.

Following the previous comment, each 50-foot band, has been assigned a C_h that corresponds to the upper limit of the band. Thus the linear distribution was converted to a SF where the C_h corresponds to the elevation at the top of the interval. This, while conservative, has a small impact on the areas in the 0-50 feet band and a negligible impact on the bands above 50 feet.

1973 – Classification Rules for MODUs revised by ABS

The ABS Special Committee on MODU, identified a great number of improvements to the Rules in force since 1968 and a revised edition of the Rules (Ref 3) was published. The following items are of note:

- The boundary layer profile is still presented in the form of an SF with a similar trend as the MMTN 6-66 but with slightly different values;
- The SF was presented only in tabular format, the analytical formulation was not published;
- The velocity reference height is 50 feet unlike the MMTN (30 feet);

- The SF is subsequently adopted by all other classification societies, incorporated into the Code of Federal Regulations, the IMO MODU Code, and most Coastal State regulations;
- The table (shown below) is extended to 256m (850') elevation – a height impossible to conceive in any MODU then but plausible today.

Table 3 – Height Coefficients per ABS 1973

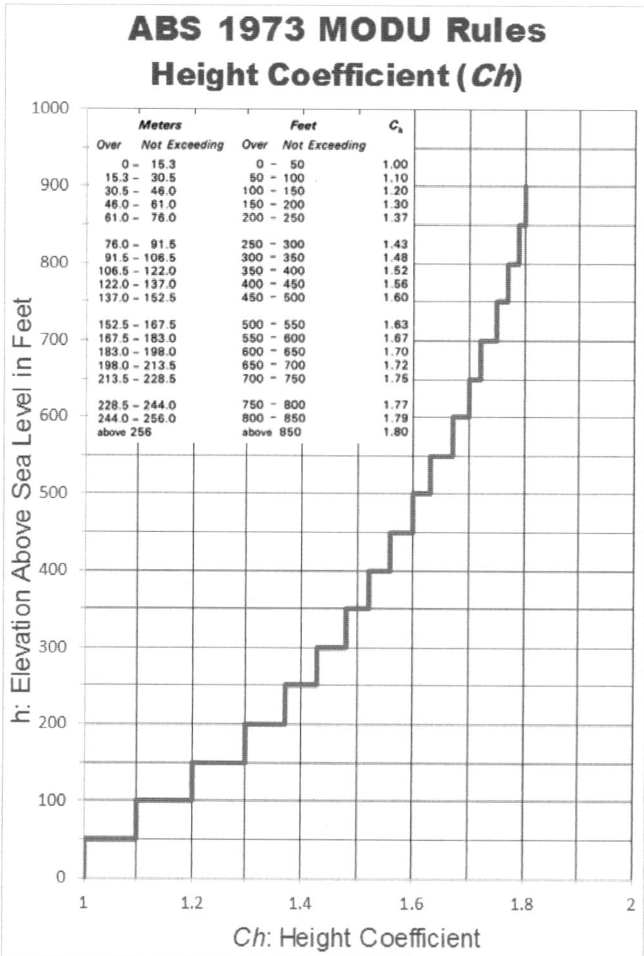

Meters Over – Not Exceeding	Feet Over – Not Exceeding	C_h
0 - 15.3	0 - 50	1.00
15.3 - 30.5	50 - 100	1.10
30.5 - 46.0	100 - 150	1.20
46.0 - 61.0	150 - 200	1.30
61.0 - 76.0	200 - 250	1.37
76.0 - 91.5	250 - 300	1.43
91.5 - 106.5	300 - 350	1.48
106.5 - 122.0	350 - 400	1.52
122.0 - 137.0	400 - 450	1.56
137.0 - 152.5	450 - 500	1.60
152.5 - 167.5	500 - 550	1.63
167.5 - 183.0	550 - 600	1.67
183.0 - 198.0	600 - 650	1.70
198.0 - 213.5	650 - 700	1.72
213.5 - 228.5	700 - 750	1.75
228.5 - 244.0	750 - 800	1.77
244.0 - 256.0	800 - 850	1.79
above 256	above 850	1.80

1973 Onward

The 1973 standards were adopted by other agencies and became an industry standard for wind load calculations. The following items are of note:

- The SF presented in the 1973 Rules is subsequently adopted by all other classification societies, the IMO MODU Code, and most Coastal State regulations, and API.
- In 1978, the USCG issued a broad set of requirements under the cover of NVIC 04-78. These

requirements were expanded and incorporated into the Code of Federal Regulations Title 46, Sub-chapter I-A and Sub-Chapter S, which contains the stability requirements for MODUs. Part of this included wind force and moments and the C_h Step Function from the 1973 ABS Rules.

- The table of height coefficients is identical to the one published by ABS but it only applies to afloat stability.
  Some agencies also incorporated a C_h power law distribution with a reference height
- = 10m (32.8') and an exponent α = 0.1
- Concurrently, API[5] developed a BLP for mooring and structural purposes with a power law having a reference height (h_0) of 10m (32.8') and an exponent α = 0.125.
- ABS and others have since published Rules for Floating Production Installations (FPI) and for mooring analysis have adopted a power law with a reference height h_0 = 10m (32.8') and an exponent α = 0.09 to 0.16 depending on the site where the FPI operates.
- ABS and others have continued with annual meetings of the special committee on MODUs to upgrade the requirements for classification. But the stability requirements are codified in the CFR Regulations and other than the incorporation of the metric system, the C_h table has not been updated in almost five decades.

METHODOLOGY TO DERIVE AN EQUIVALENT BLP

Having established the need for a continuous BLP function to use in wind tunnel testing and the obvious simplification for wind software development the authors propose the following:

Formulation Type

A power law formulation that is consistent with the step function should be used.

Comparison and Validation

The power law function and corresponding height coefficient in the formulae below shall be determined using a least-square regression analysis between the power law and the tabulated step function values.

$$\frac{}{_0} = [\frac{}{_0}] \quad (4)$$

$$= [\frac{}{_0}]^2 \quad (5)$$

Regression Analysis

The authors have considered two (2) reference heights and two distinct goals:

1. Establish the profile that best describes the basis curve for the step function;

2. Establish the profile that best emulates the results from the application of the step function using a very fine 1-foot interval up to a height of 900 feet.

Proceedings of the 22nd Offshore Symposium, February 2017, Houston, Texas
Texas Section of the Society of Naval Architects and Marine Engineers

Validation

For the purposes of validation, the step function and power law formulations were applied to the same set of geometries to compare the resultant force and over-turning moments to ensure they had similar overall results.

The following geometries were considered:

1. 100m x 1m rectangle (base-line for peer-to-peer comparisons)
2. Four different MODU models
   a. Jack-up (see Figure 1A)
   b. Moored Semi-submersible (see Figure 1B)
   c. Dynamically Positioned Semi-submersible (see Figure 1B)
   d. Dynamically Positioned Drillship (see Figure 1C)

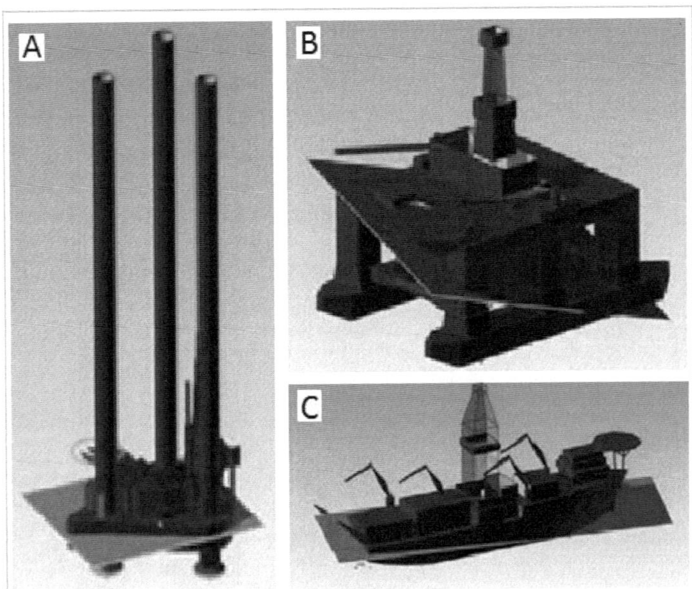

Figure 1 – Eagle Wind Models of Representative MODUs

Note, the rectangle is included for peer-to-peer comparisons of the different power law formulations and performing sensitivity studies but real MODUs will have a distribution of wind area with height and so each type of MODU must be compared separately as well.

When comparing results, the following must be understood.

a. Overturning moments are the only concern when examining stability and comparisons must be made for all angles of heel, trim, and wind direction;
b. Wind drag (force) is relevant when analyzing structural and mooring loads but only for the upright hull. Thus, forces for conditions with heel and trim are not significant for these.

***** Formatting Issue - Best Available Paper/Graphic *****

*** Formatting Issue - Best Available Paper/Graphic ***

RESULTS OF THE DERIVATION OF AN EQUIVALENT BLP

Best Fit to Original Basis Curve

When using a reference elevation h_0 = 50 feet as implied in the original step function (SF), the best fit is achieved (see Appendix I) using α = 0.1019.

The curve fit (see Appendix I) is notably close which confirms a power law with a 50ft reference height was the original source. A sensitivity analysis (Appendix II) demonstrates that decimals beyond the third place for the exponent α are irrelevant. Consequently, it can be rounded to 0.102 for engineering purposes and the final form is given by:

$$= \left[\frac{}{50'}\right]^{0.102} \qquad (6)$$

Best Fit to Resultant Loads

Of the various possible approaches to determine this profile, it was agreed that a profile that would best fit the midpoints of each step was the most rational (See Figure 2 below).

Figure 2 – Excess Caused by Downward Projection of Steps from Basis Curve

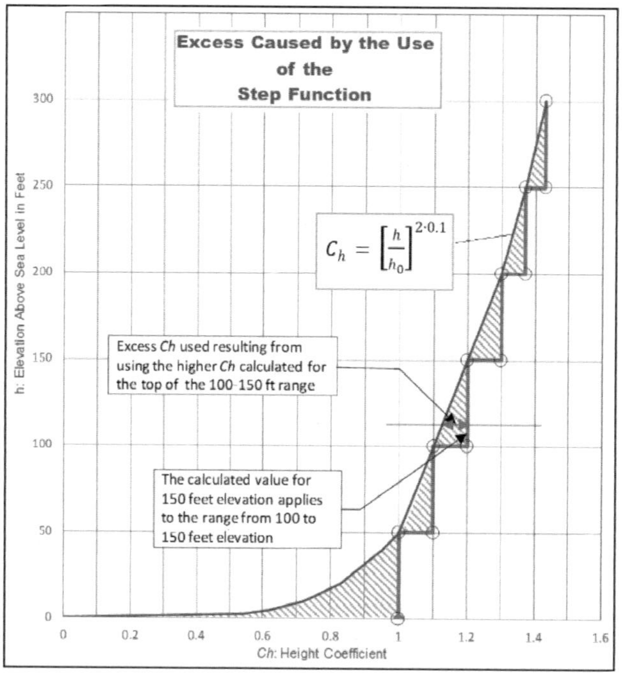

A variety of exponents were examined but when using a reference elevation h_0 = 10m the best overall fit is achieved (see Appendix I) using $\alpha = 0.0883$. This reference height is not related to the original step function but is used frequently elsewhere. For reference heights of 10m, the profile below in Equation 7 is an equivalent continuous formulation of the step function.

$$= \left[\frac{\;}{10}\right]^{0.088} \qquad (7)$$

Comparison to 100m High Rectangle

The results for the rectangle are given in Appendix II.

A height of 100m was selected to be similar to most conventional offshore units. The simplicity of a single rectangle allows an easy comparison with minimum parameters.

Some of the notable findings for the 100m x 1m rectangle from Appendix II are discussed below.

1. A comparison between the forces and moments using the h_0/α combinations from the above regression analysis (Case 1 and Case 2) shows that the SF calculated forces and moments are slightly higher than the theoretical result obtained by applying the formulation from Ref [6].

   In applying the step function, C_h is selected at the top of each 50-foot interval and it is projected down over the entire 50-foot strip which results in increased forces and moments at the lowest range; this excess is 20% in forces and 10% in moments.

2. A sensitivity analysis presented in Appendix III shows that disregarding the fourth decimal

modifies the resulting wind forces and moments by less than 0.5%. However, ignoring the third decimal can cause discrepancies of 5% or more.

3. None of the calculations using a BLP obtained by regression exceeds the values calculated with the step function. Case 1, determined to be the source of the values in the SF, falls short by 5% in in moments, a discrepancy that matches the theoretically predicted values.
4. While the step function is unquestionably based on a reference height h_0 = 50feet, a comparison of cases 1 through 5 shows that using an h_0=10m adds a conservatism that makes results close to the SF. However, as noted above, this is only applicable to a rectangular strip and not to actual MODUs.

Calculations in Appendix II were carried out in order to compare results for a simple rectangle using the Step Function (as defined in the regulations) and for five (5) different cases of proposed power law approximations:

1. BLP h_0 = 15.24m - α = 0.1019 (50'/0.1019),
2. BLP h_0 = 10m - α = 0.0864 (50'/0.0.084),
3. BLP h_0 = 15.24m - α = 0.1039 (32.8'/0.1039),
4. BLP h_0 = 10m - α = 0.0883 (32.8'/0.0883), and
5. BLP h_0 = 10m - α = 0.1 (32.8'/0.1).

The results from Appendix II show that the wind moments are best approximated with a reference height of 10m and an exponent of 0.0883.

Proceedings of the 22nd Offshore Symposium, February 2017, Houston, Texas
Texas Section of the Society of Naval Architects and Marine Engineers

Comparison to Representative MODUs

Calculations in Appendix III were carried to compare results using representative MODU geometries and the Step Functions (as defined in the regulations) and for five (5) different cases of proposed power law approximations:

1. BLP h_0 = 15.24m - α = 0.1019 (50'/0.1019),
2. BLP h_0 = 10m - α = 0.0864 (50'/0.0.084),
3. BLP h_0 = 15.24m - α = 0.1039 (32.8'/0.1039),
4. BLP h_0 = 10m - α = 0.0883 (32.8'/0.0883), and
5. BLP h_0 = 10m - α = 0.1 (32.8'/0.1).

Several comparisons were made applying the SF and the power function. Wind loads and overturning moments were calculated using the Eagle Wind Software by ABS (Ref 5).

The jack-up and the semisubmersibles were calculated for three wind directions (0, 45, and 90 degrees) while the drillship was only evaluated for the broadside wind (0 degrees). All models were calculated with two angles of heel (even keel and 15 or 30 degrees).

The sixth boundary layer with a reference height of 10m and an exponent α = 0.1 is frequently part of other applications and the results will show that this profile is more conservative than all the cases based on regression and the Step Function.

In all models, the Step Function is to some degree more conservative than the combinations of h_0/α derived from regression of the SF data. Again, this is to be expected as the conversion from the BLP to a SF incorporates a conservatism that leads to higher results.

Jack-up

The model (Figure 1-A) of a typical modern unit is the "tallest" of all models analyzed. While its height of 440 Feet (about 135m) above the water is not the most extreme but it represents most jack-ups in the world fleet. Exceptionally legs may extend as high as 600 Feet (180 m). This makes the upper 300 feet of the SF not applicable to existing jack-ups.

Given the similar results, the cases for BLP with h_0 = 10m reference heights reflected in the table can be used as an alternative to the SF for this type of unit.

Semisubmersible (DP and conventionally moored)

The model (Figure 1-B) used corresponds to a large modern semisubmersible with a "box" deck.

The results using the BLP with a reference height h_0 = 50 feet are very close on overturning moments but fall short by as much as 6% in forces (only the upright hull is relevant). The cases where h_0 = 10m exceed in the overturning moment columns and are very close on forces.

Drillship

The model (Figure 1-C) corresponds to a typical modern large drillship including dynamic positioning. The most notable aspect of a drillship is that it is lower in elevation and most of the exposed area lie close to the waterline.

Following the same trend as the above cases, the BLP with h_0 = 10m yield results close to those obtained by applying the regulations. The results for the unit at 30-degree list appear exaggerated but for these cases most regulations allow using the overturning moments for the upright hull affected by the cosine of the angle of heel. Wind forces are a small percentage lower than those obtained using the SF. This is caused by the built-in margin of the SF that is up to 20% higher in forces at the lower elevations.

Final

Based on the calculations presented in Appendix IV, a power law with a reference height of 10m and an exponent of 0.0883 (see Equation 7) provides the best overall fit for the variety of unit types found in the industry.

CONCLUSIONS AND RECOMMENDATIONS

- The Current Step Function found in various rules and regulations is the foundation for the stability analysis of all existing MODU designs and is embedded in almost all statutes. As such, it must be retained as a choice for design and approval purposes. However, a continuous function that can be used in wind tunnels must be derived.
- A reference height of 50feet and an exponent of 0.1 defines a curve which can be cut and projected downward over 50-foot intervals to create a step function. This yields the table of height coefficients versus range of heights that matches those specified in the regulations.
- Using the common reference height of 10m, an exponent of 0.0883 provides the best-fit mid-points of the height coefficients in the step function. Forces and moments calculated with this compare well to the step function for all vessel designs. Consequently, this is recommended for use in wind tunnel testing for stability purposes.

*** Formatting Issue - Best Available Paper/Graphic ***

APPENDIX I

Least Square Regression of the Height Coefficient Step Function

Table I-1: *Least Square Regression of the Power Law exponent for the Published Height Coefficients (Ch) for h_0=50 feet.*

α Exponent	Reference Height h_o (Feet)
0.1018928	50

h_1 (Feet)	h_1 (Meters)	Ch Step Function	Ch Calculated w/Above Parameters	Ch ho = 50Ft α = 0.1	Ch ho = 10m α = 0.1	Ch ho = 10m α = 0.0833	Δ Delta calculated and step	$\Delta^2 C_h$
0	0.00	0.00	0.000	0.000	0.000	0.000	0.000000	0.000000
50	15.24	1.00	1.000	1.000	1.088	1.073	0.000000	0.000000
100	30.48	1.10	1.152	1.149	1.250	1.204	0.051716	0.002675
150	45.72	1.20	1.251	1.246	1.355	1.288	0.050923	0.002593
200	60.96	1.30	1.326	1.320	1.436	1.352	0.026451	0.000700
250	76.20	1.37	1.388	1.380	1.501	1.403	0.018162	0.000330
300	91.44	1.43	1.441	1.431	1.557	1.446	0.010708	0.000115
350	106.68	1.48	1.487	1.476	1.606	1.484	0.006685	0.000045
400	121.92	1.52	1.528	1.516	1.649	1.517	0.007695	0.000059
450	137.16	1.56	1.565	1.552	1.688	1.547	0.004807	0.000023
500	152.40	1.60	1.599	1.585	1.724	1.575	-0.001231	0.000002
550	167.64	1.63	1.630	1.615	1.757	1.600	0.000125	0.000000
600	182.88	1.67	1.659	1.644	1.788	1.623	-0.010713	0.000115
650	198.12	1.70	1.687	1.670	1.817	1.645	-0.013425	0.000180
700	213.36	1.72	1.712	1.695	1.844	1.665	-0.007761	0.000060
750	228.60	1.75	1.736	1.719	1.870	1.685	-0.013517	0.000183
800	243.84	1.77	1.759	1.741	1.894	1.703	-0.010528	0.000111
850	259.08	1.79	1.781	1.762	1.917	1.720	-0.008656	0.000075
900	274.32	1.80	1.802	1.783	1.939	1.737	0.002214	0.000005
							Minimum $\Sigma \Delta^2_{Ch}$ =	0.007269

***** Formatting Issue - Best Available Paper/Graphic *****

Table I-2: *Least Square Regression of the Power Law exponent for the Published Height Coefficients (Ch) for h_0 = 10m.*

UPDATE ALPHA

HREF = 32.8084 (feet)

ALPHA = 0.0883 SUM DIFF = 2.188093

MIN (feet)	MAX (feet)	MID (feet)	CH	CH_CALC	DIFF^2	% INCR
0	1	0.5	1	0.48	0.272008	-52.2
1	2	1.5	1	0.58	0.175859	-41.9
2	3	2.5	1	0.64	0.132983	-36.5
3	4	3.5	1	0.67	0.106187	-32.6
4	5	4.5	1	0.70	0.087226	-29.5
5	6	5.5	1	0.73	0.072889	-27.0
.	.	.	.	.	.	.
.	.	.	.	.	.	.
.	.	.	.	.	.	.
50	51	50.5	1.1	1.08	0.000443	-1.9
51	52	51.5	1.1	1.08	0.0003	-1.6
52	53	52.5	1.1	1.09	0.000186	-1.2
.	.	.	.	.	.	.
.	.	.	.	.	.	.
.	.	.	.	.	.	.
200	201	200.5	1.37	1.38	3.22E-05	0.4
201	202	201.5	1.37	1.38	4.74E-05	0.5
202	203	202.5	1.37	1.38	6.53E-05	0.6
.	.	.	.	.	.	.
.	.	.	.	.	.	.
.	.	.	.	.	.	.
500	501	500.5	1.63	1.62	0.000188	-0.8
501	502	501.5	1.63	1.62	0.000173	-0.8
502	503	502.5	1.63	1.62	0.000158	-0.8
.	.	.	.	.	.	.
.	.	.	.	.	.	.
.	.	.	.	.	.	.
897	898	897.5	1.8	1.79	7.29E-05	-0.5
898	899	898.5	1.8	1.79	6.7E-05	-0.5
899	900	899.5	1.8	1.79	6.14E-05	-0.4

Figure I-1: *Comparison between the various Boundary Layer Profiles discussed*
The Jack-up and Semisubmersible profiles are a scale representation of typical large MODU

*** Formatting Issue - Best Available Paper/Graphic ***

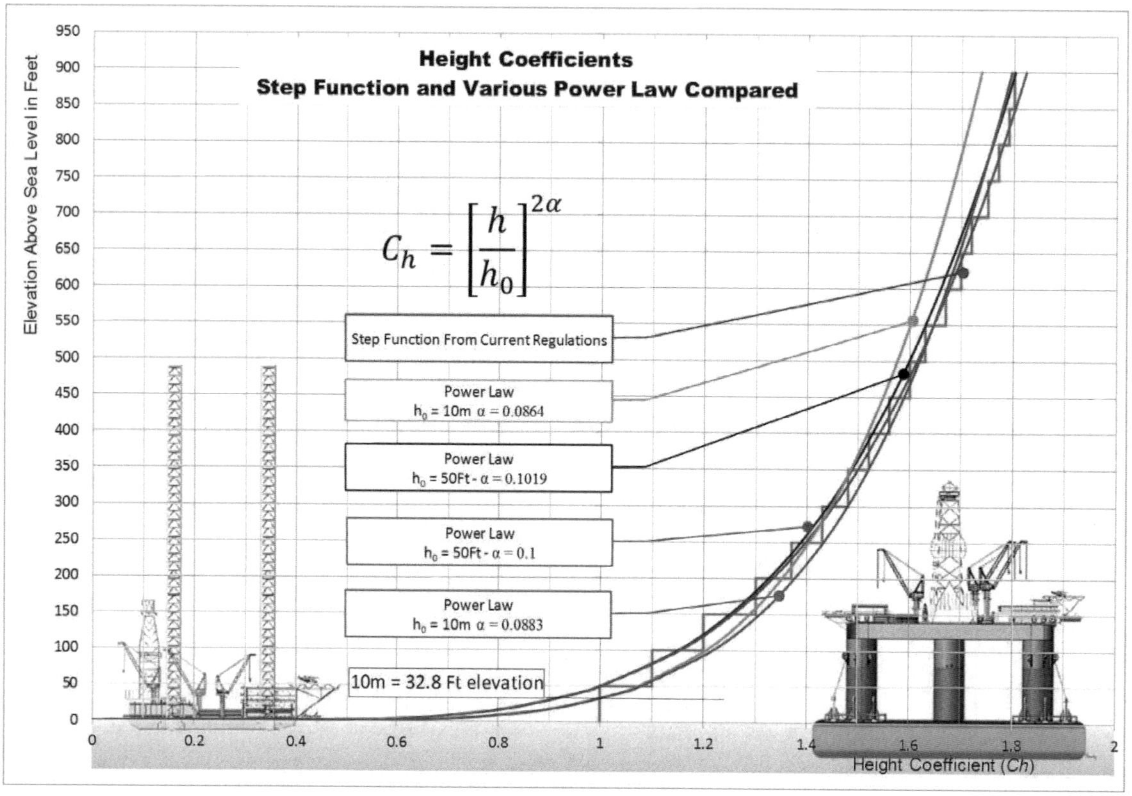

Appendix II

Comparison between Theoretical Calculations and the Step Function on a 100m x 1m Rectangle

Case 1

v_0 = 51.444m/s (100 knots), h_0 = 15.24m (50 Ft), exponent α = 0.1019

*** **Formatting Issue - Best Available Paper/Graphic** ***

100m x 1m Rectangle - 51.44m/s (100 Knots) Draft= 0 Heel=0					
Elevation Range	Area	Ch Height Coeff.	Ch x A	Arm	Moment
m	m^2	-	m^2	m	m^3
0.00-15.24	15.24	1.00	15.24	7.62	116.1
15.24-30.48	15.24	1.10	30.24	22.86	691.3
30.48-45.72	15.24	1.20	18.29	38.10	696.8
45.72-60.96	15.24	1.30	19.81	53.34	1056.8
60.96-76.20	15.24	1.37	20.88	68.58	1431.9
76.20-91.44	15.24	1.43	21.79	83.82	1826.7
91.44-100	8.56	1.48	12.67	95.72	1212.7
	100.00		138.92	50.62	7032.2

Total Force = 138.92 x 0.0623 x 51.44^2 = 22,901.1 Kg

Total Moment = 7032.2 x 0.0623 x 51.44^2 = 1,159,259.5 Kgm

Theoretical Calculations (See Reference [6])

$$ {}_1 = \frac{1}{2} \qquad \frac{{}^2}{{}^2_0} \cdot $$

$$ = \frac{1}{2 \quad + 1} \cdot \quad {}^{2\alpha+1} $$

$$ = \frac{1}{2 \quad + 2} \cdot \quad {}^{2 \quad +2} $$

$v_0 = 51.444$ m/s

$h_0 = 15.24$m $\alpha = 0.1039$

$\rho = 1.222$ kg_m/m^3

h = 100m b = 1m

Cs = 1.0

g = 9.81 m/s^2

$$ {}_1 = \frac{1}{2} \cdot 1.222 \cdot 1 \cdot \frac{51.444^2}{15.24^{0.2038}} = 928.150 $$

$$ = \frac{1}{2 \quad + 1} \quad {}^{2 \quad +1} = \frac{928.150}{2 \cdot 0.1019 + 1} \cdot 100^{1.2038} = $$

$$ = 197{,}090 \qquad = 20{,}091 $$

$$ = \frac{1}{2 \quad + 2} \quad {}^{2 \quad +2} = \frac{928.150}{2 \cdot 0.1019 + 2} \cdot 100^{2.2038} $$

$$ = 10{,}765{,}796 \qquad = 1{,}097{,}431 $$

Strict application of the SF in published regulations

*** Formatting Issue - Best Available Paper/Graphic ***

Case 2

v_0 = 51.444m/s (100 knots), h_0 = 10m (32.8 Ft), exponent α = 0.0864

100m x 1m Rectangle - 51.44m/s (100 Knots) Draft= 0 Heel=0					
Elevation Range	Area	Ch Height Coeff.	Ch x A	Arm	Moment
m	m^2	-	m^2	m	m^3
0.00-15.24	15.24	1.00	15.24	7.62	116.1
15.24-30.48	15.24	1.10	30.24	22.86	691.3
30.48-45.72	15.24	1.20	18.29	38.10	696.8
45.72-60.96	15.24	1.30	19.81	53.34	1056.8
60.96-76.20	15.24	1.37	20.88	68.58	1431.9
76.20-91.44	15.24	1.43	21.79	83.82	1826.7
91.44-100	8.56	1.48	12.67	95.72	1212.7
	100.00		138.92	50.62	7032.2

Total Force = 138.92 x 0.0623 x 51.44^2 = 22,901.1 Kg
Total Moment = 7032.2 x 0.0623 x 51.44^2 = 1,159,259.5 Kgm

Theoretical Calculations (See Ref. [6])	Strict application of the SF in published regulations

Theoretical Calculations (See Ref. [6])

$$_1 = \frac{1}{2} \quad \frac{^2}{^2_0} \cdot$$

$$= \frac{1}{2\ \ +1} \cdot \quad ^{2\alpha+1}$$

$$= \frac{1}{2\ \ +2} \cdot \quad ^{2\ \ +2}$$

v_0 = 51.444 m/s
h_0 = 10m α = 0.864
ρ = 1.222 kg_m/m^3
h = 100m b = 1m
Cs = 1.0
g = 9.81 m/s^2

$$_1 = \frac{1}{2} \cdot 1.222 \cdot 1 \cdot \frac{51.444^2}{10^{0.1728}} = 1086.202$$

$$= \frac{1}{2\ \ +1} \quad ^{2\ \ +1} = \frac{1086.202}{2 \cdot 0.864 + 1} \cdot 100^{1.1728} =$$

$= 205{,}251 \quad = 20{,}923$

$$= \frac{1}{2\ \ +2} \quad ^{2\ \ +2} = \frac{1086.202}{2 \cdot 0.864 + 2} \cdot 100^{2.1728}$$

$= 11{,}078{,}754 \quad = 1{,}129{,}333$

Strict application of the SF in published regulations

Proceedings of the 22nd Offshore Symposium, February 2017, Houston, Texas
Texas Section of the Society of Naval Architects and Marine Engineers

Case 3

v_0 = 51.444m/s (100 knots), h_0 = 15.24m (50 Ft), exponent α = 0.1039

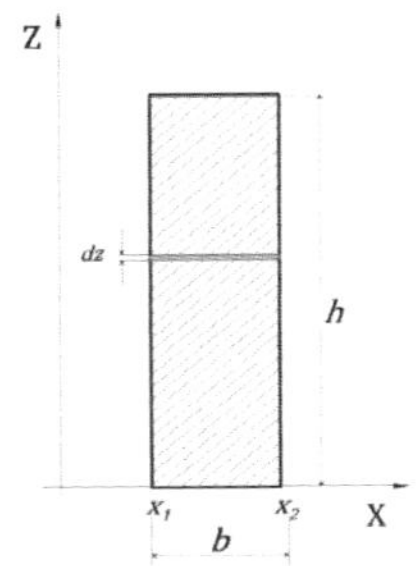

Theoretical Calculations (See Ref. [6])

$$_1 = \frac{1}{2} \qquad \frac{^2}{_0^2}\cdot$$

$$= \frac{1}{2 \quad +1}\cdot \quad ^{2\propto+1}$$

$$= \frac{1}{2 \quad +2}\cdot \quad ^{2 \quad +2}$$

v_0 = 51.444 m/s
h_0 = 15.24m α = 0.1039
ρ = 1.222 kg_m/m^3
h = 100m b = 1m
Cs = 1
g = 9.81 m/s^2

$$_1 = \frac{1}{2}\cdot 1.222 \cdot 1 \cdot \frac{51.444^2}{15.24^{0.2078}} = 918.092$$

$$= \frac{1}{2\propto+1} \quad ^{2\propto+1} = \frac{918.092}{2\cdot 0.1039+1}\cdot 100^{1.2078} =$$

$$= 197{,}920.8 \quad = 20{,}175$$

$$= \frac{1}{2 \quad +2} \quad ^{2 \quad +2} = \frac{918.092}{2\cdot 0.1039+2}\cdot 100^{2.2078}$$

$$= 10{,}827{,}462 \quad = 1{,}103{,}717$$

Strict application of the SF in published regulations

100m x 1m Rectangle - 51.44m/s (100 Knots) Draft= 0 Heel=0

Elevation Range	Area	Ch Height Coeff.	Ch x A	Arm	Moment
m	m^2	-	m^2	m	m^3
0.00-15.24	15.24	1.00	15.24	7.62	116.
15.24-30.48	15.24	1.10	30.24	22.86	691.
30.48-45.72	15.24	1.20	18.29	38.10	696.
45.72-60.96	15.24	1.30	19.81	53.34	1056.
60.96-76.20	15.24	1.37	20.88	68.58	1431.
76.20-91.44	15.24	1.43	21.79	83.82	1826.
91.44-100	8.56	1.48	12.67	95.72	1212.
	100.00		138.92	50.62	7032.

Total Force = 138.92 x 0.0623 x 51.44^2 = 22,901.1 Kg
Total Moment = 7032.2 x 0.0623 x 51.44^2 = 1,159,259.5 Kgm

Case 4

v_0 = 51.444 m/s (100 knots), h_0 = 10m (32.8 Ft), Exponent α = 0.0883

100m x 1m Rectangle - 51.44m/s (100 Knots) Draft= 0 Heel=0					
Elevation Range	Area	Ch Height Coeff.	Ch x A	Arm	Moment
m	m^2	-	m^2	m	m^3
0.00-15.24	15.24	1.00	15.24	7.62	116.1
15.24-30.48	15.24	1.10	30.24	22.86	691.3
30.48-45.72	15.24	1.20	18.29	38.10	696.8
45.72-60.96	15.24	1.30	19.81	53.34	1056.8
60.96-76.20	15.24	1.37	20.88	68.58	1431.9
76.20-91.44	15.24	1.43	21.79	83.82	1826.7
91.44-100	8.56	1.48	12.67	95.72	1212.7
	100.00		138.92	50.62	7032.2

Total Force = 138.92 x 0.0623 x 51.44^2 = 22,901.1 Kg

Total Moment = 7032.2 x 0.0623 x 51.44^2 = 1,159,259.5 Kgm

Theoretical Calculations (See Ref. [6])	Strict application of the published regulations

$$_1 = \frac{1}{2} \quad \frac{2}{\frac{2}{0}} \cdot$$

$$= \frac{1}{2 \quad +1} \cdot \quad ^{2\propto+1}$$

$$= \frac{1}{2 \quad +2} \cdot \quad ^{2 \quad +2}$$

v_0 = 51.444 m/s

h_0 = 10 m α = 0.0883

ρ = 1.222 kgm/m^3

h = 100m b = 1m

Cs = 1

g = 9.81 m/s^2

$$_1 = \frac{1}{2} \cdot 1.222 \cdot 1 \cdot \frac{51.444^2}{10^{0.1766}} = 1{,}076.740$$

$$= \frac{1}{2 \propto +1} \quad ^{2\propto+1} = \frac{1076.572}{2 \cdot 0.0883 + 1} \cdot 100^{1.1766} =$$

$$= 206{,}386.8 \quad = 21{,}038$$

$$= \frac{1}{2 \quad +2} \quad ^{2 \quad +2} = \frac{1076.572}{2 \cdot 0.0883 + 1} \cdot 100^{2.1766}$$

$$= 11{,}156{,}604 \quad = 1{,}137{,}269$$

*** Formatting Issue - Best Available Paper/Graphic ***

Case 5

v_0 = 51.444 m/s (100 knots), h_0 = 10m (32.8 Ft), Exponent α = 0.1

100m x 1m Rectangle - 51.44m/s (100 Knots) Draft= 0 Heel=0					
Elevation Range	Area	Ch Height Coeff.	Ch x A	Arm	Moment
m	m^2	-	m^2	m	m^3
0.00-15.24	15.24	1.00	15.24	7.62	116.1
15.24-30.48	15.24	1.10	30.24	22.86	691.3
30.48-45.72	15.24	1.20	18.29	38.10	696.8
45.72-60.96	15.24	1.30	19.81	53.34	1056.8
60.96-76.20	15.24	1.37	20.88	68.58	1431.9
76.20-91.44	15.24	1.43	21.79	83.82	1826.7
91.44-100	8.56	1.48	12.67	95.72	1212.7
	100.00		138.92	50.62	7032.2

Total Force = 138.92 x 0.0623 x 51.44^2 = 22,901.1 Kg
Total Moment = 7032.2 x 0.0623 x 51.44^2 = 1,159,259.5 Kgm

Theoretical Calculations (See Ref. [6])	Strict application of the published regulations
$_1 = \frac{1}{2} \quad \frac{^2}{^2_0} \cdot$ $= \frac{1}{2 \; + 1} \cdot \quad ^{2\propto+1}$ $= \frac{1}{2 \; + 2} \cdot \quad ^{2 \; +2}$ v_0 = 51.444 m/s h_0 = 10 m α = 0.0883 ρ = 1.222 kg_m/m^3 h = 100m b = 1m Cs = 1 g = 9.81 m/s^2 $_1 = \frac{1}{2} \cdot 1.222 \cdot 1 \cdot \frac{51.444^2}{10^{0.1766}} = 1{,}020.260$ $= \frac{1}{2\propto+1} \quad ^{2\propto+1} = \frac{1{,}020.26}{2 \cdot 0.1 + 1} \cdot 100^{1.2} =$ = 213,564.7 = 21,770.1 $= \frac{1}{2 \; + 2} \quad ^{2 \; +2} = \frac{1{,}020.26}{2 \cdot 0.1 + 1} \cdot 100^{2.2}$ = 11,648,982.4 = 1,187,460	

Comparison between Theoretical Calculations and the Step Function

*** **Formatting Issue - Best Available Paper/Graphic** ***

on a 100m x 1m Rectangle

All cases compared summary

Case	BLP Description	Wind Force	Wind Overturning Moment	Wind Force Percent Difference	Moment Percent Difference
		Kg_f	$Kg_f m$	Compared to Step Function Results	
	SF	22,901	1,159,260	-	-
1	50Ft/0.1019	20,091	1,097,431	(-12.27%)	(-5.33%)
2	10m/0.0864	20,923	1,129,333	(-8.64%)	(-2.58%)
3	50Ft/0.1039	20,175	1,103,717	(-11.90%)	(-4.79%)
4	10m/0.0883	21,038	1,137,269	(-8.14%)	(-1.90%)
5	10m/0.1	21,770	1,187,460	(-4.94%)	2.43%

Proceedings of the 22nd Offshore Symposium, February 2017, Houston, Texas
Texas Section of the Society of Naval Architects and Marine Engineers

Appendix III

Sensitivity to number of decimals defining the exponent
(Compare results to Case 5 above)

α = 0.0995	**α = 0.1004**	**α = 0.095**	**α = 0.104**
h_0 = 10 m v_0 = 51.444 m/s ρ = 1.222 kg_m/m^3 h = 100m b = 1m Cs = 1 g = 9.81 m/s^2 $_1$ = *1,022.611* = *21,738.1* = *1,185,268*	h_0 = 10 m v_0 = 51.444 m/s ρ = 1.222 kg_m/m^3 h = 100m b = 1m Cs = 1 g = 9.81 m/s^2 $_1$ = *1,018.382* = *21,795.7* = *1,189,217*	h_0 = 10 m v_0 = 51.444 m/s ρ = 1.222 kg_m/m^3 h = 100m b = 1m Cs = 1 g = 9.81 m/s^2 $_1$ = *1,044.024* = *21,453.3* = *1,165,729*	h_0 = 10 m v_0 = 51.444 m/s ρ = 1.222 kg_m/m^3 h = 100m b = 1m Cs = 1 g = 9.81 m/s^2 $_1$ = *1,001.638* = *22,028.0* = *1205,154*

*** Formatting Issue - Best Available Paper/Graphic ***

Proceedings of the 22nd Offshore Symposium, February 2017, Houston, Texas
Texas Section of the Society of Naval Architects and Marine Engineers

Appendix IV

Forces and Moments calculations for Various MODU and Boundary Layer Profiles
Case 1

Jack-up

Reference Height h_0 (Feet)	Exponent α	Draft (Feet)	Heel (Degrees)	Wind Direction (Degrees)	Wind Force (LT)	Wind Overturning Moment (LT-Feet)	Wind Force Percent Difference (Compared to Step Function Results)	Moment Percent Difference (Compared to Step Function Results)
Step Function (as required by regulations)		15	0	0	631.9	125253.2	0.00%	0.00%
		15	0	45	687.2	130475.7	0.00%	0.00%
		15	0	90	664.3	128228.2	0.00%	0.00%
		15	15	0	668.5	120457.4	0.00%	0.00%
		15	15	45	722.1	127596.4	0.00%	0.00%
		15	15	90	711.8	125660.0	0.00%	0.00%
50 Feet (15.24 m)	0.1019	15	0	0	596.9	123170.2	(-5.54%)	(-1.66%)
		15	0	45	651.0	128257.3	(-5.27%)	(-1.70%)
		15	0	90	630.3	126058.3	(-5.12%)	(-1.69%)
		15	15	0	623.8	118064.3	(-6.68%)	(-1.99%)
		15	15	45	677.4	125088.1	(-6.18%)	(-1.97%)
		15	15	90	664.0	123024.1	(-6.72%)	(-2.10%)
10m (32.808 Ft)	0.0864	15	0	0	625.1	126407.6	(-1.07%)	0.92%
		15	0	45	682.9	131807.2	(-0.63%)	1.02%
		15	0	90	660.9	129500.6	(-0.51%)	0.99%
		15	15	0	656.7	121480.9	(-1.77%)	0.85%
		15	15	45	713.2	128863.8	(-1.24%)	0.99%
		15	15	90	699.5	126706.9	(-1.73%)	0.83%
50 Feet (15.24 m)	0.1039	15	0	0	599.1	123922.2	(-5.20%)	(-1.06%)
		15	0	45	653.2	129018.1	(-4.95%)	(-1.12%)
		15	0	90	632.4	126811.9	(-4.80%)	(-1.10%)
		15	15	0	625.6	118746.3	(-6.41%)	(-1.42%)
		15	15	45	679.4	125791.0	(-5.92%)	(-1.41%)
		15	15	90	665.8	123719.6	(-6.46%)	(-1.54%)
10 m (32.808 Ft)	0.0883	15	0	0	628.1	127331.6	(-0.60%)	1.66%
		15	0	45	686.0	132748.3	(-0.17%)	1.74%
		15	0	90	664.0	130431.0	(-0.05%)	1.72%
		15	15	0	659.4	122329.4	(-1.36%)	1.55%
		15	15	45	716.1	129744.7	(-0.83%)	1.68%
		15	15	90	702.3	127576.7	(-1.33%)	1.53%
10 m (32.808 Ft)	0.1000	15	0	0	647.3	133234.5	2.44%	6.37%
		15	0	45	706.1	138760.0	2.75%	6.35%
		15	0	90	683.6	136375.0	2.91%	6.35%
		15	15	0	676.9	127751.1	1.26%	6.06%
		15	15	45	735.1	135371.3	1.80%	6.09%
		15	15	90	720.5	133133.6	1.22%	5.95%

*** Formatting Issue - Best Available Paper/Graphic ***

Case 2

Semisubmersible - Dynamic Positioning

Reference Height h_0	Exponent	Draft	Heel	Wind Direction	Wind Force	Wind Overturning Moment	Wind Force Percent Difference	Moment Percent Difference
Feet	α	m	Degrees	Degrees	tonnes	tm	Compared to Step Function Results	
Step Function (as required by regulations)		25	0	0	404.7	20166.2	0.00%	0.00%
		25	0	45	572.0	25924.7	0.00%	0.00%
		25	0	90	433.7	21016.1	0.00%	0.00%
		25	30	0	905.4	35764.2	0.00%	0.00%
		25	30	45	1055.9	39296.6	0.00%	0.00%
		25	30	90	880.5	32213.2	0.00%	0.00%
50 Feet (15.24 m)	0.1019	25	0	0	376.6	19495.0	(6.95%)	(3.33%)
		25	0	45	532.5	25060.9	(6.91%)	(3.33%)
		25	0	90	405.9	20503.9	(6.41%)	(2.44%)
		25	30	0	825.9	34017.2	(8.78%)	(4.88%)
		25	30	45	963.2	37196.7	(8.77%)	(5.34%)
		25	30	90	797.7	30606.2	(9.41%)	(4.99%)
10m (32.808 Ft)	0.0864	25	0	0	403.0	20402.4	(0.41%)	1.17%
		25	0	45	570.3	26276.5	(0.31%)	1.36%
		25	30	90	433.9	21333.1	0.06%	1.51%
		25	30	0	887.8	36096.6	(1.94%)	0.93%
		25	30	45	1037.9	39572.6	(1.71%)	0.70%
		25	30	90	859.3	32418.8	(2.41%)	0.64%
50 Feet (15.24 m)	0.1039	25	0	0	376.8	19525.4	(6.89%)	(3.18%)
		25	0	45	531.9	25098.9	(7.02%)	(3.19%)
		25	0	90	406.2	20535.1	(6.34%)	(2.29%)
		25	30	0	826.0	33658.5	(8.77%)	(5.89%)
		25	30	45	963.1	36912.2	(8.79%)	(6.07%)
		25	30	90	797.6	30319.0	(9.42%)	(5.88%)
10 m (32.808 Feet)	0.0883	25	0	0	403.9	20457.7	(0.20%)	1.45%
		25	0	45	571.5	26347.6	(0.10%)	1.63%
		25	0	90	434.9	21387.9	0.29%	1.77%
		25	30	0	889.2	36189.4	(1.78%)	1.19%
		25	30	45	1039.3	39662.5	(1.57%)	0.93%
		25	30	90	860.5	32496.4	(2.28%)	0.88%
10m (32.808 Ft)	0.1000	25	0	0	409.4	20805.9	1.16%	3.17%
		25	0	45	579.0	26794.6	1.22%	3.36%
		25	0	90	441.2	21732.1	1.74%	3.41%
		25	30	0	898.4	36773.6	(0.77%)	2.82%
		25	30	45	1048.1	40229.0	(0.73%)	2.37%
		25	30	90	867.9	32985.1	(1.43%)	2.40%

Case 3

Semisubmersible - Conventional Mooring

Reference Height h_0	Exponent	Draft	Heel	Wind Direction	Wind Force	Wind Overturning Moment	Wind Force Percent Difference	Moment Percent Difference
Feet	α	m	Degrees	Degrees	tonnes	tm	Compared to Step Function Results	
Step Function (as required by regulations)		25	0	0	404.7	15935.2	0.00%	0.00%
		25	0	45	566.3	20509.3	0.00%	0.00%
		25	0	90	425.6	14145.2	0.00%	0.00%
		25	30	0	905.4	33570.3	0.00%	0.00%
		25	30	45	1055.0	36556.1	0.00%	0.00%
		25	30	90	873.7	28473.8	0.00%	0.00%
50 Feet (15.24 m)	0.1019	25	0	0	376.5	15264.0	(6.95%)	(4.21%)
		25	0	45	526.9	19649.7	(6.97%)	(4.19%)
		25	0	90	397.8	13638.6	(6.52%)	(3.58%)
		25	30	0	825.9	31823.3	(8.78%)	(5.20%)
		25	30	45	958.4	34459.8	(9.16%)	(5.73%)
		25	30	90	790.9	26870.2	(9.48%)	(5.63%)
10m (32.8 Ft)	0.0864	25	0	0	403.0	16171.4	(0.42%)	1.48%
		25	0	45	564.5	20853.4	(0.33%)	1.68%
		25	0	90	425.7	14451.7	0.04%	2.17%
		25	30	0	887.8	33902.7	(1.94%)	0.99%
		25	30	45	1033.0	36824.9	(2.09%)	0.74%
		25	30	90	852.3	28668.0	(2.44%)	0.68%
50 Feet (15.24 m)	0.1039	25	0	0	376.8	15294.4	(6.89%)	(4.02%)
		25	0	45	527.2	19684.4	(6.91%)	(4.02%)
		25	0	90	398.1	13665.4	(6.46%)	(3.39%)
		25	30	0	826.0	31863.8	(8.77%)	(5.08%)
		25	30	45	958.2	34489.9	(9.18%)	(5.65%)
		25	30	90	790.7	26899.3	(9.50%)	(5.53%)
10 m (32.8 Feet)	0.0883	25	0	0	401.6	16081.8	(0.77%)	0.92%
		25	0	45	562.6	20744.9	(0.66%)	1.15%
		25	0	90	424.2	14372.0	(0.32%)	1.60%
		25	30	0	885.5	33752.6	(2.19%)	0.54%
		25	30	45	1030.8	36684.2	(2.30%)	0.35%
		25	30	90	850.7	28549.4	(2.63%)	0.27%
10m (32.8 Ft)	0.1000	25	0	0	409.4	16574.8	1.16%	4.01%
		25	0	45	572.9	21341.7	1.16%	4.06%
		25	0	90	432.5	14810.3	1.63%	4.70%
		25	30	0	898.4	34579.7	(0.77%)	3.01%
		25	30	45	1042.9	37458.8	(1.15%)	2.47%
		25	30	90	860.6	29202.9	(1.50%)	2.56%

*** Formatting Issue - Best Available Paper/Graphic ***

Drillship - Dynamic Positioning

Reference Height h_o	Exponent	Draft	Heel	Wind Direction	Wind Force	Wind Overturning Moment	Wind Force Percent Difference	Moment Percent Difference
Feet	α	m	Degrees	Degrees	tonnes	tm	Compared to Step Function Results	
Step Function (as required by regulations)		12	0	0	965.7	30284.9	0.00%	0.00%
		12	30	0	1164.9	33168.4	0.00%	0.00%
50 Feet (15.24 m)	0.1019	12	0	0	978.9	29263.4	1.36%	(3.37%)
		12	30	0	1069.4	31749.1	10.74%	4.83%
10m (32.8 Ft)	0.0864	12	0	0	957.7	31432.2	(0.83%)	3.79%
		12	30	0	1165.2	34435.1	20.67%	13.70%
(15.24 m) (50 Feet)	0.1039	12	0	0	878.9	29312.9	(8.99%)	(3.21%)
		12	30	0	1069.1	31788.6	10.71%	4.97%
10 m (32.8 Feet)	0.0883	12	0	0	947.0	30778.2	(1.94%)	1.63%
		12	30	0	1152.6	33782.2	19.35%	11.55%
10m (32.8 Ft)	0.1000	12	0	0	947.0	31339.2	(1.94%)	3.48%
		12	30	0	1163.4	34342.3	20.48%	13.40%

Case 4

REFERENCES

(1) US Coast Guard. *Merchant Marine Technical Note No. 6-66 – Floating Drilling Rigs - Appendix C*, 13 July 1966

(2) American Bureau of Shipping. *Rules for the Building and Classing Offshore Mobile Drilling Units*, 1968

(3) American Bureau of Shipping. *Rules for the Building and Classing Offshore Mobile Drilling Units*, 1973

(4) US Code of Federal Regulations, *Title 46 Shipping, Subchapter IA and Subchapter S- - §174.055 Calculation of wind heeling moment (Hm)*

(5) ABS Eagle Wind V 1.1, November 2015

(6) The Fifteenth International Conference: The Jack-Up Platform – Design, Construction and Operation, City University London, September 2015 - Boundary Layer Profile, Background and Mathematical Tools, J.A. Breuer, J. H. Rousseau, J. N. Brekke

SNAME
THE INTERNATIONAL COMMUNITY
FOR MARITIME AND OCEAN PROFESSIONALS

A COMPARISON OF FRICTION ALGORITHM PERFORMANCE IN FINITE ELEMENT SOLVERS

NATHAN COOKE
SENIOR ENGINEER, INTECSEA INC., ST. JOHN'S, NL (CAN)

ARYA MAJED
TECHNICAL ADVISOR, INTECSEA INC., HOUSTON, TX (USA)

LUCA CHINELLO
SENIOR ENGINEERING SPECIALIST, INTECSEA INC., HOUSTON, TX (USA)

DHARMA PASALA
ENGINEERING SPECIALIST, INTECSEA INC., HOUSTON, TX (USA)

XIANGLEI NI
SENIOR RISER ENGINEER, INTECSEA INC., HOUSTON, TX (USA)

Proceedings of the 22nd Offshore Symposium, February 2017, Houston, Texas
Texas Section of the Society of Naval Architects and Marine Engineers

ABSTRACT

The capability to accurately predict friction (both sticking and sliding states as well the timing of the transitions between the two states) is critical to the design of flexible risers, umbilicals, and connectors. For example, the critical curvature at which the tensile armor wires of a flexible riser transition from sticking to slipping determines the maximum bending moment and plays a key role in strength and fatigue life predictions.

Friction is complex and difficult to predict, and its simulation within multi-body structures is typically relegated to Finite Element (FE) methods. The FE solver friction algorithms are perhaps the most sophisticated and highly guarded part of the solver. This is a predominant differentiator from one solver from another – without proprietary algorithms like friction, FE packages can quickly reduce to a library of standard element formulations, which are mostly common to all solvers.

In this study, the performance and accuracy of a number of nonlinear FE solvers in predicting friction are characterized. A case study is presented involving a small scale, dynamic, friction dominated system designed to isolate and challenge FE solver friction algorithms. The arrangement of interconnected bodies within the system is analogous to an isolated section of a complex structure, such as the tensile armor and surrounding layers of a flexible, or a series links in a mooring chain. The system is exercised through each solver to produce a set of nonlinear time-history outputs using a range of solution parameters, including both lagrangian and penalty enforcement methods. The fidelity of the produced response, as well as the sensitivity to solver parameters is analyzed in detail and evaluated with respect to a predefined set of performance criteria relating specifically to frictional behavior.

A number of important findings are made relative to the friction algorithm in the different FE solvers. These findings include accuracy/performance at the recommended (default) penalty settings, as well as an expanded discussion on the observed effects. Since the case study system can be thought of as representative of interfacing components within an assembly, the findings correlate directly to internal kinematics which governs both stress levels and wear rates. The performance criteria and findings established herein highlight several deficiencies which can impact design and analysis, and encourages due diligence in selecting FE solvers and parameters for systems involving friction.

Keywords: Contact, Friction, Algorithm, Finite Element (FE), Penalty, Lagrange

INTRODUCTION

Offshore structures featuring intricate geometries and interacting components are prevalent within the industry, and play a key role in the ongoing operability of production facilities. These large multi-component structures, such as connectors, mooring chains, and flexible risers, are exposed to severe cyclic loads due to combined wind, wave, current and vessel motions, and are expected to remain in service for decades. Due to the level of complexity involved, FE methods are widely accepted as the ideal means for modeling and analysis, and are therefore routinely deployed in design activities.

One of the main design considerations for any such multi-component structure is the role played by contact mechanics, specifically friction. For example, the critical curvature at which the tensile armor wires of a flexible riser transition from sticking to slipping determines the maximum bending moment and plays a key role in strength and fatigue life predictions. It is widely accepted, however, that capturing the nonlinear kinematics associated with contact are among the most challenging aspects of numerical analysis (Christensen et al., 1998). The capability to accurately predict friction - both sticking and sliding states as well as the timing of the transitions between the two – imposes a major burden on FE solvers, even for simple problems featuring basic constitutive models. Even with the modern computational power and sophisticated algorithms available today, contact is often avoided or simplified wherever possible in favor of assumed conservatism and improved convergence (Steinbuch, 2011).

The significance and challenges associated with accurate contact modeling are such that a wide range of numerical strategies has been researched and developed, and FE solvers have established unique and highly proprietary algorithms used for tracking and enforcement. This differentiation can cause significant disagreement between solvers, even for rather simple problems. In this study these differences are explored and characterized in terms of solution accuracy for several commercial FE solvers, as well as the proprietary FLEXAS solver.

BACKGROUND

The difficulties involved in proper contact modeling within FE solvers stem from both tracking and enforcement. Regardless of any constitutive models used, contact algorithms must first be able to detect when and where to apply enforcement conditions for either normal or tangential components, both of which are highly dependent on the deformed geometries, spatial orientations and loading histories of the interacting bodies involved. Problems involving contact are thus fundamentally nonlinear, and require constant and precise tracking of the bodies in relation to one another. This is especially important when dealing with friction, where a proper model has to also track the evolution of the transitions between sticking and slipping. In other words, it must have a memory in order to achieve correct enforcement, which is a major contributor to what makes the algorithms so complex. Furthermore, both the normal and tangential contact conditions are interrelated, and must therefore be established in concert. Several different approaches have been theorized and employed to achieve this, some of which are shown in Figure 1 below to give a brief appreciation for the subtle differences at a high level.

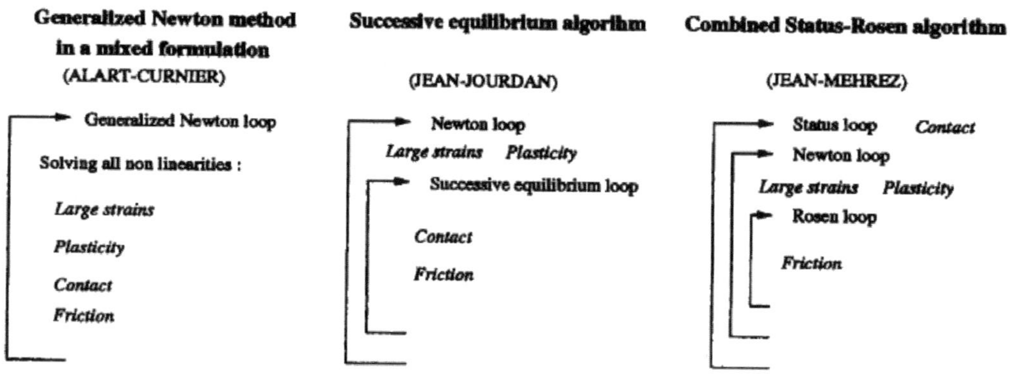

Figure 1 Example of Algorithm Structures (Jourdan et al., 1996)

In conjunction with the above, imposing precise enforcement of contact conditions becomes exceedingly difficult within a traditional FE solution scheme. The discrete changes in contact status and sharp transitions in friction states contribute to ill-conditioning and convergences issues in many cases. These long standing numerical challenges are directly responsible for the development of compromised methods, which trade accuracy for numerical stability and ease of implementation. The most commonly used and recognizable method associated with these is known as the penalty method, which introduces spring forces to oppose contact violations. With respect to friction, this results in a certain amount of allowed displacement while in the sticking condition. This is shown in Figure 2 below for a typical Coulomb friction model.

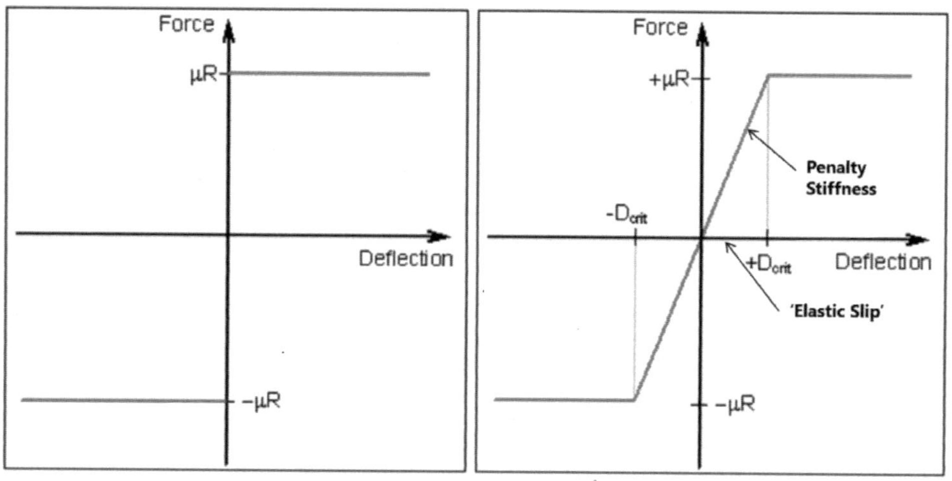

Figure 2 Coulomb Friction Model – Exact (Left) and Penalty (Right)

While the penalty method is highly attractive in terms of simplicity and robustness, the introduced elastic slip can adversely affect the solution by artificially reducing system stiffness, introducing spurious forces and moments, and interfering with the correct timing and location of stick-slip transitions by redistributing loads.

The alternative and much more accurate approach is known as the Lagrange multiplier method, which in essence treats the constraint force required to maintain exact enforcement as an additional independent variable. While this method provides much more precision, the added constraint variables increase the number of degrees of freedom, and the algorithms required are much more complex and difficult to implement properly without adversely affecting convergence. As a result, only selective commercial codes feature Lagrangian enforcement methods, mainly in an augmented or hybrid formulation, and even then its use is relegated only to special cases due to poor numerical efficiency.

METHODOLOGY

For the purpose of this comparison, a case study approach was decided as the ideal means to assess the relative performance. In order to facilitate model selection, the system requirements listed below were established. These requirements ensured that the problem remained sufficiently small to allow efficient operation of all possible friction algorithms, while retaining sufficient complexity to test each solver's capabilities. Furthermore, they de-leverage any advantage of FLEXAS large model nonlinear simulation capabilities.

1. Multi-body system – activate multiple interdependent friction states.
2. Rigid bodies – no elastic deformation.
3. Short duration irregular (time-domain) load input – activate dynamics (inertial forces).
4. Small number of DOFs – enable efficient use of optimal solution procedures.

To execute the study, the system was provided to separate analysts, each deemed to be proficient with one of the solvers of interest. The analysts were tasked with producing multiple solutions using various combinations of applicable solver parameters such as time increment control, friction enforcement (penalty / Lagrange), and solution scheme (implicit / explicit). The optimal solution(s) from each solver were then provided to the principal investigator for comparison and performance assessment. Assistance from the vendor technical support group for each solver was highly encouraged, and actively used throughout the study. Throughout this process, a consistent FE representation of the chosen system across all solvers was ensured.

In order to gauge the relative performance of each solver, a set of criteria was developed intended to capture the most fundamental aspects of frictional behavior. The criteria, described in Table 1 below, were applied on a relative, subjective basis using a point score ranging from 0 to 10. While more rigorous mathematical definitions and numerical assessments are available to quantify relative error for these criteria, the chosen method was preferred here as it suits the objective of simply highlighting the degree of variability in results.

Table 1 Solver Performance Criteria

Criteria	Description
Convergence	Ability of the solver to resolve and hold the solution response without deviation, and with minimal to no over- or under-shoot and subsequent decay time.
Numerical Spikes	Artificial introduction of a sudden peak or valley in the solution response. Graded based on amplitude and quantity.
Numerical Ringing	Artificial introduction of multiple numerical spikes in rapid succession. Graded based amplitude and persistence. Differentiated from convergence rate based on relative frequency in response.
Transition (Sticking to Slipping)	Ability of the solver to capture a clean transition from sticking to slipping, evidenced by an immediate change from an increasing friction force to a constant value matching the critical friction force and without exceeding it.
Transition (Slipping to Sticking)	Ability of the solver to capture a clean transition from slipping to sticking, evidenced by an immediate change to a reducing friction force from a constant value matching the critical friction force.
Computational Efficiency	Size of the time increment needed in order to converge on a solution.
Parameter Dependence	Degree of variability of the solution based on solver/integrator parameters.
Absence of Numerical Effects	A complete absence of any numerical noise or chatter in all response quantities. The solver produces a clean solution with no artificial portion.

CASE STUDY

The chosen system, shown in Figure 3, features three pairs of stacked rigid blocks connected by horizontal linear elastic springs resting on a flat surface. One end is sprung to ground, while at the opposite end a forcing function, shown in Figure 4, is applied to the top block only. Frictional interaction is considered between the pairs of stacked blocks only, using the weight of the blocks to provide a constant normal force. The interface between the lower set of blocks and the supporting surface is considered frictionless. This system, while relatively simple, may be thought of as analogous to a set of stroking segments within a connector, or even an isolated set of tensile armors within a flexible riser. The response of each component and in turn the overall system stiffness is heavily dependent on the individual contact states. This is consistent with the primary characteristics of the chosen system, which are listed below.

- When stuck, each pair of blocks behaves as a single rigid body, effectively representing 1 DOF and maximizing system stiffness. Equal displacements must therefore placed on both the upper and lower set of springs, generating equal forces on the adjacent pair of blocks and precluding any imposed shear force. Any oscillation is immediately transferred into the springs, and therefore should not be present in the friction force output.
- Relative displacement only occurs once the critical friction force is exceeded, which reduces system stiffness. Any residual displacement between block pairs directly results in an offset in force seen by the springs, which is transferred to adjacent blocks.
- Relative velocity between blocks must be exactly zero when blocks are stuck, regardless of any residual displacement.

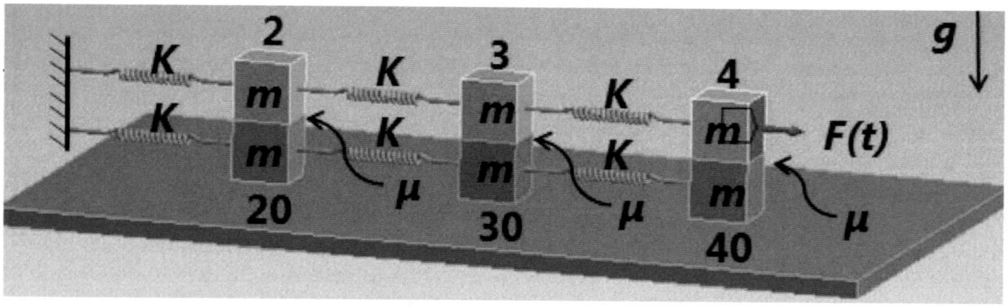

Figure 3 Case Study System

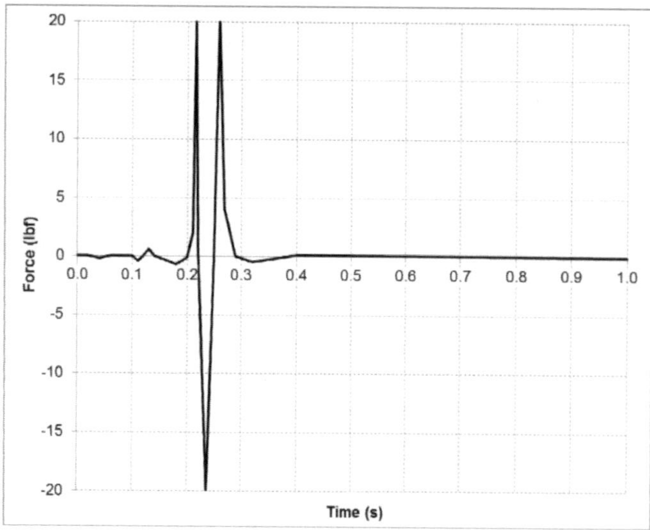

Time (s)	Force (lbf)
0.000	0.00
0.010	0.01
0.020	0.00
0.040	-0.20
0.060	0.00
0.080	0.03
0.100	0.00
0.110	-0.40
0.120	0.00
0.130	0.60
0.140	0.00
0.180	-0.70
0.200	-0.10
0.210	2.00
0.215	20.00
0.220	0.00
0.235	-20.00
0.250	0.00
0.260	20.00
0.270	4.00
0.290	0.00
0.320	-0.50
0.400	0.10
0.500	0.10
1.000	0.00

Figure 4 Forcing Function Time Signal, F(t)

The system response was measured in terms of the friction force, relative displacement, and relative velocity between each pair of blocks (2/20, 3/30, and 4/40 as labelled in Figure 3) using the input parameters listed below. Only the relative values were chosen, as these represent the primary contributions from the friction algorithm isolated from the global dynamics system response. Two friction coefficients were considered; a commonly used default value of 0.1 and a larger value of 0.2, with the latter intended to provoke an increased dynamic response at slip onset and further challenge each solver.

- Weight of each block, W = 10.0lbm (m = 10.0lbs/386in/s^2)
- Stiffness of each spring, K = 50.0lbf/in
- Coefficient of friction between each block pair, $\mu = 0.1, 0.2$

Results are presented in the following section for only a selective set which best illustrate the performance differentiators according to the guiding criteria defined above.

RESULTS & FINDINGS

Prior to presenting the results, it is prudent to take a critical look at the system to develop a high level understanding of the expected behavior. According to the system characteristics and input parameters defined above, we can think of the blocks as a series of gates, which prevent the transmission of shear forces when stuck, and freely pass them when slipped. With this in mind, we can easily see that for the first 0.2 seconds of forcing function input the friction force seen between blocks 4 and 40 will not exceed the critical value for either friction coefficient. Blocks 3 and 30, and by extension blocks 2 and 20, will therefore see exactly the same force through their respective springs, and all response quantities will remain null. At 0.215 seconds, the forcing input has reached 20.0 lbs, which far exceeds the critical friction force and causes blocks 4 and 40 to slip. This induces a friction force between blocks 3 and 30 approximately proportional to the amount of slip and therefore relative compression between upper and lower springs. Beyond this point as the forcing input direction progresses and the other blocks begin to slip, the response becomes increasingly difficult to predict due to inertial effects, dynamics, and the complex path and state dependence.

With this in mind, we first look at the predicted full time-history response of friction force between blocks 4 and 40, shown in Figure 5. From this response, we can immediately identify the initial sticking phase as discussed above for the first 0.2 seconds, for which all solvers agree. We then note the characteristic square wave response caused by the discrete transitions between stuck and slipped friction states. Nearing the tail end of the input, at 0.35 seconds, we arrive at the first significant departure between solutions. It is important to distinguish here that in prior transitions, the blocks were transitioning rapidly from slipping to sticking, but then immediately slipping again due to a change in direction. At this point, we see the first true transition from slip to stick, which highlights the difficulty involved in capturing this state change properly.

Here we see a sudden presence of significant spike amplitude and ringing from solution D, and to a much lesser extent solution B. Solution E simply fails to converge on a final residual force, showing irregular large amplitude, and large period fluctuations. This fluctuation in the trailing response signal is also apparent in solution B, but at significantly reduced amplitude, and at a moderately reduced period. It is also important to note that despite the significant amount of noise present in the solution D response, there is seemingly little impact on the converged result. Solutions A and C were free of any artificial effects.

Taking a closer look at the transition regions, presented in Figure 6, further discrepancies and flaws are revealed. In these areas, we can see minor differences in the timing of the transitions; however this can be affected to some extent by the resolution of solver time increment and output frequency. Crucially, though, solution B shows severe difficulty in resolving the transitions properly, consistently overshooting the critical friction value, and exhibiting poor convergence and occasional numerical spikes.

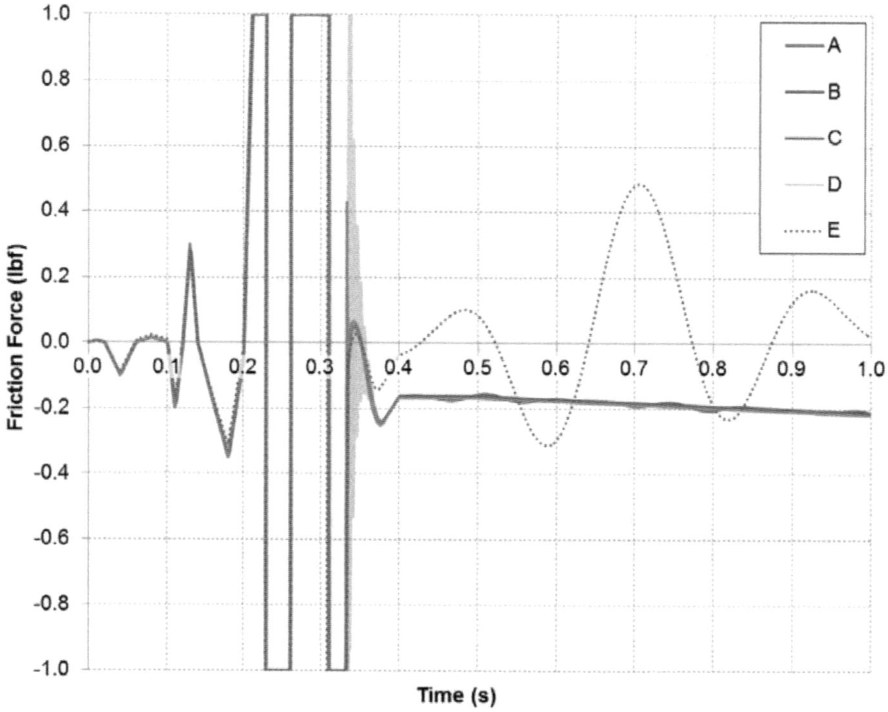

Figure 5 Friction Force | Blocks 4-40 | μ=0.1

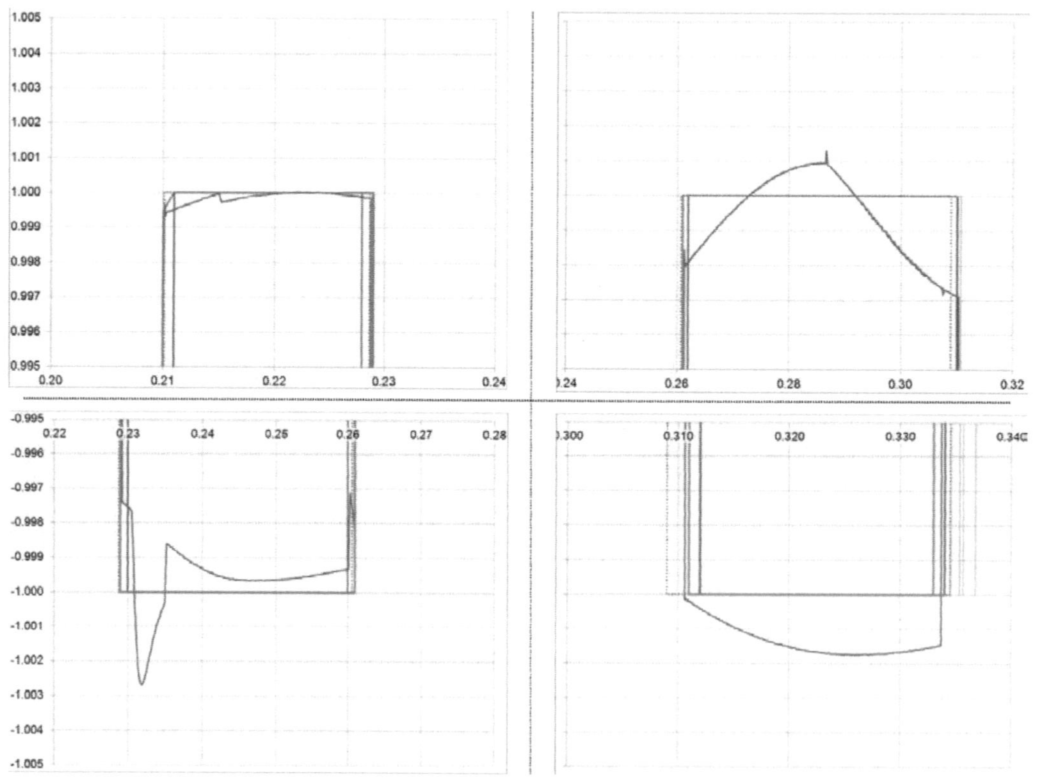

Figure 6 Friction Force at Transitions | Blocks 4-40 | μ=0.1

Looking at the subsequent responses at blocks 3 and 30, as well as blocks 2 and 20 shown in Figure 7 and Figure 8 respectively, it is apparent that the behaviors noted above are persistent, and increasingly appear to be characteristic flaws in the respective friction algorithms. Aside from this, it should be noted that the deviations in solutions are manifested in the final residual friction force, with increasing separation propagating from blocks 4 and 40 to blocks 2 and 20.

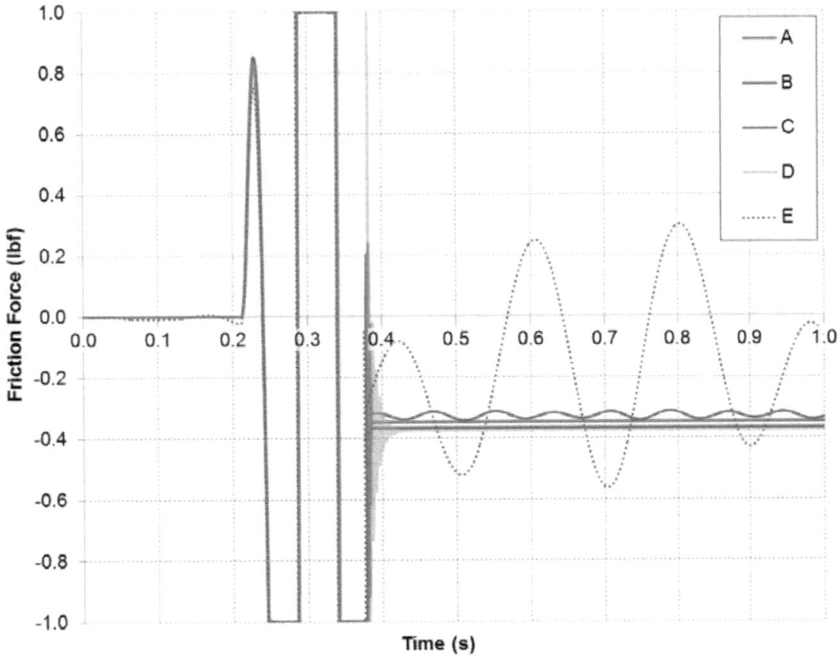

Figure 7 Friction Force | Blocks 3-30 | μ=0.1

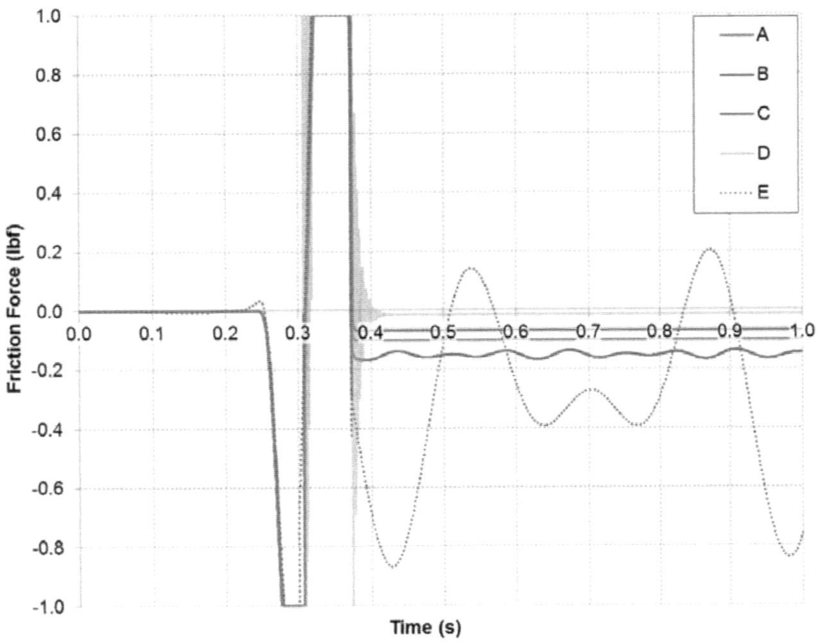

Figure 8 Friction Force | Blocks 2-20 | μ=0.1

Based on the above observation of maximum error at blocks 2 and 20, it follows then that this discrepancy is also found in the relative displacements and velocities, which is confirmed in Figure 9 and Figure 10, respectively. Here we can see that solution B consistently overpredicts the response, while solution D underpredicts. It is also noted that despite showing similar forces, solution E underpredicts the motion by a significant margin, indicating further issues possibly unrelated to the friction algorithm.

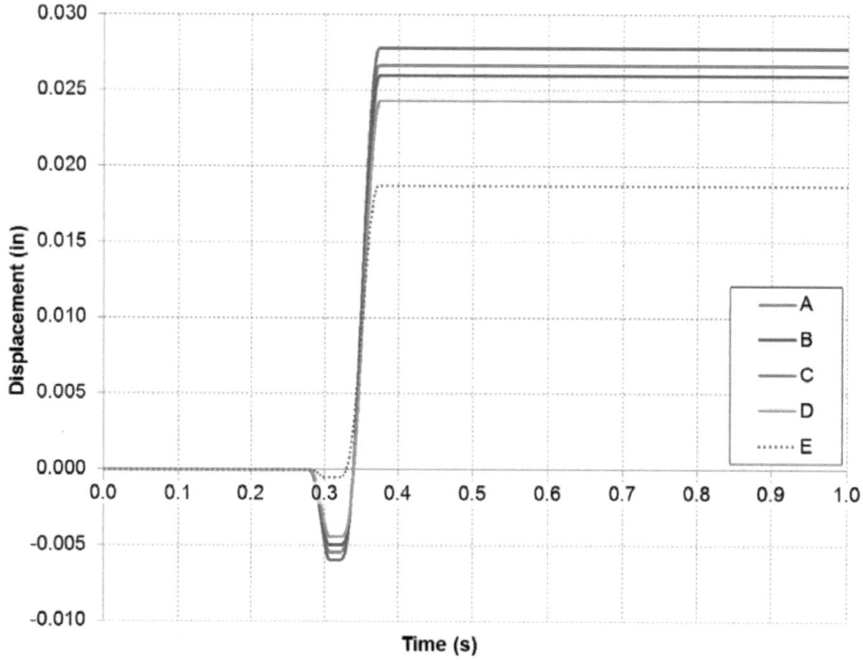

Figure 9 Relative Displacement | Blocks 2-20 | μ=0.1

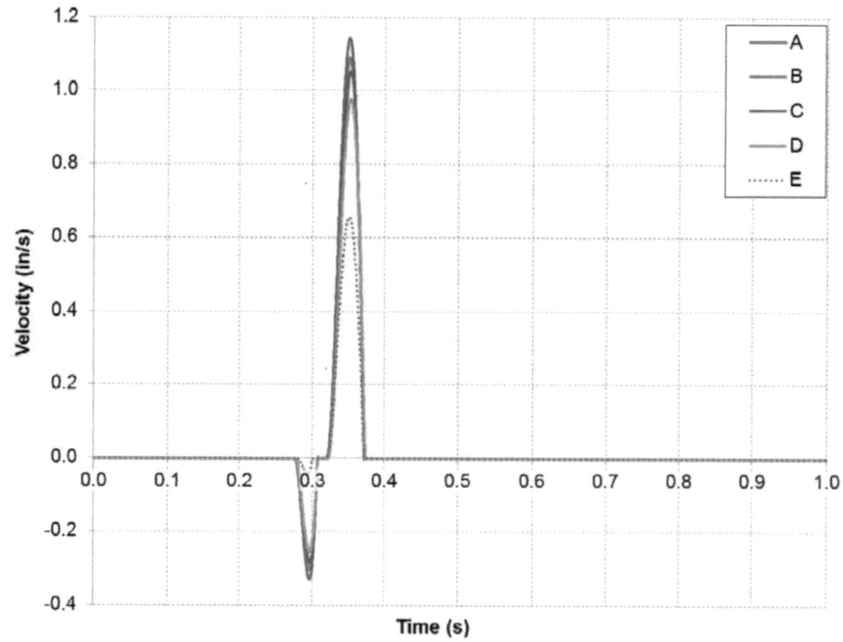

Figure 10 Relative Velocity | Blocks 2-20 | μ=0.1

We now move to the second case, using a higher coefficient of friction. Unsurprisingly, this set of results shows similar characteristic behavior for each solver, but also reveals additional flaws. First and foremost, solution B fails to deliver a complete response, with all relative quantities dropping to zero at just after 0.5 seconds. This behavior, shown in Figure 11, was thoroughly investigated, including verification of the model by the vendor, but the issue persisted across a multitude of different solver parameters. Furthermore, a closer look at the results near the transition regions, shown in Figure 12, reveals that solution A begins to exhibit mild numerical spikes, and only solution C remains free of artificial effects.

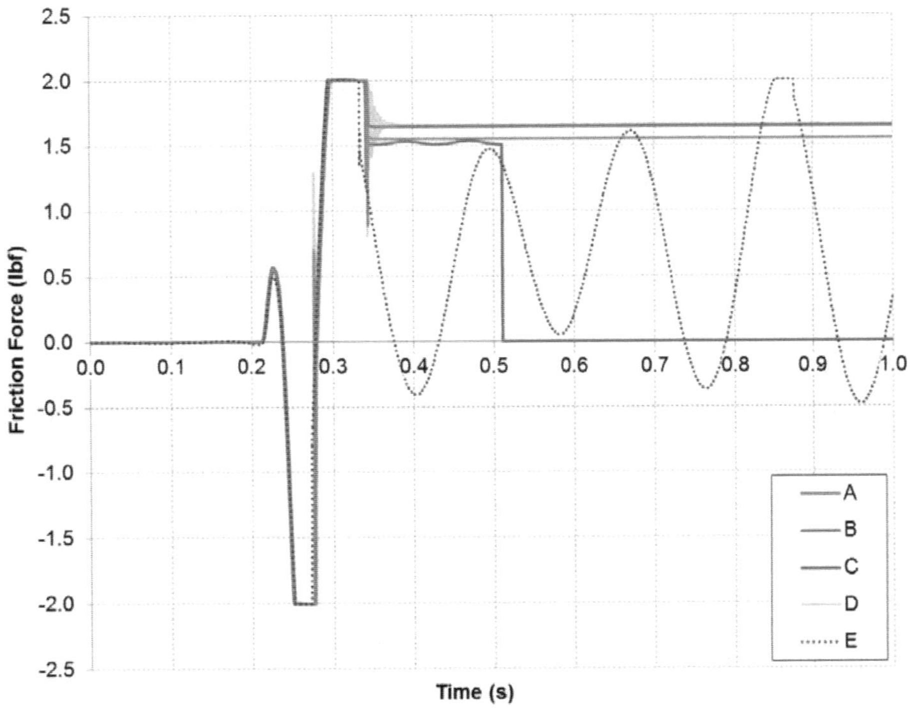

Figure 11 Friction Force | Blocks 3-30 | μ=0.2

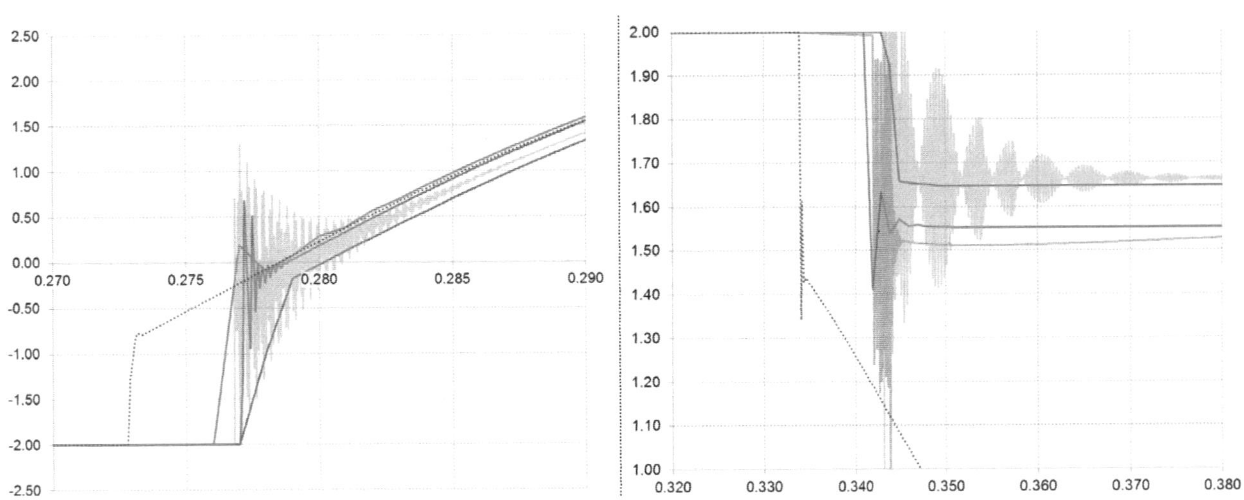

Figure 12 Friction Force Transitions | Blocks 3-30 | μ=0.2

In addition to the above, consideration was also given for how each solver arrived at its optimal solution. The common trend here was that all solvers, with the exception of solver C, were prone to similar inefficiencies when attempting to enforce friction states strictly. In particular, each required significant modification of default penalty paramters (stiffness / elastic slip) to produce a meaningful solution, resulting in increased computational cost in the form of slower convergence (more iterations per increment), or reduced increment size required. The significant effect of this parameter on solution accuracy is demonstrated in Figure 13 below, using output from solver A.

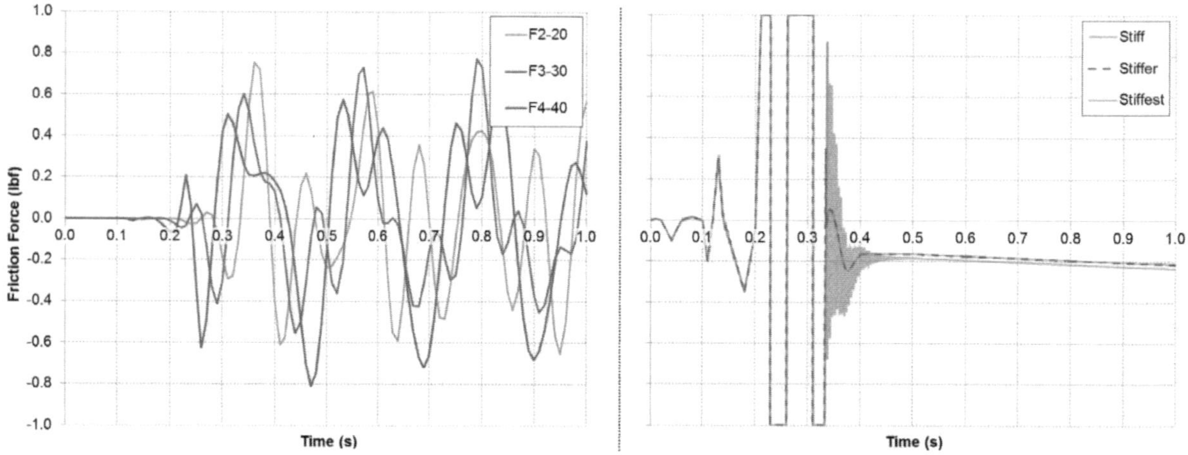

Figure 13 Friction Force | All Blocks | Default Penalty | μ=0.1 (Left)
Friction Force | Blocks 4-40 | Modified Penalty Stiffness | μ=0.1 (Right)

Only solvers A and C were able to implement Lagrangian enforcement successfully, however, the former exhibited strange time-dependence whereby numerical ringing and convergence actually worsened with a reduction in increment size. This was not the case using strict penalty enforcement.

Based on the above, the relative scores shown in Table 2 and Table 3 were established. Note that while these scores were assigned somewhat subjectively, minor differences in grading would not detract from the overarching trends, and it is evident that solver C provided the most reliable and least error prone solutions throughout, with solver A following closely.

Table 2 Performance Evaluation | μ = 0.1

Criteria		A	B	C	D	E
1	Convergence / Stability	10	4	10	7	2
2	Numerical Spikes	10	6	10	5	8
3	Numerical Ringing / Chatter	10	6	10	3	9
4	Transition - Stick to Slip	10	5	10	10	7
5	Transition - Slip to Stick	10	5	10	8	5
6	Computational Efficiency	7	7	10	5	2
7	Parameter dependence	7	7	10	5	3
8	Absence of numerical effects	0	0	10	0	0
	Overall Accuracy	**74**	**40**	**80**	**43**	**36**

Table 3 Performance Evaluation | μ = 0.2

Criteria	A	B	C	D	E
1 Convergence / Stability	8	4	10	7	2
2 Numerical Spikes	9	7	10	6	8
3 Numerical Ringing / Chatter	9	6	10	4	9
4 Transition - Stick to Slip	10	5	10	10	7
5 Transition - Slip to Stick	10	5	10	8	5
6 Computational Efficiency	8	7	10	5	2
7 Parameter dependence	7	7	10	5	3
8 Absence of numerical effects	0	0	10	0	0
Overall Accuracy	**61**	**41**	**80**	**45**	**36**

CONCLUSIONS

In this study, the relative performance of the friction algorithms implemented within several FE solvers was investigated using a case study. Despite the relative simplicity of the problem, multiple discrepancies in performance were identified, due to large variances in algorithm precision and efficiency. Although the names of each commercial solver shall remain undisclosed, we can at this point identify that solver C represents FLEXAS.

While it may be argued that the absolute differences in this case were minor, especially for the displacements and velocities, it is important to keep in mind the scale of the problem, especially in the time domain. In the much larger systems typically deployed in offshore environments, undergoing thousands of load cycles over a multi-year design life, even small to moderate errors may have substantial impacts on the predicted integrity. In a flexible riser, for example, the incremental slip and potential ratcheting could easily lead to tensile armor misalignment and subsequent impacts on stress, and even increased risk of local buckling. The most concerning aspects revealed in this study, however, were to do with the predicted friction forces, which correlate directly to stresses. Considering again the analogy of a flexible riser, one of the major contributors of fatigue stresses is those which arise from the stuck state of the wires during bending, known as "friction stress". Clearly then, accurately and reliably capturing the correct fatigue state requires precise friction algorithms, producing solutions free from artificial noise. For structures with less well defined contact states, such as connectors, this becomes increasingly important as the artificial noise may become nearly impossible to distinguish from the correct physical response.

The importance of parameter selection was also highlighted, especially with respect to penalty stiffness values. Selection of inadequate penalty parameters, which can easily occur due to mesh dependence or user ignorance, can have major implications on the analysis. The elastic slip introduced to aid convergence effectively introduces mechanical "play" or "slop" into a structure, resulting in potentially gross misrepresentations of stiffness and component response. Analysts must therefore be well aware of all parameters at their disposal and how the default values are chosen in order to make appropriate decisions regarding convergence rate and model accuracy.

ACKNOWLEDGEMENTS

The authors acknowledge INTECSEA's global technology for supporting this task.

REFERENCES

Christensen, P. W., Klarbring, A., Pang, J. S., and Stromberg, N., "Formulation and Comparison of Algorithms for Frictional Contact Problems", International Journal for Numerical Methods in Engineering 42, 145-173, 1998.

Steinbuch, R., "Contact Modelling in Structural Simulation – Approaches, Problems and Chances", Trends in Computational Mechanics, LNACM 58, 339-354, 2011.

SNAME

THE INTERNATIONAL COMMUNITY
FOR MARITIME AND OCEAN PROFESSIONALS

SEMI-SUBMERSIBLE BRACE TO COLUMN CONNECTION – A MEANS FOR IMPROVED STRENGTH AND REDUCED STRUCTURAL WEIGHT

ROY H. COTTRELL, P.E.
PRESIDENT – COTTRELL & COTTRELL, ENGINEERING AND BUSINESS SERVICES

BARTON D. GRASSO, P.E.
DIRECTOR OF ENGINEERING – 3DENT TECHNOLOGY

JOSE H. VAZQUEZ, PH.D.
PRESIDENT – 3DENT TECHNOLOGY

JUAN SANTOS
ENGINEER I – 3DENT TECHNOLOGY

Proceedings of the 22nd Offshore Symposium, February 2017, Houston, Texas
Texas Section of the Society of Naval Architects and Marine Engineers

ABSTRACT

Structural braces are provided between port and starboard columns or pontoons of semi-submersibles to reduce the bending moment at the column deck box connection created by squeeze/pry loads from hull-wave interaction. This hull-wave interaction is usually in the form of squeeze/pry action and bending caused by the relative longitudinal and vertical displacement between the pontoons caused by hull racking.

Traditionally, the braces have been designed as fixed end beam/columns for withstanding axial loads. The racking displacement of the brace ends results in brace end bending moment and induced stresses at the brace to hull connection. To handle the high bending moment, the brace wall thickness must be increased, which results in a stiffer connection. Because the racking deflection is controlled by the deck box, and not the stiffness of the braces, the magnitude of this deflection remains essentially constant regardless of brace stiffness. Thus the wall thickness increase simply attracts more bending moment, more stress and further increases in wall thickness. The additional reinforcement to resist the bending stresses typically results in substantially heavier structure than required by the compressive and tensile squeeze/pry loads alone, that they are intended to resist and control.

The authors have developed and analyzed an alternate structural arrangement that reduces the bending stiffness of the braces at their connection to the hull, allowing the braces to act more like pin ended column elements. The reduction in bending moment at the brace to hull connection results in reduced scantlings and a significant reduction in the weight of the braces and their associated hull structure.

Keywords: Semi-submersible Vessel, brace to hull structural arrangement, weight reduction

INTRODUCTION AND BACKGROUND

In the global sense, the semi-submersible hull is subject to wave loading which results in squeeze/pry/racking loading and deflection, as shown in Figures 1 and 2 below, from the Handbook of Offshore Engineering (Chakrabarti, 2005).

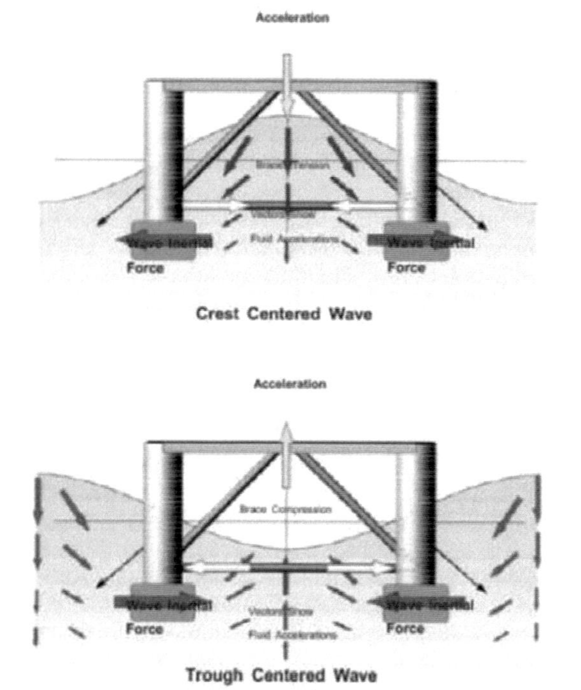

Figure 7.43 Twin pontoon semi-submersible squeeze/pry wave load system

Figure 1. Pry above and Squeeze Below

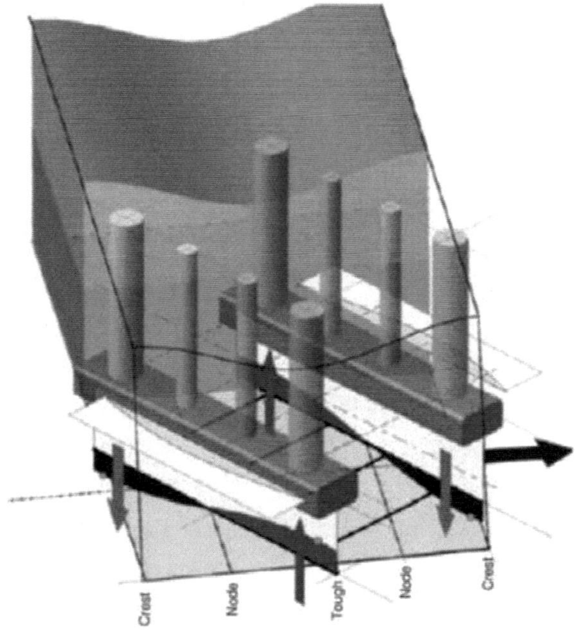

Figure 7.44 Twin pontoon semi-submersible wave racking load system

Figure 2. Racking

Many structural arrangements have been used to enable the hull to resist these loads. The structural arrangements include the following configurations, as illustrated in their respective descriptions below:

1. Column to deck box connection through "haunches" without bracing.
2. A single horizontal brace per column.
3. Two horizontal braces per column.
4. A space frame truss connecting port and starboard columns.
5. Diagonal bracing with horizontal bracing in a "K" configuration.

While the column to deck box connection, without bracing, has its advantages, concerns over structural cracking have arisen on recent implementations with the fix being to retrofit the hull with horizontal braces. The space-frame truss arrangement and the "K" brace configuration have both proved effective, but are typically heavier.

Thus the preferred arrangement, as evident by the number of designs utilizing it, is the horizontal brace shown in Figure 3. The braces provide resistance to squeeze/pry loads, while allowing the deck box to take the racking loads. The single horizontal braces, shown as item 18, features the lowest overall structural weight of the five alternatives. An iteration of the horizontal brace arrangement utilizing two braces per column is also employed in harsh environments to meet fatigue life requirements. This two brace arrangement is heavier than the single brace design.

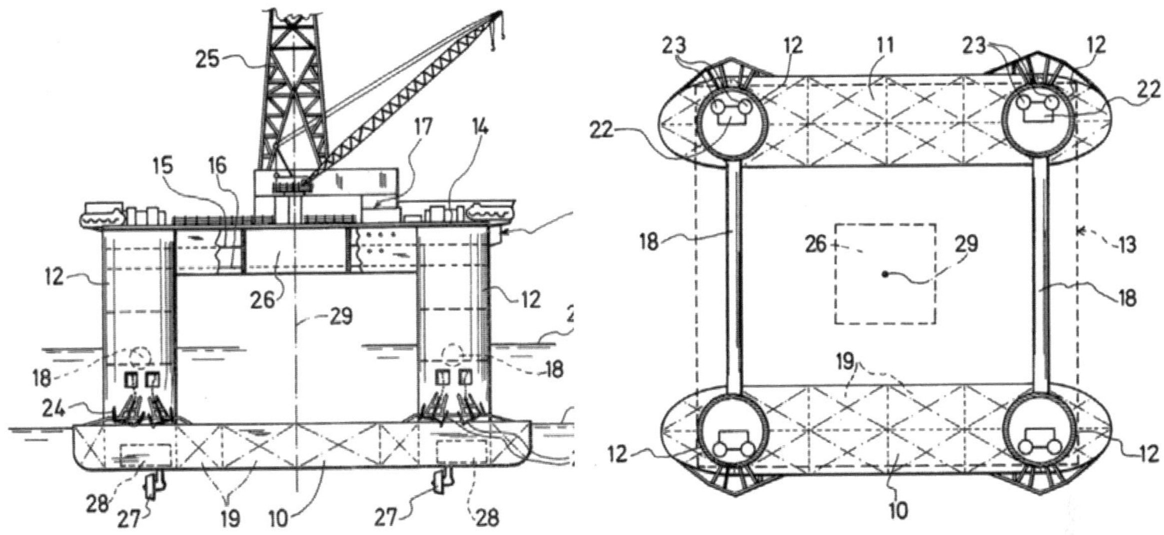

Figure 3. Configuration 2: One Brace per Column (Liden et al, 1986)

For the single brace designs, the brace ends are "built-in," resulting in fixed ends. The braces must have sufficient strength as columns to resist the squeeze/pry load, and thus a large diameter. Because the deck box is orders of magnitude stiffer than the braces, the stiffness of the brace, or braces, has a negligible effect on the racking deflection. However, this large diameter results in sufficient stiffness to attract significant bending moment and stress caused by that racking deflection. As a consequence, rather than acting as pure columns, the braces behave as beam/columns, with the magnitude of the induced bending stress roughly equal to the axial stress. This drives the brace wall thickness up to roughly twice what is needed to resist the squeeze/pry loads as simple columns. The greater the wall thickness, the greater the stiffness and bending moment. Greater bending moment results in more stress and weight is needlessly added in a vicious cycle. The structural details of this arrangement must consider fatigue, which is negatively affected by wall thickness and adds further weight.

In this paper, the authors explore a structural arrangement for minimizing the bending moment experienced by the braces of Configuration 2 to show how the structural weight can be reduced by over 20%, compared to the single horizontal braces with fixed ends.

METHODOLOGY AND DESIGN CRITERIA

Two brace configurations were examined, one with quasi-full fixity at the brace to column connection, per Traditional Design, and a New Design with a central Flex Element to minimize bending stiffness at the interface. The two configurations are shown in Figure 4.

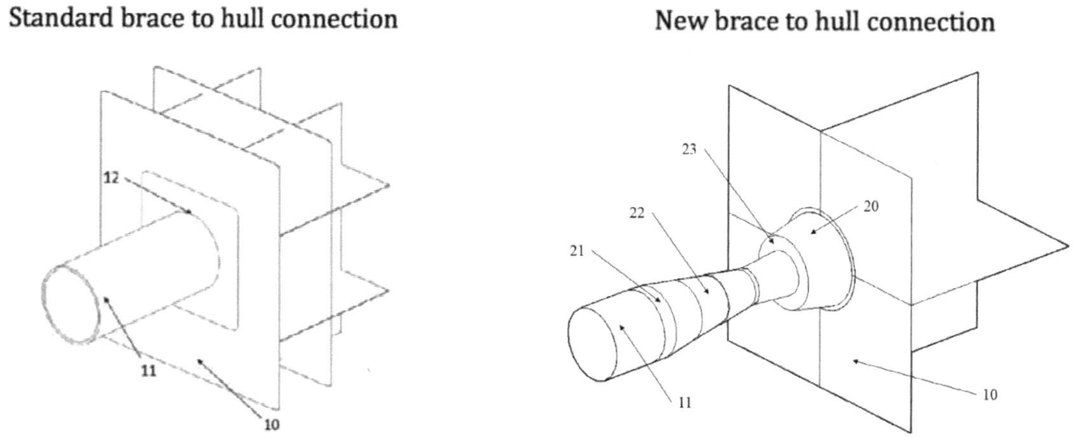

10 – Column or Pontoon Outer Plate (shell)
11 – Brace
12 – Brace to Hull Connection

20 – Outer Transition Piece
21 – Outer Band with Increase Plate Thickness
22 – Reduced Radius before Connection
23 – Warping Plate

Figure 4. Brace to Hull Connections (Cottrell & Cottrell, LLC, 2016)

A simplified diagrammatic showing the load paths provided by each configuration is given in Figure 5. As can be seen, the fixed end brace suffers bending moment and bending stress, due to its engaged large diameter, and larger stiffness, whereas the new configuration greatly reduces the bending stress component, by virtue of the Flex Element's smaller diameter. This reduction in stiffness allows the New Design brace to act more like a simple column.

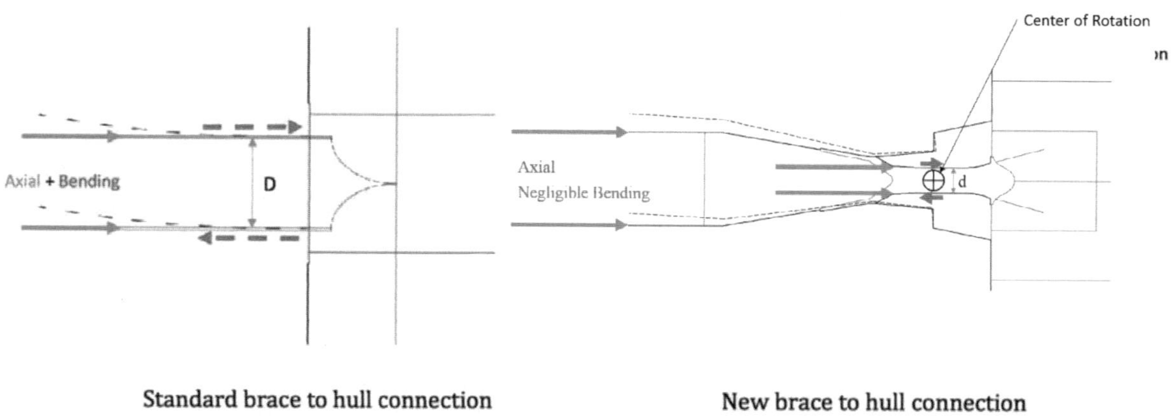

Figure 5. Load Paths: Traditional Fixed-Ended vs Pinned-Ended Brace (Cottrell & Cottrell, LLC, 2016)

When the new brace experiences end deflection, the Flex Element deforms into a "C" shape, and has a constant bending moment throughout its length. The Flex Element does not adopt an "S" shape because the warping plate acts like a gimbal, constraining the brace end to rotate (0.21° in the base case) about the mid-length point of the Flex Element. During brace end deflection, the in-plane load on the warping plate, is caused by the Flex Element end shear acting in one direction and the brace end shear acting in the opposite direction plus or minus the in-plane component of brace compression or tension axial load. The resulting net in-plane load on the warping plate is manageable and the warping plate shear stress is low. When the brace end is deflected, the warping plate deforms into an "S" shape, in the plane swept by the brace, which is in the quadrant of the warping plate perpendicular to the plane experiencing maximum shear stress. Because the warping plate load is low, it can be thin, so this deflection results in very low bending stress in the warping plate.

DESIGN LOAD CASES

To quantify the potential advantages of such a large reduction in bending stress, the authors designed a brace of each configuration, both having a column to column length of 1,640 inches, subjecting each to the same nominal representative loads and base case deflections. The strength design criteria was as follows:

1. Axial load of 10,000 kips in compression.
2. End deflection, fore and aft, or vertically of 6 inches from rest, as the base design case.
3. End deflection was then varied from 0 to 12 inches, to explore the behavior and benefits available from the New Design under different conditions.

These loads and deflections are considered representative of the loading experienced by a Semi-submersible brace when the global body is subjected to the design wave load. Obviously, the magnitude of these loads will vary dependent upon the semi's size, deck box stiffness and hull design, however for the sake of comparison of the two brace end designs, the authors feel these loads are reasonable. A diagram of the base case loads applied to each configuration is shown in Figure 6.

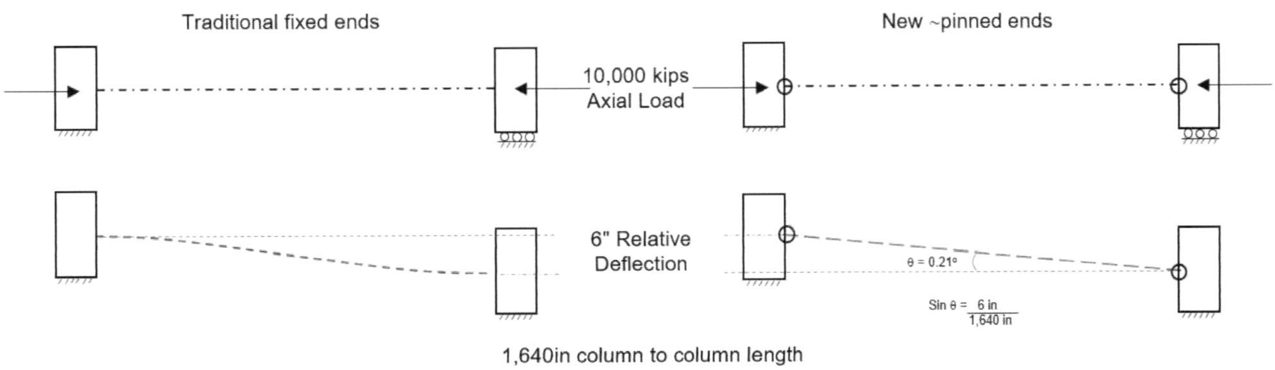

Figure 6: Load and Deflection Case Used to Size the Braces

From a global perspective, since the scantlings of the New Design are lighter than the Traditional Design, the same 10,000 kips wave loading, results in a greater axial deflection with the New Design. Consequently, the increase in angular deflection between the column and deck box structure, causes some of this load to be shared at that connection. This means that the New Design will actually see less axial load than the Traditional Design, so our results from that point of view are conservative in penalizing the New Design to a greater extent. If the column to deck box structure is governed by the net horizontal load from the pontoons, rather than the deflection of one side relative to the other, then this additional deflection does not increase the scantlings of the column to deck box structure and further analysis would reduce the New Design brace structural loading and hence its weight more than credited in this report. The reduction in load, for the New Design, is around 8% and if applied, would further reduce the weight of the New Design. If this greater flexibility of the New Design requires strengthening the column to deck box structure, then that increase in weight would reduce the benefit derived from the 8% load reduction for the New Design. For the sake of this comparison, the authors believe this effect to have a negligible net impact on the overall weight of the two structural arrangements; if anything potentially resulting in an understatement of the net benefit to be derived from the New Design. So at this time, we compare the weights derived through application of the same axial load to both designs.

Analysis has shown that because the deck box stiffness is orders of magnitude greater than that of the braces, large changes in brace stiffness does not significantly affect brace end deflection. The authors therefore believe that the application of the same 6" deflection to both designs makes for a fair and realistic comparison.

A secondary load case check was made to ensure sufficient strength perpendicular to the brace axes, by applying a distributed wave loading of 440 kips along the length of each brace. This load corresponds to the maximum wave slamming load during transit, with the columns and braces above the waterline. This load was applied independently of the survival load case above, to ensure the braces had sufficient bending and radial shear strength. This check revealed the need in the New Design for the so-called "warping plate" (piece 23 in Figure 4), to resist brace lateral loads without bending the Flex Element. The brace bending and shear stresses due to this wave slamming was found to be acceptable for both brace configurations.

BRACE SCANTLINGS DESIGN

The AISC Manual of Steel Construction (AISC 1973) Formulas for Combined Axial and Bending Loading (reproduced below for convenience), was used to establish the scantlings of the two respective braces. The allowable stresses were multiplied by 1.33 to account for the nature of the strength loads being applied.

$$\frac{f_a}{F_a} + \frac{C_{mx} f_{bx}}{\left(1 - \frac{f_a}{F'_{ex}}\right) F_{bx}} + \frac{C_{my} f_{by}}{\left(1 - \frac{f_a}{F'_{ey}}\right) F_{by}} \leq 1.0 \qquad \textbf{1.6-1a}$$

$$\frac{f_a}{0.60\, F_y} + \frac{f_{bx}}{F_{bx}} + \frac{f_{by}}{F_{by}} \leq 1.0 \qquad \textbf{1.6-1b}$$

The design approach was to begin at brace mid-span, and step outboard toward the columns. The bending moment at each section increases toward the column wall, reaching a much higher magnitude for the traditional fixed end brace than for the new pin ended configuration. Therefore, in order to establish brace wall thickness, the locations of plating thickness increases were established, as required to satisfy the above equations. At the brace ends, where the structural arrangement transitions between tubular beam-column sections and the cruciform deck and bulkhead structure of the column, combined stresses were calculated, with due consideration made for stress concentrations to establish the scantlings of the brace to column structure. To establish a proper K-factor for use in the AISC equations, an eigenvalue analysis was performed to determine the appropriate K-factor to use for the Traditional and New Designs, resulting 0.80 and 1.00 respectively.

DETAILED FE ANALYSIS

A detailed FE analysis of the resulting brace scantlings and the varying end connection designs was performed in ANSYS. The FE Analysis was used to confirm the adequacy of the design and ensure that stress levels were similar between the two configurations to establish a fair comparison. A generic semi-submersible design was used for the analysis of the brace scantlings and end connections. Figure 7 shows the FE model for both brace designs, noting that the semi was created using a generic plate thickness of ½ inches throughout the columns, pontoons, and deck box. Additional strengthening was done at the deck box to column connection.

Shell elements were used for the majority of the model, except for the Flex Element, which was modeled using solid elements. Finite element node spacing was 6 inches at the connections and 60 inches away from the connections. AH 36 material was assumed with an 88% Utilization Factor for the brace beam-columns, appropriately reduced at the ends to account for stress concentrations, depending on the respective structural geometry.

For this analysis, the boundary of the port pontoon was fixed and the starboard pontoon was deflected until the axial load on one brace was 10,000 kips and the end deflection at the brace to column connection was 6 inches, measured perpendicularly to the end of the brace at its connection to the column.

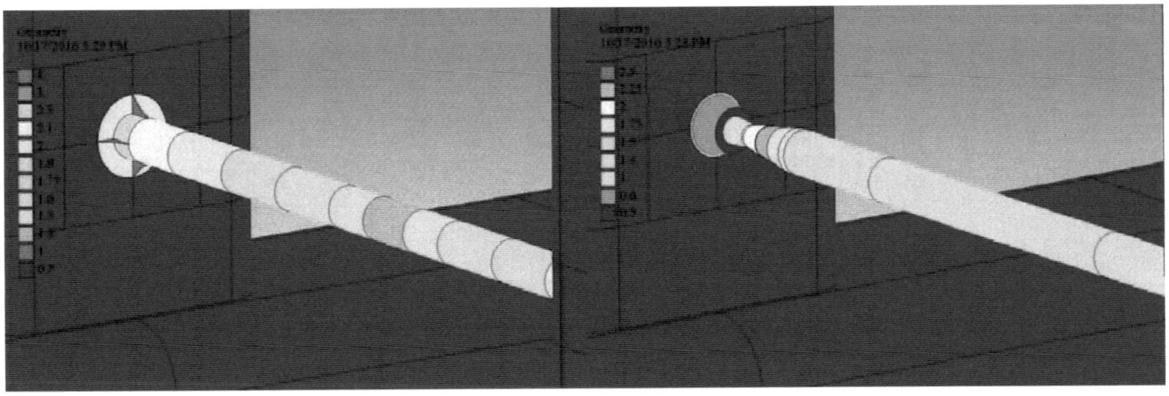

Figure 7. Brace Scantlings (Base Case): Traditional Design and New Design Models

DESCRIPTION OF RESULTING BRACES

Both the Traditional Design and the New Design have a mid-wall of 78-inch diameter. The base case Traditional Design required a 1.3" wall at mid-span, whereas the New Design satisfied the above AISC criteria with 1.4" plate thickness. At the brace ends, the Traditional Design had a 2.3" wall, whereas the New Design had a 1.4" wall. The Flex Element of the New Design is 21.4" in diameter and its cylindrical portion is 80" long.

Inside the column, both designs feature a cruciform configuration formed by the intersection of a transverse bulkhead and a deck, but with differences in the remaining column internal backing structure local to the connection.

The Traditional Design has 4 knife-edge intersections, which require transition brackets within the column in the form of a structural scalloped cylindrical skirt, of the same diameter as the brace to transfer the axial and bending stresses between the orthogonal walls. On the brace side of the connection, radial brackets inside and outside of the brace are required in alignment with the bulkheads and flats of the column. Within the column structure, the bulkhead and flat of the Traditional Design must be reinforced to transfer the knife edge axial and bending induced loads from the brace walls through the 4 knife edge crossings. This complicated structural transition is typically handled at the column wall by making the brace structure, including tubular, brackets and diaphragm as a cast piece. In this study, the allowable stress for this casting has been reduced by 10% to allow for thickness increases required for stress concentrations and casting mill tolerance.

The Flex Element of the New Design transitions inboard with a forged conical flange to the conical pieces of the brace and outboard with a forged flared bulbous portion to the column backing structure. In the New Design, the bulkhead and flat intersection must be reinforced to transfer well aligned, almost purely axial, loads at a single crossing. The Flex Element will be a more reliable and readily available forging, with generous radii at the cylindrical to flared locations, to minimize stress concentrations.

Figure 8 provides an isometric view of both the Traditional Design and the New Design and their associated plate thicknesses. Figure 9 shows a section view, looking down, taken just above the flat supporting each brace in the horizontal plane, showing the plating thicknesses. The transverse bulkhead supporting each brace in the vertical plane is similar. Full internal structural inspection access is available for both designs; from the column in the Traditional Design and through a mid-span man-way for the New Design.

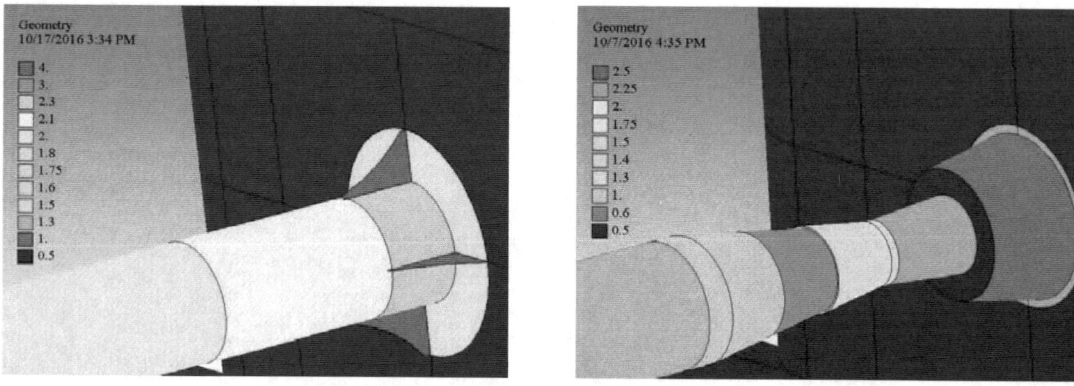

Figure 8. Detailed Brace Scantlings: Traditional Design and New Design

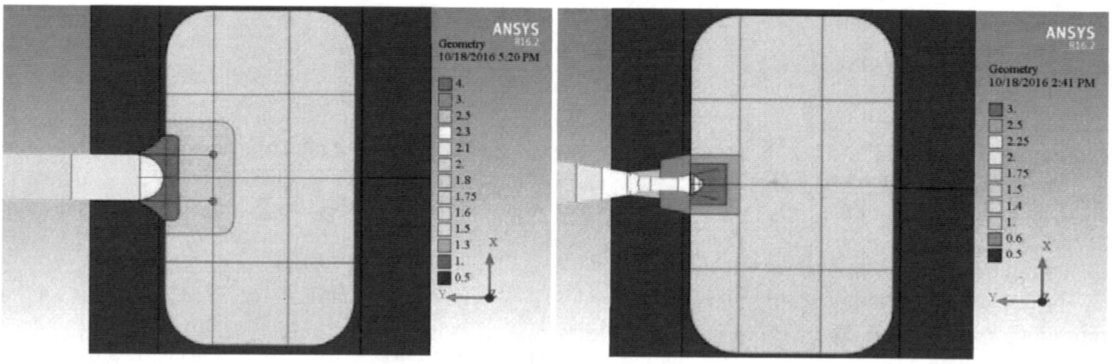

Figure 9. Brace to Column and Column Interior Scantlings: Traditional Design Left and New Design Right

A parametric analysis was performed to determine the effects of varying end deflections on brace wall sizing and flex element lengths. Varying the end deflection from 0" to 12" results in the scantlings as shown in Figure 10.

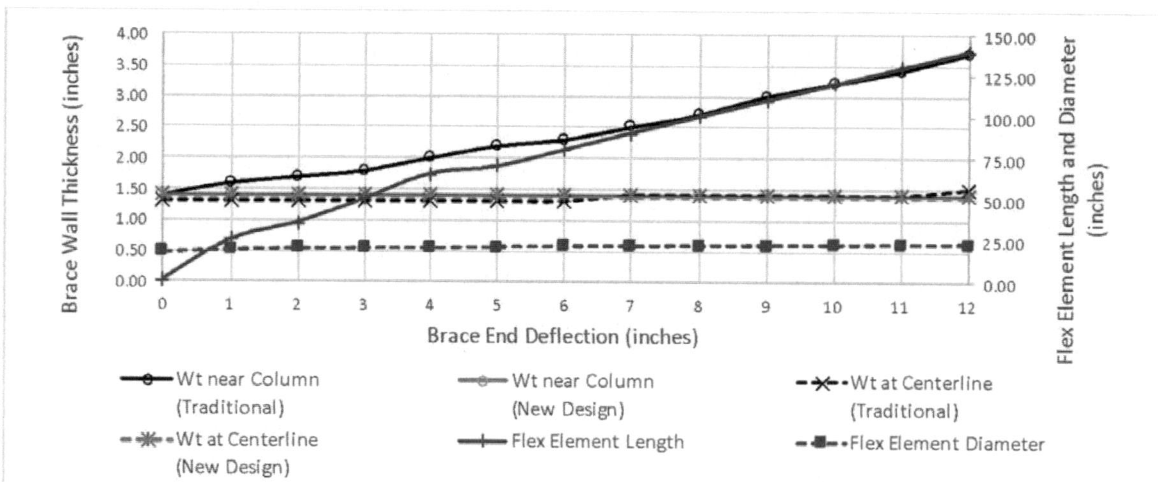

Figure 10. Brace Scantlings, Traditional and New Designs vs End Deflection

LOAD, GEOMETRY AND STRESS COMPARISON BETWEEN TRADITIONAL AND NEW DESIGNS

The following tables compare key features of each of the two designs, summarizing base load case, geometry and scantlings, and structural performance.

Table 1. Brace Comparison Summary: Base Load Case

Basic Load Case	Traditional Design	New Design
Axial Load	10,000 kips	10,000 kips
End deflection	6 inches	6 inches

Table 2. Brace Comparison Summary: Geometry and Scantlings

Geometry and Scantlings	Traditional Design	New Design
Diameter to mid-wall	78 inches	78 inches
Wall thicknesses:		
mid span	1.3"	1.4"
Brace at outboard end	2.3"	1.4" / 21.4" dia. X 80" Flex
Column interior structure thicknesses	4" to 1-3/4"	3" to 1-1/2"

Table 3. Brace Comparison Summary: Structural Performance

Structural Performance	Traditional Design	New Design
Column Ends	Quasi-Fixed – Quasi-Fixed	Quasi-Pinned – Quasi-Pinned
K	0.80	1.00
Kl/r	48 (brace)	56 (brace) 10 (Flex)
End shear, due to end bending moment	270 kips	19 kips
End bending moment	163,000 kip in	14,600 kip in
Stresses		
Global maximum	23,000 psi	22,000 psi
Global average	18,000 psi	18,000 psi
Hot spot	52,894 psi	38,752 psi

The following Figure shows the utilization factors from the AISC beam column buckling code Formula, separated into axial and bending stress components for the Traditional Design, New Design and Flex Element vs end deflection.

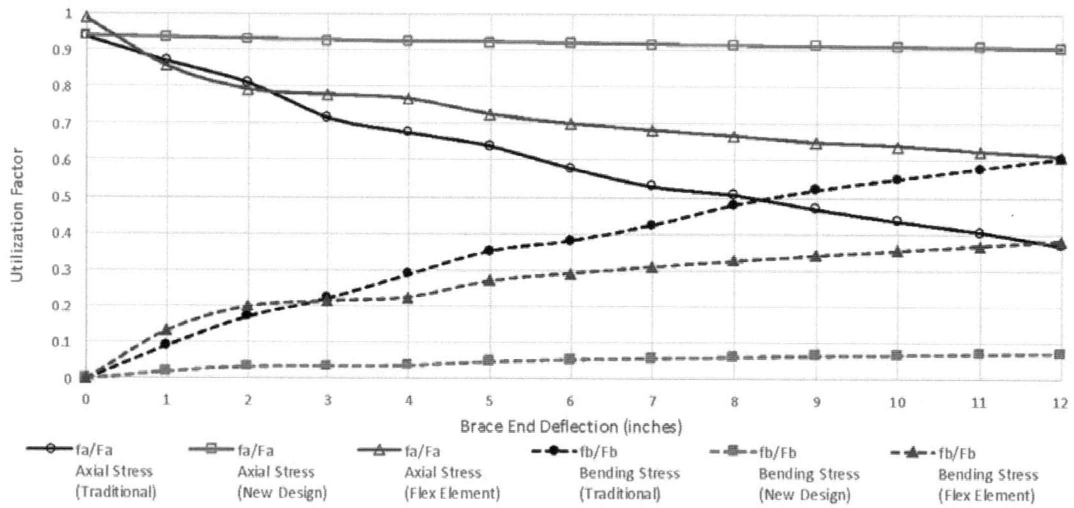

Figure 11. Utilization Factors of Traditional, New and Flex Element Designs vs End Deflection

The results of the detailed finite element analysis used to confirm the first principals calculations presented in Figure 11, are given in the following Figures. Figure 12 shows the Equivalent Stress found for the Traditional Design while Figure 13 provides a plot of the Equivalent Stress for the New Design for the same loading condition. Close-up views of the connections themselves is shown in Figure 14 and of the Flex Element in Figure 15. As is evident from the stress plots provided, both designs have a similar level of stress throughout including their maximum stresses (both located at hotspots) which are 52.9 ksi and 38.8 ksi for the Traditional and New Design respectively.

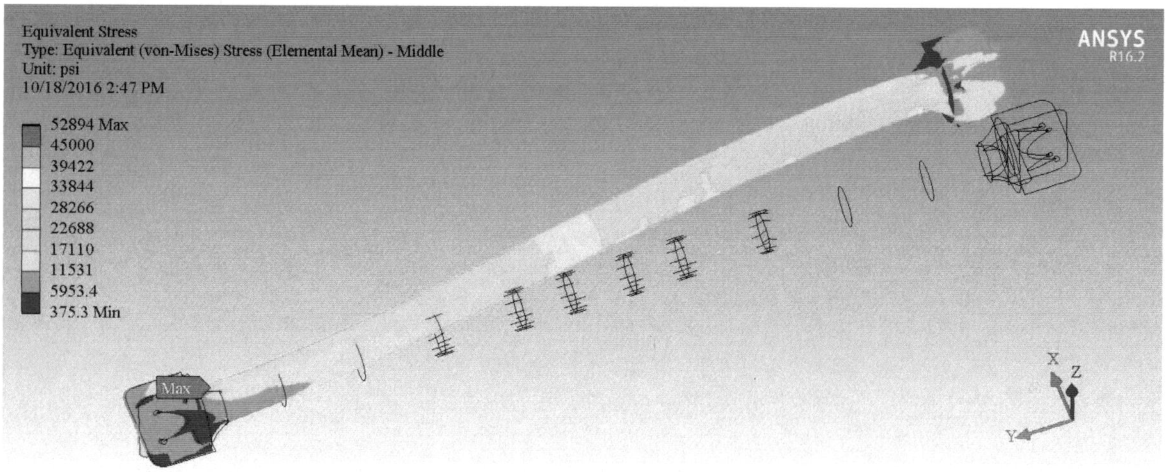

Figure 12. Equivalent Stress: Traditional Design

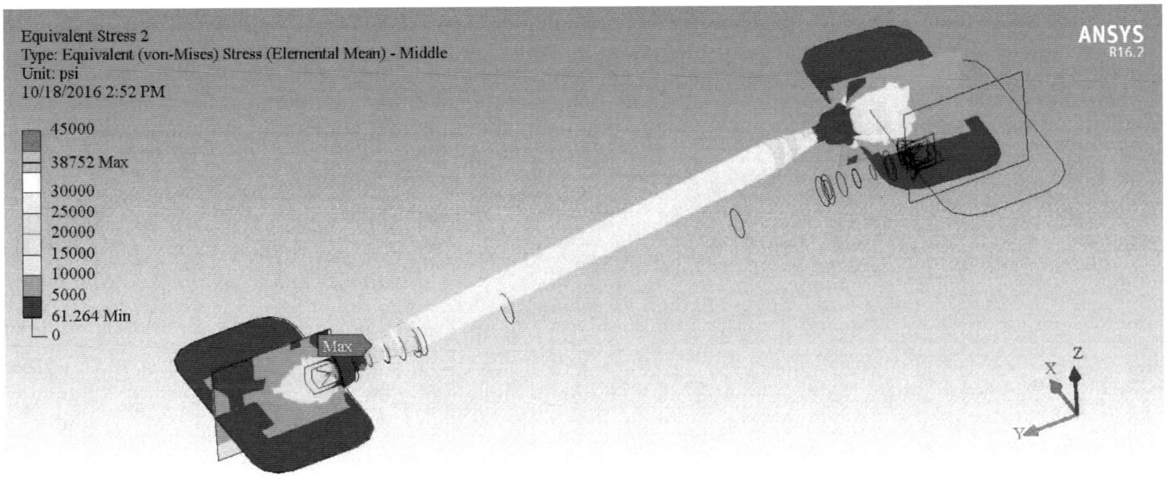

Figure 13. Equivalent Stress: New Design

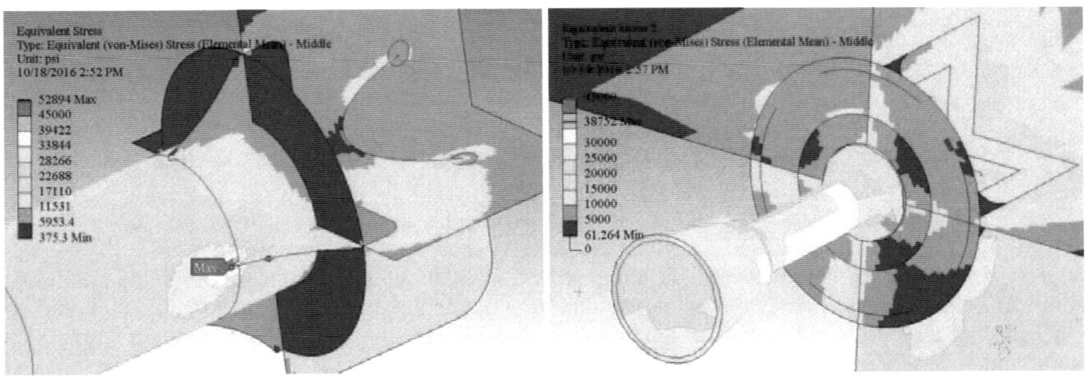

Figure 14. Equivalent Stress: Traditional Design Left and New Design Right

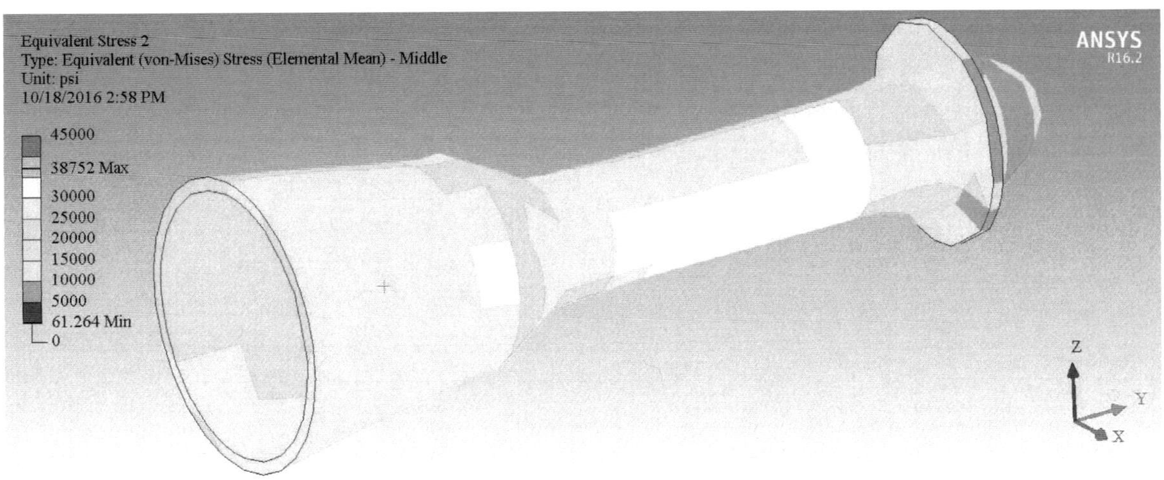

Figure 15. Equivalent Stress: New Design Flex Element

The following Figure shows a section view, looking down, taken just above the flat supporting each brace in the horizontal plane, showing the equivalent stress. The transverse bulkhead supporting each brace in the vertical plane is similar.

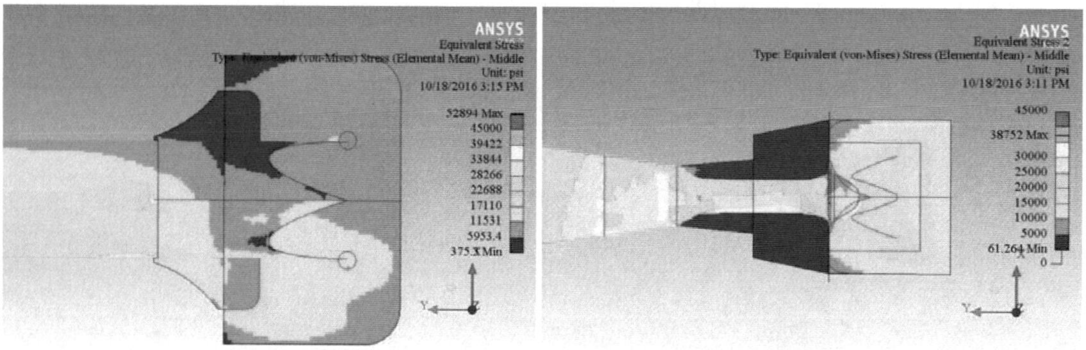

Figure 16. Brace to Column and Column Interior Equivalent Stress: Traditional Design Left and New Design Right

Having demonstrated that the stresses for each of the two brace configurations is very close when subjected to the same loading case, the authors believe that the designs provide a fair basis for comparing their structural weight.

WEIGHT COMPARISON BETWEEN TRADITIONAL AND NEW DESIGNS

The New Design is lighter. Not only is the brace itself lighter, but the attachment structure at the brace to column connection is also lighter on the column side. As mentioned above, this is because the New Design connection does not have the high induced bending moment inherent in the Traditional Design and the New Design transfers its axial load more efficiently from its central Flex Element through to the cruciform intersection line.

To quantify the weight savings available through the New Design, the weights of both designs was calculated, first by hand, then through the weight feature from ANSYS. The hand calculation includes stiffeners. The ANSYS weight compares the two models, so includes more of the semi-submersible's structure, but doesn't capture stiffener weight.

Table 4. Weight Comparison Summary (Base Case)

Hand Calculation (including stiffeners)	**Traditional Design**	**New Design**
Total	784 kips	601 kips
Savings in kips = Traditional Design – New Design		183 kips
Savings in % = Savings in kips/Traditional Design		23%
Model Weights from ANSYS (no stiffeners)		
Total	773 kips	603 kips
Savings in kips = Traditional Design – New Design		170 kips
Savings in % = Savings in kips/Traditional Design		22%

From Figures 10 and 11, it can be appreciated that with increasing end deflection, a higher component of the AISC allowable stress is devoted to resisting bending and that this results in increased scantlings for the Traditional Design. In contrast, for the New Design, due to the brace end moment reduction provided by the Flex Element, the component of AISC allowable stress devoted to bending in the brace remains below 10% for all end deflections. This results in an increased weight advantage for the New Design, where utilized for conditions of higher end deflections. This trend is demonstrated in the following Figure:

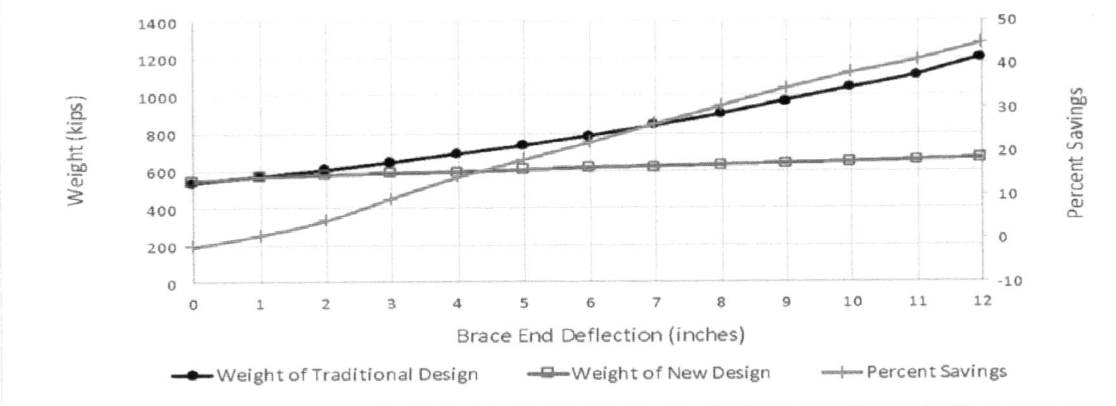

Figure 17. Brace Weight and Percent Saved vs End Deflection

POTENTIAL FATIGUE PERFORMANCE

It should be noted that a detailed fatigue analysis has not been carried out on either brace design. However, the New Design is expected to outperform the Traditional Design for the following reasons:

1. The fatigue bending stress is anticipated to be a higher component of stress than assumed in the strength bending stress. This will favor the New Design as it is less sensitive to bending.
4. The New Design flex element better aligns with the column internal structure, enabling less drastic changes in geometry, resulting in lower Stress Concentration Factors.
5. The flex element is in air, not exposed to seawater, whereas the traditional design's fatigue sensitive external brackets must use the cathodically protected S-N curve, which has a lower allowable stress range than in air.
6. The above factors are anticipated to overcome the greater thickness correction applicable to the new design's forging.

The authors expect to verify these assumptions in future work.

SUMMARY AND CONCLUSIONS

The horizontal brace structural arrangement has become the preferred method for counteracting squeeze/pry loading while complying with the racking deflection in semi-submersible hulls. The traditional horizontal brace to column connection design is typically a built-in connection that acts as a fixed-fixed connection. This connection design attracts unnecessary bending moment and thus stress. To counteract this increased stress, the scantlings of the traditional brace are increased, increasing the weight.

The authors have proposed a structural arrangement for the brace (essentially a pinned-pinned connection) that minimizes the bending moment and thereby reduces the structural weight. Not only is the brace itself lighter in the New Design, but the reduced end moments and good alignment with the column structure, results in a lower weight for the column side support structure. For a typical semi-submersible, subjected to the base case loading, the weight savings was found to be a significant 23%, representing 183 kips of weight saved for the New Design.

ACKNOWLEDGEMENTS

The authors would like to thank the employees of 3Dent Technology who have provided analysis for guidance to our understanding of the optimization and behavior of the semi-submersible brace.

REFERENCES

AISC (1973) Manual of STEEL CONSTRUCTION. New York, NY: American Institute of Steel Construction, Inc.

ANSYS Release 16.2.

Cottrell & Cottrell, LLC and 3Dent Technology, LLC, Bending Stiffness Reducer for Brace to Hull Connection. US Provisional Patent Application 62302905, March 3, 2016.

Chakrabarti, Subrata Kumar (2005) Handbook of Offshore Engineering. Plainfield, IL: Elsevier.

Liden, Hadar; Frolunda, Vastra Semisubmersible Vessel. U. S. Patent 4,436,050, Feb 11, 1986.

A RESPONSE CONDITIONED WAVE APPROACH FOR STRUCTURAL ANALYSIS OF FPSO PRESENTING RESONANT SECOND ORDER ROLL MOTIONS

GUILLAUME DE-HAUTECLOCQUE
HEAD OF HYDRODYNAMIC SECTION, BUREAU VERITAS PARIS

FLAVIA C. REZENDE
MANAGER TECHNICAL DEVELOPMENT, BUREAU VERITAS HOUSTON

FABIEN BIGOT
HEAD OF STRUCTURE SECTION, BUREAU VERITAS PARIS

QUENTIN DERBANNE
DIRECTOR OF RESEARCH DEPARTMENT, BUREAU VERITAS PARIS

Proceedings of the 22nd Offshore Symposium, February 2017, Houston, Texas
Texas Section of the Society of Naval Architects and Marine Engineers

ABSTRACT

Rolling of FPSOs is a major concern, as excessive motions may result in several operational problems including process plant downtime. Moreover, it is well known that the loads induced by roll motions may govern the design of several structural elements of the unit, such as topside supports. In order to improve the hydrodynamic performance of new FPSOs, longer roll natural periods are targeted so that the resonance within the range of periods of incident waves is avoided. Nevertheless, such a measure does not completely mitigate the resonant motions and the unit may still respond at second order (at the difference between wave frequencies). Much work has already been devoted to the accurate estimation of second order roll motions of FPSOs. However, the state-of-the-art methods for structural assessments still rely on linear assumptions and are not able to account for second-order response. In this work, we present a novel approach to obtain a design load case for structural assessment of elements whose maxima stresses are governed by second order roll. The approach is based on the concept of response conditioned wave (RCW). As the process is non-Gaussian, it is not possible to derive the RCW analytically. Therefore, first order reliability method (FORM) is used to numerically obtain the most probable wave yielding the target roll-affected load value. It is shown that the resulting wave also leads to the most probable responses associated to the target roll motion, as an example first order heave motions, demonstrating the robustness of the method.

Keywords: Second-order roll, EDW, RCW, structural assessment

INTRODUCTION

FPSOs are the preferred solution for the development of fields in various offshore regions, as they present several advantages comparing to other solutions, including large deck footprint, storage capacity, etc. Due to their ship-shape, one of the main disadvantages of this kind of unit is the hydrodynamic behavior, particularly their roll motions. Therefore, since the start of the use of FPSOs, roll motions have been one of the major concerns for designers and operators. Large motions, not predicted during the design phase, may lead to several unexpected problems, including downtime of process plant, excessive structural fatigue of the hull and risers` system, greenwater events and discomfort for the crew, among other limitations for performing marine operations. The roll motions effects become even more relevant for LNG FPSOs, which have more strict operational constraints than oil FPSOs, including smaller maximum inclinations for the efficiency of the process plant.

The reason why rolling of FPSOs may be so important is that they are resonant motions with low damping. With that in mind, one way of limiting the roll response is to move the resonant period of the vessel outside the range of periods of the incident waves, which is the region where the excitation loads are highest. This can be done by optimizing the hull geometry in such a way that the center of gravity will be as high as practically possible without compromising the stability of the unit.

Indeed, a long natural period will dramatically reduce first order roll motions, i.e. the motions that occur at the frequencies of the incident waves. Nevertheless, it does not completely mitigate the response and the vessel will continue to roll around its natural period due to second order loads, which occur at the difference-frequency between incident waves. As for every resonant response, damping is a key parameter, and it depends on the motion velocity which is very small for low-frequency motions. Thus, even though the second order excitation loads are much smaller than the first order loads, the motions are sometimes not negligible due to the low damping. So, it seems essential that the second order roll motions are considered in the structural assessments of FPSOs. The current approach for structural design which is based on regular equivalent design waves is incapable of capturing second order effects, which raises a concern with respect to integrity of FPSOs presenting that type of roll response.

The novel method proposed in this work is also based on the concept of equivalent design wave, which is defined as a wave train that leads to a target value of the load parameter that is supposed to govern the stresses of certain structural elements of the subject structure. In this case, we consider an irregular response conditioned wave (RCW). As the process is non-Gaussian, there is no analytical formula for the RCW. So, the same approach as the one presented in Jensen (2007), which is based on first order reliability method, was adopted for obtaining the most probable wave leading to the target value.

To compose a load case for structural analysis, not only the governing load parameter is needed but also all other associated loads. It can be shown that the associated loads may assume a range of different values at the moment the governing parameter reaches its design value. It is shown that the RCW resulting from the application of the method presented here leads to the most probable values of the associated loads. In order to demonstrate the application of the method and its relevance in practice, 2D beam model of a typical FPSO section is used. In such a case, it is possible to compute QTFs (Quadratic Transfer Functions) and time series of stresses which are used as benchmarks to validate the results of the RCW method.

NOMENCLATURE

FPSO: Floating Production Storage Offloading
EDW: Equivalent Design Wave
RCW: Response Conditioned Wave
QTF: Quadratic Transfer Function
RAO: Response Amplitude Operator
FORM: First Order Reliability Method

THE SECOND ORDER ROLL RESPONSE

As mentioned above, the second order roll response occurs when the vessel is excited at its natural period by the low-frequency loads resulting from the interactions between the waves in a seaway. These loads, proportional to the square of wave amplitude, are usually obtained by frequency domain diffraction-radiation computations in the form of quadratic transfer functions (QTFs) that provide the roll exciting moment per meter square of wave amplitude for different pairs of wave-frequencies. In fact, as the response is narrow-banded, not every pair of wave frequencies is of relevance; only those for which the frequency difference is close to the natural frequency of the unit.

It has been shown in Rezende *et al* (2013) that the computation of roll exciting moment QTFs presents some specificities compared to the well-known computation of the horizontal loads QTFs (used for mooring analyses). But once they are obtained, the roll motions can be computed.

The design roll motion will be the extreme value associated to a defined return period (typically taken as 100YR for FPSO designs). It should be stressed that for the estimation of the design value, one shall take into account the nonlinearity of the excitation loads and of the damping in a proper manner. Figure 1 shows how the probability distribution of second order roll motions are affected by the non-linearity of excitation loads (time-domain reconstruction). Furthermore the variability of the short term response (the fact that the extreme response varies for different realizations of the same sea state) shall be considered. The accuracy of the damping is also essential to a correct estimation of extreme roll motion, however, as this is not the focus of the paper, for the sake of simplicity, a linear roll damping model has been used in the results presented herein.

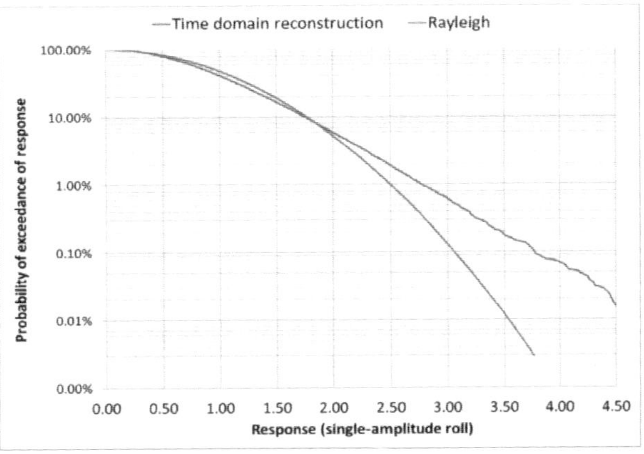

Figure 1 Probability distribution of second order roll motions (in red) in comparison with Rayleigh distribution (in blue)

FPSO STRUCTURAL ASSESSMENTS

One of the main purposes of the seakeeping analysis is to set the dynamic loads for the structural assessments of the FPSO design. The accuracy of the dynamic loads is essential since it will significantly affect the final scantlings. The usual approach for strength assessment is based on an equivalent design wave (EDW) method. Generally those EDWs are defined as regular waves aimed at reproducing the same load magnitude as the one obtained from a long term analysis of the loads. This approach provides the advantage of drastically reducing the number of structural computations compared to the full spectral analysis which would require that two calculations (i.e. real and imaginary parts) are performed for each pair of frequency and incidence to compose a RAO of stress. Furthermore the EDW approach enables accounting for non-linearity in the wave-body interaction in a more cost-effective way.

The main assumption behind the EDW method is that the stresses in every element of the structure are maximized at the same time as one of the dynamic load parameters applied to the vessel (ex. bending moment, acceleration, pressure, etc.), so that the maximum stress among all the considered design waves (likely few) will match the full spectral calculation result.

This has been extensively validated for first order (and weakly non-linear) responses, in particular by Classification Societies, who apply the method as the background for the definition of their prescriptive loads for ship-shaped vessels.

It should be noted that previous works have already demonstrated that for a tanker-type vessel like a FPSO, there are only a few waves that maximize the stresses all over the structure. Among those are the transverse acceleration and the external pressure design waves, which are both very much affected by the roll motions.

The key question is then how to define a design waves that account for 2nd order loads? One obvious way to illustrate the limitation of the design waves used for weakly non-linear response is to think of the classic regular wave. A regular wave is not able to generate any second order slowly varying response.

The following section presents a way to deal with 2nd order loads, still using the design wave methodology. To validate this approach, a realistic FPSO 2D structural model is used. In this model the stress in longitudinal stiffeners is obtained by combining the hull girder bending stress with the local bending stress due to both sea pressure and liquids in tanks. The stresses were computed on more than 400 structural elements, dispatched on 3 different sections. The mid-section is displayed in Figure 2. The loading condition is tuned so that the roll resonance period is about 45s, and thus excited mainly by 2nd order loads. The computation was performed on individual sea-states, and the short-term maximum was taken as the one with a 3h return period.

As the model is prescriptive, it is possible to explicitly write stresses as linear combination of a small number of loads (acceleration, sea-pressure and hull girder loads). Then, it is easy to generate stress transfer functions, including stress QTFs. For the QTF, only the loads directly linked to 2nd order roll have been considered. Thus, the sea-pressure, the transverse and vertical accelerations include 2nd order components, but the hull girder loads are limited to 1st order. Besides, the sea-pressure does not account for the 2nd order potential. The effects of these approximations should be negligible.

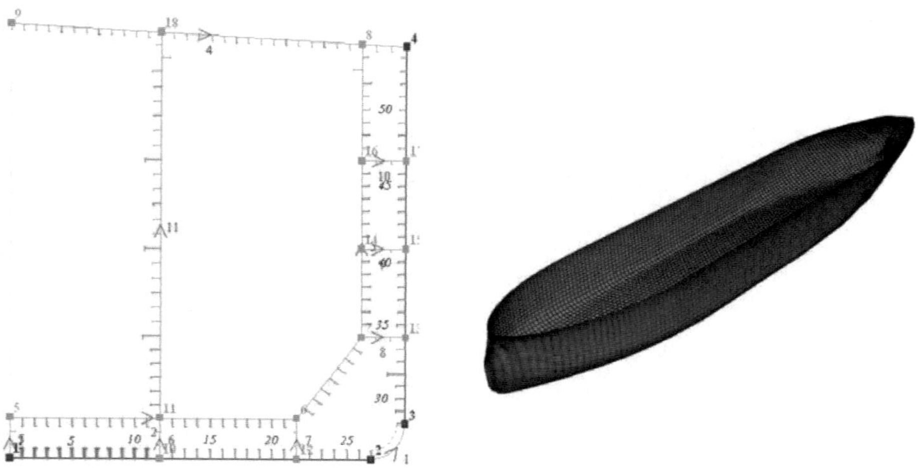

Figure 2 : FPSO section and hull hydrodynamic mesh

Once stress RAOs and QTFs are available, is possible to reconstruct time traces of all stresses, followed by an up-crossing analysis that yields the extreme value. **Figure 3** illustrates the process for one of the elements. It clearly shows a significant 2nd order contribution to the total stress. Figure 4 shows the results of that procedure for all the modeled elements, and 2nd order roll effect is significant for a substantial number of elements.

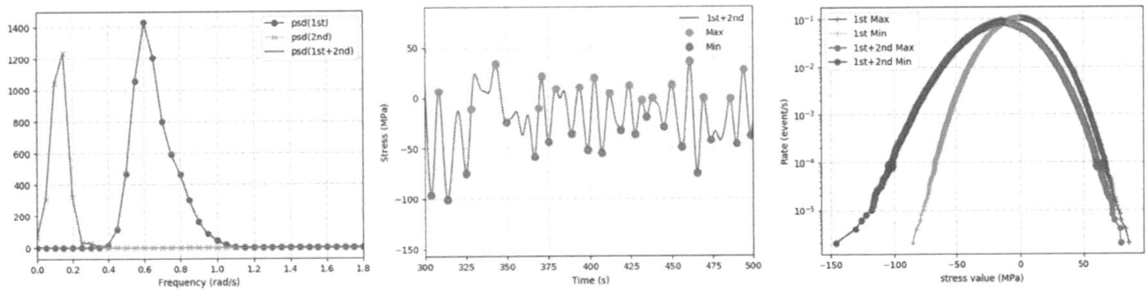

Figure 3 : Power density spectrum, up-crossing analysis and short-term distribution - Side shell elements
(Hs=10m, Tp=10s, Heading=135°).

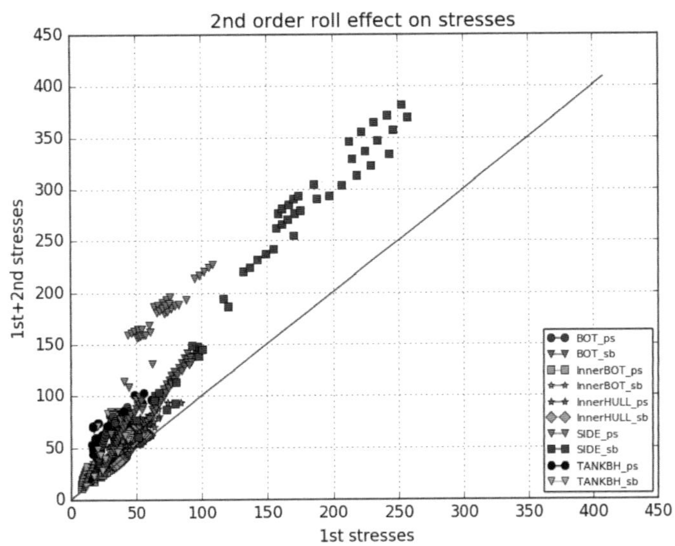

Figure 4 : 2nd order roll effect on stresses - 3h return period
(Hs=10m, Tp=10s, Heading=135°).

In practice, the above procedure is not practically feasible if the structural model is more complicated (using a FEM model for instance). Indeed, in the case of the second order response, in order to perform a full time domain analysis, one should reconstruct the QTF of stresses which requires a much larger number of structural computations than for first order responses. Furthermore, the consideration of nonlinearity in damping would be too cumbersome to handle. A simpler approach is thus necessary to provide consistent load cases.

One very simple way of including the second order effects would be to simply add them on the top of the maximized first order loads. While probably better than nothing, this approach ignores the phasing between the 2nd order roll and the other loads contributing to the stress. In order to emphasise the importance of this phasing, Figure 5 shows the comparison between the sum of 2nd order and 1st order stresses and the actual 1st+2nd order stress. It is noted that simply summing the maximized first and second order loads yield to conservative results, as one would expect. Thus, another method would be desirable to avoid over-designing the structure.

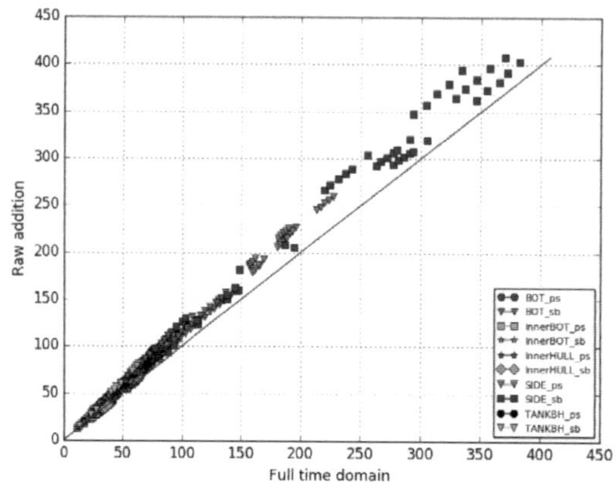

Figure 5: Comparison between the simple sum method and the time domain reconstruction of QTFs of stresses.

To illustrate the issue further, let's consider the time series of a FPSO roll motions (mainly of second order) and of the waterline pressure during the realization of a sea state as represented in Figure 6. If a very long realization of the sea state is made and we register the values of the waterline pressure at every instant the roll angle reaches a peak. We may present the distribution of roll peaks ("dom") with respect to the associated waterline pressure ("sub") as in Figure 7. It can be noticed that if the roll motion is targeted at a certain probability level (ex. 10^{-4}), there is a range of possible values of waterline pressure or other first order quantity that can be associated, which will normally result in different structural response.

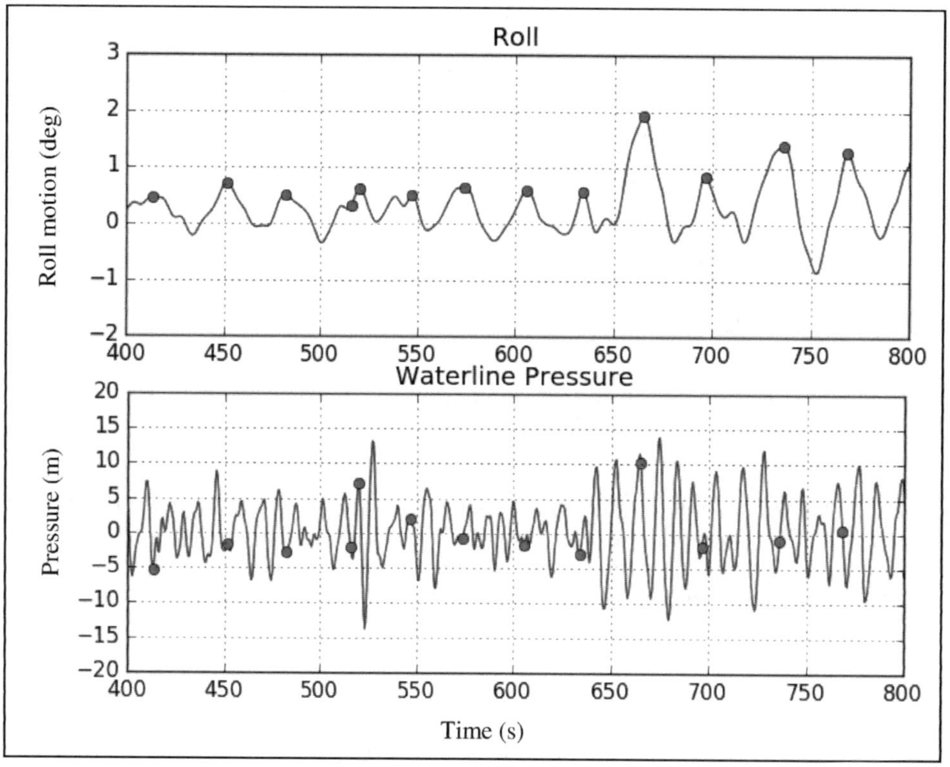

Figure 6 Time series of second order roll motion and waterline pressure. Green points in both graphs correspond to the instants where roll motion reaches a peak value.

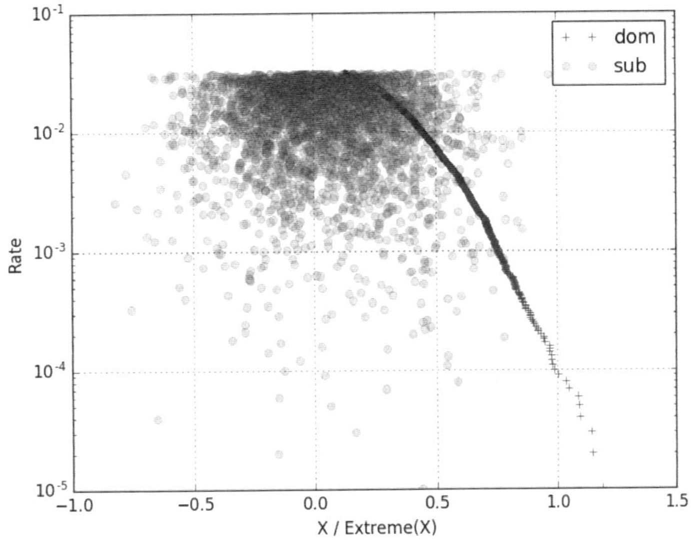

Figure 7 Distribution of roll peaks ("dom") with respect to the associated waterline pressure ("sub"). The responses are normalized by their respective extreme values for plotting.

Considering the stochastic nature of the waves, the issue comprises finding, among the waves that yield the extreme response, the most probable one. The design wave would lead to the most probable set of associated loads at the instant the governing load reaches its design value. Ultimately this design wave is expected to produce a dynamic load case that will result in a stress similar to the one obtained by a full time domain analysis.

RESPONSE CONDITIONED WAVE APPROACH

In the linear case, the most probable wave leading to a specified maximum response is called the response conditioned wave (RCW), for which an analytical formulation exists. This design wave is an irregular wave train, containing several components whose amplitude / frequency / phase are computed from the RAO and wave spectrum using the equation (1)

$$\begin{cases} A_i = S_w(\omega_i) * |RAO(\omega_i, \beta)| \dfrac{X}{m_0} d\omega_i \\ -RAO_\emptyset(\omega_i, \beta) \end{cases} \tag{1}$$

where:

A_i is the amplitude of the i-th component
ω_i is the frequency of the i-th component
S_w is the spectral density
X is the target response
m_0 is the zero-th order response spectral moment
β is the heading

The parameters to be chosen are the heading, peak period and shape of the wave spectrum. These can be selected in correspondence with the sea state contributing the most to the target response.

This approach is applied to the above presented case, where only first order is considered (both in the EDW and full time domain calculation). Figure 8 shows the excellent agreement between the results obtained from the EDW method and time domain reconstruction of stress RAOs for structural elements in different areas of the FPSO.

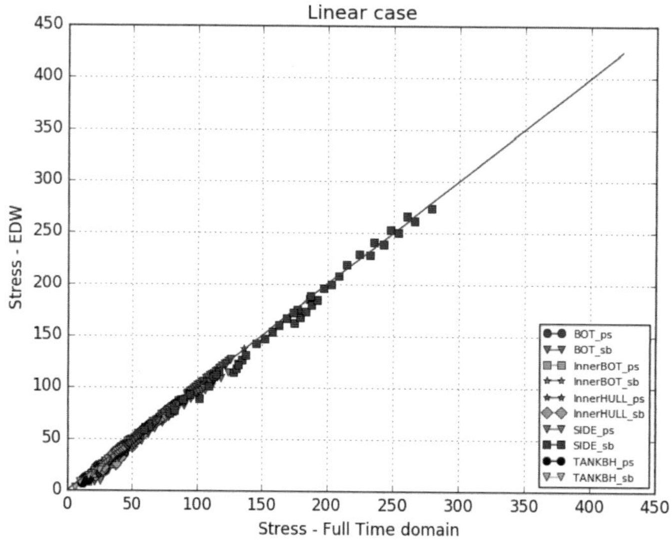

Figure 8 Comparison between the EDW method and the full time domain reconstruction of RAOs of stresses for different elements in the structure of a typical FPSO.

Here, the dominant load is not linear but of second order. Thus, the process is not Gaussian, which is the base assumption for the derivation of the RCW analytical formula. Hence, another way of deriving the design wave should be used. In this work, we adopt a method similar to the one described in Jensen (2007), which is based on first-order reliability method (FORM), well known for structural reliability problems.

In the FORM, the input process (wave elevation and associated wave kinematics) is stationary stochastic. The waves are assumed linear and long-crested. The wave elevation can be written as:

$$\eta(t) = \sum_i u_i\sigma\cos(\omega_i t) - v_i\sigma\sin(\omega_i t) \tag{2}$$

with,

$$\begin{cases} \eta(t) = \sum_i a_i\cos(\omega_i t + \phi_i) \\ a_i = \sigma_i\sqrt{u_i^2 + v_i^2} \\ \phi_i = \tan^{-1}\left(\frac{v_i}{u_i}\right) \end{cases} \tag{3}$$

where u_i and v_i are uncorrelated random variables following a standard normal distribution, and $\sigma_i = \sqrt{S_w(\omega_i)d\omega_i}$.

Having the wave elevation, the non-linear response can be computed in time-domain. Each realization will represent the response on a possible wave scenario. We are interested in the realization that will result in a target response value with the highest probability. This is in fact a limit state problem, which can be solved approximately by the use of a FORM. The reliability index for a given realization is given by:

$$\beta = \sqrt{\sum_i (u_i^2 + v_i^2)} \tag{4}$$

The calculation of the design wave requires minimizing the reliability index β, which can be done using standard constraint optimization algorithms.

It can be shown that, in the linear case, the numerical solution based on FORM and the analytical solution provide identical results as shown in Figure 9.

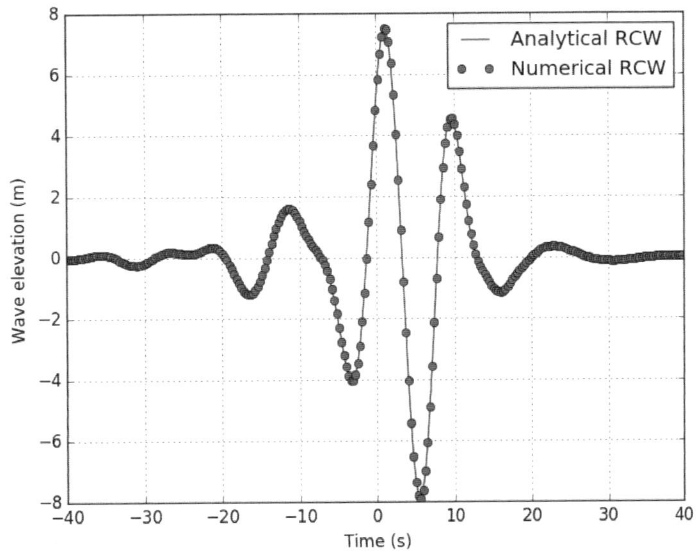

Figure 9 Comparison between numerical RCW (FORM) and analytical solution for linear response

The approach is then used on the second order roll case. Figure 10 shows the peak distribution of second order roll ("dom") with respect to the associated pressure ("sub"). The yellow squared marks are the values of pressure for RCW based on roll at different probability levels ("sub RCW"), showing that the associated pressure of the design wave corresponds to the most probable value of pressure at each probability level.

Finally, Figure 11 demonstrates that the EDW approach provides very similar results to the time domain reconstruction of the stresses, eliminating the over-conservatism obtained with the simple-sum method previously illustrated in Figure 5. The accuracy can be considered as excellent. The remaining discrepancies are possibly due to some approximation done by the EDW approach. For instance, the response is assumed to be maximized together with the selected dominant loads, which might not be 100% accurate.

As far as computation times are concerned, the determination of the EDW through FORM requires some time domain computations. Their number depends on the number of components chosen for the wave spectra, but the order of magnitude is about 1000 function evaluations for the minimization algorithm to converge. However, the computation cost of these evaluations is acceptable, since they consist in reconstructing the hydrodynamic loads at only one time step (the structural response being evaluated only on the EDW). The direct calculation of the structural response in time domain performed in the current study is only to validate the approach, and would not be required in practice.

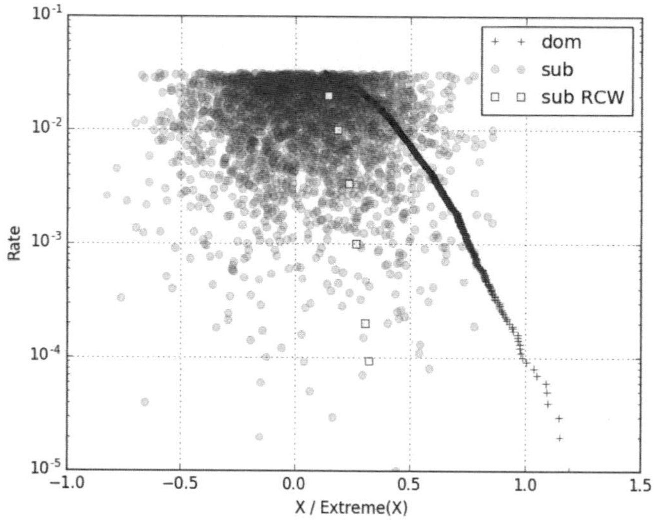

Figure 10 Peak distribution of second order roll ("dom") with the associated pressure ("sub") obtained from up-crossing analysis and values of pressure for RCW based on roll at different probability levels ("sub RCW").

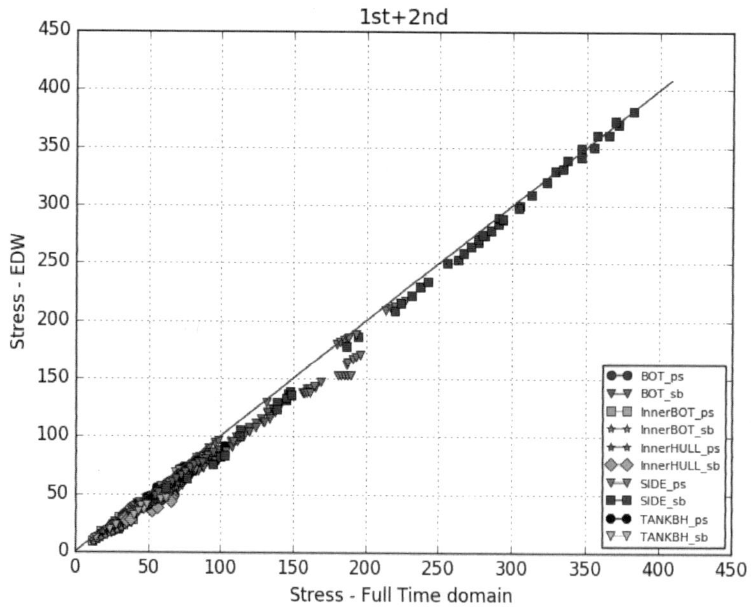

Figure 11 Comparison between the EDW method and the time domain reconstruction of stress QTFs.

CONCLUSIONS

More and more new built FPSOs present roll natural periods beyond the range of incident waves. These vessels are likely to respond at second order. Whenever second order roll is estimated to be non negligible, it should be considered in the structural design of the FPSOs. The relevance of those motions on the stresses in elements at different areas of the vessel has been demonstrated through direct computation of stress QTFs using a 2D beam model. This approach, however, is not considered practically applicable in the case of 3D FE analysis, due to the huge amount of computations required to construct the QTFs. Thus, an EDW (equivalent design wave) has been presented and demonstrated to provide results very similar to the ones obtained from the time domain reconstruction of the stress QTFs. Moreover, the EDW enables to account for the nonlinearity in damping in a very cost-effective way. Although it is validated here on a FPSO, the approach is expected to perform well on other types of units subject to 2nd order roll/pitch motion (like semi-submersible).

REFERENCES

de-Hauteclocque G., Derbanne, Q., El-Gharbaoui, A. (2012). *Comparison of Different Equivalent Design Waves with Spectral Analysis.* OMAE 2012.

Jensen, J.J., *Efficient estimation of extreme non-linear roll motions using first-order reliability method (FORM).* Journal of Marine Science and Technology, 12(4):191-202, 2007

Jensen, J.J., Capul, J., *Extreme response predictions for jack-up units in second order stochastic waves by {FORM}.* Probabilistic Engineering Mechanics, 21(4):330-337, 2006

Rezende, F., Oliveira, A.C., Chen, X.B., Menezes, F. (2013). *A Comparison of Different Approximations for Computation of Second Order Roll Motions for a FLNG.* OMAE 2013.

SNAME
THE INTERNATIONAL COMMUNITY
FOR MARITIME AND OCEAN PROFESSIONALS

CORROSION PERFORMANCE OF THERMAL SPRAYED ALUMINUM COATING IN OFFSHORE MOORING CHAINS

AINTZANE EXPOSITO
VICINAY MARINE INNOVACIÓN

JEAN-BAPTISTE JORCIN, PHD.
TECNALIA RESEARCH & INNOVATION

RAQUEL BAYÓN, PHD.
IK4-TEKNIKER

ALBERTO ARREDONDO
VICINAY MARINE INNOVACIÓN

ELENA RODRÍGUEZ, PHD.
VICINAY MARINE INNOVACIÓN

Proceedings of the 22nd Offshore Symposium, February 2017, Houston, Texas
Texas Section of the Society of Naval Architects and Marine Engineers

ABSTRACT

With 39% of mooring chain failures related to different corrosion mechanisms of Floating Production Units [6], finding a solution to delay or avoid this degradation phenomenon has become one of the offshore industry main concerns. Thermal Sprayed Aluminum (TSA), aluminum based coating, has been used as corrosion protection in chains since 2001. However, its behavior and durability in different conditions remains a question.

In the present work, corrosion behavior of the TSA is analyzed and compared to base steel material, simulating real seawater conditions, including atmospheric, splash and immersed zones. In order to evaluate TSA performance, the coating effectiveness was measured under different ageing processes and mechanical actions. Additionally, the capability of the coating to tolerate damages and its response under different cathodic potential levels were assessed. The analyses were conducted combining electrochemical techniques.

One-year tests were performed on components both in laboratory, under controlled conditions, and in real environment, using 3D scanning methods to calculate corrosion rates afterward the exposition. Components results were aligned with the previously obtained results on samples.

Results showed the benefits of applying TSA to offshore mooring chains by significantly enhancing corrosion resistance.

Keywords: Thermal Sprayed Aluminum, corrosion, offshore mooring chain.

INTRODUCTION

Thermal Sprayed Aluminum (TSA) coating has been used as a solution against corrosion in offshore industry since eighties. Many works have been published analyzing the experience with the coating in different countries such as Norway, United States, Germany, Russian Federation, etc. talking about its long term performance in marine environment. Studies show that the coating has a thickness dependent life; for example 18 years of durability for 100 micron thickness or 15 years for 125 microns unsealed TSA. Predictions suggest 30 years of life for 200 microns thickness in splash zone [5][16].

In case of the mooring chains and accessories Vicinay Cadenas S.A. coated its first chain in 2001. Starting from this year several projects have been installed with TSA coated links. Table 1.

Table 1. TSA Coating in different projects.

Chain diameter (mm)	Protection	Project	Year	Location
124	TSC/TSA	Nakika	2001	Gulf of Mexico
134	TSA	Perdido	2006	Gulf of Mexico
82/184	TSA	Escravos	2009	Nigeria
127	TSC/TSA	Gumusut	2010	Malaysia
157	TSA	Lucius	2012	Gulf of Mexico
76	TSA	West Seno	2012	Indonesia
157	TSA	Heidelberg	2013	Gulf of Mexico
157	TSA	Moho	2015	Congo

**TSC: Thermal Sprayed Carbide.*

Although aluminum is well known as a corrosion resistant material and TSA has been used in several offshore applications, it has been necessary to study its effectiveness in the protection of offshore mooring chains and see its limits under different aggressive conditions.

LABORATORY TESTS

According to ISO 12944 [12], a coating proposed to work in marine environment and immersed conditions should comply with categories C5M and Im2 respectively, and it would have a durability of more than 15 years if the high level is achieved inside this categories.

Table 2.Tests to be performed to comply with C5M and Im2 high level category and their durations.

Category	Durability ranges	ISO 12944			NORSOK M501 [14]
		ISO 6270-2 [9] Water condensation	ISO 9227 [10] Neutral salt spray fog	ISO 2812-2 [7] Immersion in water	ISO 20340 [13] Combined aged test
C5M	High	720h	1440h	-	-
Im2	High	-	1440h	3000h	-

In the first stage of the research, the validation of the coating to work in long term marine application has been performed, using ISO 12944 standard requirements.

In the second stage, the applicability of the TSA coating to mooring chains has been analyzed, in order to know as much as possible about the coating and its behavior under the conditions it would suffer during its useful life. For that purpose, different tests were performed to simulate the real conditions of the chains.

Validation of the coating for marine long term applications

To classify the coating and to analyze the resistance to humidity in water condensation conditions, to neutral salt fog and to immersion in sea water, the tests mentioned in Table 1 were performed according to the following conditions.

Table 3. Summary of the tests performed and their conditions.

<table>
<tr><th rowspan="2">Standard</th><th rowspan="2">Type</th><th rowspan="2" colspan="2">Duration of each process</th><th rowspan="2">Duration of each cycle</th><th colspan="2">Conditions in climatic chamber</th><th rowspan="2">Duration of the test</th></tr>
<tr><th>Temperature (ºC)</th><th>Relative humidity (%)</th></tr>
<tr><td colspan="8">Water condensation</td></tr>
<tr><td rowspan="5">ISO 6270-2</td><td>Constant humidity</td><td colspan="2">-</td><td>-</td><td>40(±3)</td><td>100</td><td>720h</td></tr>
<tr><td rowspan="2">Alternating temperature</td><td colspan="2">8h</td><td rowspan="2">24h</td><td>40(±3)</td><td>100</td><td rowspan="2">720h</td></tr>
<tr><td colspan="2">16h</td><td>18-28</td><td>100</td></tr>
<tr><td rowspan="2">Alternating temperature and humidity</td><td colspan="2">8h</td><td rowspan="2">24h</td><td>40(±3)</td><td>100</td><td rowspan="2">720h</td></tr>
<tr><td colspan="2">16h</td><td>18-28</td><td>Ambient</td></tr>
<tr><td colspan="8">Neutral Salt Spray Fog</td></tr>
<tr><td>ISO 9227</td><td>Salt Spray fog</td><td colspan="2">-</td><td>-</td><td>35</td><td>-</td><td>1440h</td></tr>
<tr><td colspan="8">Immersion in water</td></tr>
<tr><td>ISO 2812-2</td><td>Immersion test</td><td colspan="2">-</td><td>-</td><td>40</td><td>-</td><td>3000h</td></tr>
<tr><td colspan="8">Combined Aged test</td></tr>
<tr><td rowspan="4">ISO 20340</td><td>Ultraviolet radiation</td><td>4h</td><td rowspan="2">72h (ISO 11507[11])</td><td rowspan="4">168h</td><td>60(±3)</td><td>-</td><td rowspan="4">720h</td></tr>
<tr><td>Condensation test</td><td>4h</td><td>50(±3)</td><td>-</td></tr>
<tr><td>Salt spray ISO 9227</td><td colspan="2">72h</td><td>35(±3)</td><td>-</td></tr>
<tr><td>Low temperature</td><td colspan="2">24h</td><td>-20(±2)</td><td>-</td></tr>
</table>

After the tests, in all the samples at different conditions no pits, blisters, cracks, ferric corrosion or other defects were detected according to ISO 4628 [8], color loss was not observed and the sealant was intact. During the salt fog tests the apparition of localized white salts spots coming from the oxidation of the aluminum layer was detected. Therefore, in accordance with the ISO 12944 standard, the TSA coating can be classified with the C5M and Im2 High categories, which would correspond to a durability of more than 15 years in service without maintenance.

Condensation tests **Salt Spray fog tests** **Immersion test** **Combined aged test**

Figure 1. TSA coated samples after the completion of the tests.

Additionally, in samples tested to Salt Spray fog, X-shape incisions were made, deep enough to reach the steel underneath and after 1,440 hours of exposure adherence tests were carried out according to ASTM D3359 [2]. In the adhesion test two of three samples were classified with 5A grade, corresponding to the higher grade of the scale (best adherence), since the coating was not removed or detached from the surface after taking off the adhesive tape. The third sample was classified as 4A, corresponding to the second highest grade, where a small trace of the coating was removed or detached.

Different zones among the mooring line

The mooring line is the group of components that fix the floating structure to the sea floor. As it is possible to have chains in different positions within the same mooring line, the links are exposed to different environments. Although coated links are usually placed in the top of the mooring line, different conditions may affect them. Therefore, tests simulating those different conditions were performed.

Samples for the tests were obtained from 84mm diameter R5 grade chain after finishing all the manufacturing process. Comparison between non-coated samples, TSA coated samples and artificially damaged coated samples was done.

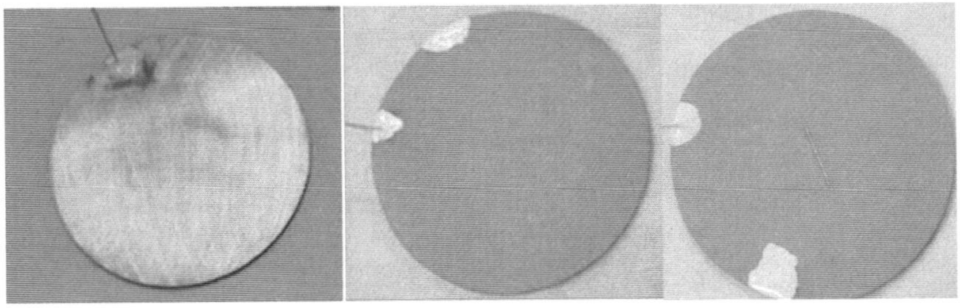

Figure 2. From left to right uncoated, TSA coated and TSA coated artificially damaged samples before tests.

Electrochemical techniques were used to analyze the corrosion during the tests: Open Circuit Potential (OCP) with Ag/AgCl reference electrode saturated with KCl and Electrochemical Impedance Spectroscopy (EIS).

Immersion

Immersion part was simulated in the laboratory by immersing the samples in a bath of synthetic sea water, aqueous solution containing 3.5% of NaCl. The samples were immersed during 720 hours.

For uncoated samples the OCP recorded was -550mV which is coherent for low alloyed steel. After 720 hours, the samples were all heavily corroded with two kinds of corrosion layers: first black top layer, which was mainly constituted by magnetite and chloride iron complexes and a second one, rusty and not adherent, mainly composed by iron hydroxides, oxo-hydroxide and hematite. The impedance spectra recorded during the immersion test shows that the low frequency impedance was lower than 100 $\Omega.cm^2$, which means that surface area was very active. See in Figure 3.

In case of TSA coated samples the OCP recorded was around -1V which is a typical OCP value measured on aluminum based material. After 720 hours no sign of corrosion or degradation was observed in the samples. The low frequency impedance modulus was around 10^4 $\Omega.cm^2$, showing the effectiveness of the coating protecting the steel.

For the TSA coated and artificially damaged samples the OCP recorded was around -800mV, which is a mixed potential between steel and aluminum based material. No sign of corrosion was found, even in the defect, after 720 hours of ageing. Sacrificial effects of the aluminum in favor of steel took place in the sample. That demonstrated that if the coating is damaged, there will be an increase in the aluminum corrosion but steel remains intact.

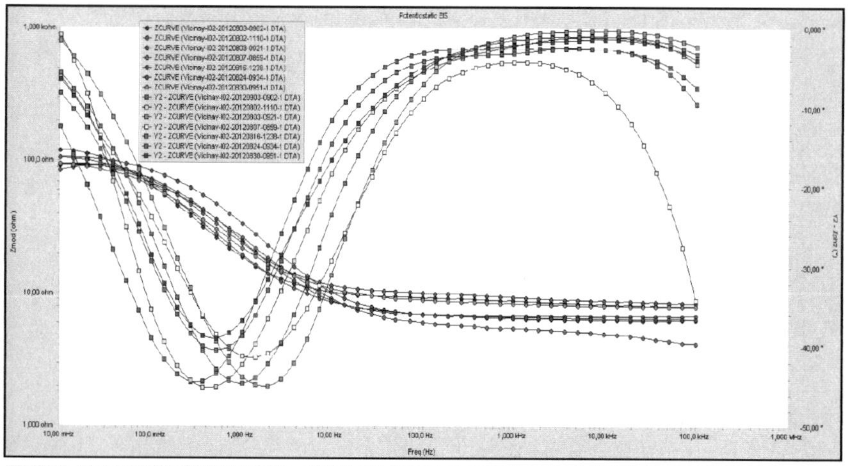

Uncoated sample after immersion test

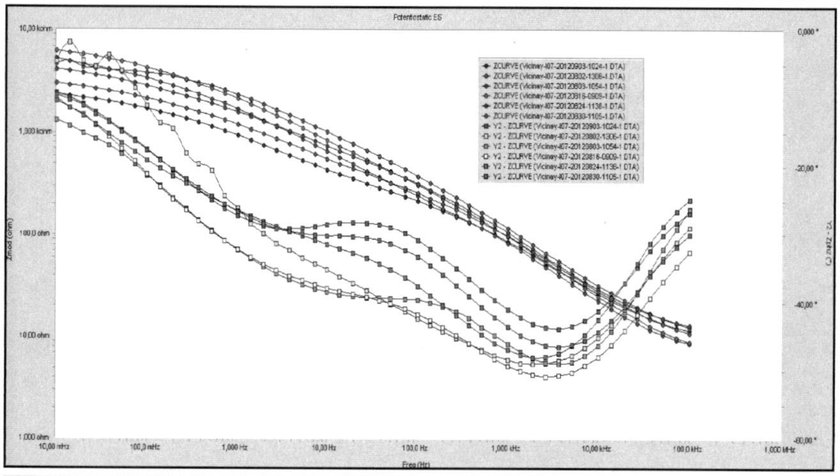

TSA coated sample after immersion test

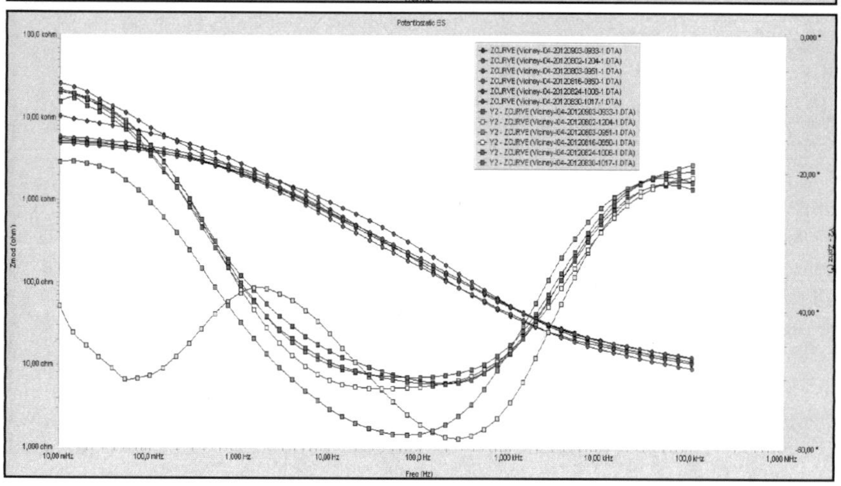

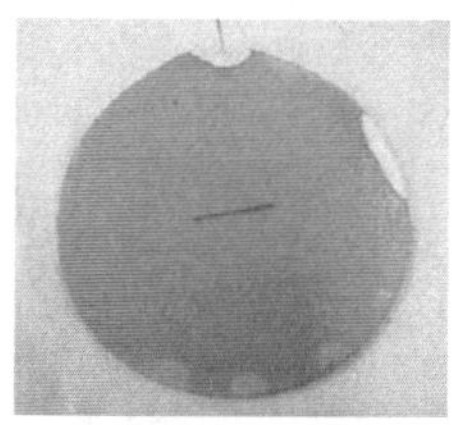

Artificially damaged sample after immersion test

Figure 3. EIS diagrams recorded during the ageing time in immersion.

Splash zone

A modified salt spray chamber was used to reproduce an environment close to the offshore splash zone. This chamber allows doing cycles between salt spray exposition, according to standard ISO 9227 standard, and drying moment. The cycle consists of 0.5 hours of salt fog (synthetic sea water containing 3.5% of NaCl) and 1 hour of drying at 35°C. The samples were exposed for 720 hours.

During the salt spray cycle, the constant addition of chloride through the fog generation increases salt concentration at the surface during the drying part which promotes to an over concentration in salt at the beginning of the following cycle leading to acceleration of the corrosion process. For uncoated samples the OCP recorded was -630mV. After 720 hours uncoated samples were corroded such as in immerse condition, though salt sprayed chamber environment seems to be more aggressive than immersion test. Coated samples, with a -1V of recorded OCP value, had no sign of corrosion or degradation. For the coated and artificially damaged samples the OCP recorded was around -800mV. No sign of corrosion was found, even in the defect, after 720 hours of ageing.

EIS diagrams obtained during the tests are similar to the ones obtained during the immersion tests with the same low frequency impedance modulus indicating a similar corrosion behavior.

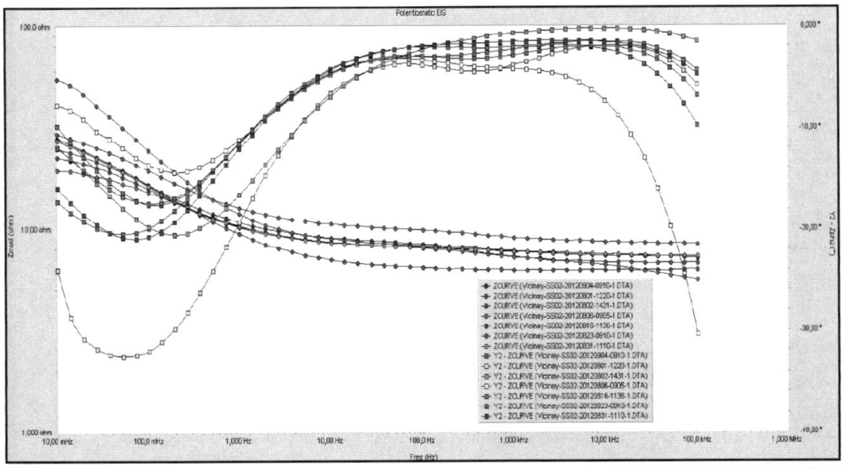

Uncoated sample after test in salt spray chamber.

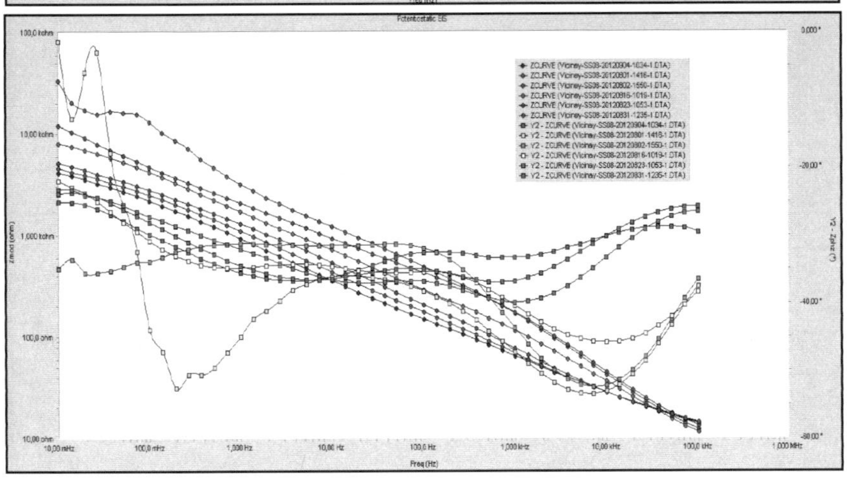

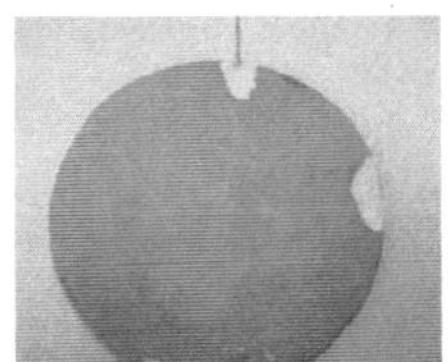

TSA coated sample after test in salt spray chamber.

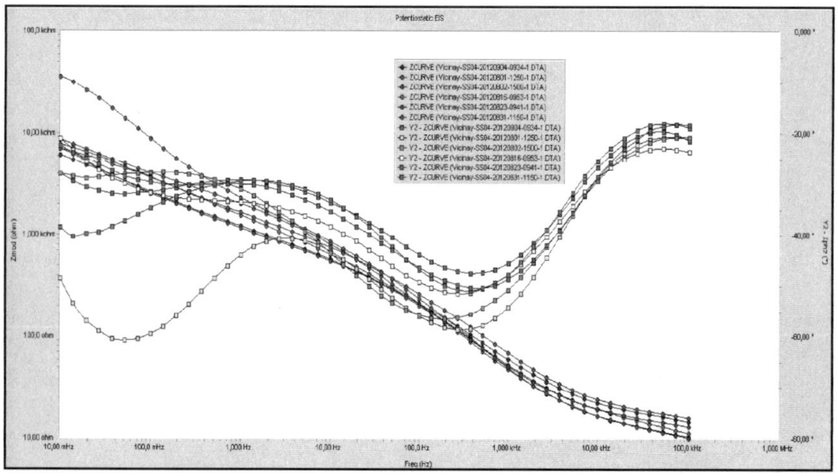

Artificially damaged sample after test in salt spray chamber

Figure 4. EIS diagrams recorded during the ageing in salt spray chamber.

North Sea environment

The third ageing method chosen for the study was the "Norsok cycle" (annex A standard ISO 20340) which was specially designed to simulate a North Sea environment. This cycle is a combination between three processes, with a total duration of 720 hours:

- UV/condensation during 3 days according to the standard ISO 11507.
- Salt spray during 3 days following standard ISO 9227 (same timing than for simple salt spray ageing).
- Climatic chamber at -20ºC during 1 day.

Norsok environment was the most aggressive of the three tested environments. The OCP recorded after the test was -630mV in uncoated samples, showing a heavy corrosion and some localized pitting corrosion. In case of coated samples, with an OCP value of -1V, no sign of corrosion appeared. For the coated and artificially damaged samples the OCP recorded was around -800mV. No sign of corrosion was found, even in the defect, after 720 hours of ageing.

EIS diagrams obtained during the tests were similar to the ones obtained during the immersion tests with the same low frequency impedance modulus indicating a similar corrosion behavior.

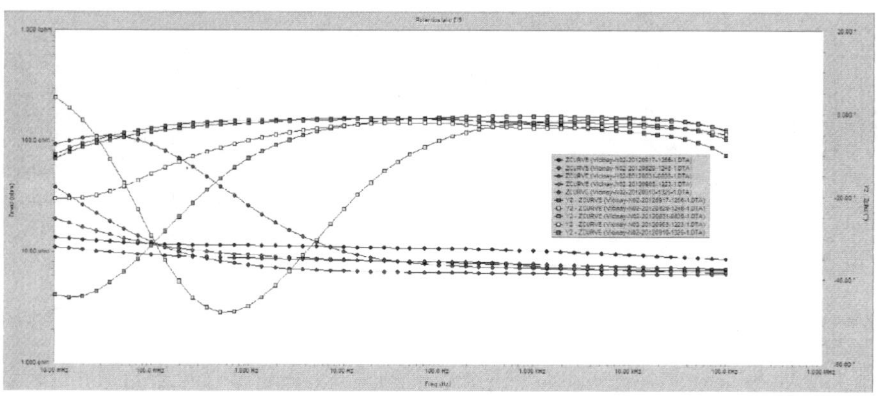

Uncoated sample after Norsok Cycle.

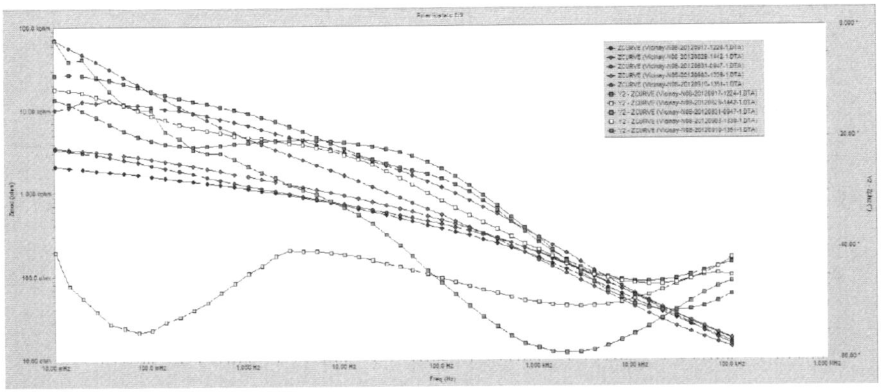

TSA coated sample after Norsok Cycle.

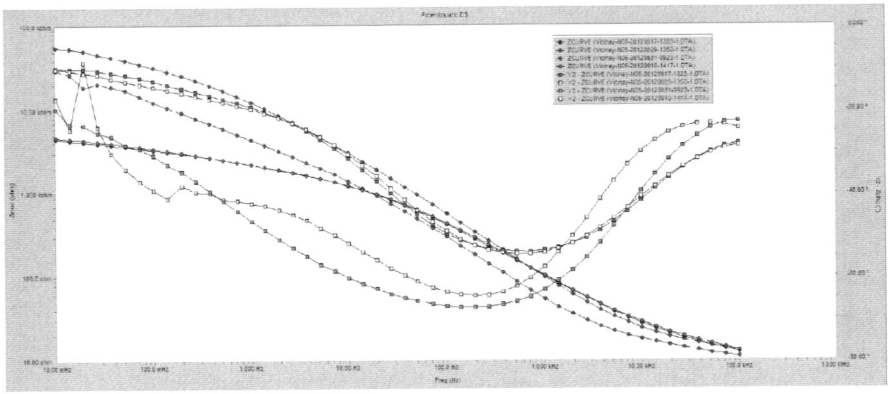

Artificially damaged sample after Norsok Cycle.

Figure 5. EIS diagrams recorded during the ageing time in Norsok Cycle.

Under cathodic protection

In marine environment, most of the important structures are under cathodic protection to avoid corrosion damage. The big structures are usually protected by impressed current systems. Some problems may happen locally if, due to a combination of morphological factor of the structure or defect of the protection system, there is an overprotection (above -950mV vs Ag/AgCl), which may lead to an overconsumption of the TSA or delamination or embrittlement problem due to the possible hydrogen evolution. For this reason, it has been necessary to analyze the TSA response under these conditions: first, without applying any load; and second, under a constant load.

TSA without load

In order to simulate the first mentioned case, samples under a cathodic overprotection without any load were analyzed. These samples, consisting of coated, non-coated and artificially damaged coupons (Figure 2), were immersed in synthetic sea water, with 3,5% of NaCl, and polarized at -1.2V vs. Ag/AgCl using a potentiostat, an Ag/AgCl reference electrode and graphite counter-electrodes.

EIS spectra of the samples obtained during the tests were in good agreement with the ones obtained during immersion tests without any cathodic protection.

Uncoated samples presented only one time constant and the low frequency impedance modulus reached $100\Omega.cm^2$. The cathodic protection pushed the steel in its immunity domain to avoid corrosion. As soon as it was switched off, the corrosion mechanism starts. A gas evolution was observed above the samples, which is normal in these conditions where the hydrogen evolution should be observed.

In case of the coated samples, they did not seem affected neither by the immersion, nor by the cathodic protection. The damaged samples had the low frequency impedance around 3kΩ.cm^2 indicating a good overall corrosion protection, taking account of the damage, even if a delamination was observed starting from the damage.

The delamination is generally due to a strong increase of the pH as a consequence of the reduction of oxygen in OH- on the steel, which is happening in high negative polarization like in this case. It generates a disbonding of the coating around the damage.

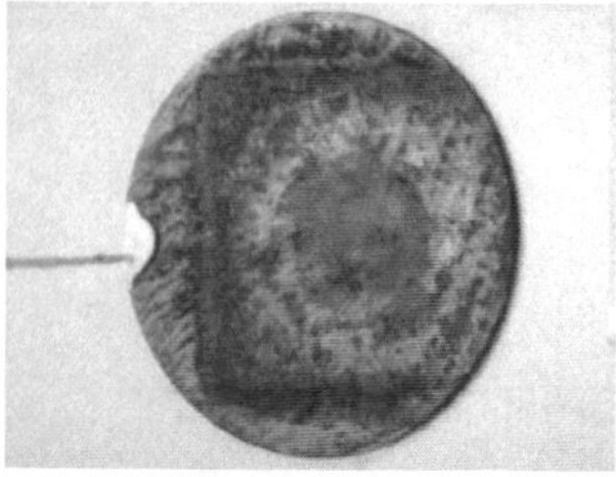
Uncoated samples.
Slightly corroded areas due to measurements system.

Coated samples.
White deposits a mix of salt and aluminum oxide.

Artificially damaged samples.
Delamination starting from the damage.

Figure 6. Samples after the tests with cathodic overprotection.

The delamination speed measured, 1-2mm in 835 hours in aggressive condition, was quite low but it led to an extra consumption of aluminum and thus a decrease of the life of the coating.

TSA under constant load

In order to assess the sensitivity of the material to hydrogen embrittlement, constant load tests according to NACE TM0177 [15] were performed.

Aqueous solutions at 3.5% NaCl, and with different levels of cathodic protection were used to complete the tests: without any cathodic protection (CP), -850mV and -1,200mV. The load applied using a calibrated ring was 80% of the yield strength of the material, in that case, R5 mooring chain grade.

In Table 4 the samples matrix consisted of non-coated samples, TSA coated samples and half-coated samples.

Table 4. Constant load tests matrix.

Samples	Without cathodic protection	-850mV	-1,200mV
Non-coated	x	x	x
Coated	x	x	x
Half-coated	x	-	-

After 720 hours of test, none of the samples broke, so according to the acceptance criteria of NACE TM0177, the material is not sensitive to hydrogen embrittlement in the conditions of the tests. In Figure 7, details of the samples during and after the tests are shown.

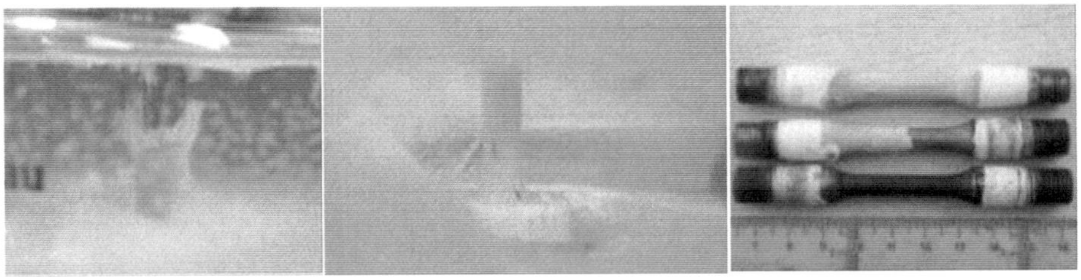

Figure 7. Detail of samples during the tests: a) without CP; b) at -1,200mV ; c) after the tests.

FULL SCALE TESTS

Full scale tests were performed in order to confirm the tendencies obtained in lab tests. Two test sites were used: one in controlled laboratory conditions and the other one in a real environment in a port. For that purpose, 84mm diameter R5 grade studless chains were manufactured, complying with the common chain manufacturing process including welding, heat treatment, proof loading, non-destructive inspection (NDT) and mechanical tests.

Full scale test in laboratory

Six links with different configurations were installed in a container with recirculated synthetic sea water containing 3.5%NaCl. The links were immersed in the container during one year with periodical measurements of oxygen concentration, temperature, links diameters and visual inspection. Before the exposition, links were artificially damaged to reproduce the worst case, as it can be observed in Figure 8.

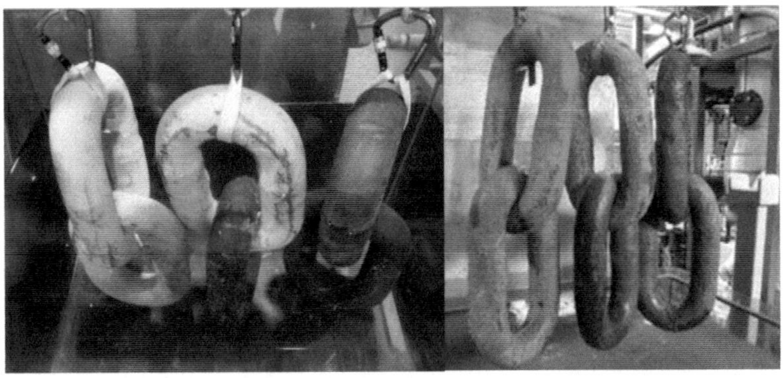

Figure 8. Left, 1st day of the links in the lab (In yellow TSA coated links). Right, after 9 months.

The environmental conditions during the different measurements showed an average in temperature of 24ºC, a stabilized pH around 8, oxygen content 80% saturated and with a 7ppm of concentration.

During the exposition, the uncoated links were very quickly covered with a black corrosion layer, compact and well adhered. On top of the black corrosion layer, a brown-orange layer of corrosion products was formed. This last layer was powdery and poorly adherent. This top layer releases very easily to the solution in tiny particles. The aspect of the chain surfaces observed was very similar to the one observed for the small scale study in immersion (Figure 3).

The TSA coated links already presented defects before the immersion and starting from them, some degradation were observed with time. The degradation of the TSA coated links was not general and homogeneous but local. Some small flake detached from the surface close to a TSA defect. The flake was taken and analyzed in the Scanning Electron Microscope (SEM) with the Energy-Dispersive X-Ray Spectroscopy (EDS) detector. The EDS spectrum obtained showed the presence of aluminum which confirms that a part of the TSA coating detached.

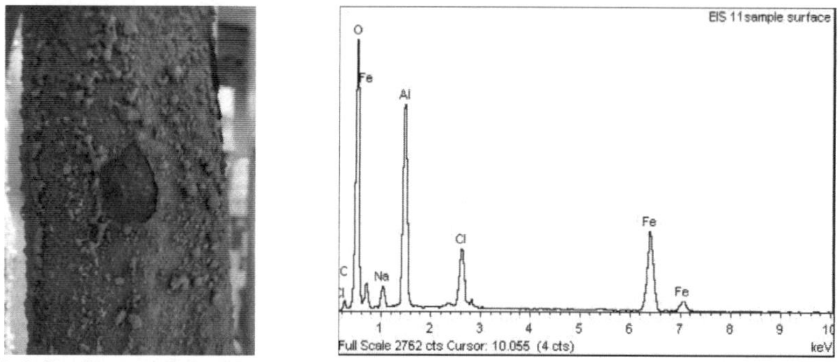

Figure 9. Link joined to uncoated link, after 3 months with a flake detached and the EDS Spectrum.

Regarding the diameter of the links, in non-coated samples an increase of the diameter occurred, due to the corroded layers that appear as a reaction of the oxidation of the steel. In case of TSA sample, the diameters increased at the beginning but then stabilized and remained in the same values.

At completion of the tests, samples were removed from water and cross sections of the links were cut and metalographically prepared to analyze them at SEM. The black corrosion product layer developed on uncoated links had a thickness around 300µm and it was composed of a complex system of various iron oxides with chemistry close to the magnetite Fe_3O_4; mix of Fe II and Fe III, with various amounts of oxygen, and the addition of other metallic cations.

In the coated link, the TSA layer appeared to be homogeneous, porous and still quite thick with a thickness of 400µm. An extra thin and bright layer was observed on top of the TSA, the sealer. Figure 10.

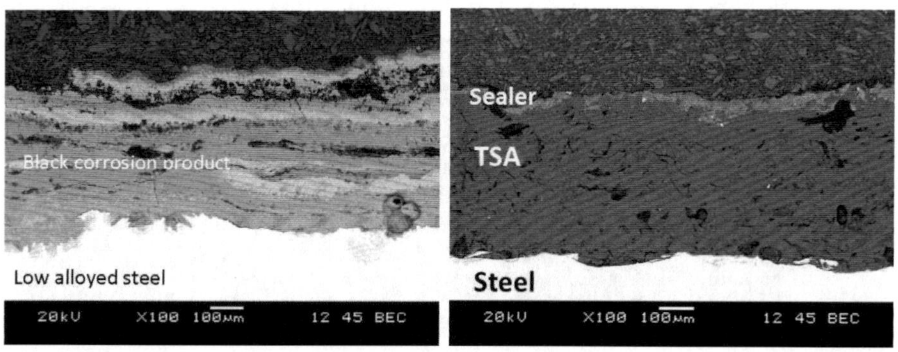

Figure 10. SEM analysis. A) Uncoated link. b) Image of the TSA layer after 1 year of immersion

Full scale test in real environment

Six links were exposed in the Port of Pasaia, in the North of Spain, for one year with periodical inspections and measurements of the diameter and the environmental parameters: temperature, pH, turbidity, oxygen content and salinity.

The environmental conditions recorded during the exposure showed a variation of the temperature and salinity measurements due to the different seasons of the year. The minimum of the temperature registered during the exposition was 10ºC, the maximum 24ºC and the average 18ºC. The minimum of salinity registered was 12g/l, the maximum 34g/l and the average 28g/l. The oxygen saturation and the pH were stable at 70% and at 8, respectively.

Due to the high content of nutrients present in the port area, a layer of biofouling appeared very quickly. By the third month, the configuration composed by TSA coated links showed a biofouling covered surface probably due to the surface roughness and non-toxicity of the aluminum. The configuration composed by coated and uncoated links exhibited less surface covered by biofouling. The pair of uncoated links presented more general corrosion, with the formation of a double layer of corrosion products with a black layer adherent to the metal covered by a powdery and poorly adherent orange layer, and a few zones of biofouling films. In subsequent inspections it was observed that the three configurations began to resemble one another: the samples with a thicker layer of biofouling were stabilized while samples with less biofouling increased the thickness of the biofouling layer.

Uncoated-uncoated **Coated-uncoated** **Coated-coated**

Figure 11. Links in the port after the year of exposition.

In all the links, a gradual increase of the diameter was observed due to the adhesion of the fouling and corroded layers.

After one year of exposition, a SEM analysis was performed showing that the uncoated steel was corroded, a double layer of corrosion product with a black layer adherent and a poorly adherent orange layer as in the case of full scale test realized in the laboratory conditions.

In case of TSA coated links, no sign of corrosion was observed on the coating, it remained intact under the biofouling layer. Tests performed using barnacles during a doctoral thesis showed a similar behavior of the coating, where after two years of exposition there was no penetration of the shell into the thermal sprayed aluminum resulting in non-loss of corrosion protection. [20]

The deposition of the biofouling and the formation mechanism of the scale in TSA coated links could be affected by the local pH close to the surface. This pH could increase by the reduction of oxygen at cathode areas, inducing the precipitation of calcium and magnesium carbonates onto the cathode surface. The scale formation mechanism on a duplex stainless steel partially covered with TSA would be similar to the links studied in this project. The formation of scale could be accelerated if the link is catholically polarized; something that can occur if the corrosion mechanism leads to a local alkalization. Figure 12.

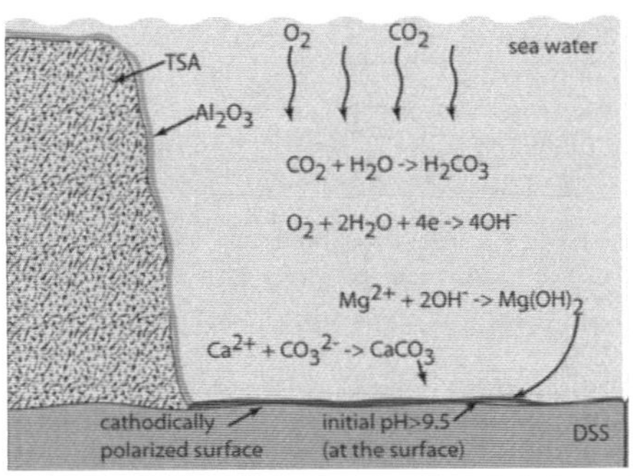

Figure 12. Formation of the scale deposit on a duplex stainless steel with TSA.[17], [18]

Additionally, a comparison between links before and after exposition was performed using 3D scanning methods. For that reason the links were shot blasted to remove all the marine growth. In case of the coated sample, also the coating was lost because of this cleaning process.

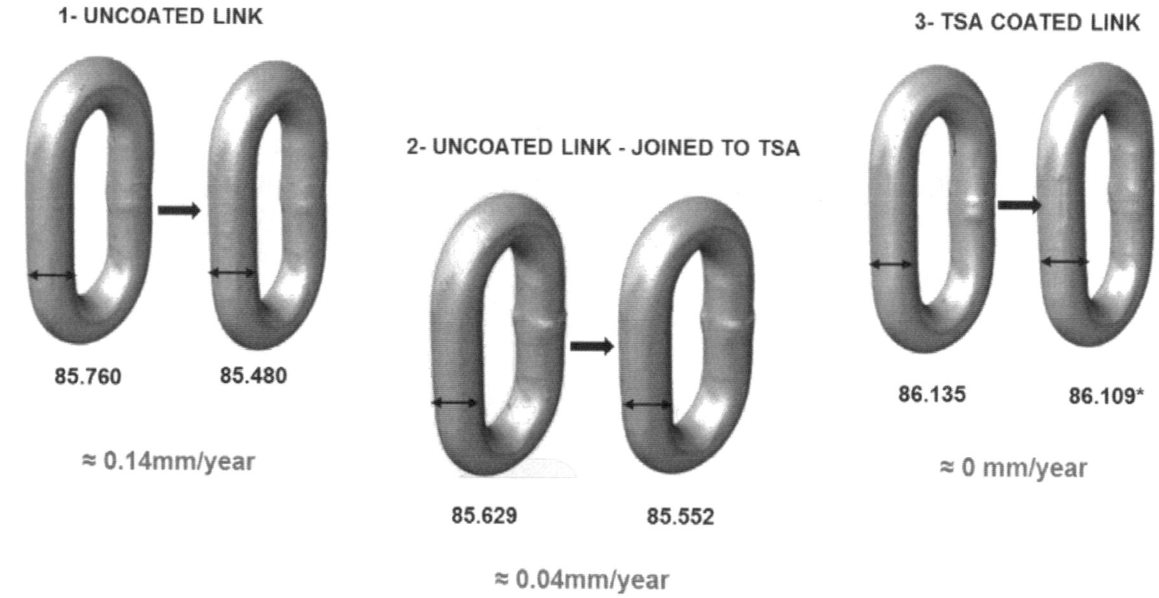

Figure 13.Reduction in diameter of the links due to corrosion.

Taking into account the average diameter of the links an approximated corrosion rate was calculated. In case of the coated link, the thickness of the coating layer was summed to the diameter in order to calculate the rate.

Table 5. Estimation of the corrosion rate in the different links.

Link	Uncoated	Uncoated (joined to a coated link)	Coated
Corrosion rate (mm/year)	0.14	0.04	0

The biggest loss in dimeter due to corrosion took place in the uncoated link, showing a steel corrosion rate in concordance with the offshore standards expected values [1], [3]. In case of the uncoated link joined to a coated link the corrosion rate observed was lower than in the previous case, showing a sacrificial effect of the TSA coated link in favor of the uncoated one, which concurs with the laboratory results. In the configuration composed only by coated links the corrosion rate is negligible.

CONCLUSIONS

Taking into account the performed tests and different inspections; it can be concluded that:

- TSA coating in mooring chain can be classified as a coating for CM5 and Im2 corrosivity categories, which would correspond to a durability of more than 15 years in service without maintenance.
- TSA gives good protection to steel against corrosion under immersed conditions and under different ageing cycles even if they are very aggressive.
- The coating handles damages well, having a sacrificial effect for the steel in case of small damages.
- The system steel/TSA coating has good behavior under cathodic over-protection, though low delamination was observed.
- Constant load tests showed that, neither the R5 grade steel nor the coating are sensitive to hydrogen embrittlement in tested conditions.
- In case of real components TSA performs well both in the laboratory and in port. Although in the cases where the coating has a previous substantial damage, the TSA is locally affected.
- In the presence of high levels of biofouling both the steel surface and the coating are covered, although the TSA layer remains intact below fouling attachments.

Thermal Sprayed Aluminum coating has demonstrated to be an optimum option for corrosion protection of mooring chains and a suitable alternative to the overthickness of the links, one of the solutions historically accepted by Class Societies and the industry in general. The reduction of this overthickness, which is different for different standards, locations and zones along the mooring line, would lead to decreasing the raw material and energy used in its manufacture as well as an improvement in handling, transport and installation processes.

ACKNOWLEDGEMENTS

Authors would like to thank to the Spanish Ministry of Economy and Competitiveness for funding part of study in the project "Float Solutions", INNPACTO 2011, IPT-2011-1083-920000. Authors would also like to thank Vicinay Cadenas S.A. the support throughout all the work.

REFERENCES

[1] API RP-2SK. Design and Analysis of Stationkeeping Systems for Floating Structures. 2005.

[2] ASTM D3359 Standard Test Methods for Measuring Adhesion by Tape Test.

[3] DNVGL-OS-E301. Position Mooring. 2015.

[4] Egtvedt.S, "Thermally Sprayed Aluminum (TSA) with Cathodic Protection as Corrosion Protection for Steel in Natural Seawater". Trondheim, Norway. 2011.

[5] Fischer, K.P., Thomason, W.H., Rosbrook, T. and Murali, J.,"Performance History of Thermal-Sprayed Aluminum Coatings in Offshore Service". USA, 1995.

[6] Fontaine, E., Kilner, A., Carra, C., Washington, D., Ma, K.T., Phadke, A., Laskoswki, D., Kusinski, G., "Industry Survey of Past Failures, Pre-emptive Replacements and Reported Degradations for Mooring Systems of Floating Production Units" OTC-25273-MS. Published at Offshore Technology conference in Houston, USA, 2014.

[7] ISO 2812-2. Paints and varnishes - Determination of resistance to liquids - Part 2: Water immersion method.

[8] ISO 4628. Paints and varnishes - Evaluation of degradation of coatings - Designation of quantity and size of defects, and of intensity of uniform changes in appearance.

[9] ISO 6270-2. Paints and varnishes - Determination of resistance to humidity.

[10] ISO 9227. Corrosion tests in artificial atmospheres - Salt spray tests.

[11] ISO 11507. Paints and varnishes - Exposure of coatings to artificial weathering - Exposure to fluorescent UV lamps and water

[12] ISO 12944. Paints and varnishes - Corrosion protection of steel structures by protective paint systems.

[13] ISO 20340. Paints and varnishes - Performance requirements for protective paint systems for offshore and related structures.

[14] Norsok standard M-501. Surface preparation and protective coating. 2004.

[15] NACE TM0177. Laboratory Testing of Metals for Resistance to Sulfide Stress Cracking and Stress Corrosion Cracking in H_2S Environments.

[16] Olaf Doble and Graham Pryde, "Use of Thermally Sprayed Aluminum in the Norwegian Offshore Industry". Norway, 1997.

[17] Paul, S., Lee, C.M., Harvey, M.D.F. "Improved Coatings for Extended Design Life of 22%Cr Duplex Stainless Steel in Marine Environments". Published at ITSC 2012 International Thermal Spray Conference, Houston, USA, 2012.

[18] Paul, S., Shrestha, S., Lee, C.M., Harvey, M.D.F. "Thermally Sprayed Aluminum (TSA) Coatings for Extended Design Life of 22%Cr Duplex Stainless Steel in Marine Environments". Published at Journal of Thermal Spray Technology, USA, 2013.

[19] Shrestha, S. and Sturgeon, A. "Characteristics and electrochemical corrosion behaviour of thermal sprayed aluminium (TSA) coatings prepared by various wire thermal spray processes". EUROCORR 2005 Lisbon, Portugal. 2005.

[20] Vreijling, M.P.W., "Electrochemical Characterisation of Metallic Thermally Sprayed Coatings". Doctoral Thesis. The Nederlands, 1998.

SNAME
THE INTERNATIONAL COMMUNITY
FOR MARITIME AND OCEAN PROFESSIONALS

SAVING RIG TIME AND MITIGATING RISK THROUGH "QUICK-RELEASE" MOORING SYSTEMS

GINNY REBECCA WHISENHUNT
STAFF ENGINEER – R&D, DELMAR SYSTEMS, INC.

JUSTIN DAVID BARROW
SENIOR STAFF ENGINEER – R&D, DELMAR SYSTEMS, INC.

JASON DAVID PASTERNAK
PROJECT MANAGER – R&D LEAD, DELMAR SYSTEMS, INC.

Proceedings of the 22nd Offshore Symposium, February 2017, Houston, Texas
Texas Section of the Society of Naval Architects and Marine Engineers

ABSTRACT

The use of quick-release devices in mooring systems shortens disconnection times, allowing MODUs to leave locations in less time, resulting in significant cost savings. Additionally, being able to disconnect the moorings without AHV assistance eliminates the need for decking connections onto an anchor handling vessel or diving an ROV. This allows for the rig to be released safely in more strenuous weather conditions, correlating to a greater window to safely disconnect from a mooring system. In addition to normal disconnect operations, quick releases allow for swift departure from a field during an emergency.

Two classed and offshore-proven quick-release devices are available on the market. The Delmar Quick Release (DQR), an in-line mooring component developed by Delmar Systems, Inc., uses a simple mechanical release feature to release while a mooring line is still under tension. The Rig Anchor Release (RAR), an in-line mooring component developed by InterOcean Systems (an affiliate of Delmar Systems, Inc.), uses acoustics to release under tension.

This paper will present how the use of either release in a mooring system can both increase mooring system safety by allowing for rapid departure in the event of an emergency, as well as reducing regular disconnection times.
Keywords: Mooring, MODU, Offshore

INTRODUCTION

Offshore Mooring Overview

A typical offshore spread mooring system is defined by the orientation of the mooring lines when viewed from above. Terms such as even spread, when lines are approximately spaced at even intervals around the circumference of the vessel, or cluster pattern, when multiple lines are clustered – typically in each of four primary corners of the vessel being moored, are used to describe mooring system types.

When an individual mooring line profile is viewed from the side, referred to as an elevation view, the shape of the catenary is used to describe the type of mooring system. Terms such as taught line, referring to a system that has no grounded component, and semi-taught, a system with a small amount of grounded component with respect to the mooring line scope (the distance from fairlead to anchor), are used to define the type of mooring system. Traditionally, moored MODUs are stocked with a set of rig anchors and the associated chain and wire component to install the MODU in what is referred to as a conventional configuration. Figure 1 illustrates a typical conventional mooring line consisting solely of a drag embedment anchor (DEA) and MODU chain and wire. These types of mooring systems are primarily used on locations free of subsea infrastructure due to the large amount of grounded chain required to soften the system and the low uplift requirements typical of drag embedment anchors.

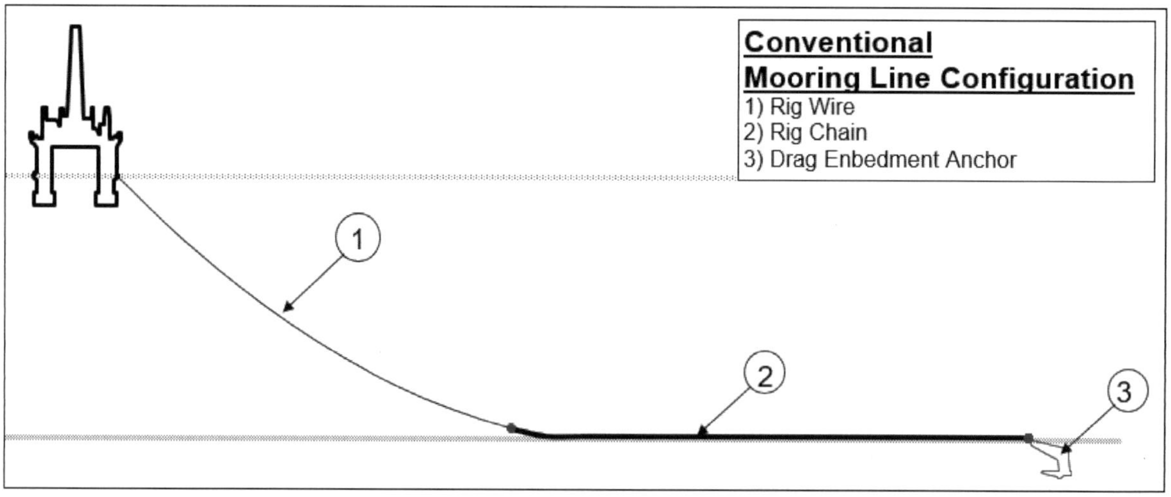

Figure 1: Conventional Mooring Line Configuration

Preset Mooring Overview

Preset mooring systems use pre-installed anchors and mooring components, allowing for the use of a larger range of anchors such as suction piles, driven piles, and both drag embedment or gravity installed vertical load anchors (VLAs). Additionally, preset mooring systems allow for the use of synthetic ropes and buoys in the mooring line configuration making this type of system ideal for locations with abundant subsea infrastructure that must be avoided during installation and operation. Preset mooring systems are also ideal for locations with deep water depths, as utilizing lighter mooring components can result in improved MODU station keeping when compared to conventional mooring systems. Two example elevations views of preset mooring line configurations are shown in Figure 2 and Figure 3.

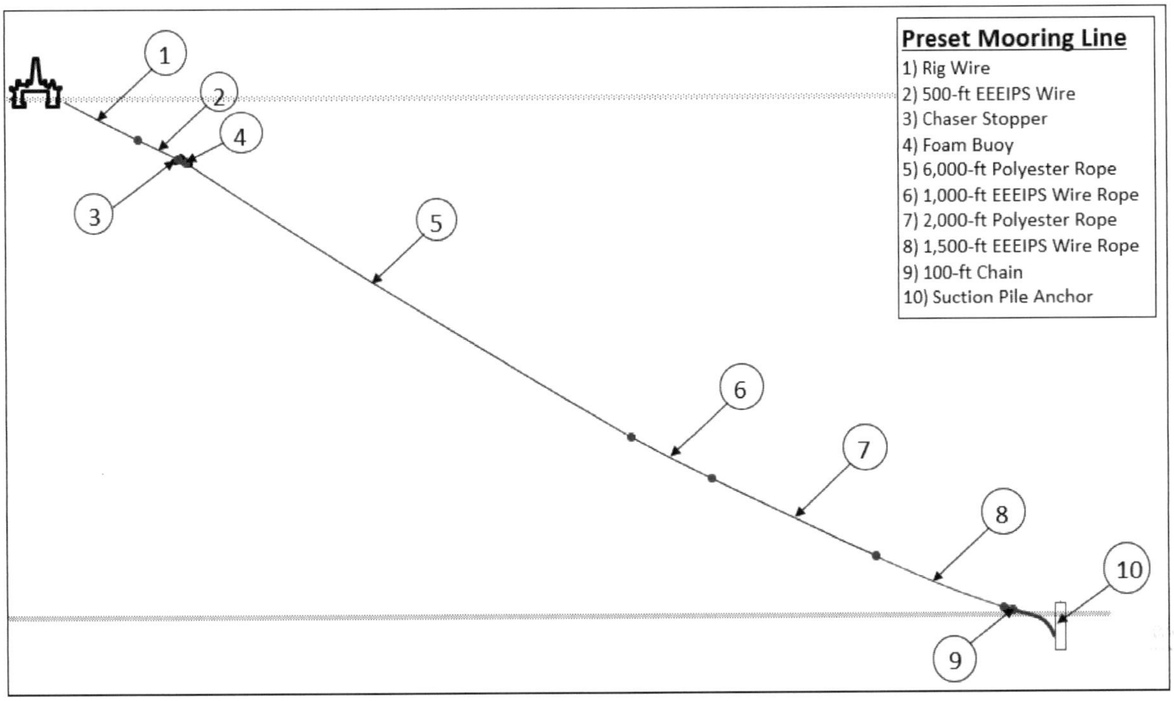

Figure 2: Example Preset Mooring Line Configuration for 7,700-ft WD

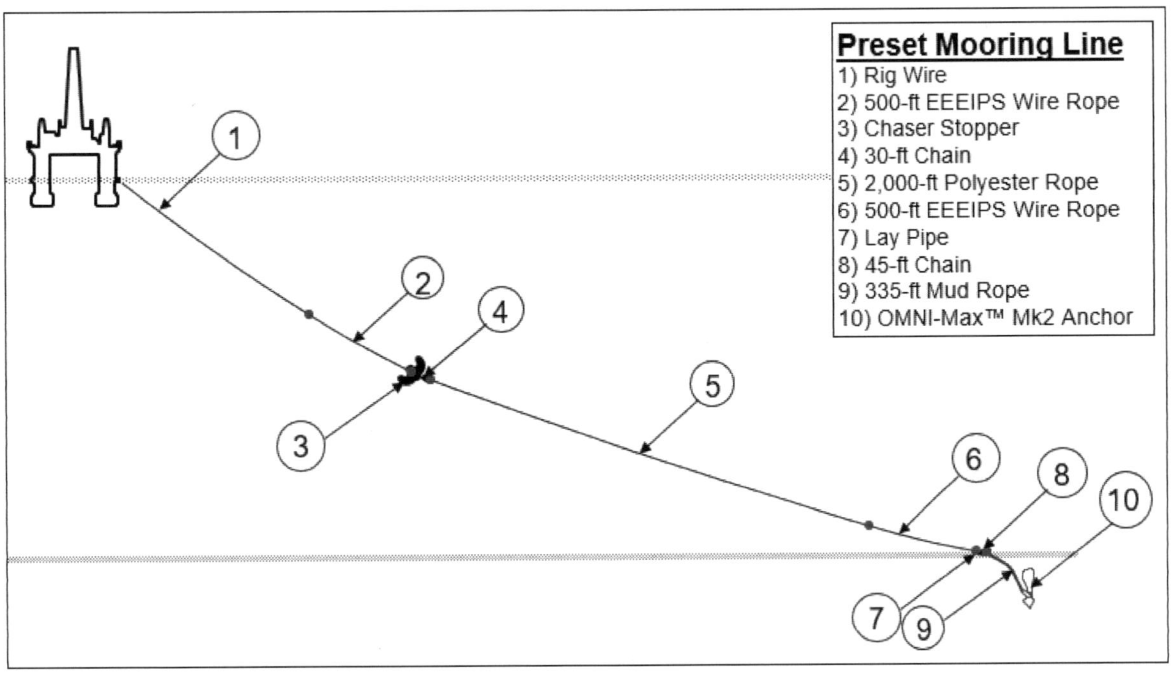

Figure 3: Example Preset Configuration for 1,100-ft WD

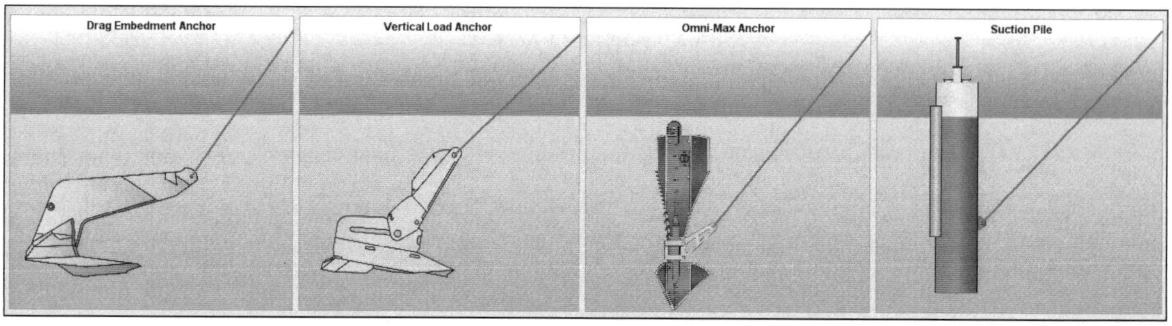

Figure 4: Example MODU Anchors (Courtesy of Delmar System's DEAP™)

Traditional Preset Mooring System Disconnection Methodology Overview

The recovery of preset mooring systems traditionally requires the decking of the mooring line prior to the breaking of a connecting link located between the rig side mooring wire or chain and the anchor side preset component. This is a time consuming process, especially in deeper water depths. Disconnection operations typically involve one or two AHVs per line, the operation of the rig crane, and the involvement of a work class ROV. Each of these requirements comes with certain operational constraints with regard to metocean criteria suitable for work. One of the key time saving benefits associated with the utilization of mooring system quick-release devices for planned disconnection is the removal of these metocean dependent operational requirements.

DQR Overview, Usage and Features

The Delmar Quick Release (DQR), shown in Figure 5, is a mechanical mooring disconnection device used for both emergency mooring system releases, and planned disconnections to improve rig move efficiency. The DQR has two different trigger methods. The fairlead release method, shown in Figure 6, involves actively winching the DQR and trigger sleeve towards the fairlead. When the trigger sleeve contacts the fairlead the shear pin contained within the DQR is broken, opening the quick release and releasing the mooring line. The second option, shown in Figure 7, utilizes an AHV and a mooring line chasing device such as a Widebody or J-Lock. The chaser is deployed in-line and chased away from the rig, towards the DQR. After contacting the trigger sleeve, the chaser then breaks the shear pin, opening the quick release. Unlike traditional disconnections, both methods do not require the lines to be recovered by a secondary vessel, or the use of an ROV, allowing the rig to depart location faster. The mooring line trajectory during release along with the recovery methodology for the released line is pre-engineered into the design of the mooring system itself with considerations for subsea infrastructure. An example mooring configuration using a DQR is shown in Figure 8.

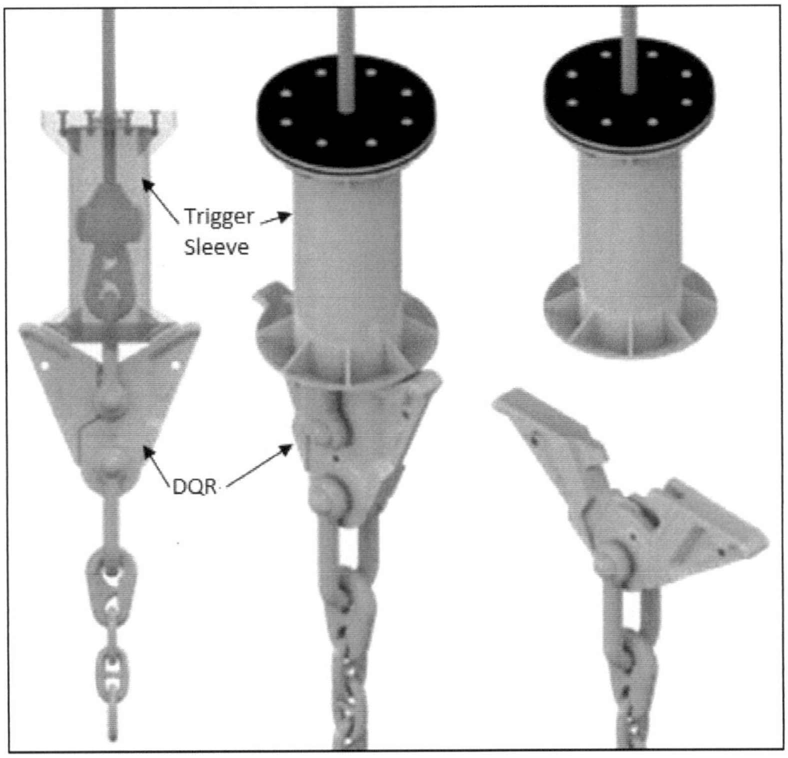

Figure 5: DQR Disconnection

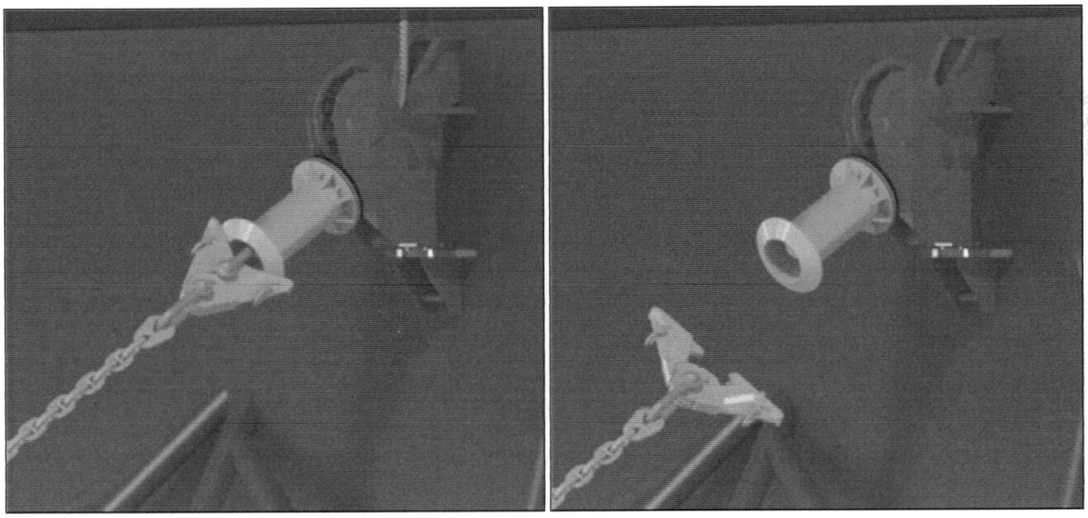

Figure 6: DQR Fairlead Release Method

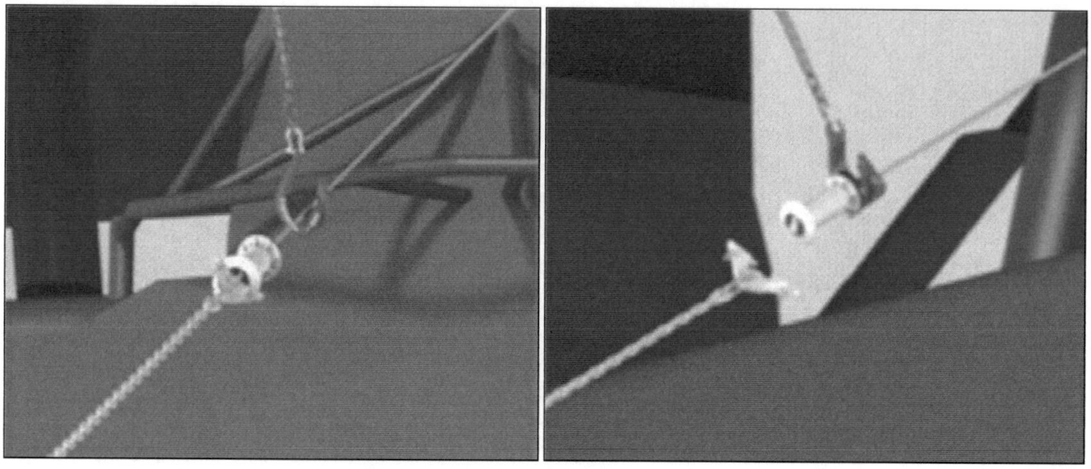

Figure 7: DQR AHV Intervention Release Method

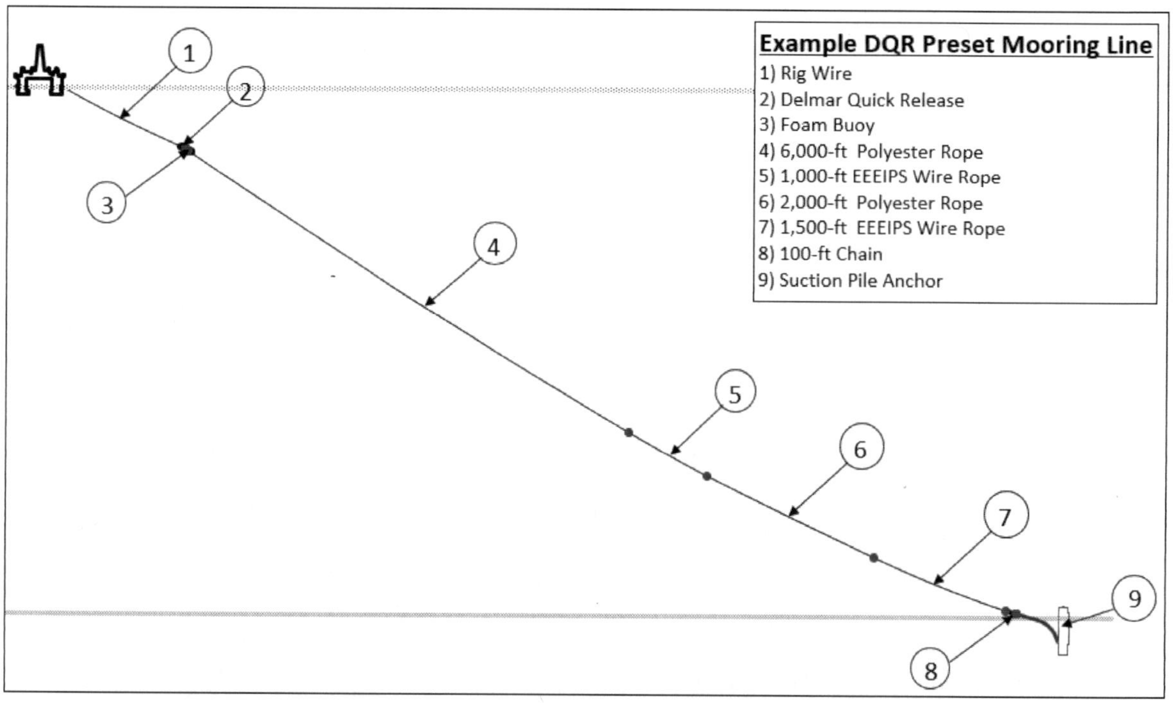

Figure 8: Example DQR Preset Mooring Line Configuration for 7,700-ft WD (Compare to Figure 2)

RAR Overview, Usage and Features

The Rig Anchor Release (RAR), developed by InterOcean Systems, is an acoustically triggered mooring disconnection device for use both in emergencies, and to increase rig move efficiency. The RAR utilizes a portable transponder deployed from a rig or other vessel which transmits an acoustic signal to the transducers on the RAR, as shown in Figure 9. An internal locking system holds the two sides of the RAR together until triggered. When triggered, the piston retracts, releasing the rig from the mooring system. The locking system and release functionality are detailed in Figure 10. The RAR operates at a low frequency to prevent interference with other acoustic operations in the field.

The RAR removes the need for an ROV to be placed in the water along with the need for the decking of mooring components during disconnection operations, often resulting in significant time savings. An example mooring configuration using a RAR is shown in Figure 11.

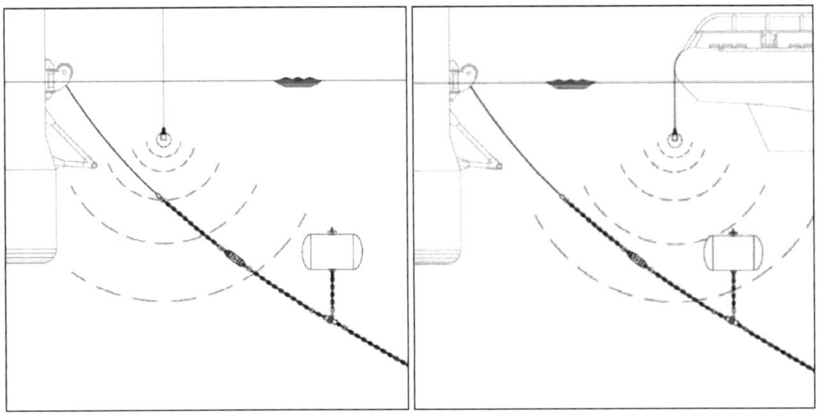

Figure 9: RAR Transponder Deployment (Left: Rig, Right: AHV)

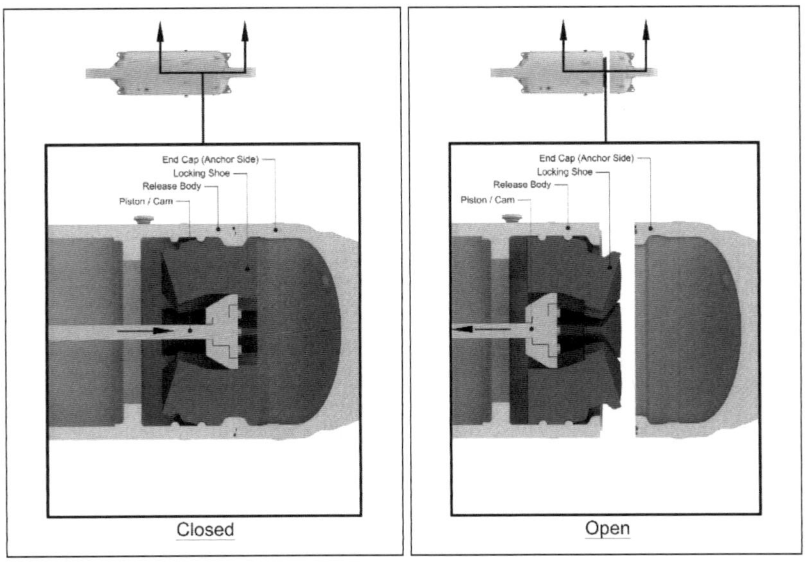

Figure 10: RAR Internal Locking Mechanism

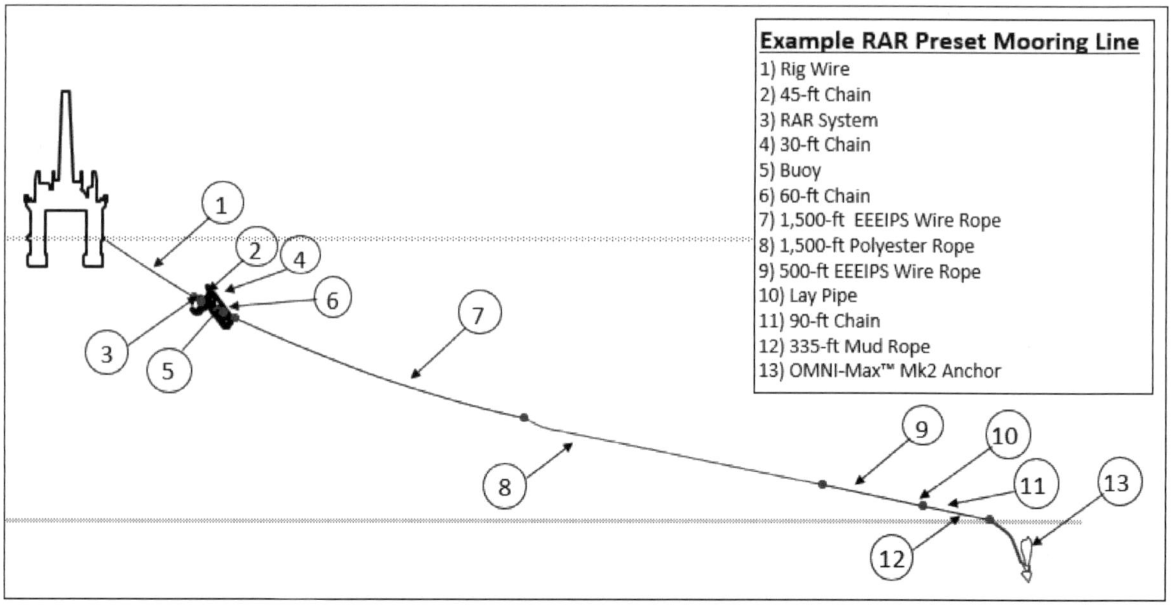

Figure 11: Example RAR Preset Mooring Line Configuration for 1,100-ft WD

OVERVIEW OF RECENT DELMAR DQR & RAR USAGE

The DQR has been field proven in the Gulf of Mexico on a taught-line preset moored semi-submersible MODU. Eight DQRs were deployed in the system for over 160 days at a water depth of approximately 7,800-ft. Seven of the DQRs were released using the rig winches to pull the trigger sleeve into the fairlead, and the remaining eighth DQR was released using an AHV and a modified chaser. Several of the mooring line releases were performed in a metocean environment that would have prevented mooring systems without quick releases from being recovered due to reliance on AHVs, ROVs, and the rig crane. The average disconnection time using the DQR was 3.3-hr per line. A similar preset mooring used at the same water depth was used as comparison. In the comparison case, the mooring line was retrieved using a secondary vessel to disconnect a subsea connector. When comparing the traditional preset system recovery to the DQR preset system, disconnection of the quick-release system saved an average of 12-hrs per line. This results in a potential time savings of 143.2-hrs when using DQRs for preset disconnection of a 12-line mooring system versus a standard preset disconnection, without the aid of mooring system quick-release devices. These time saving comparisons can be found in Table 1. It should be noted that these time savings do not account for adverse weather conditions which would add further delays (and additional time saved) to a traditional preset recovery.

Table 1: Time Savings Using DQR (Compared to Preset System in Same WD)

Type	Average WD (ft)	Avg. Disc./Leg (hr)	8-leg Total (hr)	8-leg Δ Time (hr)	12-leg Total (hr)	12-leg Δ Time (hr)
Preset	7,700	15.3	122.0	0.0	183.0	0.0
DQR	7,700	3.3	26.5	-95.5	39.8	-143.2

The RAR has a long history of worldwide use as an emergency release device. Delmar has experience utilizing the RAR to perform both emergency releases in the arctic and pre-planned efficiency enhancing releases in the Gulf of Mexico. The disconnection times from these experiences were compared to historical disconnection times of traditional preset and conventional mooring systems at similar water depths in order to quantify time savings. For quantification purposes, the disconnection times listed under the RAR type mooring system in the below table begin when the rig begins to reduce mooring line tension in preparation for the transmission of the acoustic disconnection signal, and end when the rig wire is handed back to the rig. Preset and conventional mooring system disconnection times are quantified from the time at which the PCC is passed to the AHV. Mooring system disconnection times end when the PCC is passed back to the rig in the case of a conventional system, while the secession of disconnection time occurs when both the gravity hook and ROV return to the deck of the AHV in the case of preset mooring systems.

Table 2: Time Savings Using RAR (Compared to Conventional System in Similar WD)

Mooring Type	Average WD (ft)	Avg. Disc./ Leg (hr)	8-leg Total (hr)	8-leg Δ Time (hr)	12-leg Total (hr)	12-leg Δ Time (hr)
Preset	1,500	8.8	70.7	0.2	106.0	0.3
Conventional	850	8.8	70.5	0.0	105.8	0.0
Preset	1,150	6.7	53.5	-17.0	80.3	-25.5
Preset	1,500	6.5	52.0	-18.5	78.0	-27.8
RAR	1,100	1.3	10.5	-60.0	15.7	-90.0
RAR	1,150	1.0	8.2	-62.4	12.2	-93.5
RAR	1,350	1.0	7.9	-62.6	11.8	-93.9

Based on the three examples outlined in Table 2, average mooring line disconnection times when using the RAR range from 1.0-hr to 1.3-hrs. When compared to a conventional mooring system with an average disconnection time per leg of 8.8-hrs the use of RARs in a mooring system add up to time savings in excess of 90-hrs.

Preset mooring systems innately offer the potential for savings during critical path rig moves by allowing for the mooring system at the next surface location to be installed prior to the arrival of the drilling rig. Both mooring system release devices currently available on the market allow for this type of "piggy backing" operation, where the other half of the mooring system release device is pre-installed on the next location's preset mooring system. Utilizing this methodology an AHV can be ready at the next surface location with the connection on-deck as the drilling rig is towed in. Mooring preset component can be recovered from the previous location when such operations are both operationally and economically practical based upon both current metocean and vessel availability.

EMERGENCY DISCONNECTION

The NOAA storm database comprises all tracked tropical storm and hurricane events since the year 1900. This database was cross-referenced with Delmar System's ConStat™ database of over three hundred rig surface locations, shown in Figure 12. In ninety-eight instances, a tracked storm exceeded the wind speed criteria of 64-kts for a category one hurricane while passing over a drilling rig location. The time between when NOAA began tracking each of these storms and the exceedance of category-one hurricane wind speeds on drilling locations is detailed in Table 3.

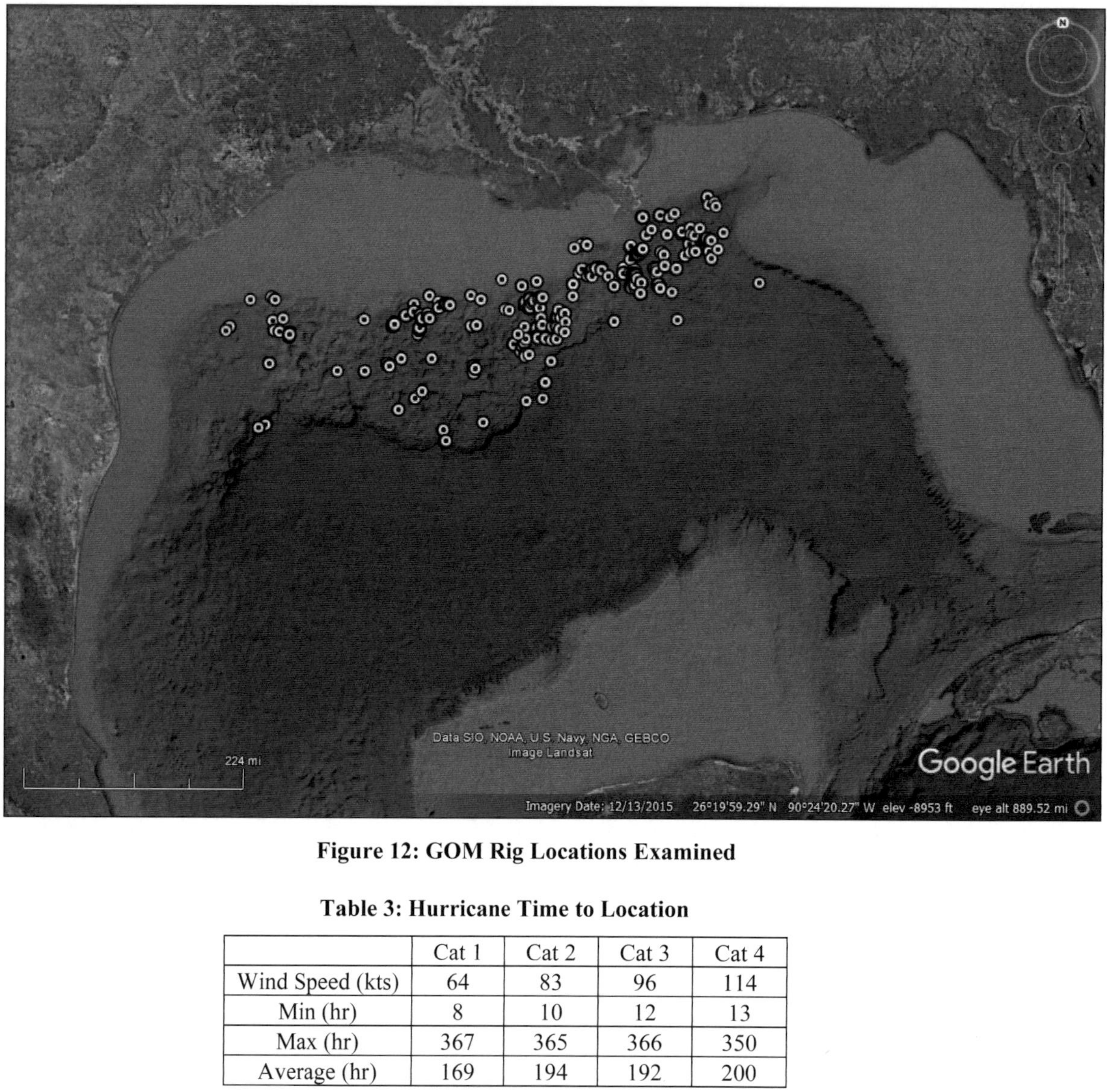

Figure 12: GOM Rig Locations Examined

Table 3: Hurricane Time to Location

	Cat 1	Cat 2	Cat 3	Cat 4
Wind Speed (kts)	64	83	96	114
Min (hr)	8	10	12	13
Max (hr)	367	365	366	350
Average (hr)	169	194	192	200

The time between when the decision to leave the drilling location is made and the actual departure of the rig from the field is referred to as a "T-time". This time includes operations such as riser recovery, and in the case of a moored system, the disconnection of the rig from the mooring system. The statistics in Table 3 indicate that the ability to reduce the T-time to below eight hours would allow for a rig to safely depart the field prior to the exceedance of hurricane force winds in all scenarios analyzed. The estimated emergency release times for mooring systems when considering a simultaneous RAR release and subsequent rig component recovery range from 3-hr for an 8-line system to 4.5-hr for a 12-line system on average. It is therefore advisable to consider the inclusion of planned mooring system quick-release devices, such as RARs, in a mooring system when operating during hurricane season.

CONCLUSIONS

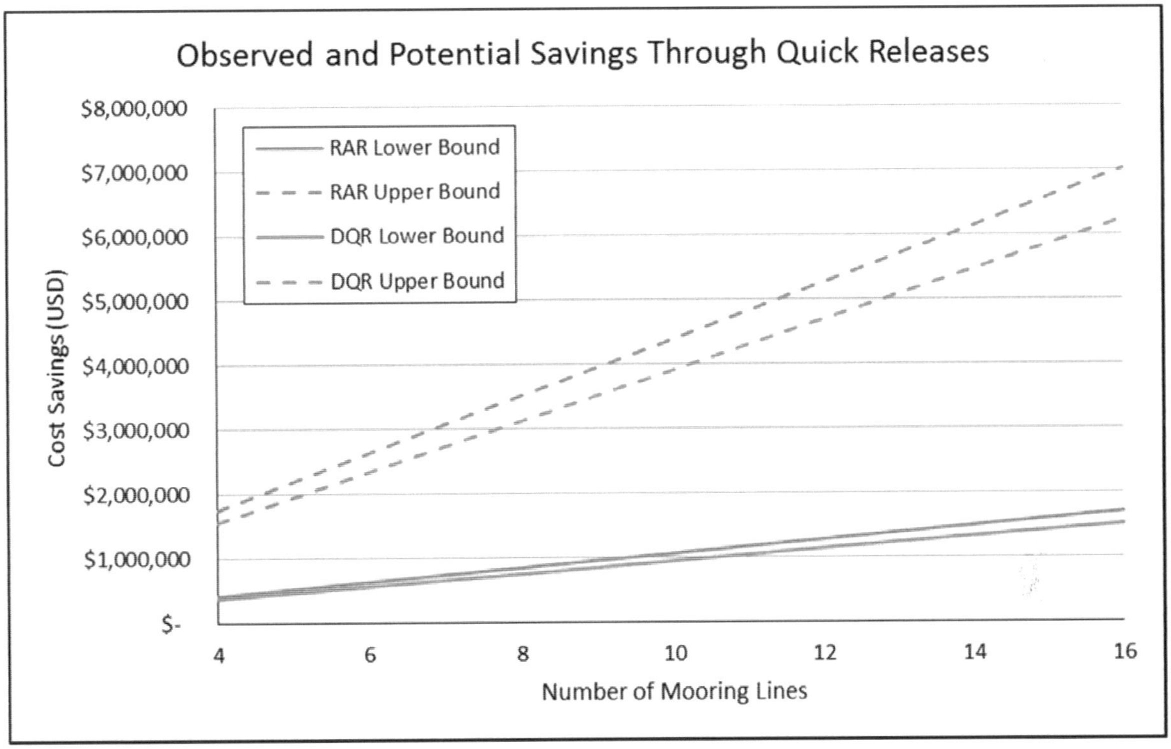

Figure 13: Cost Savings Through Quick-Release Mooring Systems

The information portrayed in the figure above was tabulated using a lower-bound day rate for a drilling rig of $130k and an upper-bound day rate of $600k. A lower-bound AHV day rate of $60k and an upper-bound of $180k per day were similarly considered. These prices were chosen to envelope lower-bound GOM prices and upper-bound GOM prices during the last ten years. The time saved using the RAR was adjusted to correlate to a water depth of 7,700-ft for comparison to the DQR. It is evident from the figure that the time saved using a mooring quick-release results in significant cost savings, even when conservatively assuming that an AHV is used during release for the recovery of the RAR. These savings do not take into consideration the cost saved from other factors such as reduced tug standby times and operational weather windows.

The use of mooring system quick-release devices presents a number of safety and time savings benefits. Rig move operations can proceed in a much broader range of metocean conditions due to the elimination of operational requirements, such as the operation of the rig crane, the diving of a work class ROV, and the maneuvering of AHVs in close proximity to each-other and/or the MODU. Drastically reduced mooring system disconnection times both increases the operational safety of the drilling campaign by allowing for moored MODUs to depart safely prior to the arrival of a hurricane or sudden metocean events, and shaves days of critical path time off of typical rig move operations.

ACKNOWLEDGEMENTS

The authors would like to thank Daniel Alonso, Steven Bowles, Mikaela Farrell, and InterOcean Systems for their assistance on this paper.

REFERENECES

"Gulf of Mexico." 26°18'12.77"N and 90°31'36.38"W. Google Earth. December 13, 2015. October 3, 2016.

InterOcean Systems LLC. San Diego, CA. <http://www.interoceansystems.com/>

Pasternak, J., Alonso, D., Garrity, R., Shelton, J. (2012). DP Vessel Passive Quick Release Contingency Mooring Systems. SNAME Offshore Symposium. Houston, TX

Pasternak, J., Hersley, M., Farrell, M. (2013). Mitigating Risk in Shallow Water and Artic Station-Keeping. SNAME Offshore Symposium. Houston, TX

SNAME
THE INTERNATIONAL COMMUNITY
FOR MARITIME AND OCEAN PROFESSIONALS

Global performance of a turret-moored large tanker in complex sea states dominated by swell and wind-seas

AUTHOR (RAZIEH ZANGENEH)
DEPARTMENT OF MECHANICAL ENGINEERING, UNIVERSITY OF MAINE, ORONO, ME, USA

AUTHOR (KRISH P. THIAGARAJAN)
DEPARTMENT OF MECHANICAL ENGINEERING, UNIVERSITY OF MAINE, ORONO, ME, USA

AUTHOR (MATHEW CAMERON)
ADVANCED STRUCTURES & COMPOSITE CENTER, UNIVERSITY OF MAINE, ORONO, ME, USA

Proceedings of the 22nd Offshore Symposium, February 2017, Houston, Texas
Texas Section of the Society of Naval Architects and Marine Engineers

ABSTRACT

Producing hydrocarbons from deeper waters in various offshore environments requires operations and survival in complex seas made up of non-collinear winds, wind-driven seas and long period swells. Such bi-directional wave conditions appear in regions like west of Africa and Brazil (Campos Basin), where the 100-year condition comprises of extreme swells in addition to local waves of varying degrees of severity. A nascent interest of the offshore industry has been on studying the global performance of FPSO and FLNG systems in complex sea conditions in deep waters. This paper presents a study on the behavior of a turret moored Very Large Crude Carrier (VLCC) in typical bi-directional sea states at two experimental scales ranging from 1:60 to 1:120.

It is still uncommon to find experimental capabilities for simulating and modeling bi-directional sea states. At the University of Maine's recently completed Alfond W^2 wind wave facility, bi-dimensional sea-states are created by systematically combining two different wave systems from varying directions. The wave generator comprises of 16 independent paddles with the capability of generating regular and random uni-directional or bi-directional waves at angles up to +/- 60 degrees relative to the basin center line. A geometrically scaled down version of a VLCC which can be considered representative of an FPSO or an FLNG, at 1:120 scale was tested at the W2, whose hull and the mooring system were similar to a 1:60 scale moored-tanker hull tested at the National Research Council's Institute of Ocean Technology wave basin facility in Canada in 2006. The mooring system comprised of four spring-loaded lines attached to a forward turret.

Results show that statistics of the vessel global motions and mooring tension at the two scales compare very favorably. When the vessel is aligned with the oncoming swell, the influence of the wind-sea increases when its heading increases, and this is particularly noticeable on the horizontal plane motions (sway and yaw) and on the mooring line tension. The power spectral density of roll motion has a prominent peak at the swell period, which exceeds the peak at the natural frequency. Our previous experiments in uni-directional long period waves identified an instability in passive weathervaning systems of FPSOs, wherein for wavelengths from $0.6 < \lambda/L < 2$ (L – ship length) the model drifted to a large angle of 45 – 67 deg with respect to head sea. This effect, noticed at both model scales, was subsequently verified by simulations and theoretical calculations. In the presence of bi-directional conditions, the wind sea appears to calm the instability, and the mean heading drift was reduced to 24 deg.

Keywords: Bi-directional sea states, FPSO and FLNG systems, W^2 wind wave facility

INTRODUCTION

The verification of the global performance in the design of floating structures and their moorings is important. Measures must be taken to ensure that their designs are appreciably reliable since such structures will be exposed to great environmental forces. One way to verify the performance in the prototype design process is model testing. A model of offshore structure can be built and subjected to the same environmental loads in the wave basin as those used in the prototype design. During testing, the responses of a floating structure to various forces due to wind, wave and current are measured and compared to those obtained in the design of the prototype floater. In recent years, Floating Production, Storage and Offloading (FPSO) and Floating Liquid Natural Gas (FLNG) vessels have proved effective for the processing and storage of hydrocarbons offshore. The effect of unexpectedly deteriorating weather conditions such as bi-directional waves and wind needs to be accounted for in the design of FPSOs to ensure the safe operation under all expected environmental conditions and pollution-free transfer of crude oil. Studies have recently been done to estimate the bi-directional wave drift forces and its influences on offshore floating structures. Pinkster (1980, 1988) calculated the drift forces of regular and irregular waves on a tanker in bi-directional waves. Nwogu and Isaacson (1991) calculated the spectral density of the second-order wave drift forces and analyzed the mooring line forces and slow-drift motions in bi-directional waves.

Kim et al. (2005) developed a coupled dynamic analysis program in time domain for the global motion simulation of a turret-moored FPSO designed for 6000-ft water depth they compared the numerical results with 1:60 model-testing results in the severe hurricane condition of Gulf of Mexico. Chen (2006) used the so called middle-field formulation to describe some results from the second-order wave forces on a liquefied natural gas (LNG) in bidirectional waves. Rezende and Chen (2009) performed numerical computations in order to evaluate the effects of directional interaction between two waves on a barge in deep water depth. Sergent and Naciri (2010) calculated wave and wave drift loads for a standard LNG carrier using the middle-field method. More recently Zhang (2012) reported that under environmental conditions in West of Africa, swell induces larger drift loads on FPSOs than wind-sea at the surge/sway natural frequency, and they both contribute to the low frequency motion.

Zangeneh and Thiagarajan (2014) published a numerical study to numerical simulations of the heading instability of this FPSO, including the decay simulation result to obtain the natural periods and damping ratios of the prototype model. Zangeneh and Thiagarajan (2015) discussed the effect of adding damping on heading instability of FPSOs in swell dominated region of west of Africa. Milne et al. (2016) obtained a reliable estimates of the heading of an operating FPSO from the equilibrium of the steady yaw moments acting at the turret for sea states where both wind and swell waves were present and non-collinear.

Pistani and Thiagarajan (2007) tested a 1:60 scale model of an FPSO hull at the ocean basin of the Institute of Ocean Technology, Canada, in bi-directional waves including swell and wind-sea. The hull was moored by means of four mooring lines connected to an internal rotating turret located at roughly one third of length from the bow of the model. In the tests, the period and amplitude as well as its direction (zero degrees) of the swell-sea were kept constant. The wind-sea direction varied between 22.5, 45 and 60 degrees with reference to the swell direction of propagation.

The particular focus of this paper is to study the responses of an FPSO to bidirectional waves at 1:120 scale and compare the result with 1:60 scale.. A geometrically scaled down version at 1:120 scale, was built at the Marine Ocean and Offshore Research (MOOR) laboratory at the University of Maine. The FPSO has an internal turret to support four mooring lines which are modeled as symmetrically positioned lines with linear springs (similar to the 1:60 scale). The experiments were conducted at the University of Maine's W2 wind wave facility. The W2 facility which is the only one of its kind in the world, equipped with an open-jet wind tunnel over a wave basin, capable of testing ocean energy devices and ship structures gives a great opportunity to do experimental research on FPSOs.

EXPERIMENTAL SETUP

Model

The tests were performed on a 1:120 scale model as shown in Figure 1. The hull and the mooring system used in this study were similar to the hull used in the experimental campaign at a 1:60 scale FPSO hull at Institute of Ocean Technology in Canada (Pistani and Thiagaragan, 2007). Froude scaling was employed for comparison of intrinsic parameters and test results between the model and the prototype. The FPSO characteristics are described in Table 1. These characteristics are described at full scale. The model was ballasted for light displacement with lead weights. The internal turret of the FPSO is located 62.4 m aft of its fore perpendicular.

Figure 1: Wave basin and FPSO model

Table.1 Parameters of prototype FPSO

Parameter	*Unit*	*Value*
Overall length	m	328.2
Length between perpendiculars *(L)*	m	318.0
Width	m	56.1
Depth	m	28.2
Draft	m	10.6
Displacement	t	145175
Turret Location (forward , aft perpendicular)	m	258.2
Longitudinal Centre of Gravity (LCG)	m	172.4
Vertical Centre of Gravity (VCG)	m	14.2

Basin Setup and Calibration

The Alfond W2 Ocean Engineering Lab is a unique facility equipped with an open jet wind machine over a wave basin. The tank dimensions are 30m x 9 m with a variable depth from 0 m to 4.5 m. A 16-paddle wave maker (Figure 2) has the following capabilities:
Maximum wave height: 0.6 m (24 in) at T = 1.65 sec, 0.8 m (30 in) at T = 2.3 sec
Regular waves and all standard spectra in addition to custom random seas
Wave angles in excess of +/- 60 degrees relative to the basin center line

Figure 2. Wavemaker comprises of 16 paddles to generate bi-directional waves

The results of the performed model tests were recorded using a motion package (Qualisys System), inline load cells to measure the mooring top tensions as well as with an accelerometer at the turret location. The initial phase of the FPSO model experiments was to calibrate the basin. Prior to installing the model, all sea-states were run with an array of five resistance wave probes in place of the model for calibrating the basin and to measure the actual wave amplitudes experienced by the model.

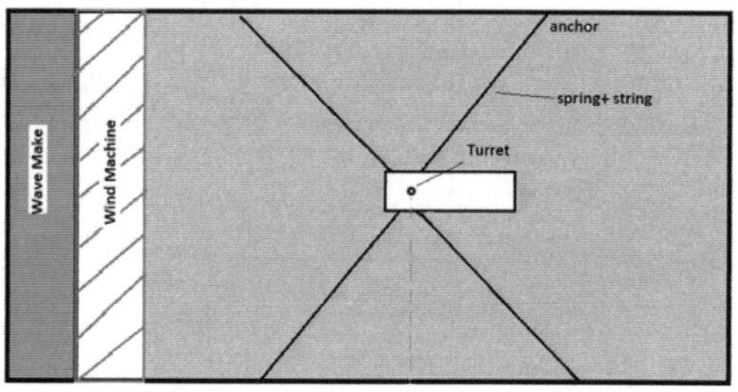

Figure 3: Basin calibration set up

Mooring System

The FPSO model was moored using 4 instrumented mooring lines designed to be compliant and provide sufficiently low natural periods in surge, sway and yaw. This compliancy was provided by an inline spring of spring stiffness 7 N/m at each mooring line and so that the mooring loads do not influence the wave frequency motions of the FPSO. The lines were attached to fairlead hooks at the turret and to the basin wall. These mooring lines were laid out symmetrically, spaced 90o from each other as shown in Figure 4.

Figure 4: Model positioning and mooring layout in W2 basin.

Static offset test was carried out to verify the static stiffness of the mooring line in the x-direction of the moored

FPSO model. In the static tests, the FPSO was moved slowly in the x-direction, and the mooring line tension was recorded. The static test results are plotted in Figure 5 versus the analytical result. Good agreement can be seen between the measurements and the target.

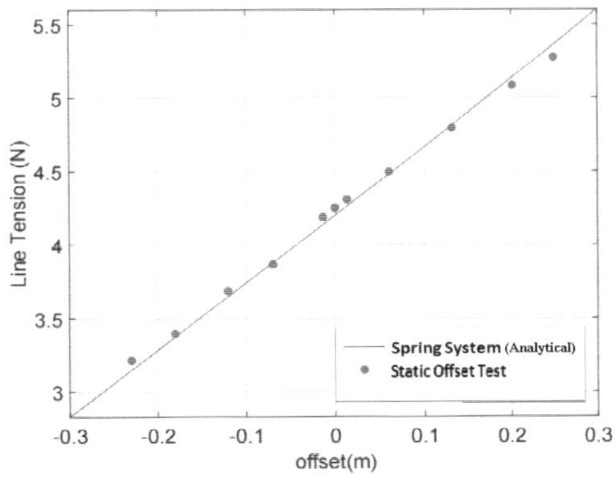

Figure 5: Comparison of the line tension in surge static offset test

RESULTS AND DISCUSSTION

Free decay test

Decay tests were conducted for surge, roll and pitch degrees of freedom to obtain the natural frequencies and the total damping of the hull/mooring system (Table 2). Heave decay tests were not carried out because to push the model in heave was physically difficult as a result of large heave stiffness. Figure 6 shows the roll decay test result. The natural periods and damping coefficients are compared with the 1:60 as well as numerical simulation results (Zangeneh et al., 2014). The comparisons of damping for various modes look satisfactory as presented in Table 2. There exist some discrepancies in the numerical

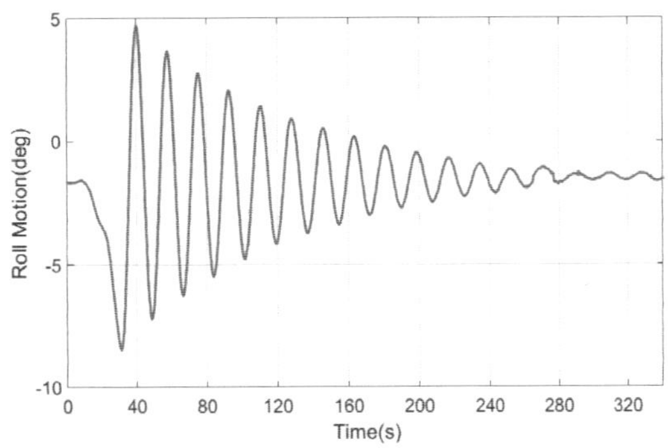

Figure 6: Model Test Roll decay

result (Zangeneh and Thiagarajan, 2014) and measured value of pitch natural period however the pitch natural period resulted from two different scales are quite close.

Table 2. Comparison of the free decay test results

	Natural Period			Damping		
	Numerical	**1/60 Scale**	**1/120 Scale**	**Numerical**	**1/60 Scale**	**1/120 Scale**
Surge	148.55	143.96	147.13	2.11	1.37	1.29
Roll	14.60	13.58	13.64	1.15	1.31	1.19
Pitch	11.80	13.31	13.45	18.9	16.2	17.2

Response in Bi-directional sea states

The wave generator at W2 has the capability of generating two different wave systems that combine at the center of the basin as if they were coming from different directions. One part of the sea-state generated in the facility was a long regular wave system that acted for our purposes as swell (Figure 5). ITTC formulation has been applied to model the wind-sea by a JONSWAP spectrum. FPSO Model tests in bi-directional Sea states were conducted as listed in Table 3. Two combinations of wind-sea and swell were tested during the model testing (Figure 7). The swell was kept constant while the wind-seas were varied in heading. The test matrix is tabulated in Table 3. For select conditions shown in Table 3, tests were conducted for 2.5 hour duration full scale.

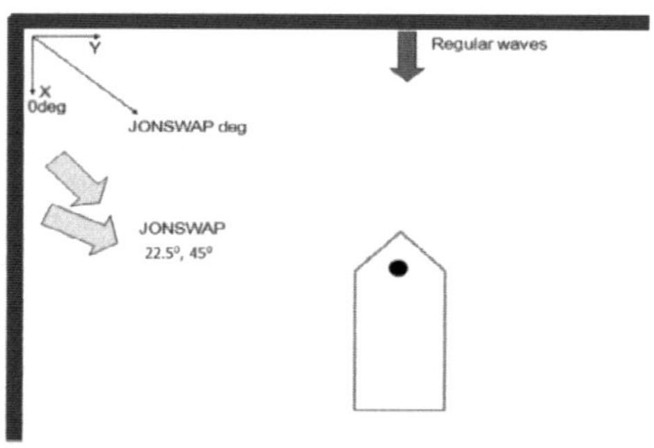

Figure 7: Initial FPSO orientations during Model tests

Table 3. Bidirectional Test Matrix

	Wind Sea			**Swell**		
Run	H_s(m)	Tp(s)	Dir (deg)	H(m)	T(s)	Dir (deg)
1	15.9	12.363	22	7.98	14.99	0
2	15.9	12.363	45	7.98	14.99	0

FPSO responses in bidirectional waves

Statistical parameters of the time series for two different wind-seas heading can be seen in Table 4. For global vessel motions, the results of two different model scales are in good agreement in view of overall trend. The differences between the results of two model scales can be attributed to many possible mismatches between two models, such as mooring system damping and differences related to wave generation. Despite the mismatches, the correlation looks very reasonable.

Table 4. Comparison of two model scales Statistical parameters results

		1/120 Scale			1/60 Scale		
Run#1	Unit	Mean	Std	Max	Mean	Std	Max
Surge at CG	m	-14.43	11.28	-42.12	-15.03	11.02	-48.13
Sway at CG	m	-6.19	11.18	-37.19	-10.89	10.26	-42.19
Heave at CG	m	-0.25	0.90	-2.53	-0.23	0.88	-2.34
Roll	deg	-0.36	1.09	-5.65	-0.41	1.29	-4.88
Pitch	deg	0.51	1.23	3.06	0.47	1.12	3.14
Yaw	deg	0.61	9.09	-23.61	0.35	8.77	23.71
Mooring Tension (Line#1)	N	7.11E6	5.10E6	8.32E6	7.19E6	5.33E6	8.87E6
Run#2							
Surge at CG	m	11.27	7.68	37.16	9.27	7.04	39.54
Sway at CG	m	-37.73	17.09	-82.90	-34.67	14.29	-81.66
Heave at CG	m	0.08	0.93	3.62	0.07	0.88	3.48
Roll	deg	0.07	0.92	3.65	.08	0.97	3.77
Pitch	deg	0.66	1.33	4.85	0.63	1.49	4.98
Yaw	deg	28.77	6.72	43.47	23.69	9.87	48.19
Mooring Tension (Line#1)	N	7.21E6	5.90E6	9.04E6	7.28E6	6.60E6	9.55E6

The results of two model tests indicates that there is a significant increase in the sway and yaw responses as the wind-sea heading changes from 22.5 degrees to 45 degrees while no significant change can observe in the surge motion. mooring line tension shows increasing trend of the mooring line tension as the heading varies from 22.5 degrees to 45 degrees indicates that the interaction of sea and swell dependent on the directionality of the wind-sea.

Figure 8 shows the time series of motions for two different model scales for Run#1 which the wind-sea direction is 22 degrees. Power Spectral density plots were used to analyze the frequency components in a time series to investigate the source of excitation. As the results low frequency slowly varying motions are dominated for the surge, sway and yaw modes, while wave frequency motions are pronounced for vertical plane modes (heave–roll–pitch). Results show that the magnitudes of motions for two model scales are in good agreements except some discrepancy in dynamics which could be because of errors in wave generations and mooring line damping. The power spectrum of the 1:120 scale model shows that the roll and pitch motions have two prominent frequency components; one at the natural frequency and the other at the swell frequency.

Surge time series for two model scales

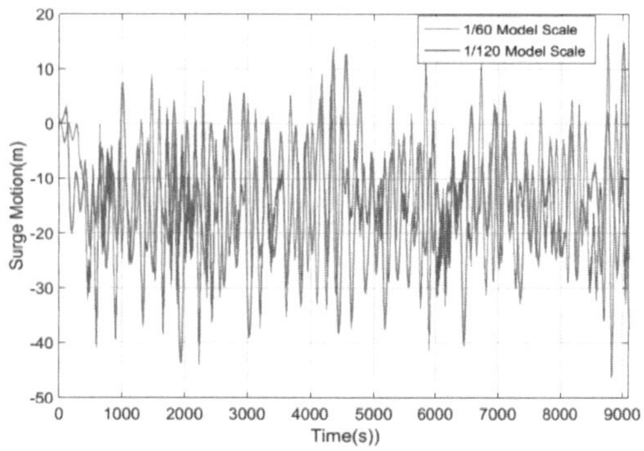

Surge power spectrum for two mode scales

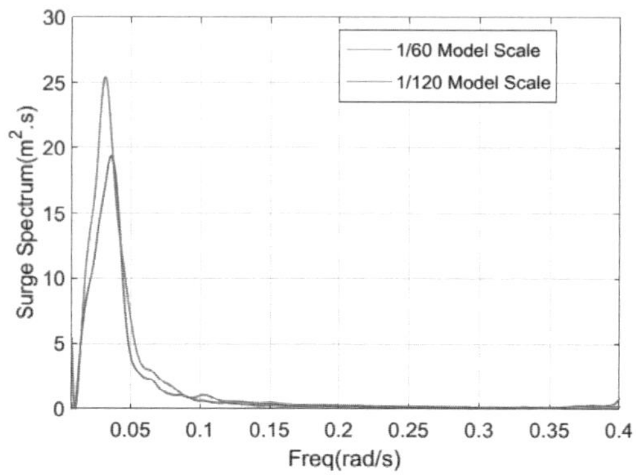

Sway time series for two model scale

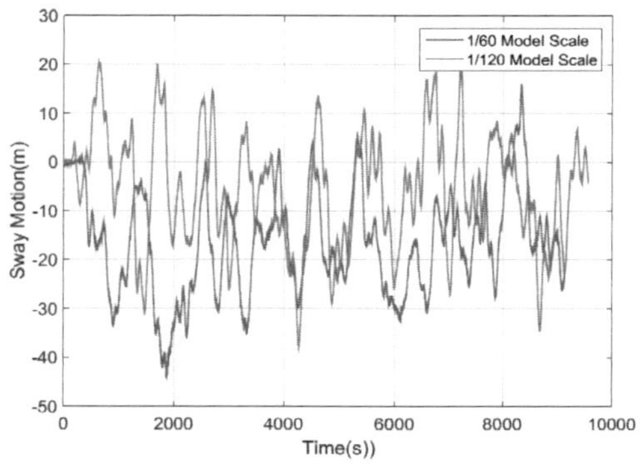

Sway power spectrum for two model scale

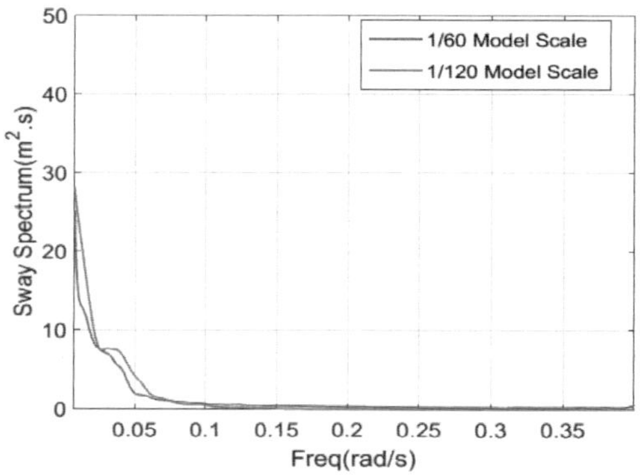

Heave time series for two model scales

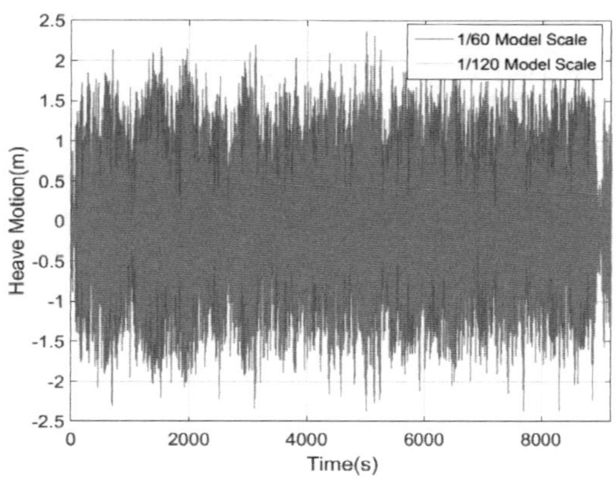

Heave power spectrum for two mode scales

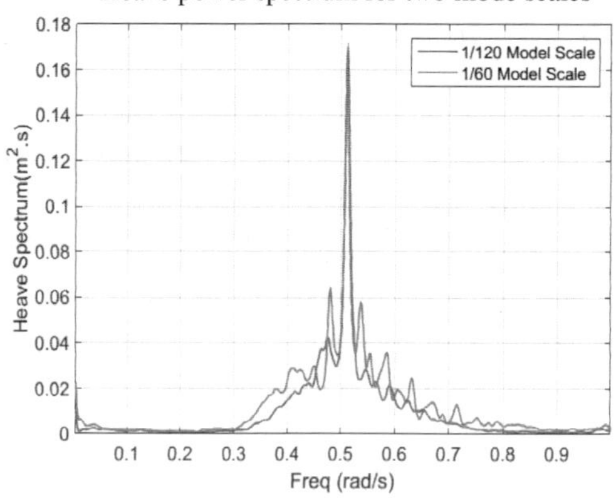

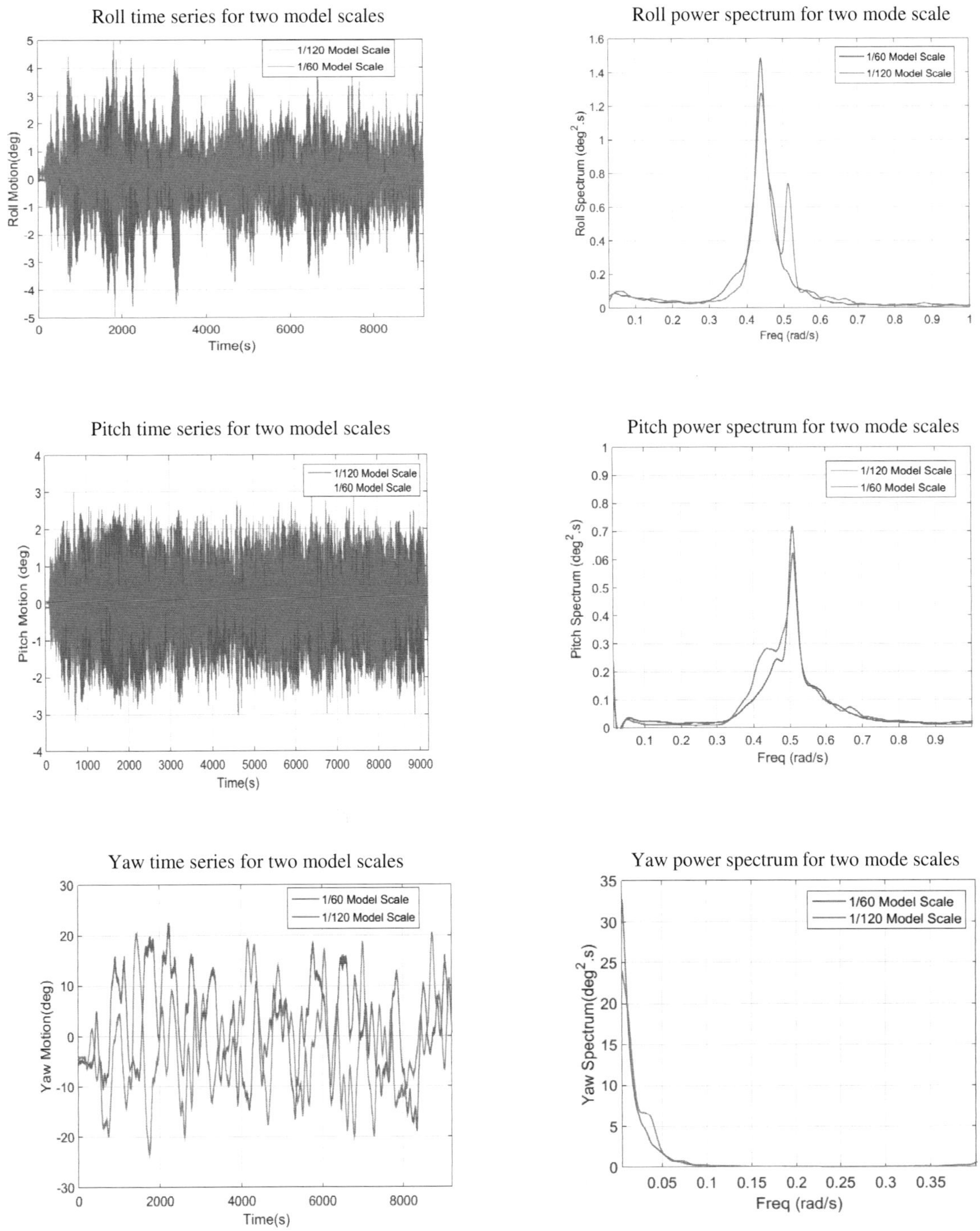

Figure 8: Summary of FPSO responses in bi-directional waves for two model scales

Because of higher roll damping at 1:60 scale model, as shown in table 2, the peak at the roll natural frequency is not visible for this mode at this scale. As the power spectrum results show, for the roll motion the swell frequency exceeds the peak at the roll natural frequency, while for the pitch mode the peak at the pitch natural frequency dominates the peak at the swell natural frequency.

Heading Instability

Previous experiments and numerical simulations in uni-directional long period waves identified instability in passive weathervaning systems of FPSOs, particularly when the wave length is comparable to the ship length (Thiagarajan and Tian, 2013). Figure 9 shows numerical and experimental prediction of the heading angle. It shows that a pitchfork bifurcation of the equilibrium predicted to occur by both experimental and simulation when wavelength/ship length ratio equals to 0.73. The numerical result predicts heading angles of 60 – 70 deg, in reasonable agreement with the maximum values seen in the experiment. In fact, the agreement is remarkably good until $\lambda/L = 1.5$. For wavelength ratio greater than 1.5, the numerical drops more rapidly while the experiment shows a second pick at $\lambda/L = 1.83$.

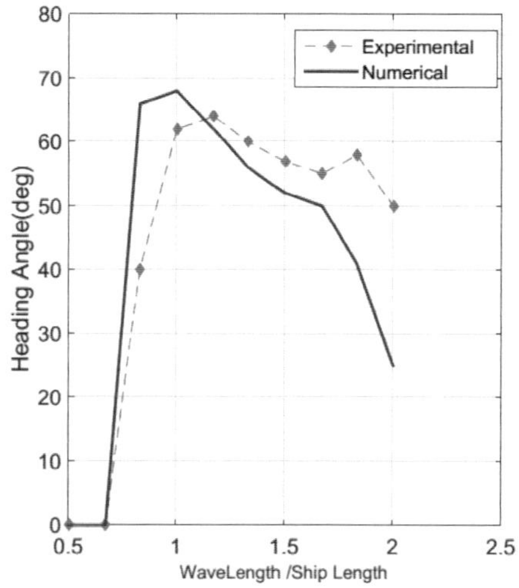

Figure 9: Equilibrium heading angle vs. wavelength/ship length

An important observation from the bi-directional waves testing is that the wind-sea appears to calm the instability significantly. The time series of heading angle for a long duration run at $\lambda/L = 1.17$ shows the heading instability with the final heading angle of 60 degrees. However as Figure 10 shows, in presence of a directional wind-sea of 45 degrees the instability decreases to about 28 degrees and with a directional wind-sea of 22 degrees the instability changes to a weathervaning with the maximum angle of 24 degrees. This indicates that in complex sea states, when two waves with different directions approach the FPSO, heading instability diminishes. However the yaw motion appears to increase.

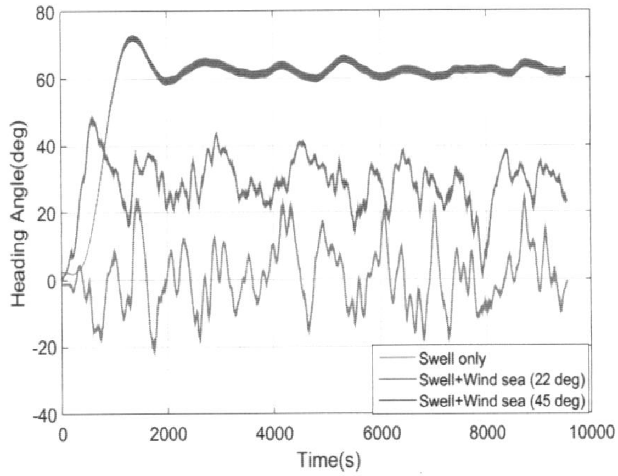

Figure 10: Heading angle in presence of Swell only and Swell+Wind-sea

CONCOLUTION

The particular focus of this paper was to study of an FPSO response to bi-directional waves in complex sea states dominated by swell and wind-seas. For this purpose a geometrically scaled down version at 1:120 scale, has been tested at W2 Ocean Engineering Lab at University of Maine. Statistical parameters of the time series for two different wind-seas heading have been analyzed and compared with the 1:60 scale model, published previously. Results show that statistical values of two different model scales are in good agreement in view of overall trend. The differences between the results of two model scales can be attributed to many minor differences between two models, such as mooring system damping and differences related to wave generation. Results indicate that the interaction of sea and swell dependent on the directionality of the wind-sea since the sway and yaw responses increase significantly when the wind-sea heading changes from 22 degrees to 45 degrees. The power spectra in vertical motions shows the excitation in roll motion is swell dominated while the pitch motion is mainly affected by the natural frequency. The paper also discusses the effect of the bi-directionality of waves on the heading instability problem of FPSOs. It has been observed that in the presence of bi-directional conditions, the wind sea appears to calm the instability, and the mean heading drift was reduced. However the yaw motion appears to increase due to the wind-sea.

ACKNOWLEDGMENTS

The experiments were conducted at The Alfond W2 Ocean Engineering Lab at University of Maine. The authors would like to thank the W2 facility operation team for their support during the model testing. The first author acknowledges the support of the Office of the University of Maine's Vice-President for Research through the Research Reinvestment Fund Graduate Assistantship scheme.

REFERENCES

Chen XB. 2006. Middle-field formulation for the computation of wave-drift loads. Journal of Eng Math. 59:61–82
Kim MH, Koo BJ, Mercier RM, Ward EG. 2005. Vessel/mooring/riser coupled dynamic analysis of a turret-moored

FPSO compared with OTRC experiment. Journal of Ocean Engineering, 32: 1780–1802.

Milne IA, Delaux S, McComb P. 2016. Validation of a predictive tool for the heading of turret-moored vessels. Journal of Ocean Engineering. 128:22-40

Nwogu O, Isaacson M. 1991. Drift motions of a floating barge in random multi-directional waves. J Offshore Mech Arct Eng. 113(1):37–42.

Pinkster JA. 1980. Low frequency second order wave exciting force on floating structures. Netherlands: NSMB Publication. No: 650.

Pinkster JA. 1988. The influence of directional spreading of waves on mooring forces. Proceedings of Offshore Technology Conference; Houston, USA.

Pistani F, Thiagaragan KP. 2007, Experimental campaign on a moored FPSO in complex bi-directional sea states, 22nd IWWWFB, Plitvice, Croatia.

Rezende FC, Chen XB. 2010. Approximation of second-order low-frequency wave loading in multi-directional waves. Proceedings of International Conference on Ocean, Offshore and Arctic Engineering; Shanghai, China.

Sergent E, Naciri M. 2010. Wave spreading and wave drift loads for a standard LNG carrier in shallow water. Proceedings of International Conference on Ocean, Offshore and Arctic Engineering; Shanghai, China.

Thiagarajan KP, Tian Z. 2013. Heading instability of moored ships in long waves. Transactions Society of Naval Architects and Marine Engineers. Alexandria, VA.

Zangeneh R, Thiagarajan KP, Urbina R, Zhigang T. 2014. Viscous damping effect on the heading stability of turret-moored ships. 33th International Conference, on ocean offshore and Arctic Engineering, Jun San Francisco, USA.

Zangeneh R, Thiagarajan KP, Urbania R, Tian Z, 2015," Effect of viscous damping on the heading stability of turret-moored tankers", J of Ships and Offshore Structures.

Zhang L, Lu H, Yang J, Tao P, Lonfei X. 2012. Low frequency horizontal motions of a moored FPSO in bi-directional swell and wind-sea offshore West Africa. J Ships and offshore structures.8: 425-440

SNAME
THE INTERNATIONAL COMMUNITY
FOR MARITIME AND OCEAN PROFESSIONALS

A New Light-Weight Semi-Submersible FPS for Offshore Marginal Oil and Gas Field Development

Mingfei Wang
STRUCTURAL ENGINNER, COTEC INC

Jim Wang
VP OF ENGINEERING, COTEC INC

Zhan Cheng
STRUCTURAL ENGINNER, COTEC INC

Yanbin Bai
NAVAL ARCHITECT, COTEC INC

Dongming Fu
STRUCTURAL ENGINNER, COTEC INC

Yong Luo
PRESIDENT, COTEC INC

Proceedings of the 22nd Offshore Symposium, February 2017, Houston, Texas
Texas Section of the Society of Naval Architects and Marine Engineers

ABSTRACT

The current low oil price environment demands innovative solutions in offshore oil & gas field development. This is especially true for deepwater marginal field development. A new concept light weight Semi-submersible floating production system (FPS) named as QX-Semi FPS has been developed for deepwater South China Sea. In this paper, the design philosophy and key features of this light weight Semi-submersible FPS are described and a FEED level design of the hull structure is presented. With an operating displacement less than 30,000MT, this multifunctional Semi FPS is designed for drilling/work-over, subsea tie-back, production, and accommodation in water depth ranging from 300m to 1000m. Compared with a conventional Semi FPS, the QX-Semi FPS is of excellent hydrodynamic performance, high variable deck load capacity, light deck structural weight and low fabrication cost due to its unique design features such as standard circular triple column configuration, webfoot pontoons, Y-shaped deck and efficient hull structures. Detailed naval and structural analyses were performed. The results show that the QX-Semi design meets all the requirements stipulated by international codes and standards. An economic analysis shows that the QX-Semi FPS can achieve more than 14% savings for an EPCI project compared with a conventional Semi FPS.

Keywords: Light weight; Semi-submersible FPS; Deepwater field; Low cost; Triple columns

INTRODUCTION

Semi-submersible FPSs (Floating Production System) have been commonly used worldwide for mid (300m) to ultra-deepwater (over 2500m) developments.Semi. Semi FPS is the second most deployed deepwater floating production floater worldwide and 47 are in operation till 2015. In general, a Semi-submersible FPS may be grouped into three categories based on its displacement, i.e. light weight (less than 30,000MT), medium weight (30,000~50,000MT) and heavy weight (greater than 50000MT). Light weight Semi-submersible FPS is mostly used for marginal oil & gas fields due to economic considerations. In particular, more than half of the existing 47 Semi-submersible FPS's are light weight Semi-submersible FPS units with displacements in the range of 15,000 to 30,000MT. However, most of the existing light weight Semi-submersible FPSs were converted from aged drilling units in the 1980s to 1990s. An example is the Nanhai Tiao Zhan Semi FPS (Figure 1) serving the Liuhua 11-1 field in the South China Sea, which was converted in 1995 from an aged 3rd generation Semi-submersible drilling unit built in the 1970s.

Figure 1 Nanhai Tiao Zhan Semi FPS

As the suitable aged Semi-submersible drilling units become less available nowadays, newly designed light weight Semi-submersible FPS's are expected to play a more significant role in developing marginal deepwater oil & gas fields. Based on the experience of the Liuhua 11-1 FPS, a newly designed low cost light weight Semi-submersible FPS can be an attractive solution for future deepwater oil and gas field development in the South China Sea. In 2014, COTEC was awarded a contract by China Nation Offshore Oil Corporation (CNOOC) to conduct a conceptual design of a low cost light-weight Semi-submersible FPS for production, development drilling/work-over and subsea tie-back in the South China Sea for mid-deepwater with ranging of water depth from 300 to 1000 meters. A novel concept Semi-submersible FPS, named as QX-Semi FPS, was developed for reducing cost and make deepwater oil/gas production profitable even in the current low oil price environment, especially for offshore marginal oil and gas field development. The new concept QX-Semi FPS has a Y-shaped deck, which is supported by three standard circular columns, with webfoot tanks between ring pontoons as shown in Figure 2.

This light weight production Semi platform is designed to provide drilling, production, accommodation, and conjunction with subsea wellheads for water depth of 300-1000m. This new Semi-submersible FPS has excellent hydrodynamic performance, efficient variable load capacity and significant reduction in cost. Typical main characteristics of the QX-Semi are listed in Table 1. For projects with specific payloads and water depth, the main dimensions can be adjusted and optimized accordingly.

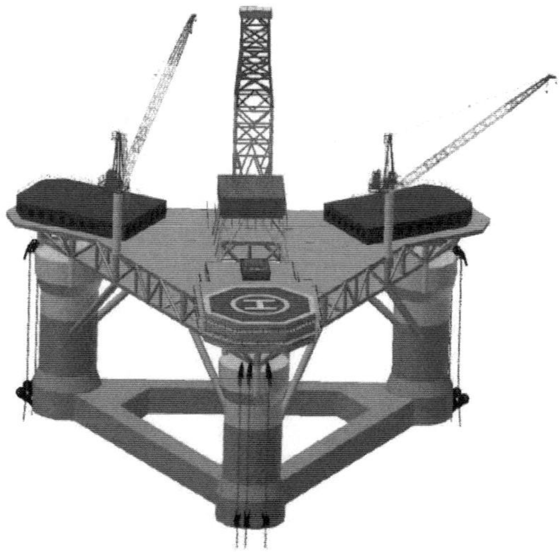

Figure 2 New Concept QX-Semi-submersible FPS

Table 1 Main Characteristics of QX-Semi FPS

Item	QX-Semi FPS
Max. displacement	30,000 MT
Max. topside weight	10,000 MT
Design draft	26 to 28 m
Mooring design water depth	300 m to 1000 m
Variable deck load	~ 5,000 MT
Deck area	> 6,000 m^2
Design life	25 years
Width over all	< 106 m
Length over all	< 110 m
Max. pontoon height	7.5 m
Min. pontoon width	6.0 m
Max. column distance (CTC)	100 m
Min. column distance (CTC)	80 m
Max. pitch (3 hrs 100 yrs. Typhoon)	10°
Max. heave motion (3 hrs 100 yrs. typhoon)	6.0 m

DESIGN PHILOSOPHY

The new concept QX-Semi FPS has an inherent cost advantage as it can be shown that a 3-column/3-leg Semi-submersible floater can be as much as 13.4% more weight efficient than a conventional 4-column Semi-submersible floater for the same displacement and draft. In addition, the QX-Semi FPS has the cost advantage of 9 mooring lines of 3x3 mooring pattern in comparison with 12 mooring lines for a conventional four columns Semi. Based on these specific advantages, the design features of the QX-Semi FPS are:

- Higher topside payloads – higher efficiency
- Less redundant structures – lower hull weight
- Improved hydrodynamic performance

- Lower overall cost

Hull configuration plays a critical role in the overall performance of a Semi FPS. The most obvious characteristic of the QX-Semi FPS is the 3-column/3-leg layout. The 3-column/3-leg layout is a popular design scheme for Jack-up platforms but it has not been used for Semi-submersible FPS platforms. There have been several 3-column/3-leg drilling Semi platforms in service for a long time, such as the Nobel Max Smith, Nobel Paul Romano and Nobel Amos Runner. On the other hand, there is no 3-column/3-leg Semi FPS either under construction or in service currently. Compared with a drilling Semi, a Semi FPS is required to be of better load-carrying capability, survival capability, and fatigue strength, especially in harsh marine environment.

The 3-column/3-leg layout is the best solution for cost control and reducing redundant structures. This layout has benefits of lower hull construction cost and mooring cost. However, the stability requirement for a 3-column/3-leg design is more challenging than that of a 4-column design. A feasible method to ensure sufficiency of stability is increasing the distance between the columns and optimizing the shape of the columns. The increased distance between the columns will affect the deck design in two ways: one is making the unsupported span of the deck greater which could cause increased deck structural weight, and another is making the deck area larger, which is beneficial for topside layout. The shape of the columns of the QX-Semi FPS is optimized by using donut ring-shaped ends at each column. The webfoot pontoon is an innovation to improve the hydrodynamic performance of the QX-Semi FPS and to reduce the unsupported span in box girders of pontoon.

Topside configuration plays another critical role in the overall design of the QX-Semi FPS. The function of the deck structure is to provide a working area for most of the functions of the platform and transfer the weight and loads of the deck to the supporting structures. For the QX-Semi FPS structure, the deck structure is also a part of platform in the global structural system connecting the top of each column of the hull. The deck area requirement may be considered a sizing factor for equipment layout. In the case of the QX-Semi FPS design, however, the large spacing of the columns should provide sufficient space for process and utility equipment mostly. In addition, a practical advantage is to extend the deck on the column outboard and place the production, utility and living quarters on top of different columns resulting in improved safety in the design as the living quarters is further away from the production area. The deck can be designed as an integrated piece structures or modularized structure. The benefit of integrated deck is that it can be installed at quayside using high capacity lifting cranes. Modular deck will be divided into a number of pieces and the size of each module can be decided accordingly to the functions and/or the availability of installation/lift equipment, especially when the size of deck is too huge to handle. Typical modular deck components can be module support frame, living quarter module, wellhead module, drilling module, production module, and utility module, etc. To answer the challenge of large span deck structure, an innovative Y-shaped truss deck layout, and column knee braces are used. The Y-shaped truss deck design makes the deck stronger for large spacing of columns and easier for the layout of the production equipment. The QX-Semi topside configuration is optimized based on in-depth analysis and evaluation of design efficiency, safety, fabrication, and integration method.

KEY FEATURES

The QX-Semi FPS is based on efficient structural design to offer a competitive FPS for deepwater marginal oil/gas field development. The general arrangement of QX-Semi FPS was shown in Figure 3 and Figure 4. The main characteristics of this new concept QX-Semi FPS are described below:

- Triple Columns / Larger Diameter Column Ends / Webfoot Pontoons

Three cylinder columns (Diameter 15m/18m) are uniformly distributed on the QX-Semi FPS. The distance between the column centers is 90m which provides enough restoring moment in terms of floating stability. By using a special design feature referred as "donut-ring", i.e. as if put a donut ring on each end of the column, which gives a larger diameter column end design as shown in Figure 5, the column water plane area can be optimized to improved hydrodynamic response for the FPS. Added-mass and damping will also be increased by using donut-ring design for the lower end. Another benefit from using donut-ring configuration is less wave run-up at the top. In addition to better hydrodynamic performance, the donut-ring design can provide increased flexibility in the QX-Semi FPS hull design with less coupled hull and topside design issues. In particular, the deck post spacing can be fixed once the column spacing and donut-ring diameter are fixed, while the column diameter can still be adjusted for changes in topside and riser payload in late stage of a project. This is a huge advantage over the conventional Semi FPS designs which often requires more coupled topside and hull design and less flexible for late design changes. The webfoot pontoon is another noticeable feature of the QX-Semi FPS design which can provide more buoyancy for pre-service floatation and make the structural connections between pontoon and column stronger. Further, it provides increased added mass and damping which gives superior hydrodynamic performance for the QX-Semi FPS. The hydrodynamic model which demonstrates the columns and webfoot configuration is shown in Figure 6.

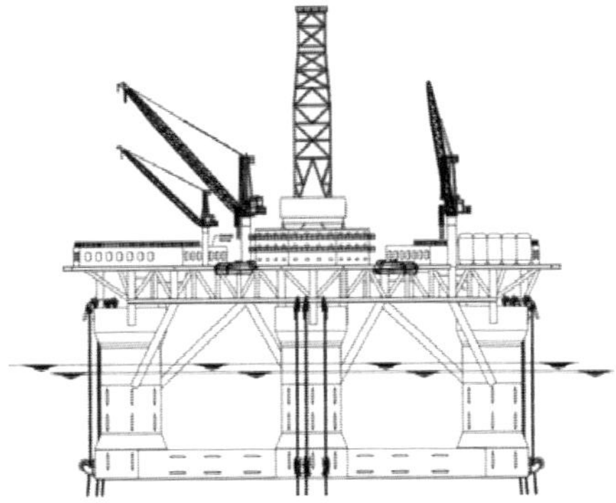

Figure 3 General Arrangement Elevation View1

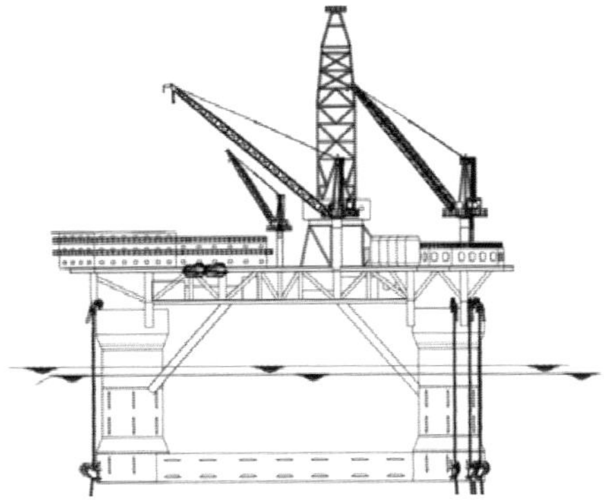

Figure 4 General Arrangement Elevation View2

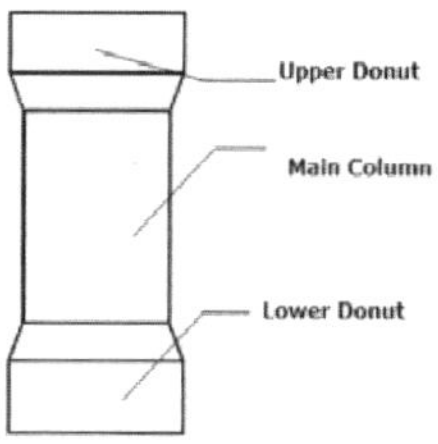

Figure 5 Donut-ring Column Profile

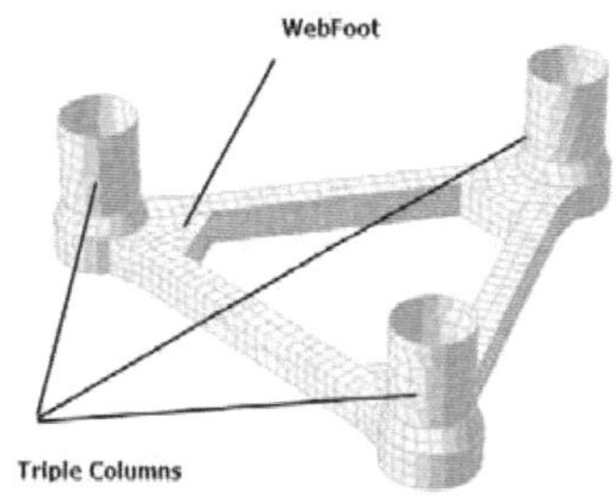

Figure 6 Hull Column Webfoot Model

- Y-Shaped Truss Deck

For heavy weight Semi FPS, modularized topside is preferred for flexible fabrication and installation, while integrated topside is more common for light weight Semi FPS. A truss type configuration in general provides a more efficient support for the topside loads. The truss type deck is most advantageous when used for 2-level topside configuration which makes a stronger space frame and carrying more loads with lighter weight.

Figure 7 Y-shaped Truss Deck

Triangular deck area is not commonly used for layout of the deck equipment. It requires smart thinking for deck equipment layout. As a result, a Y-shaped truss deck as shown in Figure 7 is developed. The knee bracing configuration for the QX-Semi topside structure is also an important feature since it reduces the span of the truss girder. The weight of the braces itself is relatively small compared to other components of the structure, and the knee braces provide effective reduction of the deck span and reduce the bending moment acting on the deck girder. It is designed for easier arrangement of drilling equipment, power plant, and processing equipment. The shape of the deck has a larger and regular (square) area on the top of the column which brings in a strong support for the topside facility arrangement. The truss type deck and center bracing of this configuration are also the optimized parameters for the QX-Semi FPS structure. While the big span of deck is a challenge for the 3-column/3-leg QX-Semi FPS, the Y-shaped truss frame with knee brace provides smart engineering solutions for this challenge.

- Buckling Control Scantling Design

The scantling design of hull structure is an important aspect of the QX-Semi FPS design, as it has direct impact on the efficiency of the weight of the entire structure. An effective way to reduce hull construction cost is to reduce number of the stiffeners for hull columns. The QX-Semi FPS uses circular columns which have the advantage of using buckling control scantling design [2]. This gives a closer-ring-girder-spacing and wider-stiffener-spacing framing pattern, which is in contrast to the conventional stiffened flat panel design for the column outer shell framing design. For the same reason, ring girder spacing is much less than flat panel design on access shaft shell and no stiffener is arranged on it. By the comparison with traditional bending control scantling design, we found the weight of the structure is almost the same as traditional design, but the length of shape steel (girders and stiffeners) can be reduced significantly [3]. This is a great benefit for reducing the cost of welding.

- Stability Analysis

To ensure the stability of the 3-column/3-leg QX-Semi FPS, extensive stability analyses were performed. Any configuration must pass strict stability checks for all loading conditions enforced by governing regulations and codes and even meet more requirements requested by the owner. Both pre-service and in-place conditions are considered in QX-Semi FPS stability design. For each loading condition, intact and damaged stability are checked respectively. Taking the benefits from the design of large column spacing and donut-ring column, the stability analysis results show the QX-Semi FPS design can meet the requirements of ABS and DNV rules. An example of stability curve is shown in Figure 8.

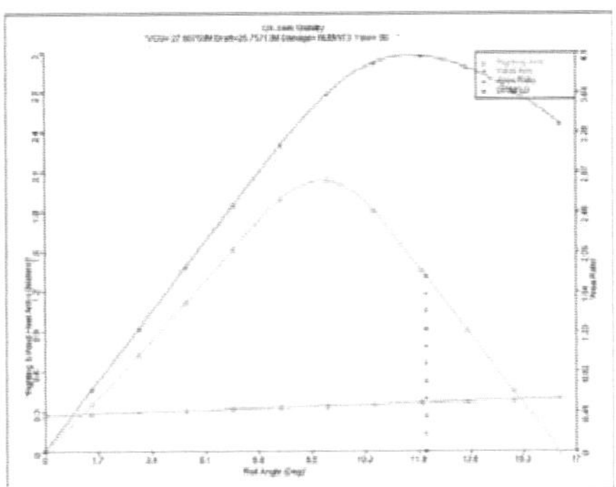

Figure 8 Stability Curve of triple column FPS

- General Arrangement

Topside arrangement includes processing facility, drilling equipment, associated drilling components, living quarters, rig, and processing liquid tanks etc. Preliminary topside arrangement is performed to assure deck space is sufficient for the QX-Semi FPS and the configuration design satisfies the design specifications. An example of topside arrangement on the main deck and the production deck is shown in Figure 9 and Figure 10, respectively. A key feature of the QX-Semi FPS topside arrangement design is that all heavy weight equipment is allocated above the columns and supported by posts. On the center of drilling deck, there are only truss structures and requisite equipment for drilling operation.

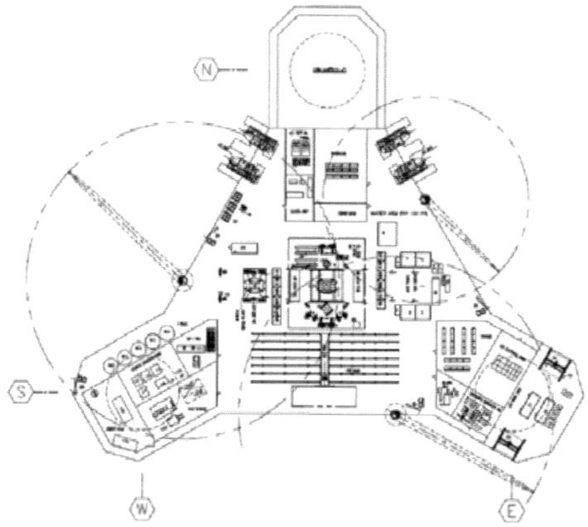

Figure 9 Main Deck Arrangement Plan View

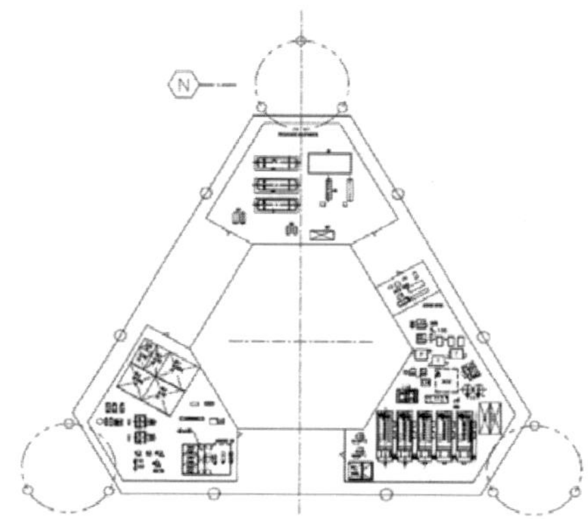

Figure 10 Production Deck Arrangement Plan View

FEED DESIGN

The QX-Semi FPS was designed complying with international industry standard. Front End Engineering Design (FEED) has been performed for the future development and detailed cost economic analysis. The FEED work is summarized as following:

- Topside Design (Structure, Process, Outfitting etc.)
- Hull Design (Structure, Marine, Outfitting etc.)
- Mooring Design
- Stability Analysis
- Hydrodynamic and Global Performance Analysis

- Mooring Analysis
- Wave Basin Model Test
- Hull Scantling Design
- Design Loads (Design Wave) Analysis
- Hull Global Strength Analysis
- Hull Global Fatigue Analysis
- Deck in-place Strength analysis
- Transportation Analysis

The goal of the FEED is to verify the feasibility of the QX-Semi FPS concept by going through every aspect of the design. Based on the FEED work, the whole concept is not only can be designed, but also can be the project executed and the platform delivered.

Global structural analysis was performed by design wave method [1]. Two hull global design load cross sections were set up as shown in Figure 10. A typical design wave curve, which plotted design wave amplitudes verse periods, is shown in Figure 11.

The global structural model including hull and topside were created by SESAM/GeniE software. The wave loads and inertial loads were provided by SESAM/HydroD software. The inundation loads generated by the wave pressure above the still waterline were calculated and combined to wave loads by in house software. The hydrostatic and hydrodynamic wave pressure on the hull structural model are shown in Figure 12 and Figure 13.

The stress contour of the hull structure from one of the critical load cases is shown in Figure 14. The hull structure has sufficient strength capacity to withstand the environmental and functional loads and can meet requirements of international codes and standards [4].

The mooring system is designed to have a 3 x 3 pattern as shown in Figure 15. Time domain mooring analysis and wave basin model test were performed (Figure 16). The model test and numerical simulation results show that the new concept QX-Semi design is of excellent hydrodynamic performance and the mooring system is strong enough for the design criteria.

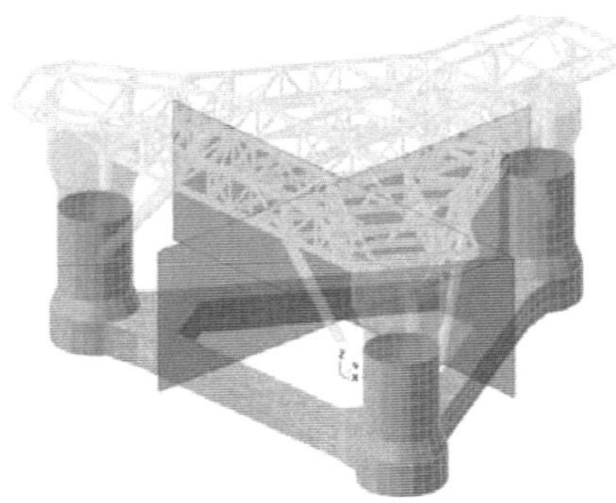

Figure 11 Design Loads Sections Definition

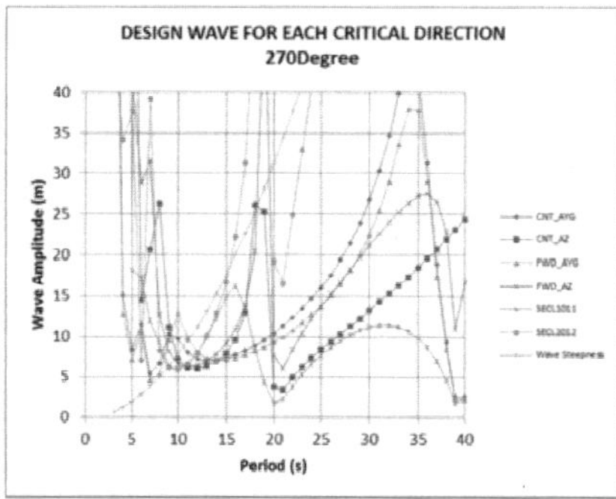

Figure 12 Design Waves Selection

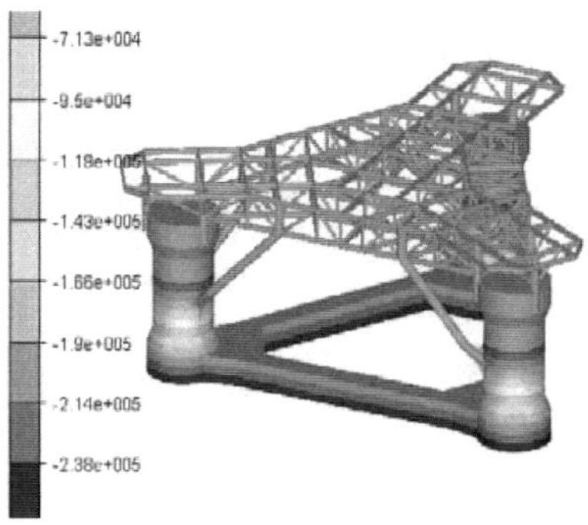

Figure 13 Hydrostatic Pressure on the Hull

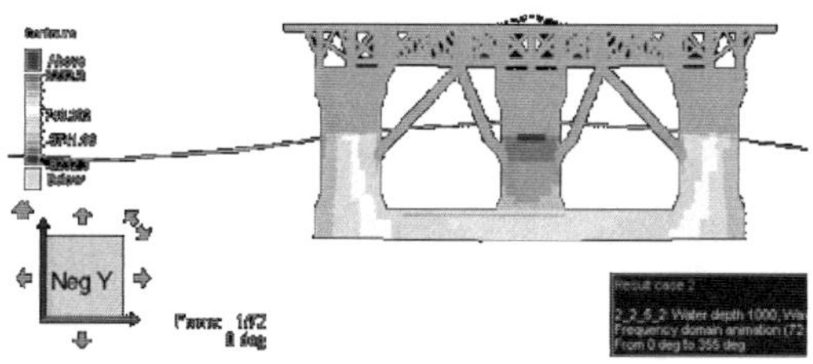

Figure 14 Wave Pressure on the Hull

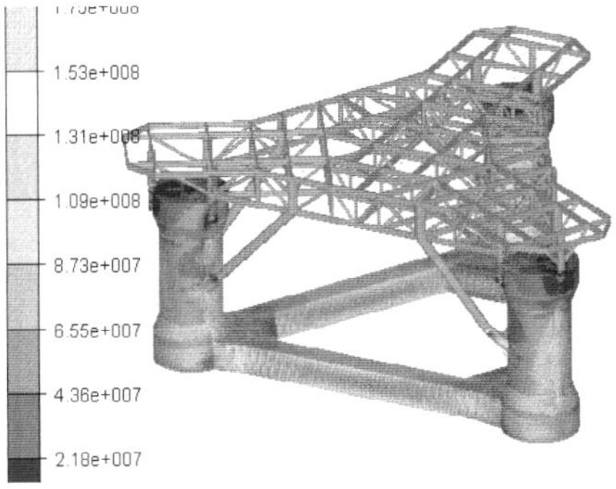

Figure 15 Stress Contour of the Hull Structure

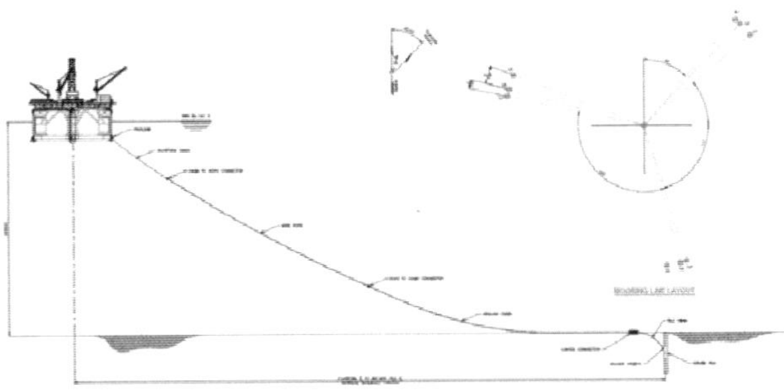

Figure 16 Mooring System Layout

Figure 17 Wave Basin Model Test

ECONOMIC ANALYSIS

There are various factors affecting the cost of a Semi FPS. The cost of facilities, such as oil production equipment, power generating plants etc., is determined by the FPS functions and market prices. For this new concept Semi FPS development, reducing the cost of the hull structure is the primary goal. The cost of the hull structure consists of two parts: material charges (mainly steel) and manufacturing charges (mainly labor cost and processing cost of cutting steel and welding). Due to China's large steel production quantity and low steel price, the material cost itself is generally less than 25% of total cost of the hull construction. Reducing fabrication cost through smart design can clearly lower the construction cost.

Compared with a conventional 4-column Semi FPS, the 3-column/3-leg QX-Semi FPS only has three columns and three node connections (a node connection is the intersection of a column and two pontoons). The node connection is often subjected to high stress concentrations that can be a governing factor for the hull structural strength, so that it is usually of very complexity in design. Fabrication of the node connection is a critical factor taken into consideration in the fabrication of a Semi FPS. Obviously, the 3-column/3-leg QX-Semi FPS has many advantages in comparison with a conventional 4-column Semi FPS. For example it has three columns which require less work for block construction and block installation; another is that it has three node connections which mean less workloads and complexity for the fabrication of the node connections. Although the displacement of the 3-column/3-leg QX-Semi FPS is almost the same as a 4-column conventional Semi FPS, the QX-Semi structure is simpler and easier to fabricate. Especially the construction workloads for building blocks and node connections, which play a great role in cost, can be distinctly economized. These are the reasons the QX-Semi FPS can give significant cost savings for hull construction.

For scantling design of the QX-Semi hull structures, the DNV buckling rules are used for designing the detailed structures, such as thickness of the plates, the sizes of girders and stiffener. In particular, this design method can take full advantage of hoop capacity of cylinder structures to reduce the sizes of ring girders and numbers of stiffeners. Compared with conventional method based on flat panel design of the hull structures, this resulted in as much as a 20% reduction in the length of profiles, and subsequently the fabrication processing cost for welding and cutting can also be significantly reduced.

In summary, the reduced cost of the 3-column QX-Semi FPS is mostly due to the following aspects:

- Only three columns which can reduce workloads during block construction and block installation;
- Only three node connections which can save complicated node fabrication cost;
- Scantling design based on buckling control method which makes the length of profiles less than conventional scantling design and cuts down welding workloads by 20%.

A preliminary economic analysis based on the hull material takeoff has been conducted. According to cost analysis results, the QX-Semi FPS design achieved about 12% CAPEX reduction for the hull construction. In addition, the 9 mooring configuration represents a nearly 25% cost savings for the mooring system. Overall, it is estimated the QX-Semi FPS can achieve a cost saving of more than 15% for a hull and mooring EPCI project in comparison with a conventional Semi FPS.

CLOSING REMARKS

The low oil price environment demands innovative solutions for the offshore industry, particularly for deepwater oil and gas development. A new conceptual low cost light-weight Semi-submersible FPS named QX-Semi FPS has been developed for deepwater South China Sea which is supported by CNOOC. In this paper, the design philosophy and key features of the QX-Semi FPS are described and a FEED level design of the hull and deck structure is presented.

With an operating displacement less than 30,000MT, this multi-functional FPS is designed for drilling/work-over, subsea tie-back, production and accommodation in water depth ranging from 300m to 1000m for the South China Sea environment. It can also be used in other deepwater regions with similar or even rougher metocean conditions, such as the Gulf of Mexico. Compared with a conventional Semi FPS, the QX-Semi FPS is of excellent hydrodynamic performance, high variable deck load capacity, light deck structure weight and low fabrication cost due to its unique design features such as standard circular triple columns configuration with variable donut ring ends, webfoot pontoons, Y-shaped deck layout and efficient hull structures. The results show that the QX-Semi FPS design meets all the requirements stipulated by international codes and standards. An economic analysis is performed which shows that the QX-Semi design can achieve more than 14% CAPEX savings for an EPCI project compared with a conventional Semi FPS.

ACKNOWLEDGEMENT

The development of the QX-Semi concept was supported by China Nation Offshore Oil Corporation (CNOOC). The authors are grateful to CNOOC's support and to many CNOOC and COTEC colleagues who participated in the project and contributed expertise that greatly assisted the QX-Semi development work.

REFERENCES

DET NORSKE VERITAS, "Recommended Practice, Column-Stabilized Units", DNV-RP-C103, 2012.

DET NORSKE VERITAS, "Buckling Strength of Plated Structures", DNV-RP-C201, 2010.

DET NORSKE VERITAS, "Buckling Strength of Shells", DNV-RP-C202, 2010.

DET NORSKE VERITAS, "Structural Design of Offshore Units (WSD Method", DNV-OS-C201, 2011.

IMPROVEMENTS ON THE CFD TOOLS FOR GREEN WATER ANALYSIS OF OFFSHORE STRUCTURES

YOANN JUS, DANIEL BARCAROLO, NICOLAS COUTY
HYDROCEAN

LAURENT CHIRON
NEXTFLOW SOFTWARE & ECOLE CENTRALE DE NANTES

PIERRE-MICHEL GUILCHER, JULIEN CANDELIER, PIERRE BIGAY
NEXTFLOW SOFTWARE

FLAVIA REZENDE
BUREAU VERITAS

Proceedings of the 22nd Offshore Symposium, February 2017, Houston, Texas
Texas Section of the Society of Naval Architects and Marine Engineers

ABSTRACT

Green Water and Wave impacts are amongst the most severe and dangerous loads that effect offshore structures. Even if many design procedures are well established, they can over or underestimate the structural loading in complex conditions as they are outside of their application scope. This therefore leads to over designed and expensive structures, or to under designed structures leading to dangerous situations. The calculation of complex structure loadings is therefore a key issue for engineering companies.

HydrOcean in the previous SNAME symposium presented a methodology to evaluate green water phenomena developed in cooperation with the French offshore industry, namely Total S.A. and Technip. This methodology is based on the use of complementary tools from potential linear flow solvers in order to perform the screening up to the very precise CFD simulations to simulate the wave impact over the platform topside structure.

More precisely on this latter aspect, HydrOcean in cooperation with NextFlow software is continuously improving its CFD numerical tools. Recently, in the SPH-flow software the adaptive particle refinement technique has been implemented which allows to improve substantially the quality of the results at the same time reducing the associated CPU costs.

In the previous article it was shown that, it was necessary to increase the number of particles to 24 million in order to start detecting with a reasonable precision the impact pressure (in comparison to experiments). Here, it will be shown that this number of particle can be decreased and the quality of the pressure signal obtained numerically is increased.

This technique is then validated and put to proof by performing the same test case as in the previous article. Then, an industrial case of a FLNG in a harsh irregular sea-state is shown. The images below illustrate the simulations and comparisons to experiments with this new technique.

INTRODUCTION

During storms and severe sea states, damages suffered by ships or offshore structures can be particularly important, possibly putting human lifes in danger, destroying equipment on board, or leading to environmental pollution. Examples of such events regularly occur and the offshore industry, as well as engineering teams and classification societies are fully aware of the need in improving the evaluation of extreme loads. These extreme loads cover a wide range of events, like the violent impact of a wave onto the ship hull –slamming impact-, the water flooding of ship decks, blowing away all the equipment –greenwater phenomenon-, or the sloshing inside the oil tanks.

Such events on ships result from an interaction between a wave train and a structure. From a numerical point of view, one has to solve two different kinds of problem with very different typical spatial lengths and time durations. It involves wave generation and propagation onto the structure, with highly nonlinear waves, irregular sea states, whose typical length and duration are respectively 300m and 10s. Eventually, the interaction between the liquid and the structure leads to impacts, with typical length and duration 5 orders smaller (1mm and 1ms). Moreover, some compressibility effects could occur in the gas, and even in the liquid during impact. Hydroelasticity (deformation of the structure under the fluid loads) should also be addressed. The full problem involves a lot of physics, and the present paper is a first step towards the simulation of all phenomena occurring during impact.

Regarding the impact simulation, special attention has been paid to the Smoothed Particle Hydrodynamics (SPH) method. Indeed, the SPH method offers advantages over classical numerical methods when simulating impact type problems. No connectivity is required for the free surface, enabling the simulation of violent flows with possible fragmentation and interface reconnection. The Lagrangian formulation cancels the interface diffusion, resulting in a sharp definition of interfaces between gas, liquid and structure. Moreover, SPH method can be applied to any continuum description, resulting in an ability to easily approach multi-physics. Therefore, SPH method can theoretically solve in a fully coupled way, the compressible multi-phase structural interaction problem occurring during impact phenomenon.

As the SPH method is best suited for fast dynamics, a careful attention need to be paid for slow dynamics like wave propagation. It would be completely inadequate to use the SPH method or any CFD software in order to simulate the propagation of a wave generated in a wave flume and then the impact phenomenon (complete problem). The propagation phase can be solved accurately by software based on potential theory (BEM for example), much faster than by CFD. However, such potential-theory-based software can obviously not simulate the local interactions between liquid, gas and structure that happen during impacts. Therefore, the strategy developed is based on a coupling between two codes sharing the different tasks. A potential-theory based code is in charge of the incident flow (non-perturbed wave train) and the SPH method is in charge of the local interactions simulations during the impact in the vicinity of the impacted structure. A specific forcing algorithm between potential solvers and SPH-flow was developed in order to reduce as much as possible the numerical diffusion of the wave trains between the wave generator and the structure. As a consequence, this method allows to simulate open sea problems, providing the incident wave field from a potential solution, and providing the fact that the diffracted wave field from the structure does not reflect on the inlet/outlet wave boundaries.

In the previous SNAME Symposium, HydrOcean presented an article with the methodology developed in cooperation with the French oil & gas industry to analyze such extreme events with CFD. This methodology has been recently improved by including the particle refinement principle. This article presents hence in a great part the methodology already presented previously but also the new improvement.

I. SOLVERS

A. *SPH-flow*

SPH-Flow is a multi-purpose, multi-physics CFD Software. It has been developed by ECN and HydrOcean through several PhD works (see [1], [2], [4], [5], [6], [8]). The solver was first developed for fluid flow simulations dedicated to complex non-linear free surface problems. SPH-Flow has now been extended to multi-fluid, multi-structure, fluid/structure interaction and viscous flows. It relies on state-of-art meshless algorithms, such as Riemann solver upwinding, MUSCL reconstruction and gradient renormalization for increasing the order of the formulation.

SPH-Flow solver is parallelized to reduce the elapsed time of a computation by use of cpu resources available on clusters. This distributed memory parallelization relies on a domain decomposition: the computational domain is divided in as many sub-domains as allocated processes. The communications between processes are achieved through the MPI library.

Through this paper, only the liquid and rigid structures are considered. The gas phase is not taken into account, and the liquid phase is considered as incompressible. Thus no compressibility effects (gas or liquid) can be addressed with the current numerical modeling.

The system of conservation laws corresponds to the isentropic Euler equations and is written in conservative form:

$$L_{v_0}(\phi) + div\left(F_E(\phi) - v_0\phi\right) = S \qquad (1)$$

where Φ is the vector of conservative variables, FE the Eulerian flux matrix and S the source term per unit of volume. The system of equations is written in an arbitrary moving reference frame (Arbitrary-Lagrange-Euler (ALE) description) where denotes the transport velocity field.

To close the system, a Tait Equation of State relating pressure to density is used:

$$p = \frac{\rho_0 a_0^2}{\gamma}\left[\left(\frac{\rho}{\rho_0}\right)^{\gamma} - 1\right] \qquad (2)$$

where ρ0, a0 and γ are respectively the nominal density, the nominal speed of sound and the isentropic constant for the considered fluid.

Fluid domain is discretized by a set P of particles. Each particle i has a location xi, a volume of control wi and carries its own properties (velocity, pressure, etc.). A weak formulation associated to (1) by use of SPH spatial derivative operators leads to the following scheme for Euler equations:

$$\frac{d}{dt}(w_i\phi_i) + w_i\sum_{j\in P} w_j 2G\left(\phi_i, \phi_j, n_{ij}\right) B_{ij}.\nabla W_{ij} = w_i S_i \qquad (3)$$

Upwind numerical flux G is given either by exact or approximate Riemann solvers in the moving framework. Extension to second order with MUSCL algorithm [10] is performed with linear reconstruction on pressure and velocity. The term Bij stands for the symmetrised renormalization matrix [11], [12].

Ordinary Differential Equations are marched in time with classical 4th order Runge-Kutta scheme, or 3rd order Strong Stability Preserving Runge-Kutta scheme. Time step is restricted by a CFL-like condition on acoustic waves.

The complex shapes of ships are solved with a general accurate boundary formulation [7].

B. Particle Refinement

The dynamic refinement algorithm used in this article is based on the article by Barcarolo et al In this algorithm, when a particle needs to be refined (called mother particle), it is divided into a finite number of smaller particles (called daughter particles) following a predefined pattern. This refinement is achieved in such a way that local density errors are minimized. The image bellow illustrates the refinement process, in which as seen once particles are created the mother particles are kept in the flow:

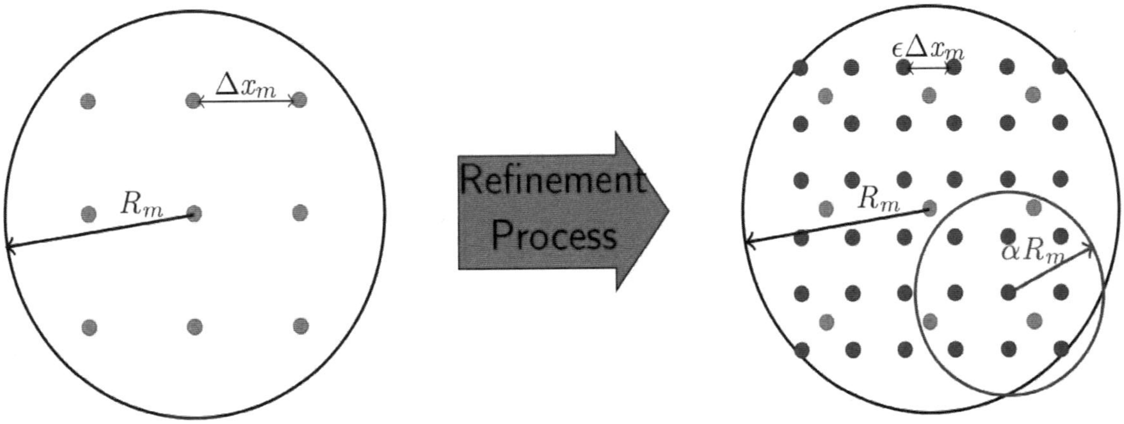

Figure 1. Refinement process

Once particles are created they evolve in time and space. The mother particles are kept in the simulations and are simply transported by convection based on the solution computed at the daughter particles. However, once particles are not needed anymore, one should be able to erase them. The methodology used here, which is based on spatial refinement criteria, consists into erasing the daughter particles and switch the flow solution to the mother particles.

C. Potential solutions

Several potential solutions can be used, depending on the complexity of the generated wave train. Two possibilities are mainly used for the coupling algorithm with SPH-flow, depending if the wave train is regular or irregular, and 2D or fully 3D. The two solutions rely on a fully nonlinear theory.

1) Fully nonlinear regular wave train solution (Rienecker and Fenton)

Rienecker and Fenton derived in 1981 a numerical solution to the problem of fully nonlinear wave trains, restrained to the regular case [9]. This solution relies on the stream function and a Fourier series decomposition. A simple solver based on this formulation was developed at Ecole Centrale Nantes, and provides a fast solution for regular waves, up to the wave breaking limit. As an output, this solver provides velocity and pressure on any point of the fluid domain.

2) Fully nonlinear irregular wave train solution (HOS solver)

The HOS solver developed by the LHEEA Labs of Ecole Centrale Nantes is a spectral potential method. HOS stands for Higher Order Spectral. It relies on a decomposition of the solution onto a set of basis functions. The spectral formulation provides quasi analytic solutions for the wave propagation problem with high accuracy and small cpu times. No simplification is involved in the system of equations, allowing solving highly nonlinear transient wave trains [3].

Typical applications are the generation and propagation of regular waves, irregular waves, 2D and 3D waves, the numerical reproduction of experimental waves from wave tank, the simulation of oceanic waves based on statistical approaches, and the propagation of focused and freak waves.

The main limitation of the HOS solver relates to the introduction of a body into the fluid domain. Thus, the solver is restricted to the problem of generation and propagation, without any possibility to compute directly the interaction with a body. A second limitation relates to the univocal definition of the free surface, avoiding the simulation of wave trains beyond wave breaking (neither fragmentation nor reconnection of the free surface).

As an output, this solver provides also velocity and pressure maps on a regular grid, on the whole fluid domain.

II. SPH-potential solver weak coupling

Once the HOS or Rienecker&Fenton solution is known for the wave propagation, this solution is used as an input for the SPH-flow computation.

A view of the coupling principle is presented in Figure 1. The wave solution is solved with the potential solver for the complete physical domain, without body. This solution corresponds to the unperturbed wave train, and is denoted as the incident wave train. The physical domain can be either a 2D/3D numerical wave tank, or an open sea domain. The wave profile can be generated through a wavemaker motion history to reproduce an experiment, or through an ocean wave spectrum with periodic conditions to reproduce an irregular open sea state.

The local interactions between the wave train and the structure are solved with SPH-flow, only for the truncated domain. The coupling between the two solvers is set up by imposing the potential incident wave solution onto the fluid domain boundaries of the SPH calculation (dotted lines). Then the waves propagate numerically from the outward boundaries to the impact area.

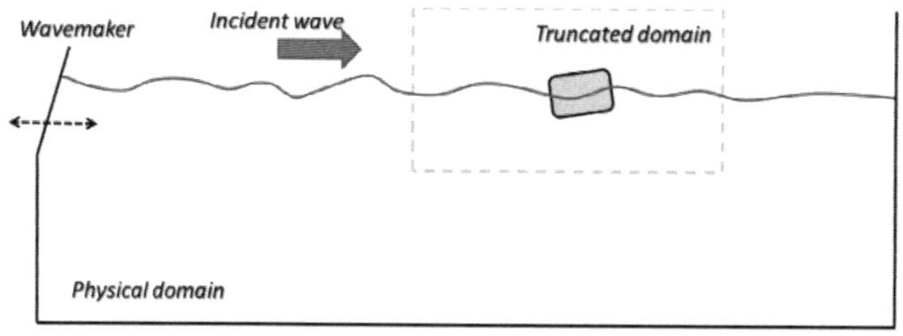

Figure 1. Coupling between potential flow solvers and SPH-flow.

These boundaries must be far enough of the area of interest, or the temporal window must be selected to ensure that no diffracted field will perturb the upstream and downstream boundaries. Despite these constraints, it is possible to simulate only a small part of the initial wave tank and reduce the overall cpu time. During the SPH computation, the inlet and outlet boundaries are continuously updated from the potential solution.

A. Inlet/Outlet boundaries for the SPH computation

As mentioned, the SPH fluid domain can be a smaller domain around the structure of interest (ship, oil rig, etc), in order to reduce the total number of particles, and then the cpu time. The main problematic is the treatment of these boundaries, which are no more wall boundaries, but inlet or outlet boundaries. Actually, in a Lagrangian framework, particles near the inlet/outlet boundaries will move with the wave velocity field, continuously going inside and outside following an elliptic trajectory (see Figure 2). The Lagrangian framework makes the treatment of particles more complex, because the position x_b of the inlet/outlet boundary is kept fixed during the whole computation (Eulerian framework). At any time, a particle can be considered as "free", in the computational domain, or "forced", in the inlet/outlet area. The (green) forced particles are updated at each time step with the incident field, and advected in time with this velocity field. The (yellow) particles inside the computational domain are considered as "free" and evolve according the classical SPH scheme, interacting with free and forced particles. As the velocity and pressure fields are fully known over the inlet/outlet area, it is possible to build MUSCL gradients and renormalization matrices also on the forced particles, to keep the same numerical scheme as in the interior domain. Particles move following the velocity field, and can be alternatively marked as free or forced, even inside a time step. The inlet/outlet area is meshed once at the beginning of the computation: the width of the inlet/outlet area must be sufficient to ensure at least a 2-Radius completion and to avoid the emptying of the area because of the Stokes drift.

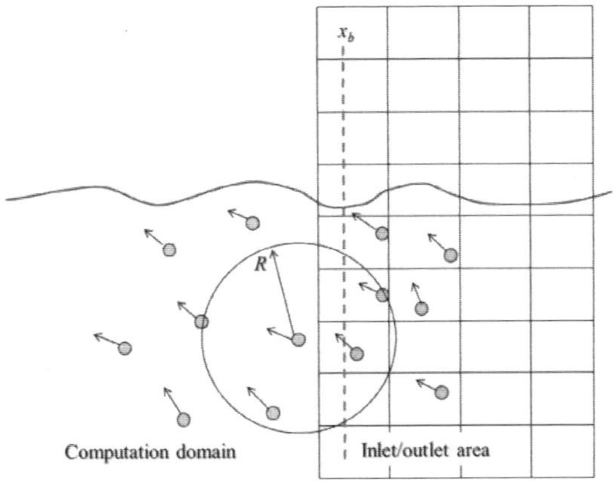

Figure 2. Interpolation of potential flow solution on a regular grid. Free particles in the computational domain are solved with classical SPH scheme. Particles in Inlet/Outlet area follow prescribed incident wave field.

A consistent formulation for the treatment of compressible inlet/outlet boundaries should rely on subsonic inlet/outlet conditions. In general, a constant velocity is imposed at inlet, and a constant pressure at outlet. Other properties of the fluid at boundary are determined with compatibility relations between the interior solution and the boundary state. The simplest solution adopted here is to impose the potential flow solution in terms of velocity and pressure. The underlying hypothesis is the fact that the perturbed area located around the impact location, does not have the time to diffract and propagate on the inlet/outlet boundaries. One must keep in mind that this solution is not fully consistent, with too much variables imposed at boundary, but is sufficient from an industrial point of view.

In practical terms, the incident fields are computed in two different ways. For the Rienecker&Fenton solution, the coefficients of the Fourier decomposition are stored then used by SPH-flow so as to reconstruct the velocity and pressure on each particle. In the case of the HOS solution, the reconstruction step is cpu time consuming, and a faster solution is adopted. The velocity and pressure fields are computed on a Cartesian grid during the potential calculation. This grid covers the area of the inlet/outlet boundary (Figure 2). This grid is saved on a file, for several time steps.

During the SPH calculation, the computation of variables on the particles is performed in two steps. A first time linear interpolation gives the incident grid at the current time. Then, a space bilinear interpolation gives the velocity and pressure at any required position. The horizontal space step can be different from the vertical one, allowing mesh refinement for the free surface area.

III. Simulation of extreme waves impacts on a FLNG

The objective of this part is to demonstrate that the SPH method coupled with HOS forcing is an efficient method for simulation of structures under extreme sea and also to demonstrate the use of the particle refinement procedure. The simulation presented hereafter is compared to the experiment conducted at Ecole Centrale Nantes.

A. Experimental set-up

Experiments were performed on a stiffly fixed FLNG model (see Figure 3) in the oceanic wave tank of Ecole Centrale Nantes, which is 50 meters long, 30 meters wide, and equipped with a wavemaker able to generate multidirectional waves. Green water effect was studied under a diversified set of regular front waves. Regular wave profiles were chosen to reduce numerical and experimental uncertainties, in order to validate the numerical tool.

The model is instrumented using a series of wave probes located in the fore deck in order to measure elevation of shipping water, as well as a set of pressure transducers, installed through the wavebreaker to observe slamming events (see Figure 4). The model deck is 1.1m wide.

Figure 3. Set-up in the Ecole Centrale Nantes oceanic wave tank.

Incident water wave is chosen in order to be numerically reproducible and to not surge, its wave length of about 7.3m and amplitude 0.44m at model scale.

Figure 4. a) Position of pressure probes on the breakwater b) Position of wave probes on the fore deck.

B. Experimental behavior of greenwater event

Those environmental conditions lead to greenwater events (see Figure 5). Time separating two frames is T/8 where T is the period of incoming waves (T = 2.2s model scale).

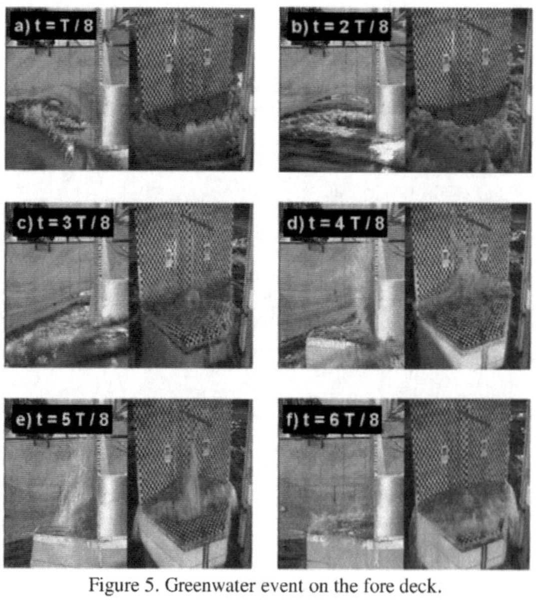

Figure 5. Greenwater event on the fore deck.

The greenwater can be described into 6 steps:

(a) Wave elevation at the bow exceeds effective freeboard.

(b) A large amount of water is plunging onto the fore-part of the deck and some water invades the deck sideways.

(c) A large amount of water at the centre of the deck results from convergence of flows from the sides and from the bow. Over 20cm (model scale) of shipped water is measured.

(d) Flows from the sides are meeting at the centre of the wall, creating a vertical water-jet. Just after, water from the front impacts the wall. Some water escapes from the deck at the edge of the wall.

(e) The centred water column on the wall starts to collapse, the shipped water is mostly close to the wall.

(f) Water first escapes through deck's sides, before the water evacuates the deck until next incident wave. Two maxima of water height are thus measured on the deck.

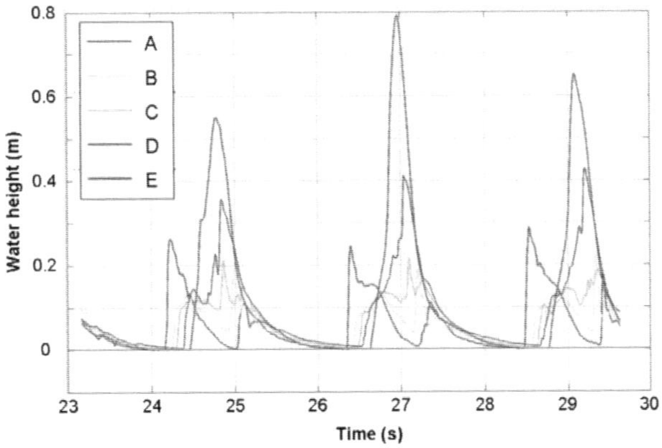

Figure 6. Time history of water height on deck (model scale).

C. Experimental measures

1) Water on deck elevation

Measures of water elevation on deck (see Figure 6) show very good repeatability at the frequency of incident wave. It gives also valuable qualitative and quantitative information about incident and returning flows (two maximum are observed for the probes located near the bow). Time of presence and spatial distribution of the amount of greenwater are also measured.

2) Pressure impacts on the wavebreaker

Pressure measures are presented in Figure 7. These results stress out the repeatable qualitative behaviour of pressure pattern during an impact. The pressure experiences a rapid (in the range of a few milliseconds) rise towards a maximum value, followed by a slower drop, related to dynamic pressure caused by the running-up of water along the wall. Finally, dynamic pressure due to the collapse of water along the wall is recorded.

The symmetric response of the sensors located at port (B1, B2, B3) and those located at starboard (T1, T2, T3) argues in favour of a precise experimental set-up. As observed, values of pressure peaks are not fully repeatable, due to incident conditions, but timescales as well as local maximum of dynamic pressure following the impact are reproducibly determined. These abilities ensure experimental results to be leveraged for comparisons with SPH. In this respect, average pressure peaks are computed (see Fig. 9), corresponding to the first incident waves (before any reflexion from the basin).

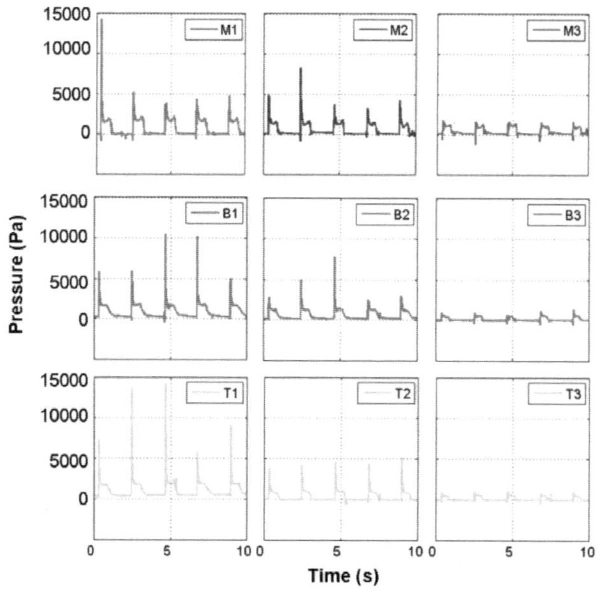

Figure 7. Time history of pressure impacts on the wavebreaker (model scale).

These experimental measures confirm that altitude from the corner formed by the wavebreaker and the deck is a key factor for pressure impact assessment in greenwater events. The pressure is maximum at bottom of the wavebreaker, where the incoming deck wave hits the wall at first. This pressure peak is related to the rapid change in momentum, from horizontal to vertical direction, and is comparable to the incompressible pressure peak occurring during a Wagner-type impact (wedge drop test).

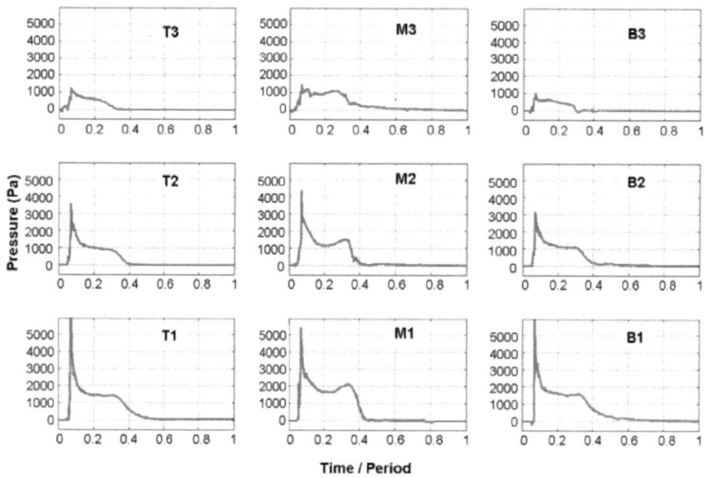

Figure 8. Average pressure peaks for pressure sensors during a period (model scale).

Table 1 sums up quantitative pressure statistics on impacts that occurred during the experiment. <Ppeak> stands for the average value of pressure peaks, σpeak its standard deviation, Ppeak+ and Ppeak- respectively the maximum and minimum measured value of pressure peaks. The maximum of pressure, at probe M1, is fairly good measured, with a ratio σpeak/<Ppeak> equals to 15%.

Table 1. Peak pressure statistics at model scale

	$\langle P_{peak} \rangle$ [Pa]	$\sigma_{P_{peak}}$ [Pa]	P_{peak}^{+} [Pa]	P_{peak}^{-} [Pa]
M1	5 440	810	6 170	4 630
M2	4 380	1 830	6 100	2 440
M3	1 510	330	1 920	1 120
B1	6 630	2 280	9 510	4 150
T1	9 780	3 080	12 230	5 350
B2	3 140	650	3 940	2 460
T2	3 590	1 540	5 490	2 240
B3	990	270	1 340	680
T3	1 250	250	1 470	900

D. Simulation

1) Set-up

SPH calculation is performed at model size in order to compare the results obtained with the experiments, and avoid any possible scaling effect. The configuration of the simulation domain is sketched in Figure 9. As the problem is symmetric, only the port side of the simplified FLNG is simulated. The particles discretization is space variable with a typical interparticle size ranking from 2mm to 5cm with 3% of growth rate, leading to approximately 2 million particles with the particle refinement solution, which with a classical approach would result into 25 million particles. The fine area is located just in front of the fore deck, in order to capture adequately the greenwater event. The adopted discretization ensures a resolution of 150 points per wave length and 100 points within the deck length.

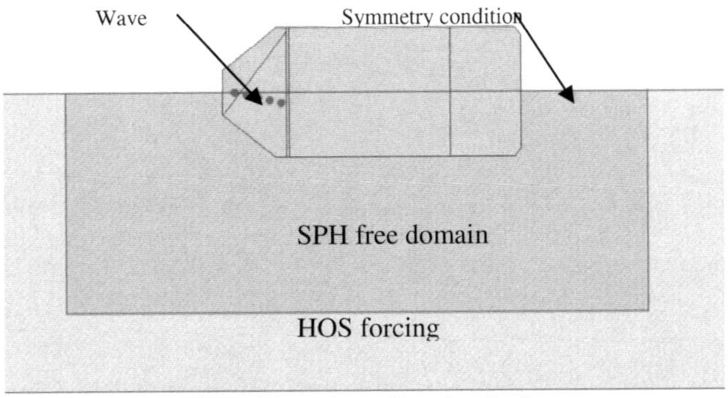

Figure 9. Domain configuration sketch.

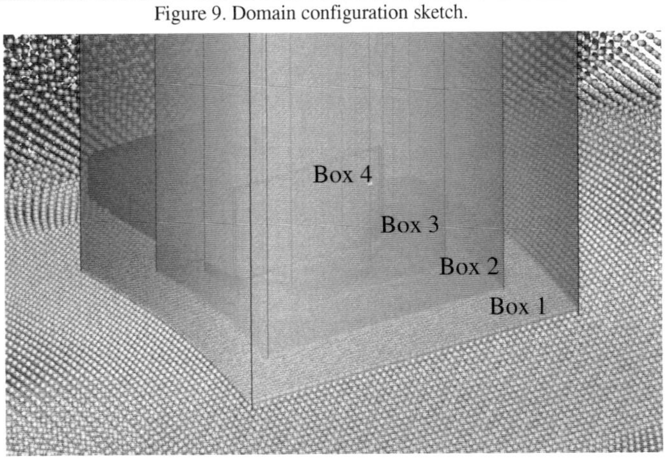

Figure 10. Refinement boxes

The speed of sound is chosen in order to fulfil both a maximum of 1% density variation at tank bottom, and a low Mach number in jets on the free surface. The hydrostatic condition appears to be the most prohibitive, leading to a speed of sound equal to 45m/s for 2m of water depth. The kernel chosen is 4th order Wendland with ratio Radius/Δx equal to 4. The time advance is solved with Strong-Stability Preserving Runge-Kutta (8,3) scheme with CFL number equal to 1.5. Four boxes of refinement are used as shown in the image below.

2) Results

The numerical incident wave in the free domain is compared with the theoretical wave forcing computed by HOS at the same probe position on **Error! Reference source not found.**. The agreement is fairly good in terms of period and synchronisation, without important phase shifting. The amplitude is also well captured, with a damping of the amplitude inferior to 10%. The SPH simulation, as well as the HOS incident field, starts from rest, and the incoming wave must propagate from the inlet boundary to the structure. In consequence, half part of the computation is dedicated to the transient propagation of the incoming wave train. However, the use of a truncated domain helps to reduce the global elapsed time by two ways: the overall number of points is reduced according to the reduction of volume simulated, and the diminution of distance between the wave generation point and the structure allows to decrease the number of points per wavelength because the numerical damping is less important.

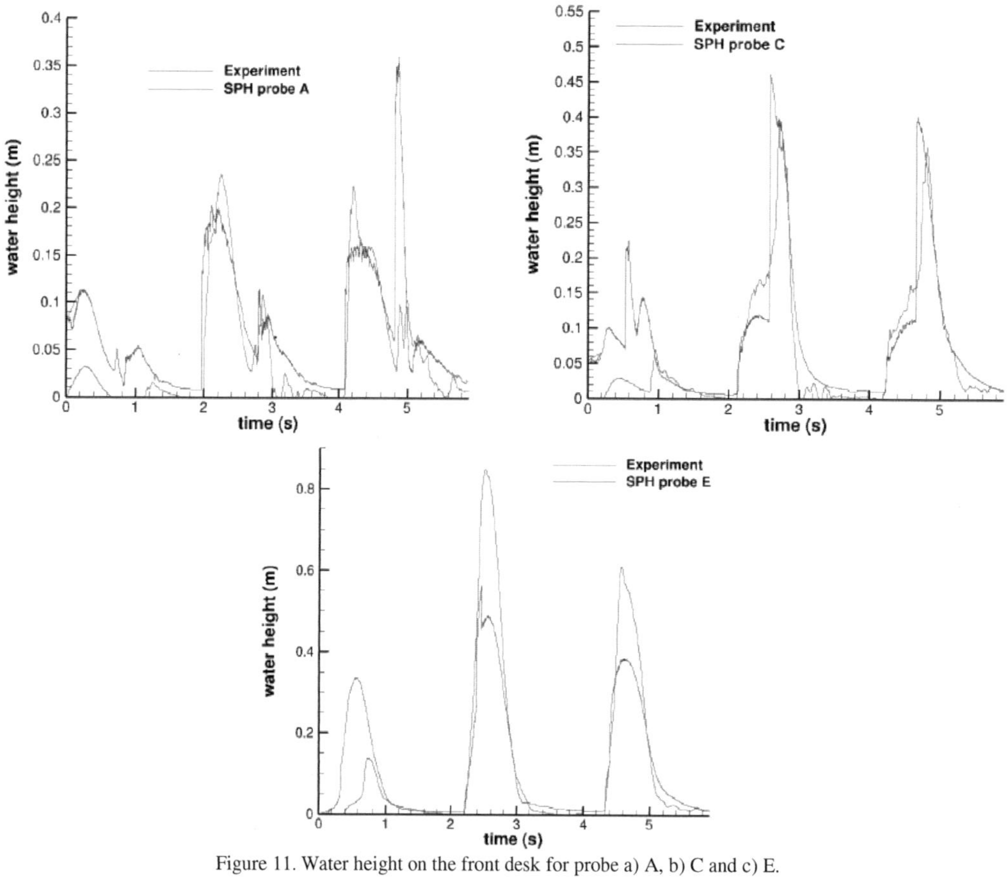

Figure 11. Water height on the front desk for probe a) A, b) C and c) E.

Three greenwater events are simulated. Comparisons are done for wave probes A, C and E respectively located on the deck from the bow to the breakwater, see Figure 4. It is possible to observe that the greenwater events are well predicted, there is a good comparison between the water elevation measured in experiments and the one computed with CFD. A good agreement is observed which confirms the fact that the adaptive particle refinement does not affect the flow behavior during the deck wetting phase of the greenwater.

The global flow is qualitatively captured, as figured on Fig. 12, with the 6 stages described in the experimental section. High number of neighbors, 256 for ratio Radius/Δx equal to 4, is very useful to propagate the swell but it leads fluid to be numerically hard to scatter. As a result, some water remains on the deck between two greenwater events.

Figure 12. Free surface colored with velocity magnitude at approximatively same times as Figure 5

The pressure value measured on the probe M1 was also compared to the experiments results. The figure below presents this comparison. A good synchronization between the experiments and the CFD results is observed, especially for the pressure peaks (impact). However, SPH tends to overestimate the pressure after the impact by 20%, which may indicate the need for an extra refinement boxes.

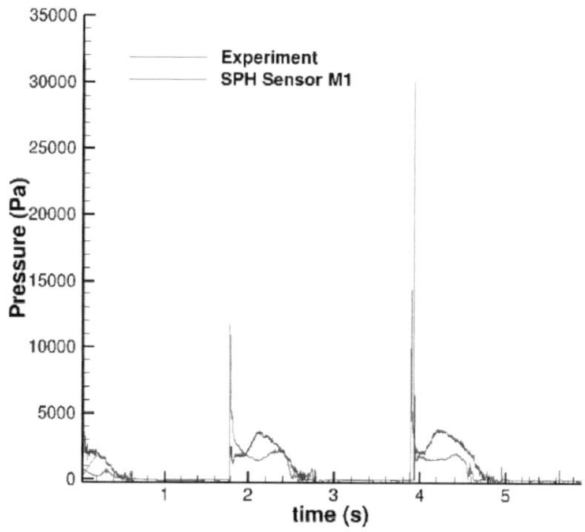

Figure 13. Comparison of the pressure forces measured at the M1 sensor.

CONCLUSIONS

An SPH monofluid version of the SPH-flow software is being developed by HydrOcean, ECN and NextFlow Software, for the special use of liquid impact simulations in the case of wave-structure interactions. An important step forward was achieved recently with the development of a coupling between potential solvers for the wave generation problem, and SPH-flow for the impact simulation. The potential solution relies on state-of-the-art solvers, to be able to take into account irregular nonlinear open sea states representative of real conditions.

The coupling between solvers is simply enforced by imposing the incident unperturbed wave solution on the boundaries of the SPH computational domain. Special care is required to manage the Lagrangian motion of particles at the inlet/outlet boundaries. This coupling allows simulating only a small part of the fluid domain around the impacted structure, as far as the diffracted flow does not reach the inlet/outlet boundaries. Further developments are under consideration, to combine general subsonic inlet and outlet conditions with the incident wave solution. This would enable the diffracted wave field to evacuate through the open boundaries, allowing larger time windows and smaller computational domains.

An extra step has been taken allowing to improve the solution obtained by the use of such approach. The use of adaptive particle refinement technique has been validated in view of the results presented in this article. By using such a technique the efficiency of the CFD simulation is increased: faster simulations, lower restitution time, more precision.

These good comparisons with experimental results bring credibility to the numerical simulations, at least with their ability to capture the relevant phenomena. The water height on deck is well captured in terms of temporal synchronization as well as maximum amplitude. Since these validations, HydrOcean has been successfully applying this methodology on projects for the offshore industry.

Nonetheless, some path of improvement remain. HydrOcean is working on integrating a dynamic particle refinement procedure which will allow maintaining the same level of precision of the simulations presented below but with less particle. Also, HydrOcean is working on developing and validating an equivalent dam-break approach that could be applied also as a first approach to assess the forces acting on some structures of the deck.

ACKNOWLEDGEMENTS

HydrOcean and ECN would like to acknowledge the support of TOTAL and TECHNIP, within the sponsorship of CITEPH R&D Program, dedicated to research and innovative projects in Exploration and Production. The results presented are part of CITEPH-66-2010 Project. In addition, the contributions of the following individuals and organizations are recognized and greatly appreciated:

- F. Bonnefoy for his invaluable contribution for the experimental overview and his wise recommendations.
- J.-L. Toularastel and S. Lambert who helped for experimental preparation as well as conception of the structure.
- Use of IFREMER Caparmor cluster for set-up of numerical model and initial validations.
- This work was supported by a grant from the Swiss National Supercomputing Centre (CSCS) under project ID s398.
- We acknowledge PRACE for awarding us access to resources on Cray XE6 "HERMIT" based in Stuttgart, Germany at GCS in HLRS.

REFERENCES

Crastto, G. (2007). *Numerical Simulations of the Atmospheric Boundary Layer.* Cagliari: Facoltà di Ingegneria.

Easom, G. (2000). *Improves Turbulence Models for Computational Wind Engineering.* Nottingham: University of Nottingham.

Hackett , J. E., Lilley, D. E., & Wildsen , D. J. (1979). Estimation of Tunnel Blockage from Wall Pressure Signatures: A Review and Data Correlation. *NASA CR.*

Ida, S. A. (2012). *Wind tunnel model testing of offshore platforms.* Norwegian University of Science and Technology.

Maskell, E. C. (1987). A Theory of the Blockage Effects on, Bluff Bodies and Stalled Wings in a Closed Wind Tunnel. *Aeronautical Research Council Reports and Memoranda.*

Parente, A., & Benocci, C. (2010). On the RANS simulation of neutral ABL flows. *The Fifth International Symposium on Computational Wind Engineering.* Chapel Hill.

Pope, A., & Harper, J. (1954). *Wind-Tunnel Testing.* New York: Wiley.

(2000). *Review of model testing requirements for FPSO's.* BMT Fluid Mechanics Ltd, Teddington.

Sumner, J., & Masson, C. (2010). k-epsilon Simulations of the Neutral ABL: Achieving Horizontal Homogeneity on Practical Grids. *AIAA Aerospace Sciences Meeting Including the New Horizons Forum and Aerospace Exposition*, (p. 1373). Orlando.

Zhang, X. (2009). *CFD simulation of neutral ABL flows.* Roskilde: Technical University of Denmark.

SNAME
THE INTERNATIONAL COMMUNITY
FOR MARITIME AND OCEAN PROFESSIONALS

VALIDATING COMPUTATIONAL FLUID DYNAMICS (CFD) MODELS IN THE ENERGY INDUSTRY – WHY AND HOW?

UDAY GODSE, PHD, PE
SENIOR ENGINEER, ADVANCED ENGINEERING GROUP, WILD WELL CONTROL, INC.

JOHN EDDIE, CENG
ENGINEER, ADVANCED ENGINEERING GROUP, WILD WELL CONTROL, INC.

ALISTAIR GILL, PE
VICE PRESIDENT, ADVANCED ENGINEERING GROUP, WILD WELL CONTROL, INC.

Proceedings of the 22nd Offshore Symposium, February 2017, Houston, Texas
Texas Section of the Society of Naval Architects and Marine Engineers

ABSTRACT

Computational fluid dynamics (CFD) tools are employed more and more frequently today to predict the behavior of fluid flow (movement of hydrocarbons, water, air, sand, etc.) in the oil and gas industry. While these tools are sophisticated enough to model complex physics, predominantly multi-phase, the accuracy and reliability of the resulting predictions are strongly dependent on how well the models used have been validated against test. Testing can present a major hurdle to validating these tools due to the cost and technical challenges associated with simulating the operational scenarios under investigation, which often involve the combination of deep water conditions and hot hydrocarbons. While the challenges should not be underestimated the importance of validating CFD predictions cannot be overstated.

Case studies including the validation effort are discussed and the challenges and importance associated with the testing and CFD predictions highlighted. Case studies include sand erosion in pipe networks, freezing of water in a pipe to form a pressure barrier, and the dispersion of gas in the water column following a subsea release.

Keywords: computational, CFD, sand erosion, subsea gas dispersion, validation, multi-phase, freezing, resin

INTRODUCTION

Computational fluid dynamics (CFD) is a computer based simulation technique which is used to predict the behavior of fluid flow phenomenon. In CFD partial differential equations are used to model the continuous fluid and their discretized form (algebraic equations) are solved to predict the flow phenomenon (Versteeg et al., 1995). The process of solving the discretized form of PDE's is a computationally intensive process and hence for a long time, the use of CFD was limited to the aerospace and automobile industry where the required resources and application requirements were closely matched. With the advent of high-speed computers in recent years, the use of CFD in other industries has increased significantly. Applications which were once too computationally intensive to model practically in a CFD simulation have now become feasible. Today's CFD code is enabled with models that emulate a range of physics that caters to applications in various industries, and one such industry is the oil and gas industry.

Applying the CFD technique to the physics at work in a typical oil and gas application is not without its challenges. Flows in the industry are highly complex and often comprised of multiphase (oil, gas, water, sand etc.), in highly turbulent conditions at elevated pressures and temperatures. Pipeline and offshore wells can be expected to transport these flows for thousands of feet, which is not ideally suited when applying the CFD approach to such large length scales. Special CFD modelling techniques have to be utilized to adequately capture the physics associated with such flows.

The list of CFD applications in oil and gas industry is diversifying as people recognize the benefits of computational analysis over traditional testing. The shorter time to perform analysis, full control over experimental conditions and the overall reduced cost to run multiple configurations make an approach such as CFD an attractive alternative to traditional methods. In recent times some most commonly studied areas include: sand erosion prediction and mitigation, performance evaluation of multiphase separators, subsea gas dispersion, conjugate heat transfer studies, vortex and flow induced vibration studies in offshore vessel and pipes, predicting explosion generated overpressures on a rig platform following a blowout, radiant heat study following accidental fire, and phase change due to pressure/temperature variations within a flow system. Each of these applications is highly diverse in the way it is modelled, and there is no one size fits all modelling approach which can fully capture the physics involved in each application. The modelling engineer, therefore, plays a critical role in selecting the appropriate models and sub models suitable for the application in hand. This bespoke modelling requirement means that each application should be thoroughly compared against real data to give the modelling engineer confidence in the fidelity of the modelling approach they have selected before the results can be accepted.

The accuracy and reliability of the resulting predictions are strongly dependent on how well the models used have been validated against tests. Testing can present a major hurdle to validating the CFD tool due to the cost and the technical challenges associated with simulating the operational scenarios under investigation. For example, a direct comparison may require the construction of a test which utilized volatile and toxic hydrocarbons at pressure and temperature, presenting a hazard to personnel and the environment. Although obtaining test data such as this is a challenge it is a vital step in the development of CFD within oil and gas and is the topic of the work presented here.

The primary objective of the current work is to present some of the CFD validation work which has been carried out to date, so as to illustrate the validation process and its importance. A total of three case studies with their validation findings will be presented: sand erosion study, the cryogenic freezing of water and subsea gas dispersion following an uncontrolled subsea hydrocarbon release.

CASE STUDY # 1: SAND EROSION ANALYSIS

Flow from a producing oil and gas well generally contains solids (sand or other particulates) along with hydrocarbons (fluid). In such flows the solids have a potential to impact the inner wall of the flow path which over time can cause erosion of the pipe wall. Accurate erosion prediction is thus important to help better design the pipe carrying the fluid-solid mixture or better predict the life of the pipeline losing material due to erosion. CFD provides distinct advantages over other prediction methods:

- It can tackle any flow geometry such as trees, flow lines, chokes and general pipe networks.
- It can tackle any flow condition such as low sand volumes during production conditions, high sand volumes during fracking/well kill conditions or high flow rates during blowouts.

The type of flow condition dictates the choice of type of modelling approach to be considered since the underlying physics are different. For low sand loading conditions (less than 5% solid by volume) where the sand particles are not expected to interact with each other to any significant degree a simple discrete phase model (DPM) can be employed. In a DPM study, the pressure and velocity field is first resolved for the dominant phase (hydrocarbon in a production scenario). The particles are then released and tracked through the existing field to determine the velocity and angle of impact. Erosion models are then employed to predict the erosion rates based on the predicted solid velocity and angle of impact.

In other cases such as fracking or a well kill scenario, a high solid content slurry is pumped through the pipe network and in such cases the particle to particle interaction becomes important and the solids have to be treated as a new n^{th} phase. The high concentration of solids also means that the erosion model has to consider not just impact, but also the shearing of solids against the wall surface which can contribute towards the overall erosion of wall material. These high solid flows are better captured with the use of the Eulerian multiphase (EMP) model rather than the DPM approach.

Validation cases have been analyzed in order to demonstrate the different approaches when considering the use of CFD in an erosion application. The first validation case involves an unsteady DPM approach with low solid loading to simulate conditions in an inlet manifold during a fracking operation. The wall movement due to erosion with time was accounted for using moving-deforming-mesh (MDM) algorithm in Fluent (Ansys Fluent's documentation, 2016). Haugen's erosion model (Haugen et al., 1995) was utilized to predict the erosion rates. This model has a general form:

$$E = M_p\, K\, F(\alpha) V_p^n \tag{1}$$

where E is the weight loss of the target material, M_p is the mass of sand impacting the target material, V_p is the particle impact velocity, α is the particle impact angle, K and n are constant that depend on the physical characteristics of the materials involved and $F(\alpha)$ is a functional relationship describing the dependence of erosion on the particle impact angle.

Figure 2 compares the erosion pattern and the eroded depth for CFD predictions to the test after 24 days of operation in the field. As can be seen, the patterns are similar in nature. The predicted eroded depth is within a factor of safety of 2 as compared to test. This is considered very good given the nature of the simulation carried out and that the dynamic wall movement due to erosion was accounted for. Note that when dynamic wall movement was not considered the predicted depth was found to be two orders of magnitude higher than test.

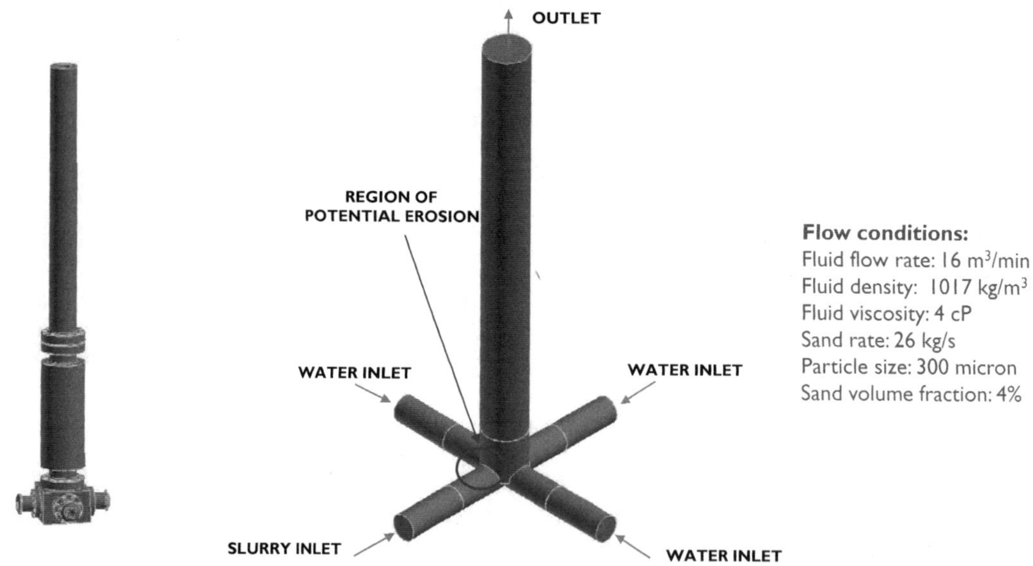

Figure 1 Erosion in an inlet manifold – Geometry and flow conditions

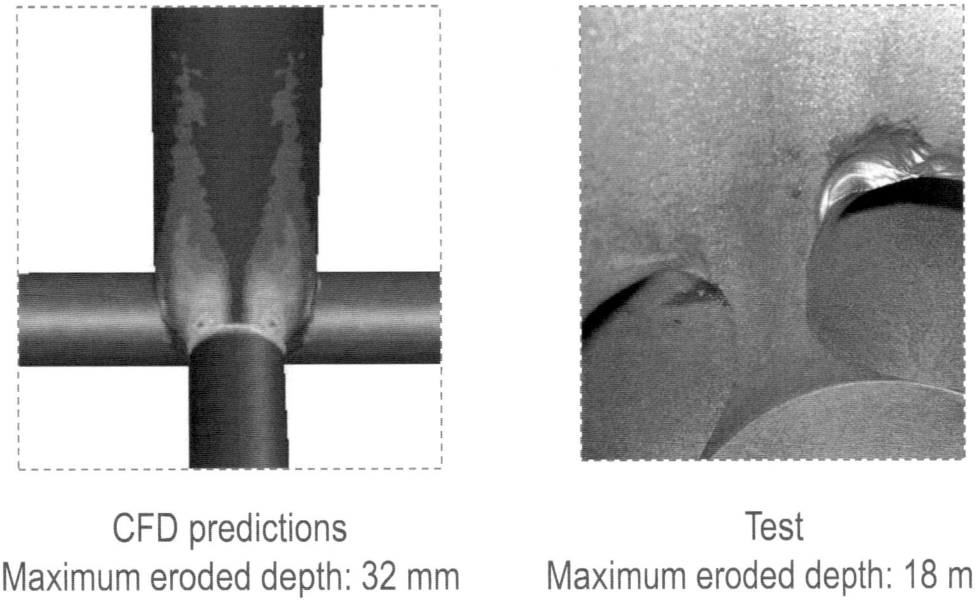

Figure 2 Erosion in an inlet manifold – CFD predictions vs. Test

The second validation case is a work in progress and involves high solid loading (~40% by volume) through an elbow that emulates a well kill job. The Eulerian granular phase model was employed in this situation to account for the particle-particle and the particle-fluid phase interactions. The water-based mud was used as the primary phase and barite with 40% volume was utilized as the secondary phase in the simulations to emulate the test conditions. Similar to the earlier erosion validation case, dynamic wall movement due to erosion was taken into account by utilizing the moving deforming mesh algorithm. Shear-based erosion mechanism was employed in addition to impact-based erosion. The shear based erosion model has a similar general form as compared to Equation (1) and can be seen in Equation (2) below. The angle of impact becomes less relevant when the particles are travelling almost parallel to the wall surface when shearing and was replaced with a constant instead. Note that the constant "K" and exponent "n" have different values than those utilized in an impact based erosion model. It was assumed that the shear stress "τ" has a linear dependence on erosion rate.

$$E = K \, \tau \, V_p^n \tag{2}$$

Figure 3 depicts the flow conditions and the geometry utilized for the study. The CFD predictions (deformed geometry) following a few hours of operations are also shown. The results have been normalized intentionally since the work is ongoing. Sensitivities around different solid fractions and flow rates have to be carried out to better understand the validity of the proposed shear based model. Figure 4 provides a comparison between impact based and shear based erosion rates after 1 hour of operations. It can be seen that the magnitude of erosion due to impact and shear are similar initially though shear based erosion is spread over a bigger area. It was found that the relative magnitude of impact based erosion decreases as time progresses with shear based erosion becoming a dominant contributor to overall erosion. The relative magnitude change and shear being a primary contributor to erosion was anticipated considering high solid fractions and the relatively lower difference in specific gravity of fluid and solid. For the tested flow rate and solid fraction, the predicted erosion pattern was found to be similar to the test (comparison not shown).

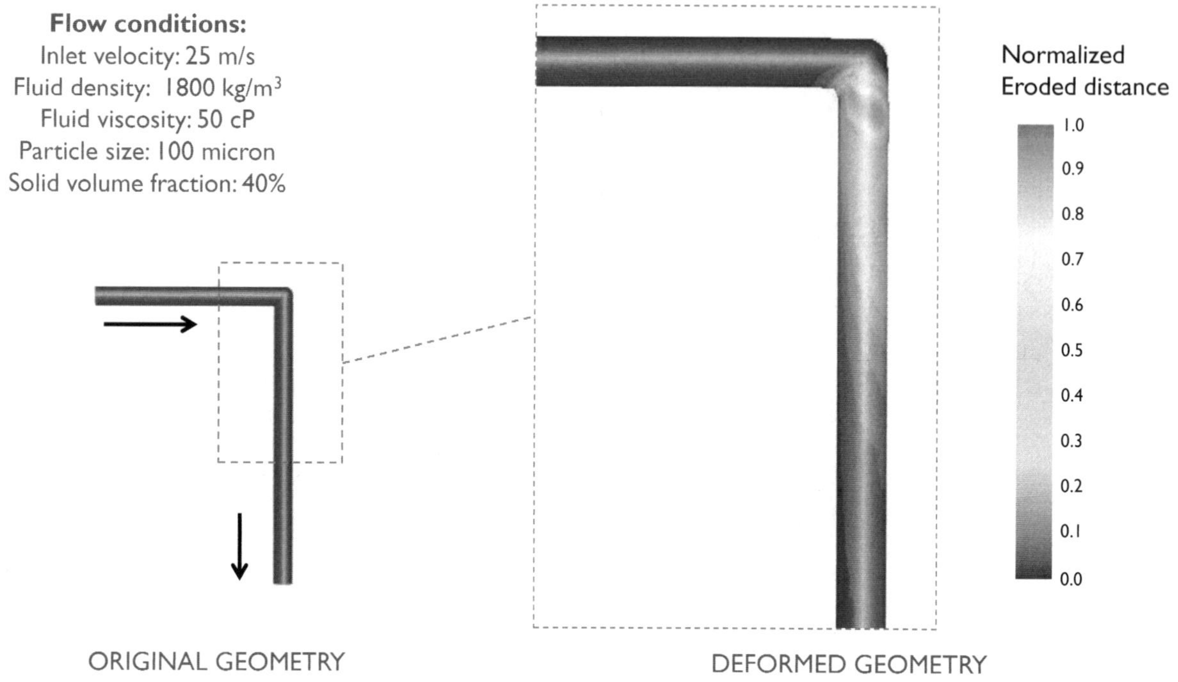

Figure 3 Erosion in an elbow – Flow conditions and CFD predictions

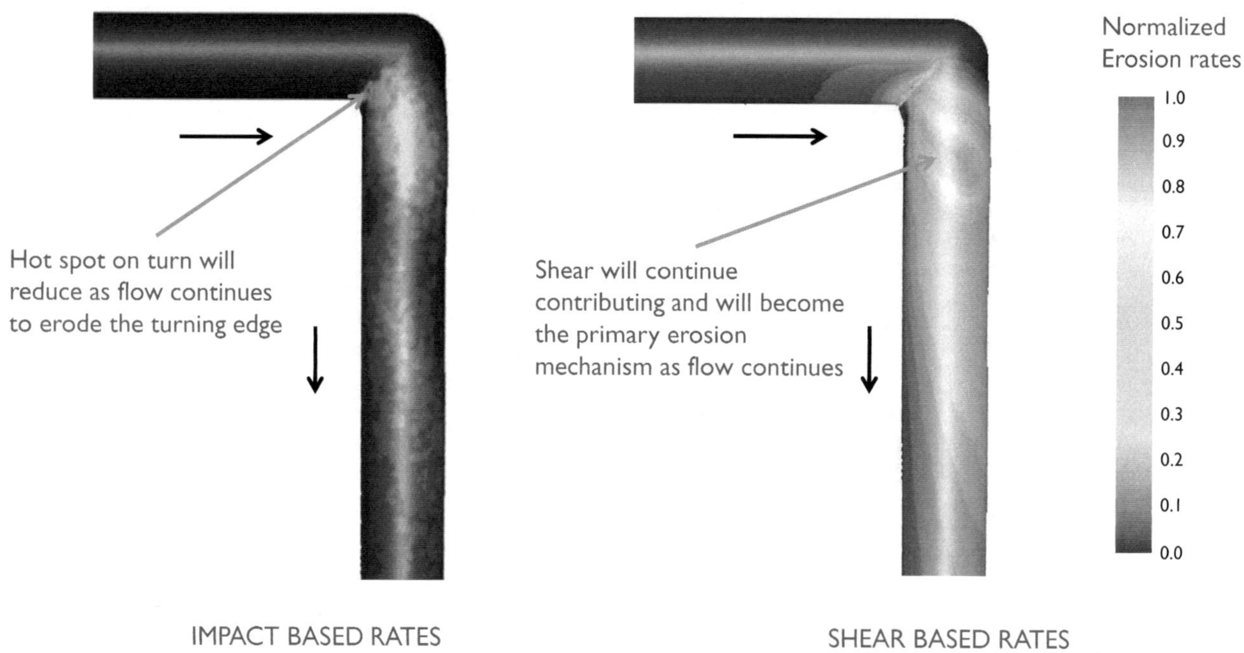

Figure 4 Erosion in an elbow – Impact based erosion vs. Shear based erosion

CASE STUDY # 2: FREEZING OF OIL FIELD PIPEWORK

The cryogenic freezing of pipework is a tried and tested technique used in the oil and gas industry to place a temporary ice plug in pipe systems to provide a temporary pressure barrier. The ice plug is generally set in place to act as a method of isolating or closing of the inside diameter of a pipe for maintenance or repair. The most frequent use for cryogenic freezing occurs when a previous operation (wire line, cementing, fracturing) has resulted in a situation wherein isolation from the wellbore cannot be achieved. An example is where wellhead and casing valves have become inoperable and need to be replaced. In such a case performing a fluid kill operation may not be feasible because of either logistics or cost, leading to the use of a temporary ice plug as a solution.

While the reason for using a cryogenic freeze may vary and the technique might have to be adapted to an extent, the fundamental process behind the operation remains the same. As depicted in the schematic below (Figure 5) the freezing operation is performed by using a fluid-tight metal jacket (as a heat exchanger). This jacket is attached to the area of the pipe to be isolated and is then precooled with nitrogen gas then filled with liquid nitrogen to freeze the water based fluid along the inside surface of the pipe near the jacket. As the cooling process continues inside the pipe, more fluid is frozen and more ice is produced along the axial plane of the pipe. Further cooling produces a highly stable ice plug that can be sustained with careful monitoring of the temperatures and pressures present at the heat exchanger by trained personnel. Note that liquid nitrogen possesses fatal characteristics and hence was always controlled and monitored during testing according to the requirements set forth in safety standards.

CFD has been used in this application to accurately predict the amount of time required to establish a solid ice plug. The test performed for CFD validation work used water as fluid (see Figure 6) but this technology can easily be used to cater to other fluid systems and environments outside those at which test was conducted. Multiple test using different ID's of piping and ambient temperature were carried out. The underpinning physics includes conjugate heat transfer (conduction, convection) and phase change from liquid water to solid ice plug. The time to freeze is sensitive to the size of freeze (ID and depth of isolation), ambient temperatures, initial fluid temperature, quality of insulation, and adjacent structures which are a potential heat source that slow down the freezing process. Note that the energy required to be taken out to change the phase of water to ice is usually order of magnitude higher than cooling water from initial/ambient to 32 °F.

Figure 7 shows the predicted time to freeze two different ID's (9-5/8" and 30" pipe) at two different ambient temperatures (50 °F and 105 °F). With an increased surface area available for heat exchange with the ambient, the large ID pipe takes much longer to freeze at higher ambient temperatures as compared to lower ID pipe. The CFD study used CD Adapco's Star CCM+ software and assumed that the section of the pipe is insulated from adjacent structures and hence in some cases was seen to be underpredicting the time to freeze. Overall, the predictions were within 10% of the available test data.

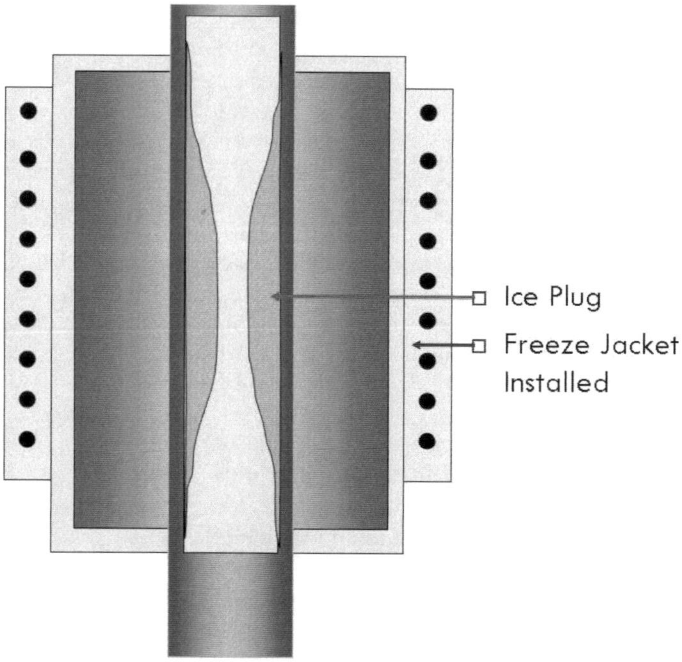

Figure 5 Schematic showing a typical freeze process

Figure 6 Freezing test using a 24" Casing and liquid nitrogen

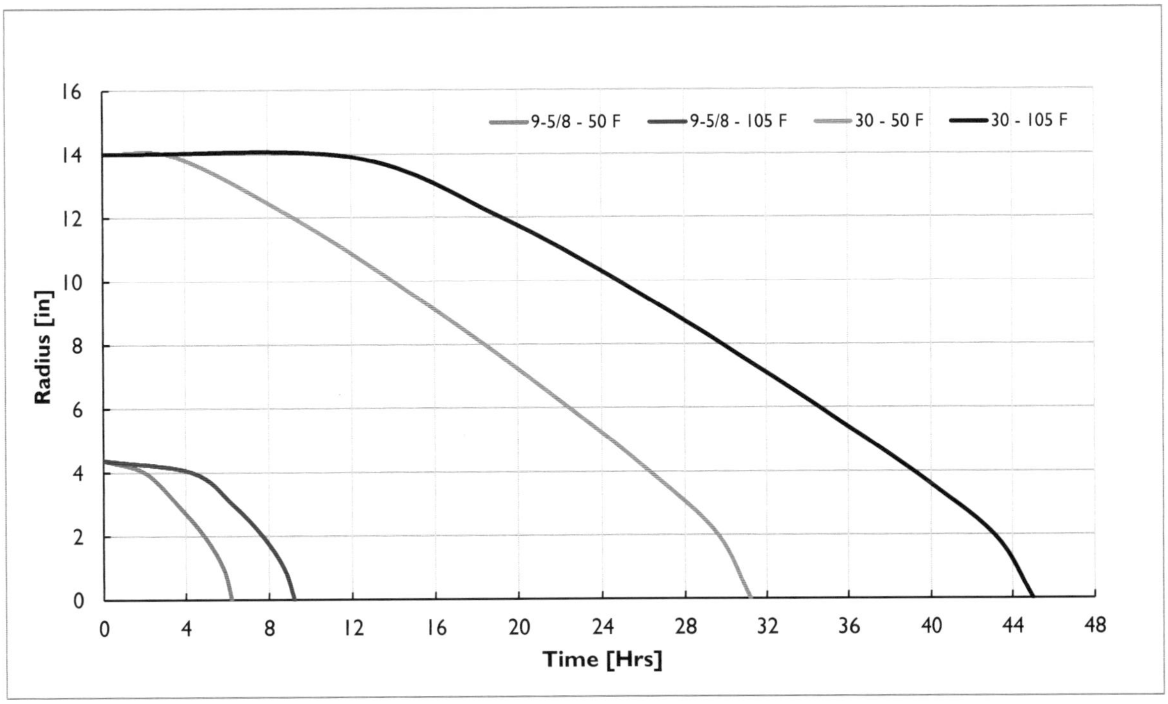

Figure 7 Time to freeze predictions using CFD

CASE STUDY # 3: SUBSEA GAS DISPERSION

An unplanned hydrocarbon release is generally a highly undesirable event that can lead to a large-scale safety, environmental and economic disaster. For example, a subsea oil and gas blowout not only has an obvious environmental consequence but also can pose a significant safety issue for the surface activities associated with the well control response actions. The rising hydrocarbons carry a potential for a flammable hazard; in some cases there will also be a toxic hazard if the release contains any hydrogen sulfide or other toxic gases. To improve safety, there is a need for reliable estimates of how much gas flows to the surface, where it arrives at the surface, and how is it dispersed into the atmosphere in the event of such a blowout. A multi-physics modelling approach (Gill et al., 2015) as part of a compressive risk assessment of the operation (whose findings can then be utilized to formulate an emergency response plan) can be utilized to predict the fate of gas rising up the water column following a release.

CFD techniques are extensively used to model such flows and recently have been extensively studied including testing for validation. Wild Well Control, Inc. is part of a Joint Industry Project which is focused on how a subsea release of gas flows to the surface. The project is managed by SINTEF Material and Chemistry (Norway). As part of the project scope detailed large-scale experiments have been carried out in both sheltered water (Trondheim Fjørd) and the open ocean in water depths up to 1200 ft. The test findings have helped understand the accuracy and limitations of current CFD techniques and further improve them. A most recent account of the project findings was published by SINTEF (Olsen et al., 2016) where an Eulerian-Lagrangian CFD model was utilized to model the gas rise in water column. Prediction was also compared to experimental data available in literature. Figure 8 shows gas surfacing in one of the experiments carried out offshore as part of the project.

As the JIP progresses, the CFD modelling approach is becoming more sophisticated based on the findings. The first generation CFD model used a RANs modelling technique to predict the turbulent dispersion within the bubble column. However, during the validation work it has been discovered that a more intensive, high fidelity mass transfer and turbulence modelling approach is required in order to accurately predict the turbulent dispersion of the bubble column and gas dissolution. This can be seen in Figure 9 where the sonar data shown in blue and red lines is compared to the CFD data. The CFD predicts the rise time accurately, however it can be seen that the turbulent dispersion of the plume is not captured using the 1st generation approach. In Figure 10 however the first generation RANS approach is compared with the latest modelling technique, which was developed and validated against the industrial scale experiments. This new generation model shows a marked improvement on the plume shape when compared with experiments.

This work has shown that the validation is critical when selecting the sub-models used in an oil and gas CFD application. The CFD models here have been developed in conjunction with the real data, allowing the CFD users to have confidence that their sub model selections are suited to the application in hand.

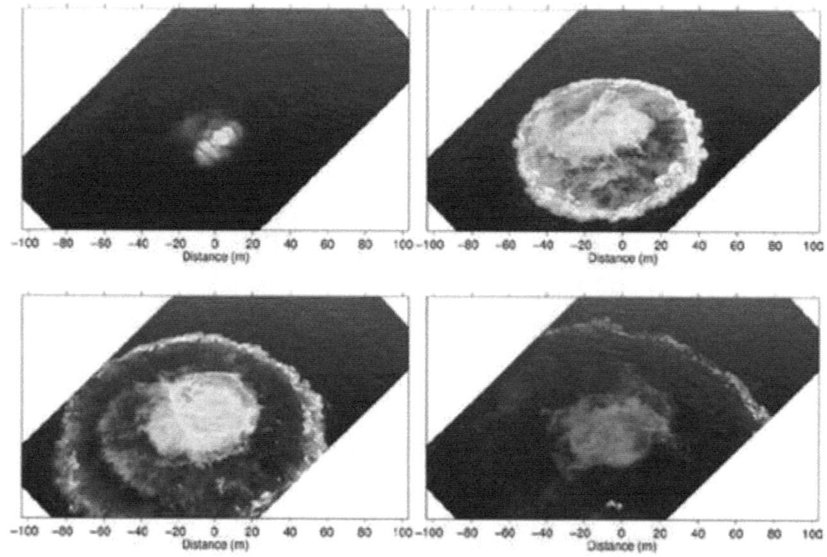

Figure 8 Surfacing gas in offshore release experiments

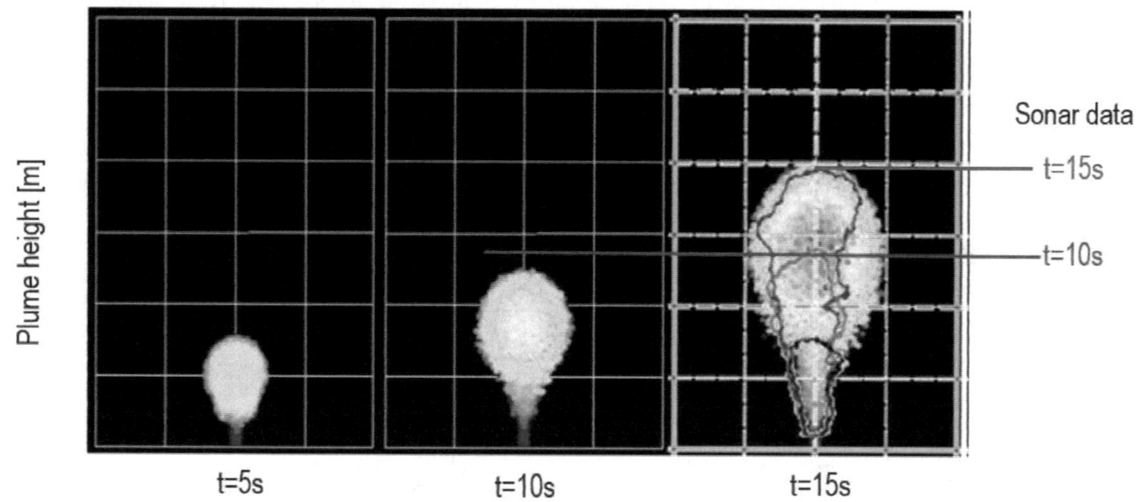

Figure 9 Sonar data (red, blue lines) vs. 1st generation CFD Plume

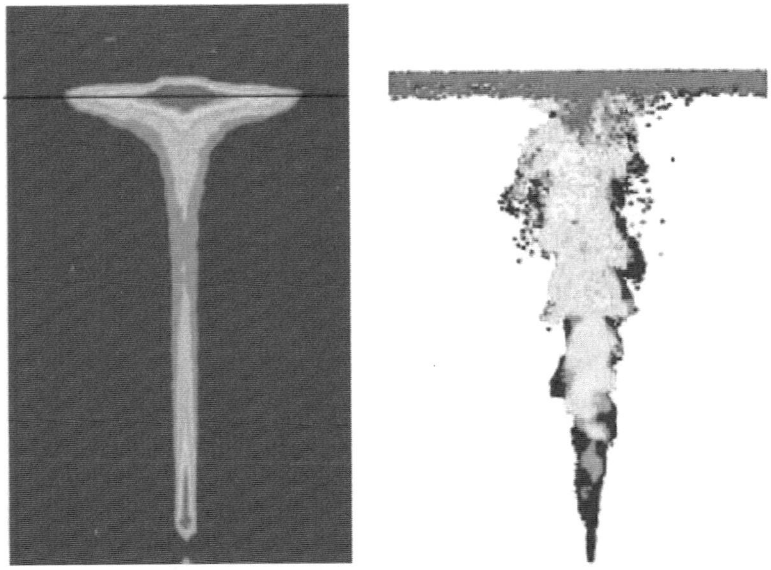

Figure 10 1st generation turbulence approach vs. JIP development turbulence approach

CONCLUSIONS

The following conclusions can be drawn from the three case studies presented:

- The use of CFD is growing in the oil and gas industry;
- CFD applications are becoming more diverse, and as such more models are available to CFD engineers;
- More options means there are more places to perhaps select inappropriate models for the application under consideration;
- Models must be validated to ensure accuracy and reliability;
- Test data can be difficult to generate for highly complex flows, but this can be done with industry initiatives and willingness from operational partners as demonstrated here;
- CFD has been modelled against test data for subsea gas release, and it was found that a higher fidelity approach is required;
- CFD has been modelled against a cryogenic freeze operational data to good accuracy;
- Erosion model selection is critical in the accuracy of the result – the correct model varies per application
- Skilled CFD users are required with access to both models and test data – no one model fits all is available.

ACKNOWLEDGEMENTS

The authors are grateful to Wild Well Control, Inc. for supporting this study. Special thank you to our colleague, Carlo Mazzella for his valuable input and guidance.

REFERENCES

Ansys documentation, Fluent 16.0 User's guide (www.ansys.com)

CD Adapco documentation, Star CCM+ 11.06 User's guide (www.cd-adapco.com)

Gill A., Godse U., Maghsoudi E. and Bellamkonda G., "Advanced analysis improves emergency response to subsea hydrocarbon releases", World Oil, 236-1, 2015.

Haugen K., Kvernvold O. and Sandberg R.R., "Sand erosion of wear-resistant materials: Erosion in choke valves", Wear 186-187 (179-188), 1995.

Olsen J. E. and Skjetne P., "Modelling of underwater bubble plumes and gas dissolution with an Eulerian-Lagrangian CFD model", Applied Ocean Research 59 (193-200), 2016.

Versteeg H.K. and Malalasekera W., "An introduction to computational fluid dynamics – The finite volume method", Published by Longman Scientific & Technical, Textbook ISBN: 0-582-21884-5, 1995

*** Formatting Issue - Best Available Paper/Graphic ***

SNAME
THE INTERNATIONAL COMMUNITY
FOR MARITIME AND OCEAN PROFESSIONALS

INTERACTION ANALYSIS FOR MODULE MARINE TRANSPORTATION

YAN ZHANG
EXMAR OFFSHORE COMPANY, HOUSTON, TX

AIMIN WANG
EXMAR OFFSHORE COMPANY, HOUSTON, TX

ALISON SHEPPARD
EXMAR OFFSHORE COMPANY, HOUSTON, TX

OTTO DASILVA
EXMAR OFFSHORE COMPANY, HOUSTON, TX

Proceedings of the 22nd Offshore Symposium, February 2017, Houston, Texas
Texas Section of the Society of Naval Architects and Marine Engineers

ABSTRACT

The modularized design and construction philosophy is becoming a common trend for large onshore and offshore projects due to its advantage in schedule and cost-savings. Multiple modules are fabricated at different construction yards across the world and then transported to service sites by marine transportation vessels. Proper and accurate sea fastening design and analysis play an important role in order to ensure the integrity of both the modules and the transportation vessel. Finite Element Analysis (FEA) has been widely used for the structural design and verification in the transportation analysis. Common practice is based on decoupled analysis where separate FE models are created for the module and vessel structure respectively.

This paper presents a more advanced and accurate approach where interaction FE analysis is utilized to analyze the complete FE model of both the module and vessel structure simultaneously. Formulations derived for this analysis will be presented along with the methodologies of applying boundary conditions and environmental loads from hydrodynamic analysis. Results from a sample module transportation analysis are presented.

Keywords: Marine Transportation, Interaction, Hydrodynamic

INTRODUCTION

A modularized construction strategy is becoming a common trend for large onshore and offshore projects. This strategy provides an economically attractive alternative to on site construction and shorten construction schedule as multiple modules can be simultaneously fabricated at different yards and transported to the service site.

During marine transportation, modules have to withstand the transportation loads and simultaneously the vessel has to be strong enough to support the modules in the rough seas. It is essential that adequate structural strength be achieved. Forethought has to be given to the marine transportation conditions in the earliest stage of module design. The structure must possess a reasonable strength for the environmental loads imposed on it during transportation.

The success of the transportation hinges on the determination of the environment loads to which it will be

Proceedings of the 22nd Offshore Symposium, February 2017, Houston, Texas
Texas Section of the Society of Naval Architects and Marine Engineers

exposed and the specification of a reasonable probability level for the environment it encounts. The environmental loads imposed on the structure will have to be determined. It is necessary to define a number of deterministic load conditions which can be used to evaluate and quantify the module and vessel capacity through the Finite Element Analysis. The transportation route needs to be studied and planned at the earliest stage of the project. An analysis is performed to identify the route and design metocean criteria.

The accurate determination of structural loads can have significant impact on the module design and transportation. Underestimation of the loads can result in structural integrity failure or schedule delays. Overestimation of loads, on the other hand, can cause the vessel to carry excess structural weight, unnecessary module/vessel strengthening, transportation difficulty or more stringent vessel selection also resulting in cost increase and project delays. Hence, the need for a rational approach to determine the appropriate loads is evident.

Common practice generally uses the decoupled vessel module model for design criteria. This approach does not consider the module's stiffness contribution to the vessel/module integrated system, therefore it can be overly conservative. In this paper, a more accurate integrated vessel/module approach is presented. A typical module transportation analysis from a selected route is performed. The resulting deflections and reactions are compared with traditional decoupled approach to validate the preference of conducting interaction analysis for module design. The results are also compared with module in-place condition to address the transportation condition is more limiting than in-place condition for module design.

The interaction analysis is an enhanced and more precise quantification of structural integrity. Additionally, it is more realistic for evaluating and developing plans. Transportation cost could be reduced based on the coupled vessel/module interaction analysis.

METOCEAN CONDITION

There are different parameters that may influence the success of the transport. One of the key factors is the metocean condition. Direct and indirect routes are simulated based on navigation hours and possibility to mitigate any environmental threshold.

Once the route has been selected, the appropriate environmental conditions along the transportation route should be established. The transportation design criteria is based on the most severe time of the year along the route in order to accommodate any schedule changes. The metocean variables of interest involved in the analysis are the significant wave height, peak wave period and wind speed with a given exceedance probability. For example, if the exceedance probability of the maximum significant wave height encountered for a tow started in a particular month is 1%, then one out of 100 tows during that month will exceed the design criteria. This is similar to the concept of return periods for fixed offshore platform design criteria as the 100 year waves. For module transportation, a metocean design threshold of during one out of every 100 voyages or the peak monthly 10-year extreme is recommended.

The range of wave spectral peak period corresponding to the significant wave height is compiled using the 3 hourly enroute statistics. The shorter wave periods, particularly those close to the natural period of motion, should be investigated as they could cause higher motion and accelerations when in resonance, even at lower wave heights.

MAXIMUM EXPECTED RESPONSES

During marine transportation, the module and vessel will experience static loads as well as dynamic loads induced by waves and wind. The waves produce external dynamic pressures, induce vessel motions in translational and rotational degrees of freedom and generate inertial forces through acceleration. It is acknowledged that a key issue is to determine the most critical load cases which can evaluate and quantify the module and vessel capacity through the Finite Element Analysis. The following critical load cases are considered.

- Vertical bending moment
- Vertical shear force
- Horizontal bending moment
- Horizontal shear force
- Torsional moment

- Vertical acceleration
- Lateral acceleration

All above load cases are assessed for both positive and negative values. For example, the vertical bending moment is assessed for both hogging and sagging condition. For each load case, the extreme value is assessed to determine the maximum response the vessel will experience during transportation.

Maximum Expected Motions The maximum expected motion responses are commonly used to determine the "observed" response of a given sea state and the maximum responses during a three hour storm. These values are calculated using the root mean square (RMS) value of the motion that is obtained by integrating the sea state power spectrum assuming a Rayleigh distribution of the peaks. The probability that a single wave amplitude is less than a specified value $a_{\max}$ is:

$$(\quad) = 1 - \quad \left(\frac{-\ \ ^{2}}{2\ \ _{0}}\right) \text{ for } a_{\max} \text{ greater than zero.} \qquad (1)$$

For example, if during a three-hour storm, there are 1000 waves on average. The maximum expected response is obtained from the following equation

$$. \ _{.} = 3.85 * \sqrt{\ _{0}} \qquad (2)$$

where, $\sqrt{\ _{0}}$ is the RMS of the response power spectrum density function (PSD).

Maximum Expected Accelerations To derive acceleration data, a 90% percentile in the extreme value distribution for short term response was used. This means there is a 10% probability that the vessel will have a larger acceleration than the given value in a monthly 10 year return wave/wind condition. The corresponding value is 2.12 multiplied by the significant acceleration response.

The maximum expected acceleration results are also based on a combination of gravity vector components and wave-induced dynamic accelerations in the complex domain. Rationality behind the determination of extreme accelerations using this approach lies in the accounting of relative phases in the components concerned, thus adding precision to the results. Note that acceleration calculations do not take into account components due to static heel and trim.

Based on the first order motions calculations, the acceleration response amplitude operator (RAO) is derived as shown below:

$$'' = - \ ^{2} * \quad * \ ^{-} \qquad (3)$$

The equation above yields the dynamic components of the acceleration at the origin of the coordinate system. To obtain the total dynamic acceleration at any given location these components have to be combined as follow:

$$'' = - \ ^{2} * \quad (\ _{1} + \quad * \ _{5} - \quad * \ _{6}) \qquad (4)$$

$$'' = - \ ^{2} * \quad (\ _{2} - \quad * \ _{4} + \quad * \ _{6}) \qquad (5)$$

$$'' = - \ ^{2} * \quad (\ _{3} + \quad * \ _{4} - \quad * \ _{5}) \qquad (6)$$

where

x, y, z represents the point at which accelerations are calculated

x'', y'', z'' are the x, y, and z accelerations

ξ is the RAO with subscripts 1-6 representing surge, sway, heave, roll, pitch, and yaw respectively.

The predictions of expected maximum motions are statistical values based on the acceleration RAOs, the given sea state, and the defined sea spectrum. Once these results are calculated, the phase relationship between each acceleration component is lost. The loss of phase relationship precludes the correct combination of these acceleration components after their significant or expected maximum values have been calculated.

The gravity components are also added to account for their phase relationships. As the dynamic components of the acceleration are fixed in the vessel coordinate system, the acceleration must be transformed for that coordinate system using the following equation:

$$'' = - \quad * \quad (\ _{5}) - \ ^{2} * \quad (\ _{1} + \quad * \ _{5} - \quad * \ _{6}) \qquad (7)$$

$$'' = \quad * \quad (\ _{4}) * \quad (\ _{5}) - \ ^{2} * \quad (\ _{2} - \quad * \ _{4} + \quad * \ _{6}) \qquad (8)$$

$$'' = \quad * \quad (\ _{4}) * \quad (\ _{5}) - \ ^{2} * \quad (\ _{3} + \quad * \ _{4} - \quad * \ _{5}) - \qquad (9)$$

DESIGN WAVE

An equivalent design wave is a regular sinusoidal wave that simulates the extreme value of the critical load cases under consideration. Critical wave selection is based on first principle and derived from WAMIT motion results. The response analysis determines the critical wave period that generates the maximum hydrodynamic loads. It further finds an equivalent regular wave height, which will generate the same amount of the hydrodynamic loads as calculated in the response analysis. The selected wave period and wave height, noted as "Design Wave", can be used for the deflection and strength check of a vessel.

The vertical bending moment generally governs the deflection of a ship shaped vessel. When hydrodynamic pressure and inertial force distributions are readily available, a rigorous approach to determine the design wave length is to integrate over the structure including both hydrodynamic loads and inertias loads to determine force RAO, and then determine the design wave by stochastic approach. The stochastic method is then used to find the regular design wave amplitude and period based on the maximum responses of various load characteristics. The method for deriving the critical wave period and the regular design wave amplitude for structural analysis is described in steps as follows.

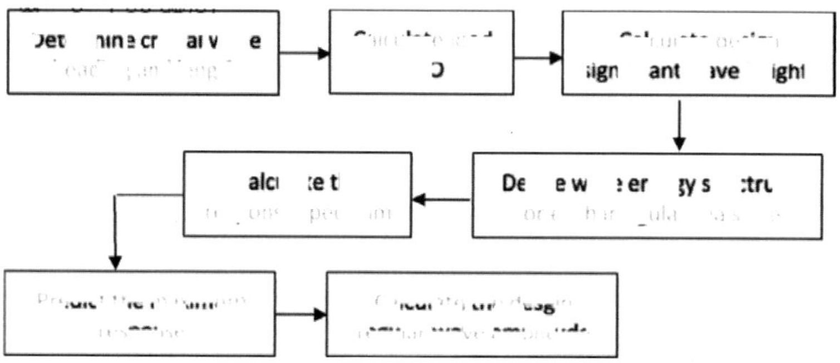

Figure 1 Design Wave Selection Flow Chart

VESSEL/MODULE INTERACTION ANALYSIS

The calculations of hydrodynamic pressures and inertial loads are performed using WAMIT. The vessel geometry is modeled by a "wire frame" mesh of panels of the wetted area. The result of the WAMIT program calculations is the set of RAOs for the six degrees-of-freedom of the vessel, hydrodynamic pressures on the vessel, slow drift forces and inertial loads. The number of panels is adjusted appropriately to achieve convergence in the calculation of main hydrodynamic properties. The pressure distribution over a hydrodynamic panel model is then linearly interpolated to obtain the pressures on the structural FE model.

An example module and vessel are used for this case study. The module and vessel particulars are listed in Table 1. The global FE model of the module and vessel are generated according to structural scantling drawings. Plate elements are used to model the vessel deck, outer shell and bulkheads. Beam elements are used to model the module, stiffeners, girders and stanchions of vessel. Grillage is simulated as beam elements and rigid connected to vessel deck. The connection reflects actual seafastening, e.g. pinned or fixed conditions. A mass distribution representing the fuel and ballast arrangement during transportation is added to the model. The hydrodynamic load representing a range of different headings, hydrostatic and wind loads with respect to the specific route metocean data are applied to the model. It should be noted that direction of wind force is coincident with design wave heading to induce the greatest structural response. A set of three points is constrained to prevent rigid body motions. The combined vessel and module FE model is shown in Figure 2.

*** Formatting Issue - Best Available Paper/Graphic ***

Table 1 Module and Vessel Particulars

	L	W or B	H	Weight/ Lightship Weight	Draft
	m	m	m	Mt	m
Module	85	30	30	5000	-
Vessel	114	37	6	4700	3.2

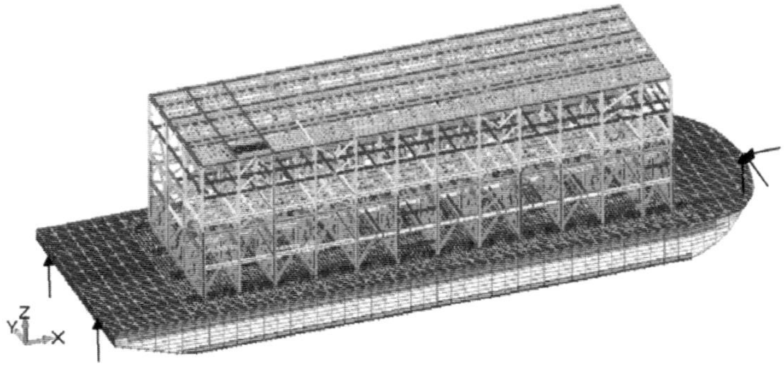

Figure 2 Combined Vessel-Module System

The resulting stresses, deflections and reaction forces are examined. Both the module and vessel's strength capacity can be determined by comparing the results to maximum allowable values. Two failure modes are analyzed in this paper to assess the module and vessel strength, buckling and yielding based on the classification requirements. The high buckling and yielding utilization members can be easily located as illustrated in Figure 3 for module and Figure 4 for vessel. Tailored modification plan can then be carried out based on the results. One or more of the mitigations may be required such as stowage plan change, boundary constraints change to reduce the column reactions, local reinforcement of the module and/or vessel.

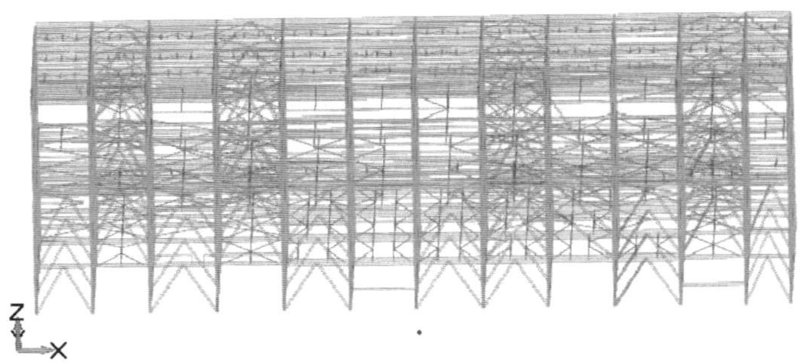

Figure 3 High Buckling Utilization Members

Proceedings of the 22nd Offshore Symposium, February 2017, Houston, Texas
Texas Section of the Society of Naval Architects and Marine Engineers

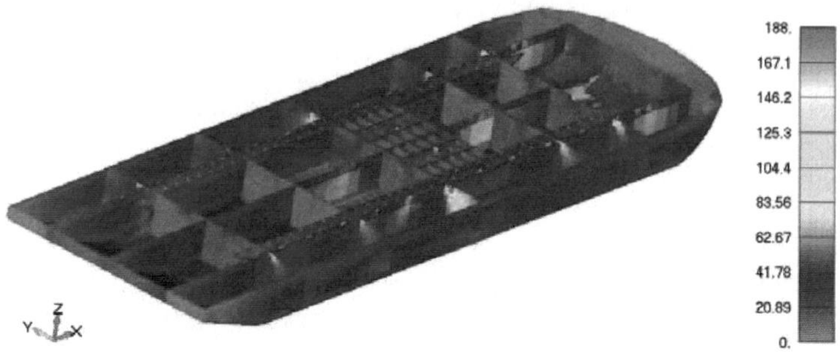

Figure 4 Vessel Envelope Von Mises Stress Contour (in MPa)

Common practice generally uses the decoupled vessel module model where module is simulated as mass elements and distributed to vessel. Only vessel's deflections and accelerations under wave conditions are examined and then imposed to module for module design. The reaction forces of module columns are then extracted and used for vessel strength capacity check and grillage design. This approach does not consider the module's stiffness contribution to the integrated vessel/module system. The examined deflection is therefore overly conservative. Figure 5 compares the vessel deflections between the decoupled and coupled approaches under hogging, sagging and stillwater conditions. It is noted that the decoupled model indicates approximately 65% more deflection than the coupled model. The maximum reaction force of single column from interaction analysis is 421Mt. The maximum reaction from decoupled approach is 688Mt of which 39% is needlessly transportation loads. The overly conservative design may result in module weight increase, more stringent vessel selection and transportation difficulty. Each situation can cause overall project cost increase and schedule delays.

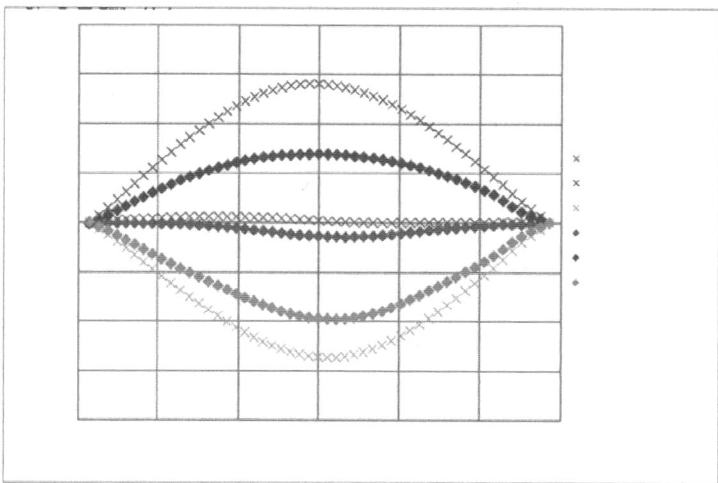

Figure 5 Deflection

Marine transportation needs to be considered in the earliest stage of the module design even though the module will spend most of its life onshore. Onshore module design considers in-place loading conditions such as wind, seismic and self-weight. However, the magnitude of the reactions depends on a large number of influences which include module/vessel length, weight distribution and metocean design criteria. The hydrodynamic induced reactions loads can be on the order of 10 times that of the in-place load borne by the same column (Bai et al., 2016). The transportation condition gives a more limiting requirement than in-place condition for module design. If the loads are not properly considered in module design, it could result in safety issues, damage of module and/or vessel, or long delays of the project.

CONCLUSIONS

From the module design standpoint, both the in-place loading condition and transportation conditions need to be considered. Accurate determination of structural loads can have significant impact on the module design. Underestimation of the loads can result in structural failure or schedule delays. Alternatively, overestimation of loads, can cause the vessel to carry excess structural weight, resulting in unnecessary module/vessel strengthening, transportation difficulty or more stringent vessel selection. Hence, the need for a rational approach to load determination is evident.

The specific objective of the study reported herein is to develop a more accurate and cost effective approach for module design and predict the vessel/module capacity during transportation. This paper presents a fundamental procedure of vessel/module interaction analysis which consists of environmental conditions, determination of representative loading cases, selection of design loads, vessel/module interaction analysis using rational design loads.

The deflection and reaction forces from decoupled approach is over conservative and undesirable. It can impose cost increases and schedule delays. The interaction analysis, on the other hand, can predict rational loads for module design and be used as the basis of modification plan if strengthening is required. It is recommended to use coupled interaction analysis for module design and marine transportation.

REFERENCES

ABS, "Guide for Dynamic Loading Approach for Floating Production, Storage and Offloading Installations", 2014

American Petroleum Institute, "Recommended Practice for Planning, Designing and Constructing Fixed Offshore Platforms", API RP 2A, 2005.

Bai, J., Koskins, S., Hodapp, D., Ma, W. and Wisch, D., "LNG Facilities Module Design Considerations During Marine Transportation", OTC-26971-MS, 2016

Det Norske Veritas, "Environmental Conditions and Environmental Loads", DNV RP C205, 2010

SNAME
THE INTERNATIONAL COMMUNITY
FOR MARITIME AND OCEAN PROFESSIONALS

VISVE, a Vorticity Based Model Applied to Unidirectional and Alternating Flow around a Cylinder

ZHIHAO LI
GRADUATE STUDENT, THE UNIVERSITY OF TEXAS AT AUSTIN

SPYROS A. KINNAS
PROFESSOR, THE UNIVERSITY OF TEXAS AT AUSTIN

Proceedings of the 22nd Offshore Symposium, February 2017, Houston, Texas
Texas Section of the Society of Naval Architects and Marine Engineers

ABSTRACT

The flow around a riser can be modeled as the flow around a cylinder, which is an extensively studied problem within the field of fluid mechanics.

In our previous work, VIScous Vorticity Equation (VISVE) method has been proposed and implemented by Tian [1], in order to model the effects of the Leading Edge Vortex (LEV) on propeller performance. The VISVE method is designed to be both spatially compact and numerically efficient in comparison to the commonly used RANS-based models. The method was recently extended in the case of cylinder subjected to uniform inflow [2].

In this study, the VISVE method is applied to alternating flow around a cylinder. The gridding scheme is changed to allow vortices to travel farther on both sides of the cylinder. In addition, the utilized grid is orthogonal and that improves the quality of the results. The vorticity, pressure, and force calculation is improved and the predicted values are compared to those from a commercial CFD software (ANSYS Fluent) in the case of unidirectional as well as alternating flow. It is found that the results from VISVE agree very well with those from Fluent at earlier time steps in the simulation in the case of unidirectional flow, while they were qualitatively similar at later time steps. In the case of alternating flow, VISVE can predict results which are very close to those from Fluent with small Keulegan-Carpenter numbers, while they will deviate at higher KC numbers.

Keywords: Viscous Vorticity Equation (VISVE), Computational Fluid Dynamics (CFD), Reynolds-averaged Navier–Stokes (RANS), Alternating Flow around Cylinder, Morison's equation

INTRODUCTION

The VISVE code was developed by Tian (2014) [1] aiming at solving the viscous vorticity equation in both 2D and 3D flow field around hydrofoil. VISVE captures the essential flow features such as leading edge vortex and tip vortex in the steady case and the results are comparable to those from RANS, as shown in Figures 1 and 2.

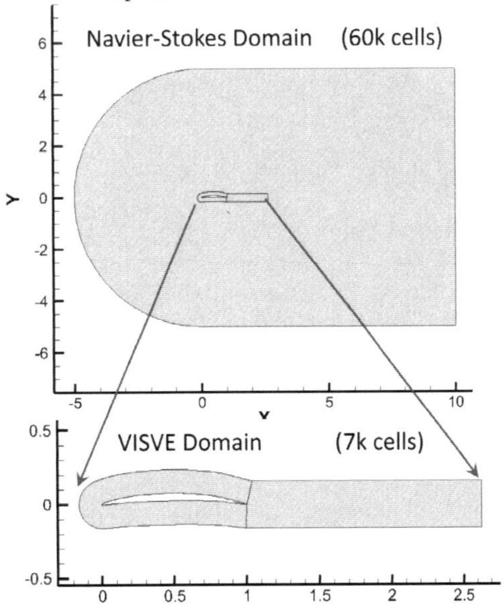

Figure 1: Comparison of the computational domain between N-S and VISVE in the case of a 2D hydrofoil [1]

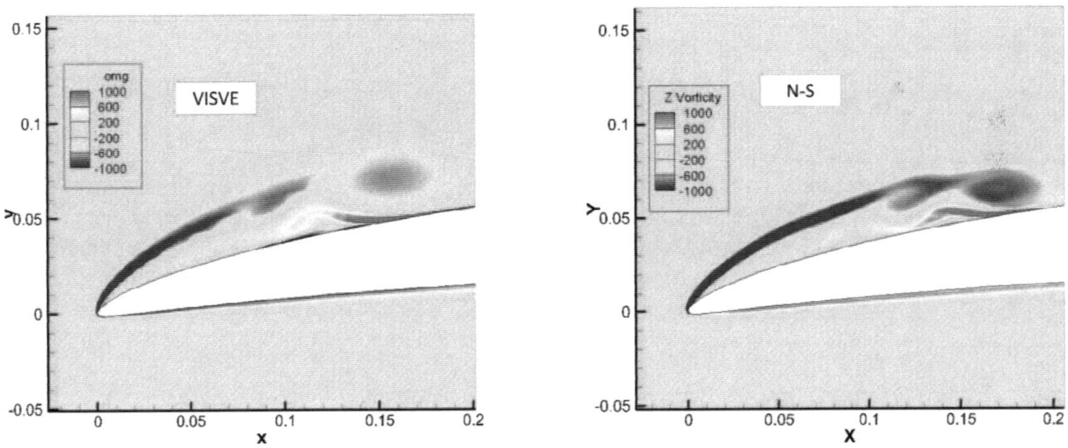

Figure 2: VISVE result of a 2D hydrofoil at AOA=10°, t = 2s comparing with RANS [1]

In the case of unidirectional flow around cylinder, we want to keep the compactness and efficiency of VISVE, at the same time, the calculation domain needs to cover all the regions where the vorticities concentrate. New gridding scheme should be developed for a more general geometry such like that for a cylinder flow. Most of the grids should be distributed close to the solid surface, where the vorticity generated. We also require sufficient number of cells in the wake region where the vortex shedding patterns are developed. In Wu's work (2016) [2], the following gridding scheme was developed, as shown in Figure 3.

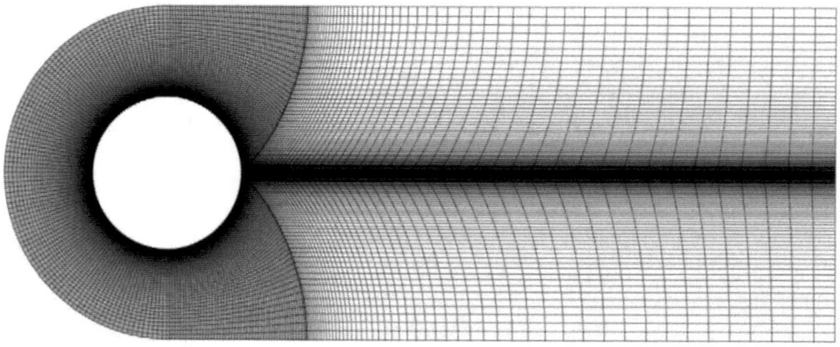

Figure 3: Computational domain and grid of VISVE in the case of unidirectional flow around cylinder [2]

From our previous studies, Wang (2015) [3] used RANS simulation of a cylinder subject to alternating flow to assess the validity of Morison's equation [4]. It is found that for Keulegan-Carpenter number less than or equal to 12, the results from Wang can match those provided by API and Sarpkaya [5] very well for both drag coefficient inertia coefficient. The computational domain, grid and boundary conditions are shown in Figure 4.

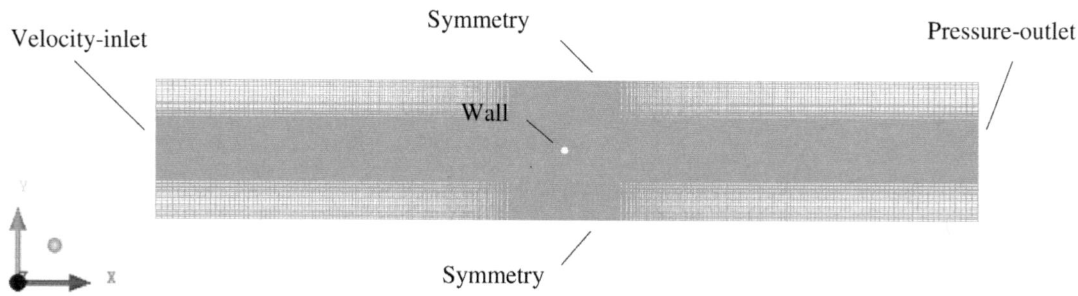

Figure 4: Computational domain, grid and boundary conditions for flow inside a flume via Fluent [3] [6]

In Wu's work, due to the curved contact surface of the front and back block, un-orthogonal mesh needs be used, which will lead to failure in the case of alternating flow, since the flow is changing more intensely near the wall. At this point, it is worthwhile to simplify the gridding scheme and use an orthogonal mesh with only one O block as is shown in Figure 5, which will be appropriate for both the unidirectional and alternating flow around cylinder. A similar mesh with the same first layer height and expansion ratio but a larger domain is generated by using Fluent's ICEM CFD.

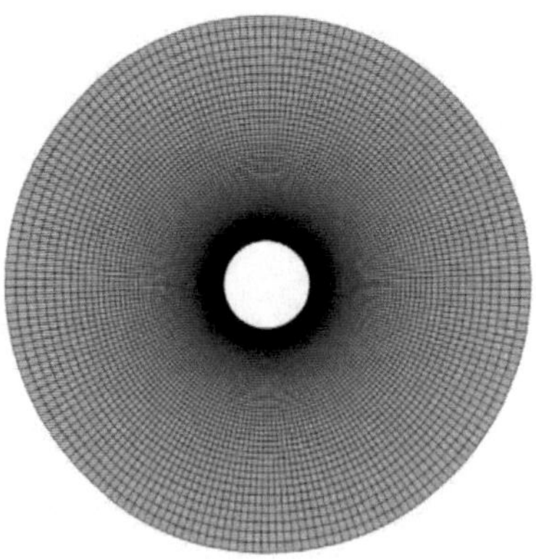

Figure 5: Computational domain and grid of VISVE used in this study

METHODOLOGY

1. Governing equation

For an incompressible viscous flow, the Navier-Stokes equations in 3D can be written as:

$$\frac{\partial \boldsymbol{q}}{\partial t} + (\boldsymbol{q} \cdot \nabla)\boldsymbol{q} = -\frac{\nabla p}{\rho} + \upsilon \nabla^2 \boldsymbol{q} \tag{1}$$

in which t is time, $\boldsymbol{q}$ denotes the velocity vector, p and ρ represent the pressure and density, respectively, and υ is the kinematic viscosity. By taking the curl of the above equation, we get the vorticity equation:

$$\frac{\partial \boldsymbol{\omega}}{\partial t} + (\boldsymbol{q} \cdot \nabla)\boldsymbol{\omega} = (\boldsymbol{\omega} \cdot \nabla)\boldsymbol{q} + \upsilon \nabla^2 \boldsymbol{\omega} \tag{2}$$

in which $\boldsymbol{\omega}$ denotes the vorticity vector, defined as the curl of the velocity vector $\boldsymbol{q}$.

It can be shown that in 2D, the vortex stretching term, $(\boldsymbol{\omega} \cdot \nabla)\boldsymbol{q}$, vanishes. The vorticity vector $\boldsymbol{\omega}$ is simplified as a scalar ω, and the vorticity equation becomes:

$$\frac{\partial \omega}{\partial t} + (\boldsymbol{q} \cdot \nabla)\omega = \upsilon \nabla^2 \omega \tag{3}$$

By introducing the Finite Volume Method into our governing equation and using Gauss divergence theorem, after discretizing the cell surface, we have:

$$A\frac{\partial \omega}{\partial t} + \sum_j (q_j \omega_j \Delta l_j) = \sum_j \left(\upsilon \frac{\partial \omega}{\partial n_j} \Delta l_j\right) \tag{4}$$

in which $q_j \omega_j \Delta l_j$ is called the convective term and $\upsilon \frac{\partial \omega}{\partial n_j} \Delta l_j$ is called the diffusive term.

Quadratic Upstream Interpolation for Convective Kinematics (QUICK) scheme by Leonard (1979) is implemented and the Alternating Direction Implicit (ADI) method by Peaceman and Rachford (1995) is used to solve the discretized vorticity equation.

2. Boundary conditions

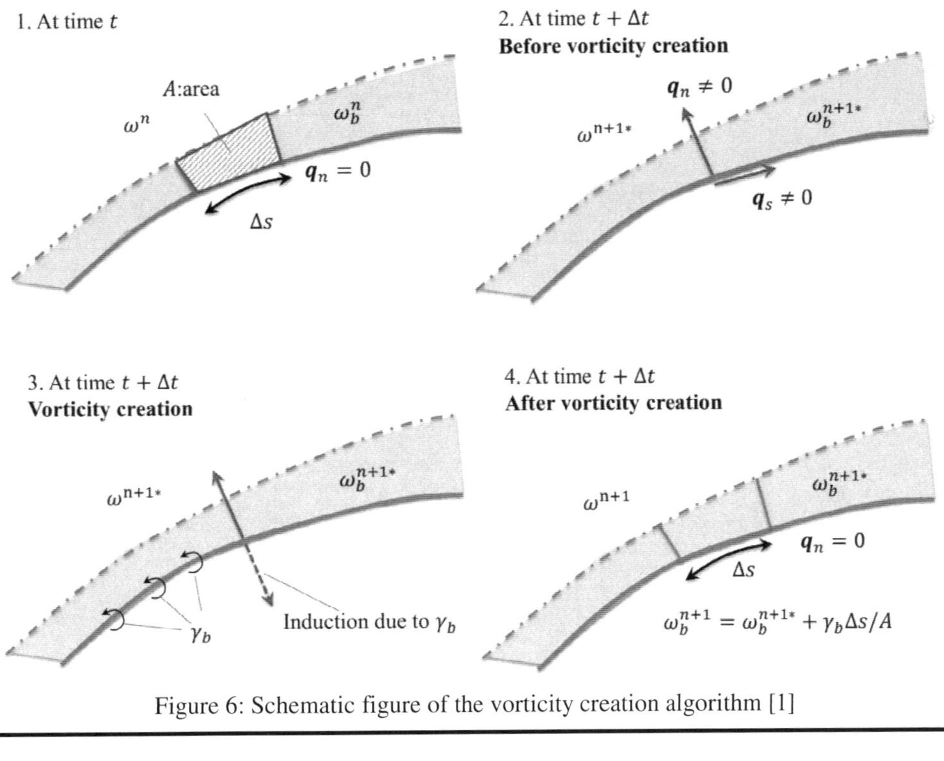

Figure 6: Schematic figure of the vorticity creation algorithm [1]

In order to understand the boundary conditions, we first need to know what happens at the boundary through our numerical treatment, as depicted in Figure 6.

1. At time t, the tangential velocity $\boldsymbol{q}_n$ is zero (non-penetrating) and the normal velocity $\boldsymbol{q}_s$ is also zero (non-slip). We have a vorticity distribution ω^n in the whole domain and we denote the vorticity in the cells next to the boundary as ω_b^n.

2. After solving the VISVE, we march to the next time level $\mathrm{t} + \Delta\mathrm{t}$, and we have ω_b^{n+1*}. However, velocities calculated from ω_b^{n+1*} using Biot-Savart law will not satisfy the boundary conditions on the wall.

3. A vorticity creation scheme based on the Boundary Element Method (BEM) is designed to eliminate $\boldsymbol{q}_n$ and $\boldsymbol{q}_s$ at $\mathrm{t} + \Delta\mathrm{t}$.

4. After assigning the newly created vorticity into cells in the first layer, the non-slip boundary condition is satisfied.

At the far field, the flow is considered as irrotational, which is consistent with external free stream. As long as we have a large enough domain, the concentrated vorticity will never reach the outer boundary. However, due to diffusion, vorticity may still penetrate the boundary, so the Neumann boundary condition is used at the boundary.

3. General solving strategy

Since velocity q appears in the governing equation, the first step of solving the VISVE is to calculate the corresponding velocity field q of a given vorticity distribution. We start the flow with the potential flow, which is a good guess for the VISVE. The flowchart in Figure 7 shows the solving strategy of VISVE. In the case of alternating flow, inflow velocity is no longer uniform, so it needs to be updated after each time step.

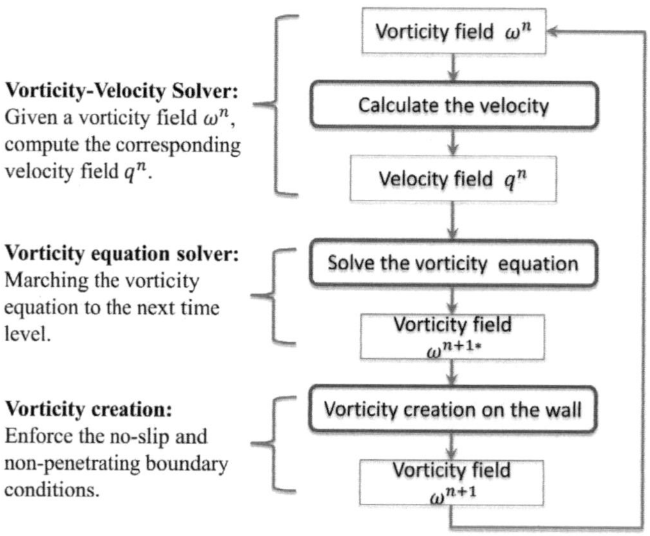

Figure 7: Flow chart of general procedure of VISVE [1]

4. Post-processing

4.1 Pressure, drag and lift coefficient

Due to the non-slip boundary condition, $\frac{\partial q}{\partial t}$ and convection term $(\boldsymbol{q} \cdot \nabla)\boldsymbol{q}$ vanish on the wall. By introducing a local coordinate system $s - n$ on the cylinder surface, s indicates the tangential direction and n is the normal direction, we have:

$$\frac{\partial p}{\partial s} = -\mu \frac{\partial \omega}{\partial n} \tag{5}$$

where μ is the dynamic viscosity.

By assuming that the stagnation point stays at the leading edge of the cylinder, the pressure can be easily determined using Bernoulli Equation. With the stagnation pressure, surface pressure along the boundary can be calculated through integration.

To find the drag and lift force, we still need the viscous shear stress τ on the surface, which is usually written as:

$$\tau = -\mu \frac{\partial u}{\partial n} \tag{6}$$

in which u is the tangential velocity parallel to the surface. The above equation can be simplified according to the definition of vorticity ω , which results into (given that $u = 0$ on the surface of the cylinder):

$$\tau = \mu\omega \tag{7}$$

With the surface pressure and viscous shear stress, the drag and lift coefficient can be calculated as follows:

$$C_d = \frac{R \oint_{bs} \left(\mu R \frac{\partial \omega}{\partial n} - \mu\omega\right) \sin\theta \, d\theta}{\rho U_\infty^2 R} \tag{8}$$

$$C_l = \frac{R \oint_{bs} \left(\mu R \frac{\partial \omega}{\partial n} - \mu\omega\right) \cos\theta \, d\theta}{\rho U_\infty^2 R} \tag{9}$$

where R is the radius of the cylinder, U_∞ is the inflow velocity, ρ is the density of the fluid, and θ is the angle of a point on the cylinder with respect to the axis in the direction of the flow.

4.2 Morison's equation

The Morison's Equation, equation (10), is a semi-empirical equation which is widely used in industry to estimate the inline force F on a body in oscillatory flow with velocity $U = U_m \cos(\omega t)$, where U_m is the maximum velocity, ω here is the angular frequency, $\omega = \frac{2\pi}{T}$, with T being the period of the flow.

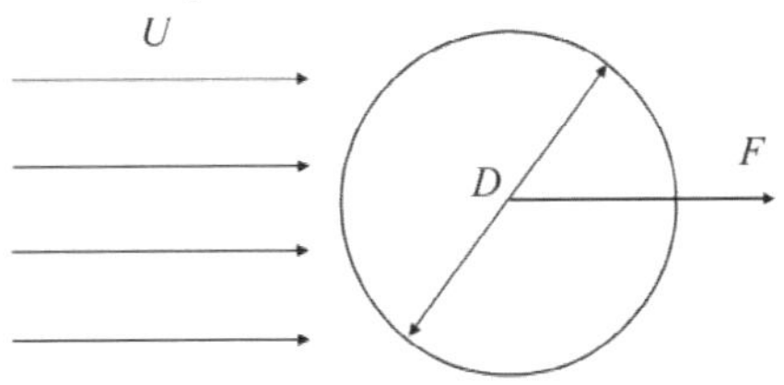

Figure 8: Cylinder is subject to an oscillatory flow

$$F = \frac{1}{2}\rho C_D D|U|U + \frac{1}{2}\rho\pi D^2 C_M \frac{dU}{dt} \tag{10}$$

The first term on the RHS of the above equation is call the drag force (viscous force) F_D and second called inertia force F_i. C_D is also called the drag coefficient (Note: This C_D has different meaning than the C_d mentioned in the post-processing part of VISVE, to distinguish them, C_d is now referred to as C_x) and C_M is the inertia coefficient.

There are two important numbers: Reynolds number ($\mathrm{Re} = \frac{U_m D}{\upsilon}$) and Keulegan-Carpenter number ($\mathrm{KC} = \frac{U_m T}{D}$).

4.3 Evaluation of C_D and C_M

The time history of C_x can be obtained directly from VISVE and from Fluent. From Morison's equation, C_x can be given as

$$C_x = C_D \cos(\omega t)\,|\cos(\omega t)| - \frac{\omega\pi D C_m \sin(\omega t)}{2U_m} \tag{11}$$

Then the integral of C_x is

$$\int_0^T C_x \sin(\omega t)\, dt = \int_0^T C_D \sin(\omega t)\cos(\omega t)\,|\cos(\omega t)| - \frac{\omega\pi D C_m \sin^2(\omega t)}{2U_m} dt \tag{12}$$

From the above integral, C_M can be calculated as

$$C_M = -\frac{2U_m}{\pi^2 D}\int_0^T C_x \sin(\omega t)\, dt \tag{13}$$

Then, the second integral is performed,

$$\int_0^T C_x\cos(\omega t)dt = \int_0^T C_D \cos^2(\omega t)\,|\cos(\omega t)| - \frac{\omega\pi D C_m \sin(\omega t)\cos(\omega t)}{2U_m} dt \tag{14}$$

From the second integral, C_D can be calculated as

$$C_D = \frac{\int_0^T C_x\cos(\omega t)dt}{\int_0^T \cos^2(\omega t)\,|\cos(\omega t)|dt} \tag{15}$$

The above integral method is used by Sarpkaya [4]. We will take the average of 5 periods to get C_D and C_M.

RESULTS

To compare the calculation time, VISVE and Fluent are run on one Stampede node of Texas Advanced Computing Center using 16 CPUs, and the parameters are shown in Table 1. VISVE is proven to be compact and efficient. Time step is set as $\Delta t = 0.001\, s$ in both VISVE and Fluent, and they are run for 2000 steps, so we get results for the first 2 seconds. CPU time is calculated simply by taking average of several cases. It should be noted that, to extract data from Fluent on the same points in VISVE, we defined 900 points in Fluent, outputting the velocity, vorticity and pressure from them, and that can take some time.

Table 1: CUP time comparison

	Total Cell Number	CPU time
VISVE	30,000	30 min
Fluent Laminar	100,000	5.5 h
Fluent RANS	100,000	5.5 h

1. Grid and domain independence studies

In Wu's work, VISVE is well validated by using domains with different sizes and different element numbers around the cylinder. It is found that a suitable and efficient domain size is around 2-4 diameters height with element number between 200-400. Based on these findings, for the new gridding scheme, we use 200, 300 and 400 elements on the cylinder to test the grid independence, and outer domain size diameter D=6 and D=8 to test the domain independence. We output the vorticity around the cylinder for each case, and the results, shown in Figure 9, match well, so it is reasonable to choose 300 elements on the cylinder. Since the domain size also does not affect the results significantly, it will be determined according to how far we want to track the vortices. The expansion ratio in the vertical direction is set as 1.05, which is the same as Wu [2]. For the Fluent runs, the convergence study can be found in Wang's MS thesis [7]. Detailed tests on the sensitivity of convergence of the solution with respect to time step is investigated and will be presented in Li's MS thesis [8].

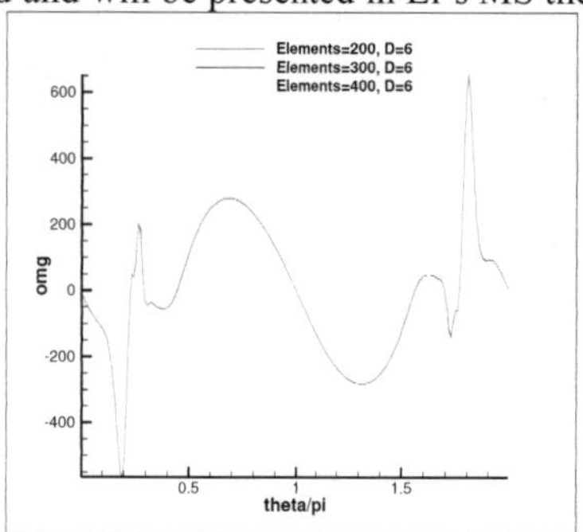

(a) Grid independence study

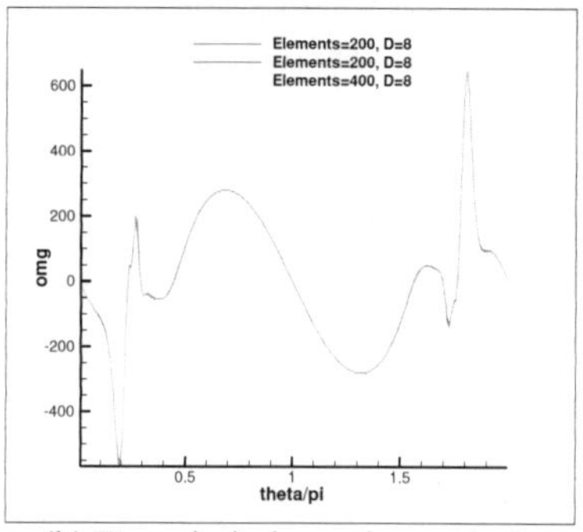

(b) Domain independence study

Figure 9: Grid and domain independence study: $Re = 9500, t = 1s$

2. Unidirectional flow

2.1 Vorticity contours

Since VISVE starts with the inviscid flow, it is best if we also initialize the flow field in Fluent with the inviscid result, which can be done easily through custom field function in Fluent. Time is made dimensionless by using the diameter of the cylinder. Vorticity contours predicted from VISVE and Fluent are shown in Figures 10-15.

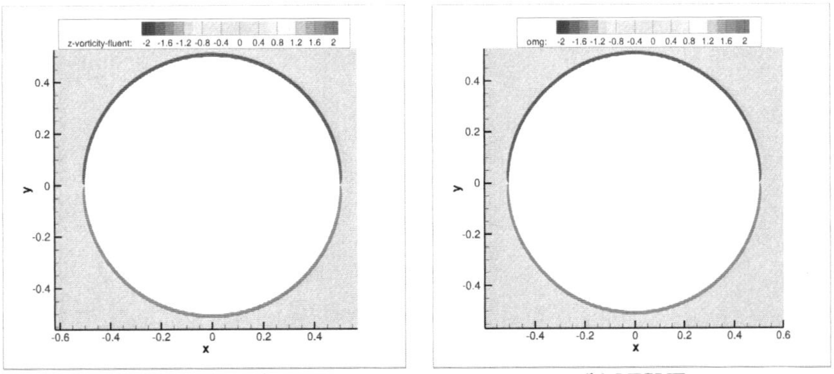

(a) Fluent-Laminar (b) VISVE

Figure 10: Vorticity contours: $Re = 1000, t = 0.01s$

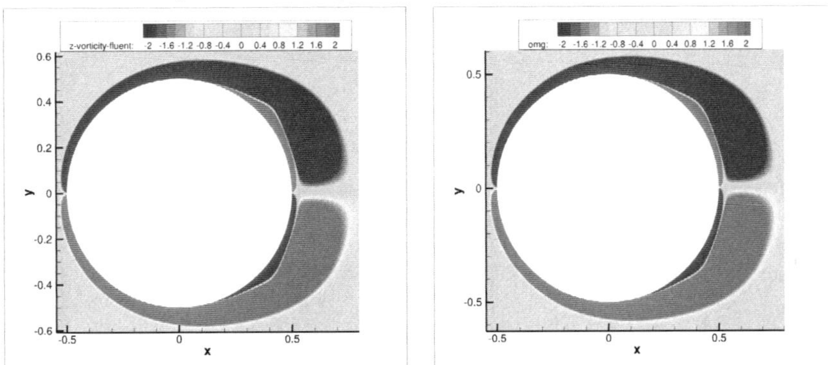

(a) Fluent-Laminar (b) VISVE

Figure 11: Vorticity contours: $Re = 1000, t = 1s$

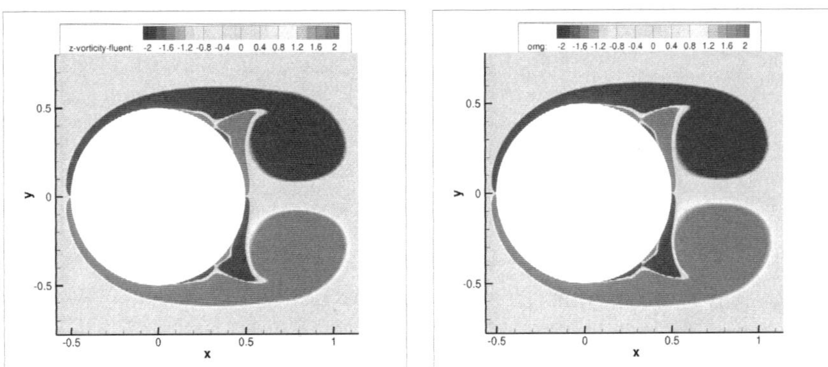

(a) Fluent-Laminar (b) VISVE

Figure 12: Vorticity contours: $Re = 1000, t = 2s$

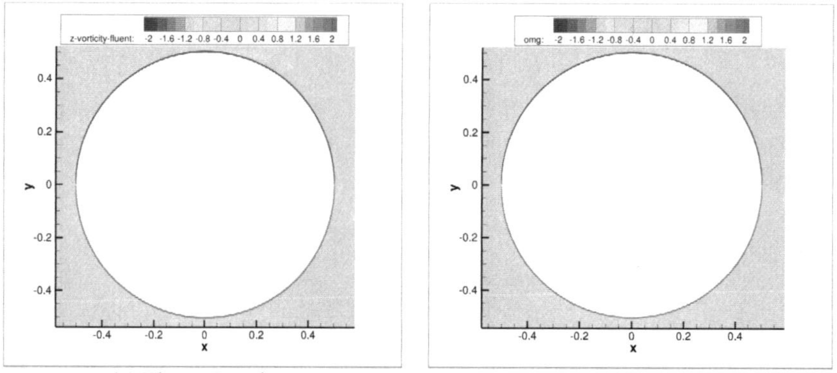

(a) Fluent-Laminar (b) VISVE

Figure 13: Vorticity contours: $Re = 9500, t = 0.01s$

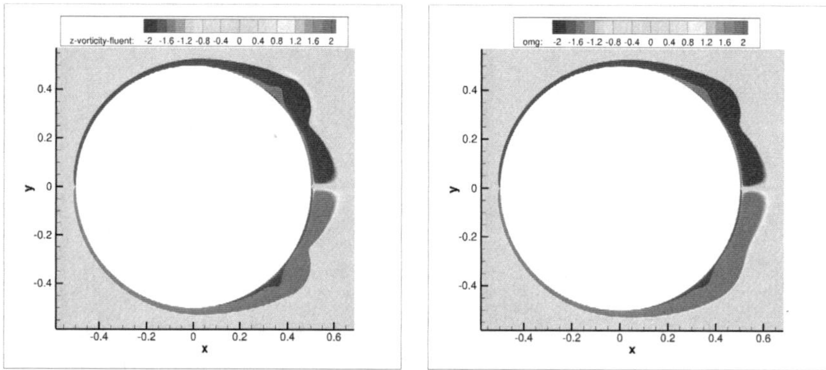

(a) Fluent-Laminar (b) VISVE

Figure 14: Vorticity contours: $Re = 9500, t = 1s$

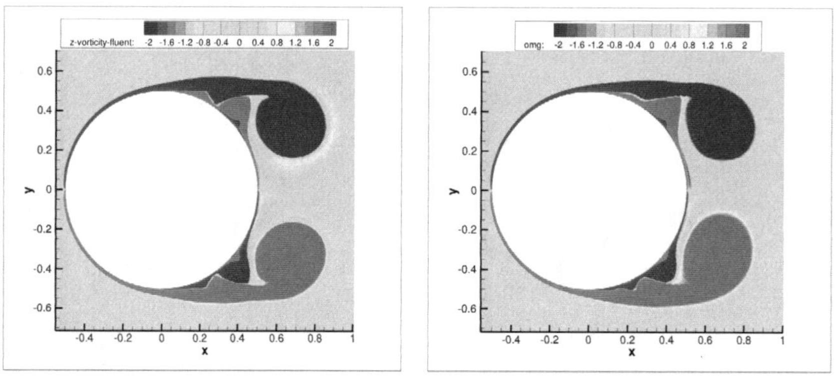

(a) Fluent-Laminar (b) VISVE

Figure 15: Vorticity contours: $Re = 9500, t = 2s$

2.2 Velocity distribution

Velocity magnitude along 8 evenly distributed radial directions j1-8 are extracted from VISVE and compared with those from Fluent, and shown in Figures 16-18. The first direction, j1, starts from the right-most point of the cylinder. In the x-axis, x means the distance between the point and the center of the cylinder. The 8 directions are numbered in the clock-wise direction.

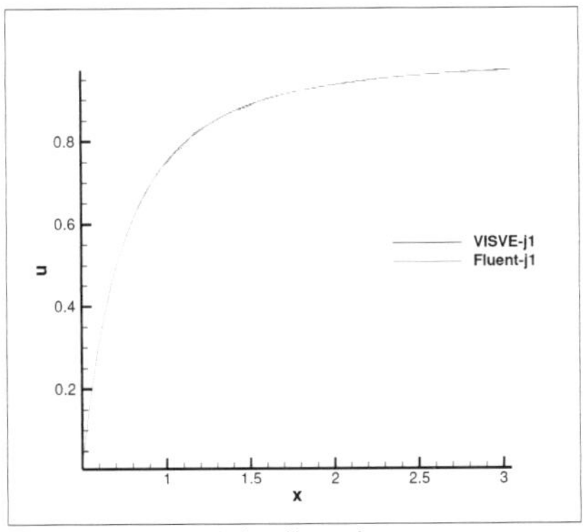

(a) j1 direction

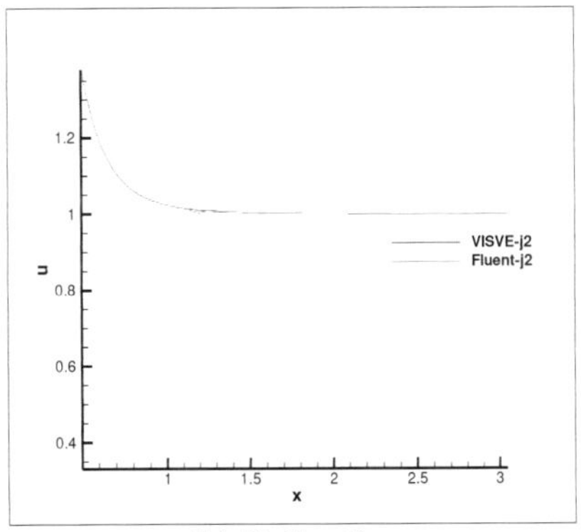

(b) j2 direction

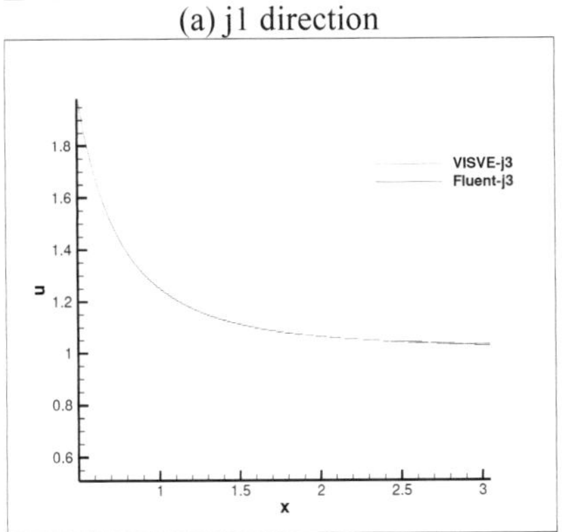

(c) j3 direction

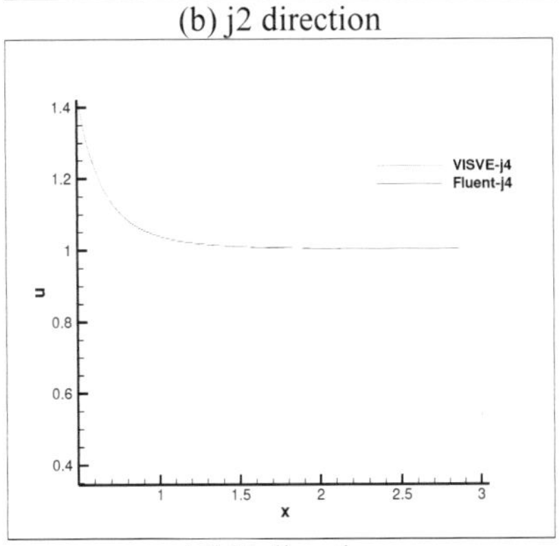

(d) j4 direction

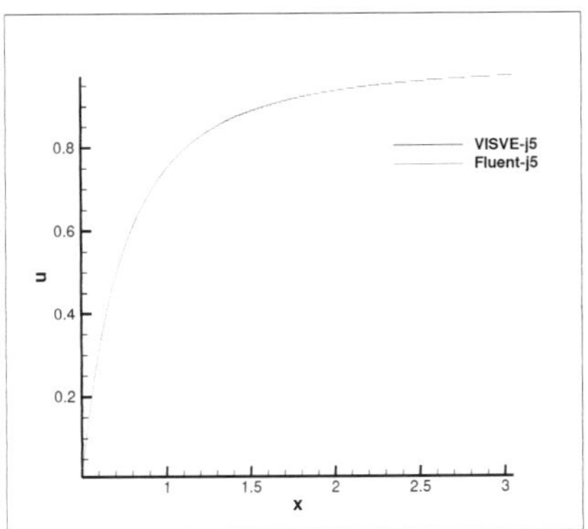

(e) j5 direction

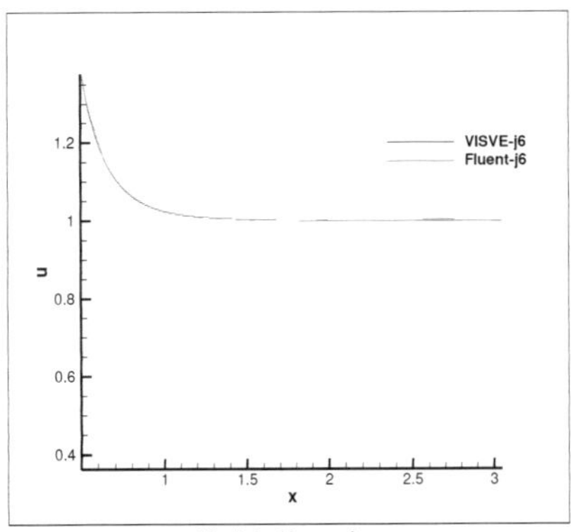

(f) j6 direction

Figure 16: Velocity magnitude along 8 vertical directions: $Re = 9500, t = 0.01s$

Velocity magnitude in j7 and j8 directions are very close to those in j3 and j2 direction, so they are not shown. We can see from Figure 16, at the beginning, the velocity profiles predicted by VISVE are very close to those from Fluent.

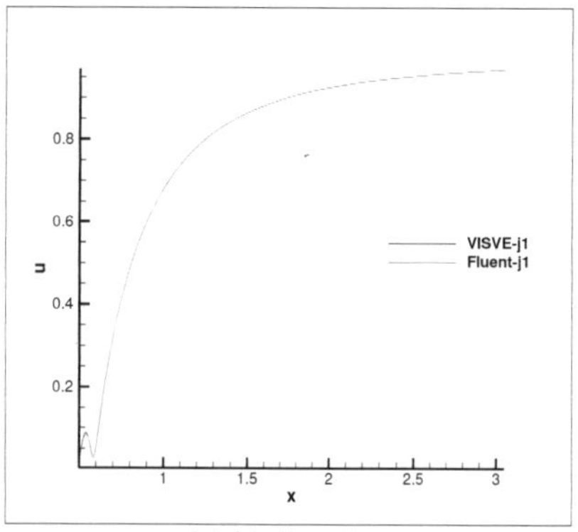

(a) j1 direction

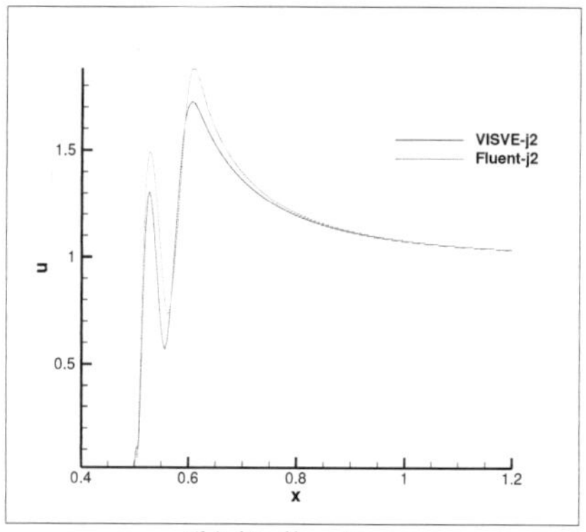

(b) j2 direction

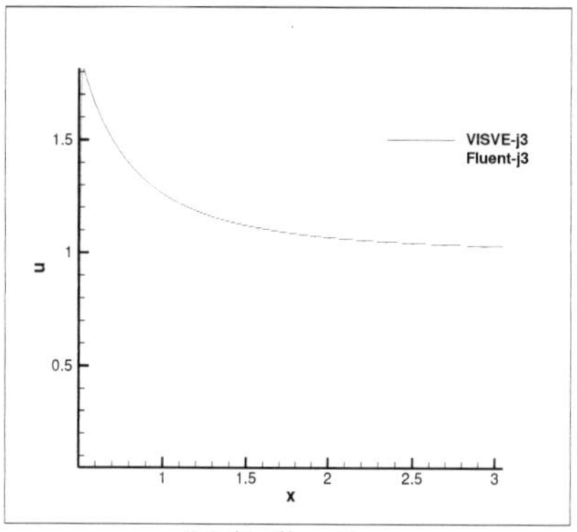

(c) j3 direction

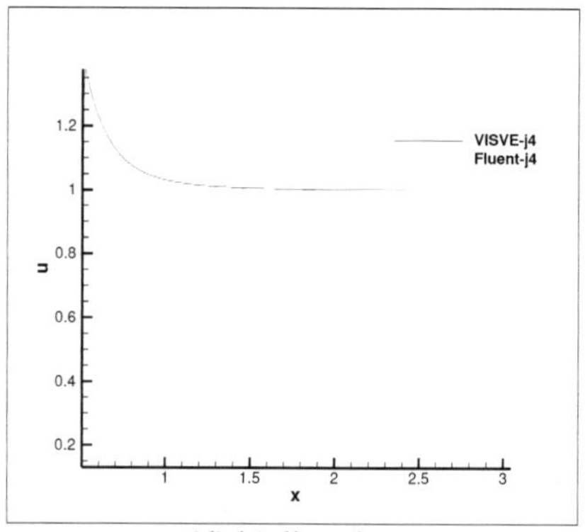

(d) j4 direction

Figure 17: Velocity magnitude along 8 vertical directions: $Re = 9500, t = 1s$

At $t = 1s$, the velocities shown in Figure 17 still match very well, except for the j2 direction.

At $t = 2s$, the velocity predicted by VISVE in j2 direction, shown in Figure 18, are deviating from those from Fluent.

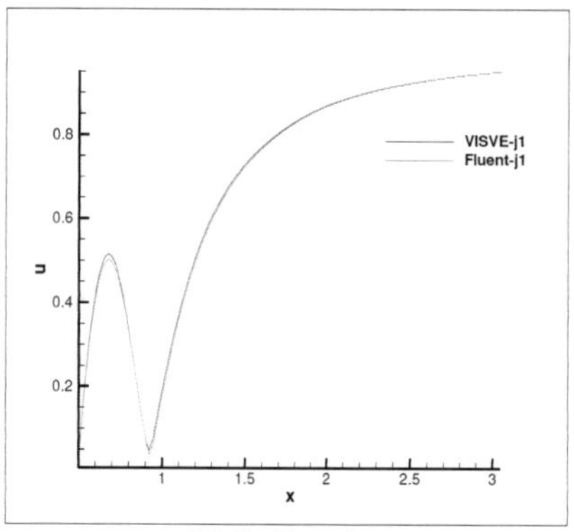

(a) j1 direction

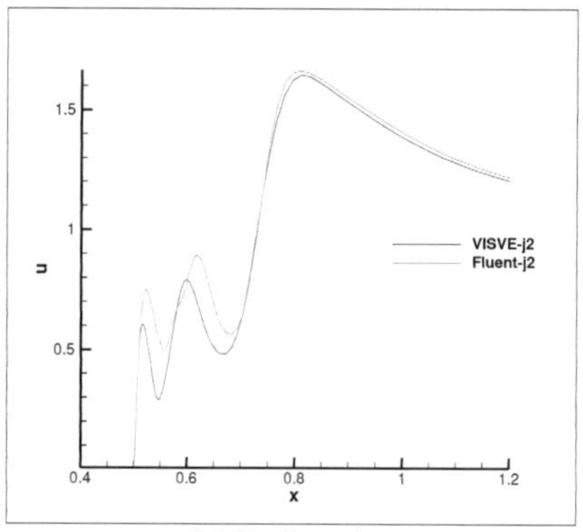

(b) j2 direction

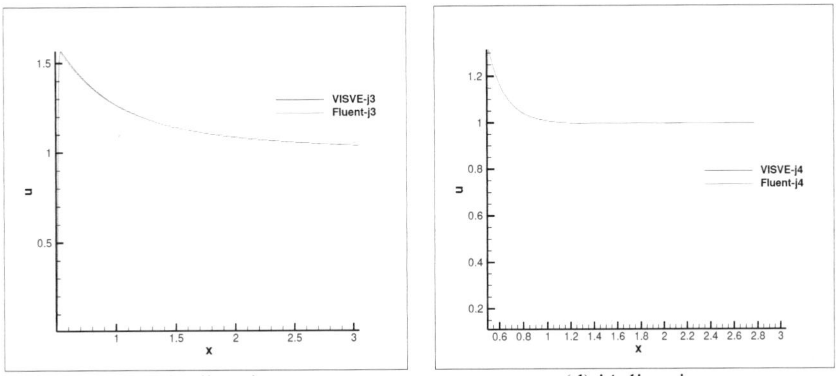

(c) j3 direction (d) j4 direction

Figure 18: Velocity magnitude along 8 vertical directions: $Re = 9500, t = 2s$

2.3 Pressure distribution

The predicted pressures from VISVE and Fluent are shown in Figures 19 and 20. Since we assume a stagnation point at the front of the cylinder, which is not always the case in Fluent, the magnitude of pressure may be slightly translated from Fluent. However, the most evident difference is the asymmetry of pressures on the top and bottom of the cylinder. Pressures on the two sides predicted by Fluent almost overlap with each other.

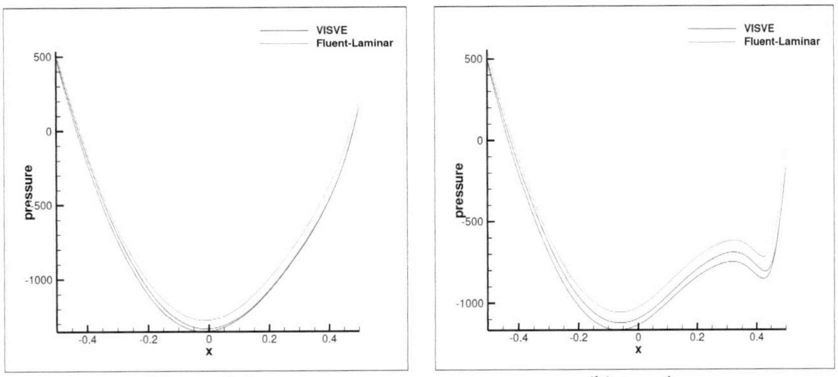

(a) $t = 0.5s$ (b) $t = 1s$

Figure 19: Pressure distribution: $Re = 1000$

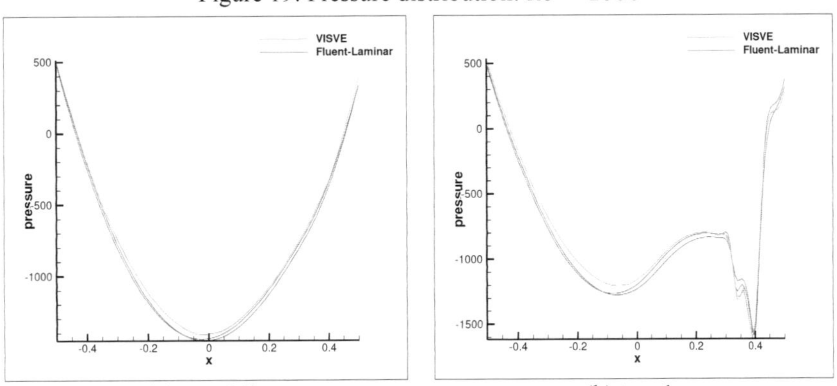

(a) $t = 0.5s$ (b) $t = 1s$

Figure 20: Pressure distribution: $Re = 9500$

2.4 Drag and lift coefficients

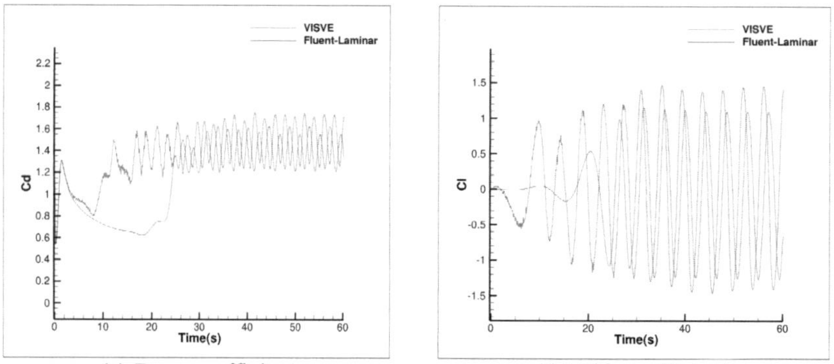

(a) Drag coefficient (b) Lift coefficient

Figure 21: Drag and lift coefficients: $Re = 1000$

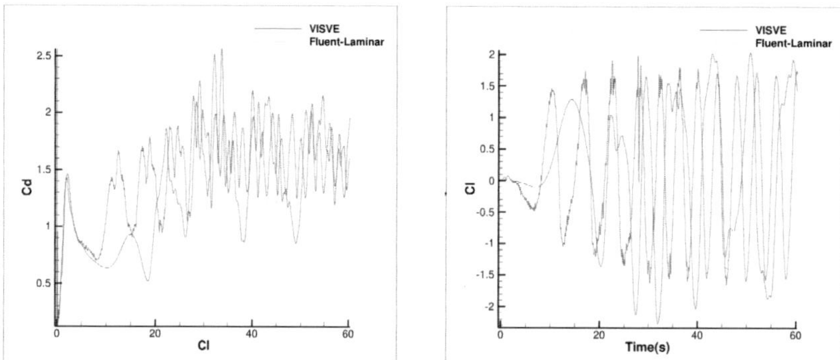

(a) Drag coefficient (b) Lift coefficient

Figure 22: Drag and lift coefficients: $Re = 9500$

The predicted drag and lift coefficients are shown in Figures 21 and 22. At the beginning, the drag coefficient predicted by VISVE is very close to that of Fluent. As time goes on, the VIV (Vortex Induced Vibration) starts earlier in VISVE. Note that the mean drags from VISVE and Fluent at later times are close to each other, while the instantaneous values differ. Given that the drag coefficient from VISVE at the first few seconds is very close to that from Fluent, which will be useful in the case of alternating flow.

3. Alternating flow

In VISVE and Fluent the alternating flow is imposed at the outer boundary of the domain. In Fluent, we used the domain shown in Figure 4, and the RANS turbulent flow was solved by using the k-omega SST model.

3.1 Vorticity contours

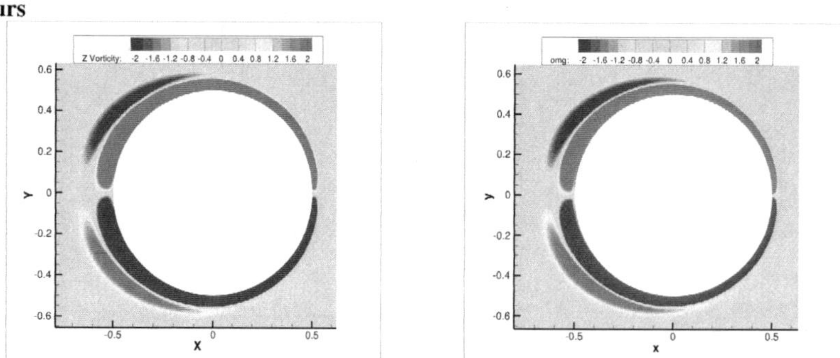

(a) Fluent-kωsst (b) VISVE

Figure 23: Vorticity contours: KC = 2, $Re = 994, t = 1s$

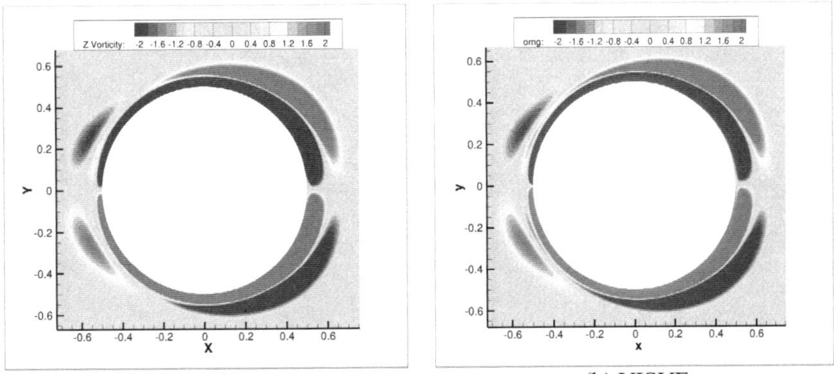

(a) Fluent-Laminar (b) VISVE

Figure 24: Vorticity contours: KC $= 2,\ Re = 994, t = 2s$

The predicted vorticities from VISVE and Fluent are shown in Figures 23 and 24. For KC=2, the vorticities predicted by VISVE are very to those by Fluent.

3.2 Drag coefficient C_x

Following the experiments conducted by Sarpkaya [5], we run the following cases. β is the frequency parameter, defined as $\beta = \frac{Re}{KC}$. The time is made dimensionless by period, which is the x-axis. The drag coefficients are shown in Figures 25-27.

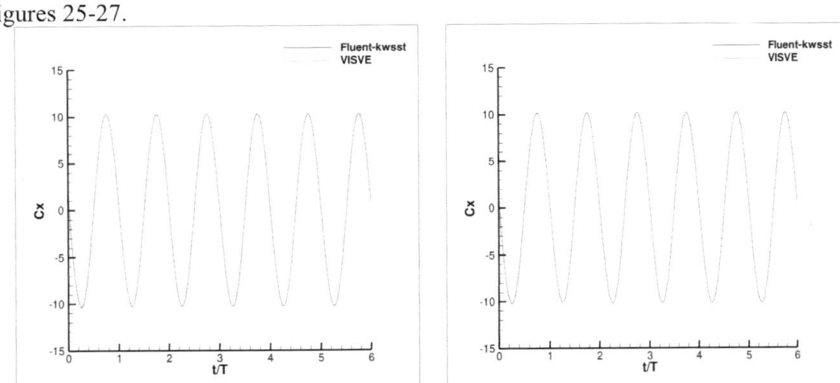

(a) $Re = 994, \beta = 497$ (b) $Re = 1568, \beta = 784$

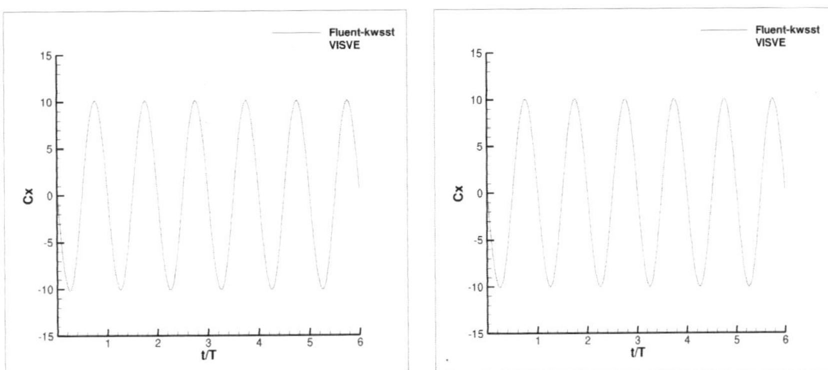

(c) $Re = 2214, \beta = 1107$ (d) $Re = 3970, \beta = 1985$

Figure 25: Drag coefficient: $KC = 2$

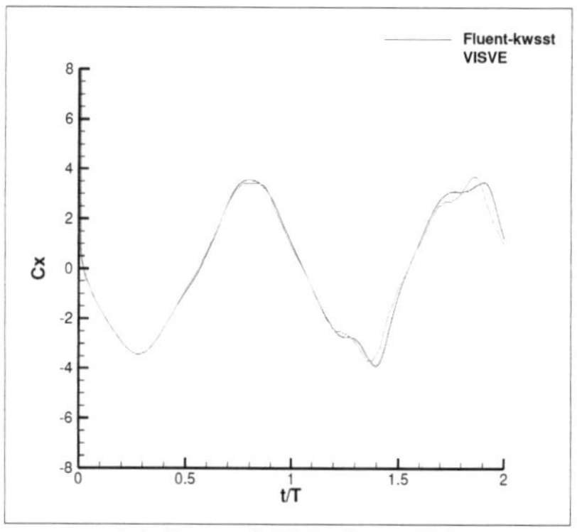

(a) $Re = 2982, \beta = 497$

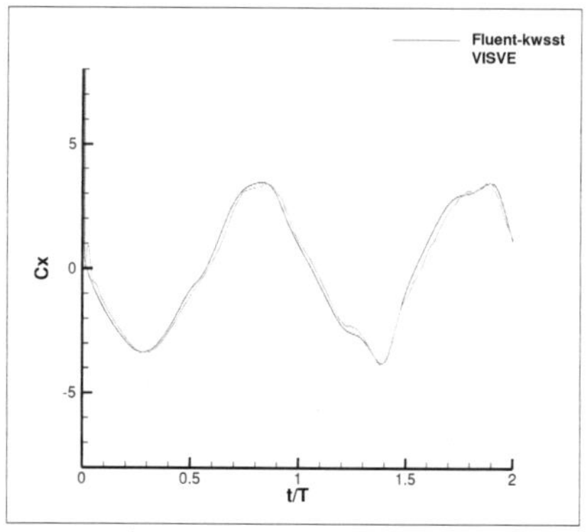

(b) $Re = 4704, \beta = 784$

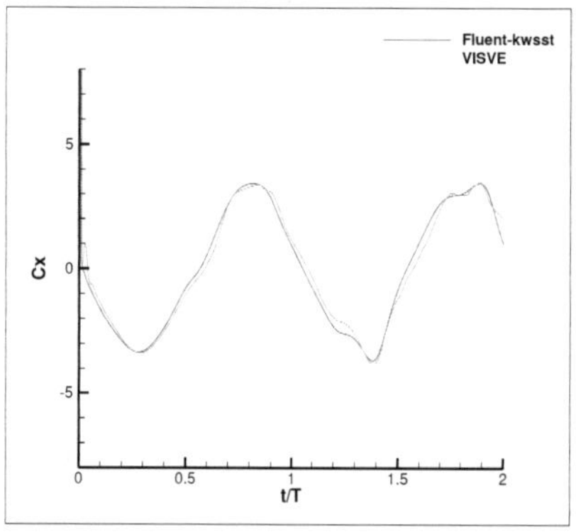

(c) $Re = 6642, \beta = 1107$

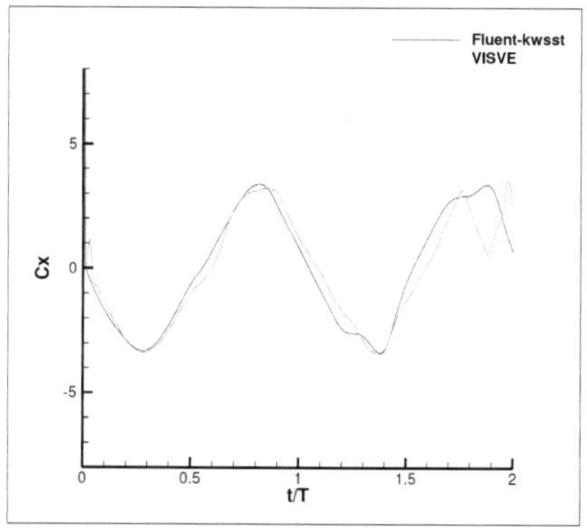

(d) $Re = 11910, \beta = 1985$

Figure 26: Drag coefficient: $KC = 6$

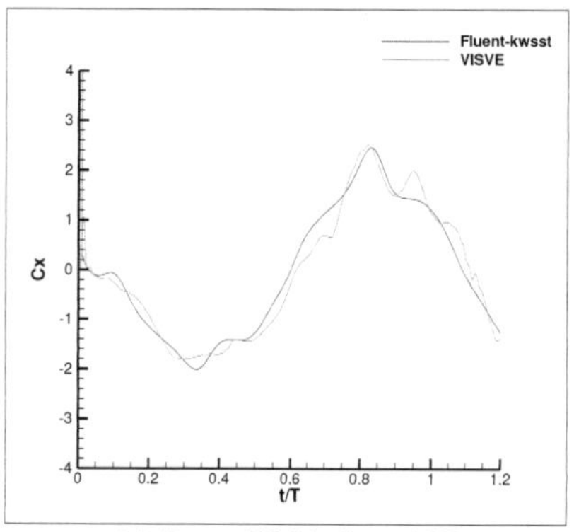

(a) $Re = 5964, \beta = 497$

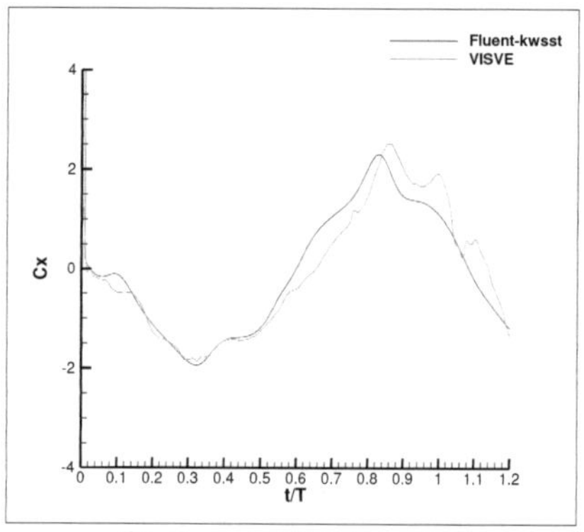

(b) $Re = 9408, \beta = 784$

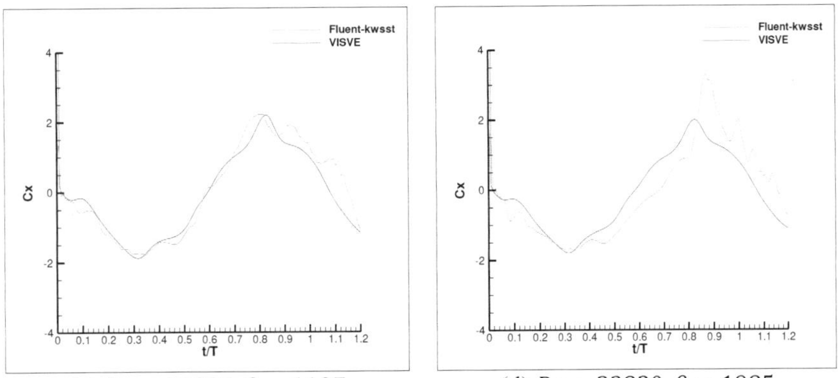

(c) $Re = 13284, \beta = 1107$ (d) $Re = 23820, \beta = 1985$

Figure 27: Drag coefficient: $KC = 12$

As KC gets larger, Re also gets larger, and the drag coefficient from VISVE starts to fluctuate between adjacent time steps and becomes wavy. The original results are processed by a small MATLAB code to eliminate the noise. The same thing was done when producing the drag and lift coefficients in the unidirectional case, to get a relatively smooth curve.

3.3 Evaluation of C_D and C_M

In Table 2, only results at $KC = 2$ are compared. When KC gets larger, the period of the alternating flow gets larger, which means the amount of calculation time will increase if we want to keep the same total number of period for the evaluation of C_D and C_M.

Table 2: Comparison of C_D and C_M for KC=2

		Fluent	VISVE			Fluent	VISVE
$Re = 994$	C_D	0.85	0.84	$Re = 2214$	C_D	0.63	0.63
	C_M	2.08	2.08		C_M	2.05	2.05
$Re = 1568$	C_D	0.71	0.71	$Re = 3970$	C_D	0.52	0.52
	C_M	2.06	2.06		C_M	2.03	2.03

By substituting C_D and C_M in Morison's equation, it will give the inline force with respect to time. Since for $KC = 2$, the predictions of Morison's equation are very good, only the case with $Re = 994$ is shown in Figure 28.

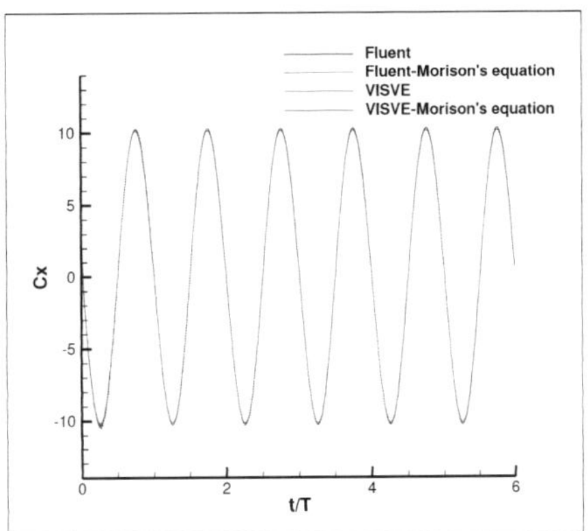

Figure 28: Comparison of C_x: $KC = 2, Re = 994$

CONCLUSIONS

At first, the gridding scheme in VISVE was changed to one O block to deal with both the unidirectional flow and alternating flow around cylinder. The pressure model was carefully examined, and a second order extrapolation method to get the derivative of vorticity normal to the cylinder surface was used to improve the precision of pressure calculation. By following this lead, the boundary condition involving the diffusive flux was also fixed.

Then, the unidirectional flow case was run in VISVE and Fluent to compare the vorticity, velocity and drag and lift coefficients. It is found that, at the beginning, velocity, vorticity and drag and lift coefficients predicted by VISVE are in good agreement with those of Fluent. As time goes on, results from VISVE starts to deviate from those from Fluent. However, it is worthwhile to note that they are qualitatively similar and the magnitude of mean drags are close. The differences are to be justified through experimental results.

Finally, in the case of alternating flow around cylinder, inline forces predicted by VISVE when KC=2 is found to be satisfying and Morison's equation is found to be valid. Since the inflow velocity is updated each time step, the stagnation point will move with time, a better pressure model for VISVE is needed. In this study, KC number is limited to 2, 6 and 12, while it can reach up to 100 in the real world. However, larger KC means longer calculation time, which is beyond current capabilities of VISVE. In the future, turbulence model will be incorporated in VISVE in addition to other improvements in the overall scheme.

ACKNOWLEDGEMENTS

We wish to thank the people who helped make this work possible: the members of Ocean Engineering Group (OEG), especially, Ms. Yunjun Wu and Dr. Ye Tian, former graduate students. We also wish to thank Ms. Lu Xing and Mr. Chunlin Wu for their efforts in improving VISVE.

REFERENCES

[1] Ye Tian. Leading Edge Vortex Modeling and its Effect on Propulsor Performance. PhD thesis, Ocean Engineering Group, The University of Texas at Austin, 2014

[2] Yunjun Wu. A Vorticity Based Model and its Application to Flow around an Impulsively Started Cylinder. MS thesis, Ocean Engineering Group, The University of Texas at Austin, 2016

[3] Guangyao Wang, S. A. Kinnas, Ye Tian. Uni-directional and Alternating Flow around 2-D Rigid Cylinder - Experimental and Computational Study, 21st SNAME Symposium, 2016

[4] Morison, J. R., O'Brien, M. P., Johnson, J. W., Schaaf, S. A., The Force Exerted by Surface Waves on Piles, Petroleum Transactions (American Institute of Mining Engineers) 189: 149–154, 1950

[5] Turgut Sarpkaya, In-line and transverse Forces on Cylinders in Oscillatory Flow at High Reynolds Numbers, Offshore Technology Conference, OTC 2533, 1976

[6] ANSYS, Inc., Fluent Documentation, Release 17.1, www.ansys.com/products/fluids/ansys-fluent, 2016

[7] Guangyao Wang, PIV Experiment and CFD Simulation of Flow around Cylinder, MS thesis, Ocean Engineering Group, The University of Texas at Austin, 2015

[8] Zhihao Li, VISVE, Applied to Unidirectional and Alternating Flow around a Cylinder – Improved Results (unfinished), MS thesis, Ocean Engineering Group, The University of Texas at Austin, May 2017

SNAME

THE INTERNATIONAL COMMUNITY
FOR MARITIME AND OCEAN PROFESSIONALS

ULTIMATE STRENGTH ASSESSMENT FOR CORRODED SUBSEA FLOWLINE

GWO-ANG CHANG
SENIOR MANAGING PRINCIPAL ENGINEER, AMERICAN BUREAU OF SHIPPING

MENG-LUNG LIU
SENIOR ENGIEER, AMERICAN BUREAU OF SHIPPING

Proceedings of the 22nd Offshore Symposium, February 2017, Houston, Texas
Texas Section of the Society of Naval Architects and Marine Engineers

ABSTRACT

The latest American Petroleum Institute Recommended Practice API RP1111, 2015 recommends repair capability for subsea flowlines and strength assessment for damaged flowlines before and during repair. In this paper, ultimate strength assessment for corroded pipe is carried out for two limit states. One is the collapse limit state resulting from external overpressure. Another is the buckling/collapsed limit state resulting from combined bending and external overpressure. A large range of corroded conditions as well as D/t (diameter to thickness) ratios are selected for the assessment including lower D/t ratios.

The assessment was carried out using Finite Element Analysis (FEA). In the FEA, ultimate capacities of both intact and corroded pipes subject to bending and/or external pressure are captured. In the Finite Element Modeling (FEM), a perfect elastic-plastic stress-strain relationship is selected to simulate the pipe's nonlinear material properties. An initial pipe ovality imperfection is also included in the FEM.

The captured ultimate capacity from the FEA is compared with ultimate strength results calculated from equations of the two limit states given in API RP 1111, 2015, API STD 2RD, 2013 and ABS Subsea Pipeline Guide, 2014. The purpose of comparison is to examine the applicability of these equations to lower D/t ratio in both intact and corroded conditions. For cases that the limit state equations are not applicable, the FEA method introduced in this paper can be applied as an alternative approach for the ultimate strength assessment of subsea flowline.

Keywords: Subsea, Corrosion Flowline, Ultimate Strength

INTRODUCTION

In the latest American Petroleum Institute Recommended Practice API RP1111, 2015 (API RP1111), repair capability for subsea flowlines is recommended, which leads to the strength assessment for damaged flowlines before and during repair. However, in the API RP 1111, there is no strength assessment criteria for damaged pipe. *The ABS Guide for Building and Classing Subsea Pipeline Systems, 2014* (ABS Subsea Pipeline Guide) addresses strength assessment criteria for corrosion damaged pipe as well as for pipe in intact condition.

Corrosion damage can occur at the pipe's inner side, outer side, or both sides. In this paper, ultimate strength assessment for inner corroded pipe is chosen for two limit states. One is the collapse limit state resulting from external overpressure. The other is the buckling and collapsed limit state resulting from combined bending and external overpressure. A large range of corroded conditions as well as D/t ratios are selected for the assessment including lower D/t ratios.

The assessment was carried out using finite element analysis (FEA). In the FEA, the ultimate capacity of inner corroded pipes subject to bending and/or external pressure are captured. In the finite element modeling (FEM), a perfect elastic-plastic stress-strain curve is selected to simulate the pipe's nonlinear material properties. An initial pipe ovality imperfection is also included in the FEM.

The captured ultimate capacity from the FEA is compared with ultimate strength results calculated from equations of the two limit states given in the API RP1111, API STD2RD, 2013 (API STD2RD), and the ABS Subsea Pipeline Guide. The purpose of comparison is to examine the applicability of these equations to lower D/t ratios of in both intact and corroded conditions. For cases that the limit state equations are not applicable, the FEA method introduced in this paper can be applied as an alternative approach for the ultimate strength assessment of subsea flowline.

METHODOLOGY

Since the FEA method is used to verify the applicability of the two limit state equations given in the API RP1111, API STD2RD and ABS Subsea Pipeline Guide for lower D/t ratios, the accuracy of the FEA results from the method is to be verified first.

The FEA verification is carried out for pipes in both intact and corroded conditions. For pipe in intact condition, the FEA results are compared with those calculated from API RP1111, API STD2RD and ABS Subsea Pipeline Guide, for collapse capacity due to pure external pressure and for moment capacity due to pure bending. For pipe in corroded conditions, the FEA results are compared with those determined from experimental tests (Naoto et al. 2008) for collapse of corroded pipe under pure external pressure. The comparison demonstrated very good agreement, therefore, the same FEA method is applied to assess pipe ultimate strength of the two limit states for a selected range of D/t ratios for both intact and inner corroded conditions. Detailed description of the FEA method and the verification results can be found in the section of Finite Element Analysis Method of this paper.

The ultimate strength equations given in the API RP1111, API STD2RD and ABS Subsea Pipeline Guide of the two limit states for pipe in intact condition are expressed as follows:

1. Collapse limit state due to external pr

a) API RP1111:

$$p_c = \frac{p_{el} p_p}{\sqrt{p_{el}^2 + p_p^2}} \quad \text{: collapse pressure} \qquad (1)$$

where:

$$pe = \frac{2 \cdot E}{1 - v^2} \cdot \left(\frac{t}{D_o}\right)^3 \quad \text{: elastic buckling pressure} \qquad (2)$$

$$pp = SMYS \cdot \frac{2 \cdot t}{D_o} \quad \text{: yield pressure at collapse} \qquad (3)$$

D_o : pipe outer diameter

t : pipe wall thickness

E : modulus of elasticity

υ : Poison's ratio, and 0.3 is used.

SMYS : Specified Minimum Yield Strength

b) ABS Subsea pipeline Guide:

$$p_c^3 - p_{el} p_c^2 - \left[p_p^2 + p_{el} p_p f_0 \left(\frac{D}{t}\right) \right] p_c + p_{el} p_p^2 = 0 \qquad (4)$$

where:

f0 = (Dmax – Dmin)/D

D : pipe average diameter

P_{el} : the same as equation (2), except D is used to replace D_0

P_p : the same as equation (3), except D is used to replace D_0

2. Buckling and collapse limit state due to combined bending and external pressure

a) API STD2RD: with zero effective tension

$$\left| \frac{M}{M_p} \right| = \sqrt{1 - \left(\frac{p_e}{p_c}\right)^2} \qquad (5)$$

where:

M_p is the plastic moment and can be analytically expresses as:

$$M_p = \frac{SMYS}{6} \cdot \left(D_0^3 - (D_0 - 2t)^3\right) \tag{6}$$

$$M_p \approx SMYS \cdot (D_0 - t)^2 \cdot t \tag{7}$$

b) ABS Subsea Pipeline Guide: with zero applied tension

$$M = M_p \cdot k \cdot \cos\left[\frac{\pi}{2} \cdot \frac{\alpha \cdot p_e}{p_c \cdot k}\right] \tag{8}$$

where:

$$\alpha = \frac{\pi \cdot D^2}{4} \cdot \left|\frac{p_c}{F_l}\right| \tag{9}$$

$$F_l = \frac{\pi \cdot ((D+t)^2 - (D-t)^2)}{4} \cdot SMYS \tag{10}$$

$$k = \sqrt{1 - (1 - \alpha^2) \cdot \left(\frac{p_e}{p_c}\right)^2}$$

For pipe in corroded conditions, inner corrosion cases are considered in this paper. There is no ultimate strength criteria in API documents for pipe in corroded conditions. The ultimate strength equations given in the ABS Subsea Pipeline Guide for the two limit states in corroded conditions are expressed as follows:

1. Collapse limit state due to external pressure for inner corroded pipe:

$$p_c^3 - p_{e}p_c^2 - \left[p_p^2 + p_{e}p_p f_0\left(\frac{D}{t_{cor}}\right)\right]p_c + p_{e}p_p^2 = 0 \tag{11}$$

where:

Equation (11) is the same as equation (5), except

$$pe = \frac{2 \cdot E}{1 - v^2} \cdot \left(\frac{t_{cor}}{D}\right)^3 \quad \text{: elastic buckling pressure} \tag{12}$$

$$pp = SMYS \cdot \frac{2 \cdot t_{cor}}{D} \quad \text{: yield pressure at collapse} \tag{13}$$

t_{cor} : minimum remained wall thickness

2. Buckling and collapse limit state due to combined bending and external pressure for inner corroded pipe: with zero applied tension

$$M = M_p \cdot \{\sin(\psi)\cdot k + 0.5\cdot(1-k_2)\cdot\sin(\beta)\cdot\left(\alpha\cdot\frac{p_e}{p_c} - k\right)\} \tag{14}$$

where:

$$\psi = \frac{\pi + (1-k_1)\cdot\beta}{2} - \frac{\pi - (1-k_1)\cdot\beta}{2}\cdot\Delta \tag{15}$$

$$\Delta = \left(\frac{\alpha\cdot p_e}{p_c\cdot k}\right)$$

$$k_1 = \left(1-\frac{d}{t}\right)\cdot\left(1+\frac{d}{D}\right)$$

$$k_2 = \left(1-\frac{d}{t}\right)\cdot\left(1+\frac{d}{D}\right)^2$$

d : corrosion depth

β : half of corroded width in radian or degree

$$\alpha = \frac{\pi\cdot D^2}{4}\cdot\left|\frac{p_c}{F_l}\right| \tag{16}$$

Equation (16) is the same as equation (9), except that Pc is to be determined from equation (11) and Fl is to be determined by the following equation as:

$$F_l = SMYS\cdot\{\pi - (1-k_1)\beta\}\cdot D\cdot t \tag{17}$$

Please see Figure 1 for example of inner corroded pipe.

Figure 1: Example of inner corroded pipe

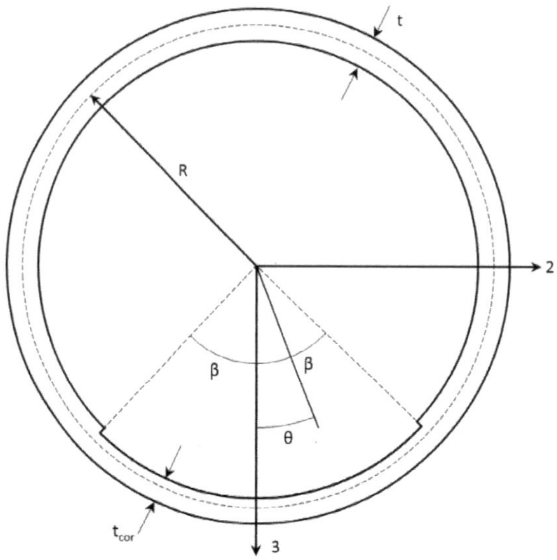

FINITE ELEMENT ANALYSIS METHOD

In the FEA, Abaqus software is used to determine pipe ultimate capacity for each of pipe D/t ratios in both intact and corroded conditions under three loading conditions: pure external pressure, pure bending moment, and combined external pressure and bending moment. For each of these three loading conditions, the ultimate capacity curve is established using the displacement control method, and the ultimate capacity is captured from the peak value of the curve. A typical ultimate capacity curve is shown in Figure 2. For pure external pressure or bending moment case, the applied load is step by step increased until the peak value is captured.

For combined external pressure and bending moment case, a pre-determined external pressure, which is between 0 and the ultimate collapse pressure, is applied first. Then, the bending moment is gradually increased step by step until the peak value of bending moment is captured. By repeatedly applying the pre-determined external pressures the corresponding ultimate capacity curve and peak bending moment can be captured for the combined external pressure and bending moment loading conditions.

Figure 2: Typical reaction bending moment and max load

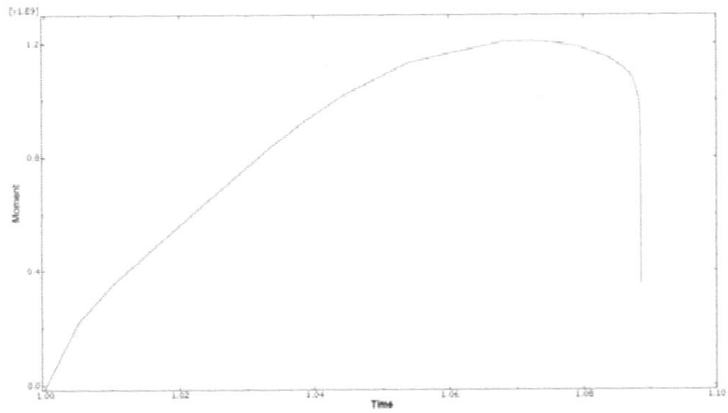

Finite Element Modeling:

For pipe in an intact condition, a half pipe model of 10 diameters long is created to take advantage of pipe symmetry. An initial ovality of 1.5% is also included in the modeling. For pipe of higher D/t ratio, 8 layers of C3D8R solid elements type or 4 layers of SC8R continuum shell element type are used along radial direction, while for pipe of lower D/t ratio, 8 layers of C3D8R element are used. About 32 elements are deployed along the circumferential direction of the half pipe.

For pipe with interior corrosion, designated meshes are taken out from the intact model to simulate corresponding material loss in corroded conditions. Figure 3 below illustrates a corrosion condition of 2β =20o and with 50% deduction of wall thickness.

Figure 3: mesh arrangement for corroded pipe

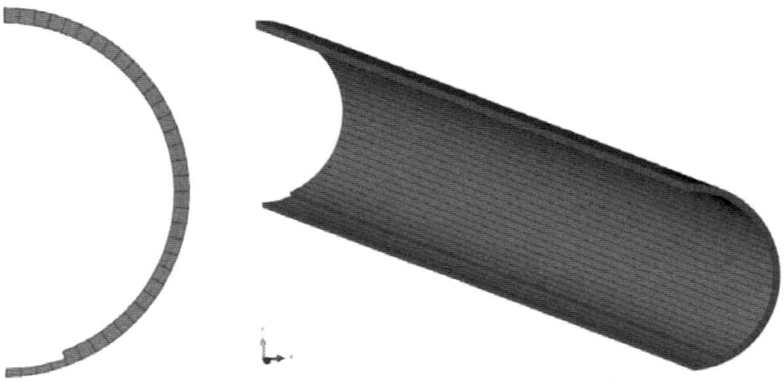

Material of X65 typically used for flowline is selected with parameters as shown in Table 1, and material nonlinearity of elastic-perfect plastic stress-strain relationship is adopted in the FEA modeling.

Table 1: material parameters

Young's modulus	Poisson's ratio	Yielding stress
200 GPa	0.3	448 MPa

FEA Verification:

Element type selection:

The case of D/t=25 with intact pipe for combined external pressure and bending moment case was selected to verify the selection of element type. In the FEA verification, both C3D8R solid element and CS8R continuum shell element types are modeled, and the results of ultimate capacity are shown in Figure 4. Visible in Figure 4, alignment between the methods was achieved with the largest deviation at one end of the curve by 2.63%. Since CS8R can effectively reduce calculation time, it was selected for cases of D/t > 25.

Figure 4: ultimate capacity curve comparison between solid and continuum shell elements

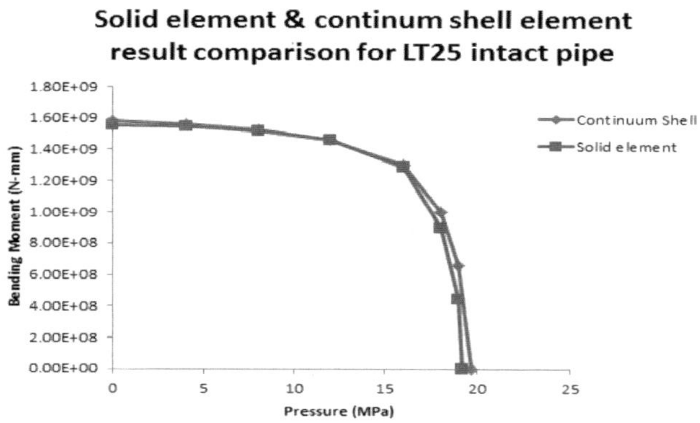

FEA verification for pipe in intact conditions:

For pipe in intact condition under pure bending moment load, FEA results were compared with analytical solution of fully plastic moment (equation 6). The results are shown in Figure 5 for D/t ratios from 5 to 45.

Figure 5: Bending moment ultimate capacity comparison

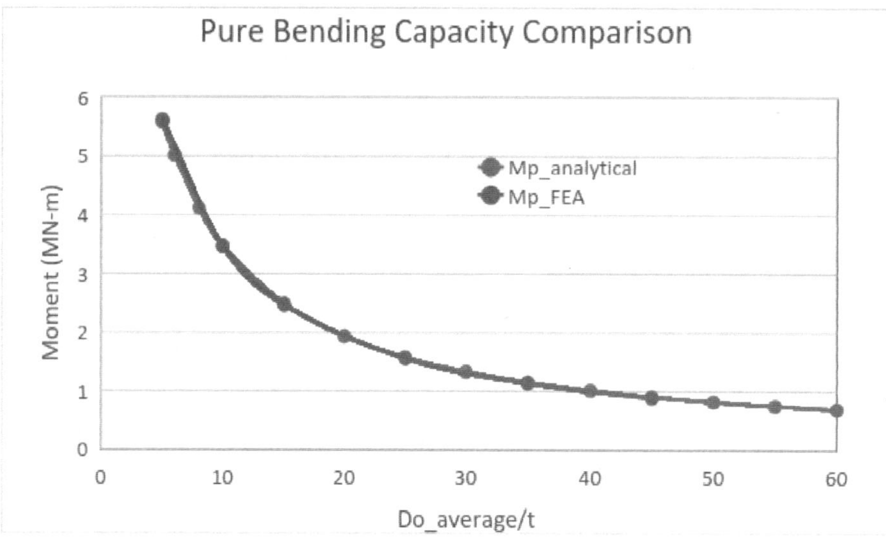

Alignment between each solution was achieved with the largest deviation at D/t=45 with 2.6%, for lower D/t ratios, the deviation is less than 2%. It indicates that current API RP1111, API STD2RD and ABS Subsea Pipeline Guide are adequate for lower D/t ratios on pure bending strength assessment.

For pipe in intact condition under external pressure load, FEA results were compared with those obtained from API RP1111 (equation 1) and ABS Subsea Pipeline Guide (equation 4). The results are shown in Figure 6 for D/t ratios from 5 to 45.

Figure 6: Comparison of collapse ultimate capacity due to External Pressure

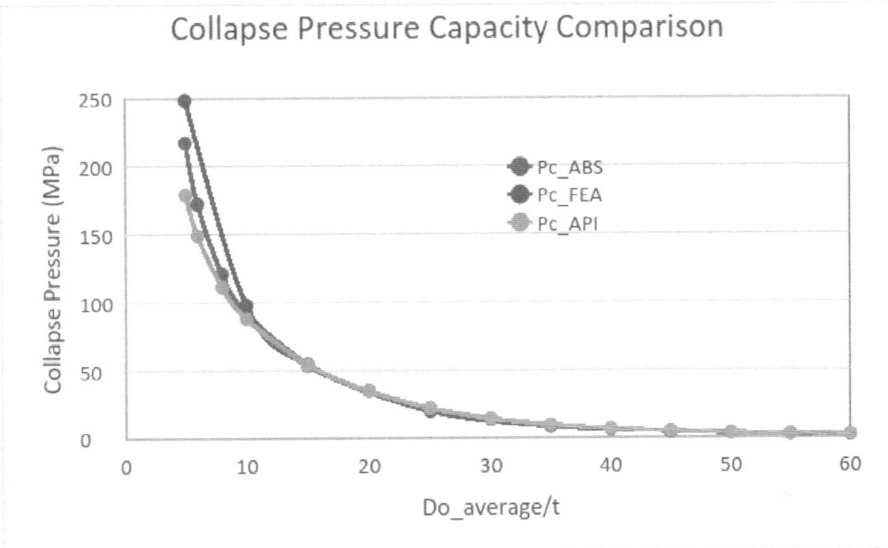

From Figure 5, it is observed that, for D/t ≥ 15, the ultimate collapse pressure calculated from API RP1111 and ABS Subsea Pipeline Guide agree very well with FEA results. But for D/t =5, the ultimate collapse pressure are underestimated up to 12% from ABS Subsea Pipeline Guide and up to 28% from API RP1111 based on the FEA results. Since collapse failure for pipe of thick wall (small D/t) is close to fully plastic failure instead of elastic buckling failure for thin wall, FEA results are higher than those predicted from the equations in API RP1111and ABS Subsea Pipeline Guide for D/t ≤ 10 are expected (see Adilson et al., 2012). However, due to conservativeness, the current API RP1111 and ABS Subsea Pipeline Guide on ultimate collapse pressure can still be applied for D/t ≤10.

FEA verification for pipe in corroded conditions:

To verify the applicability of the above FEA methodology to pipe in corroded condition, FEA verification by comparing FEA results with several selected experiment test results given in the Naoto et al., 2008 were carried out. The results of the comparisons are shown in Table 2. From the comparison of normalized collapse pressure ratio (collapse pressure of corroded pipe to that of intact pipe), very good alignment with minimal differences was observed.

Table 2: Verification of FEA and experiment tests

Case #	1	2	3	4
D/t = 21	Intact pipe	Corroded pipe		
Model Description:	Ovality=0.19%	Ovality=0.36% tcor/t=0.743 2β =20°	Ovality=0.21% tcor/t=0.737 2β=30°	Ovality=0.33% tcor/t=0.735 2β=60°
	Collapse pressure (MPa)	Normalized pressure ratio		
Experiment	19.92 (corrected for end constraints in the test)	0.799	0.801	0.781
FEA	19.57	0.789	0.774	0.765

RESULTS

Since the collapse limit state due to external pressure for pipes in intact condition is addressed in the FEA verification section, this section presents the FEA results of collapse limit state due to external pressure for pipe in corroded condition as well as the limit state of combined bending and external pressure for pipes in both intact and corroded conditions. Assuming the FEA results are the most accurate, all comparison relative to the FEA results is applied.

From the FEA, pipe failures due to pure external pressure and combined bending and external pressure can be observed in terms of lateral collapse, transition, and local buckling failures. Typical failures for a corroded pipe are shown in Figure 7.

Figure 7: Typical failure modes for corroded pipe.
Left to right: outward collapse, transition, and local buckling failure modes.

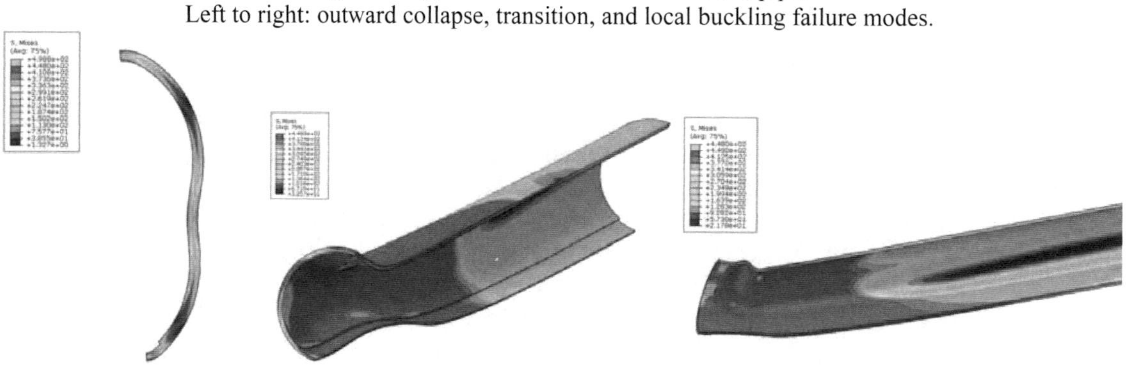

The FEA results of ultimate capacity due to combined bending and external pressure for pipe in intact condition are discussed first and shown in Figure 8 for different D/t ratios. In the figure, ultimate strength results calculated from API STD2RD (equation 5) and ABS Subsea Pipeline Guide (equation 8) are also included.

Figure 8: Ultimate capacity of combined bending and external pressure for pipe in intact condition

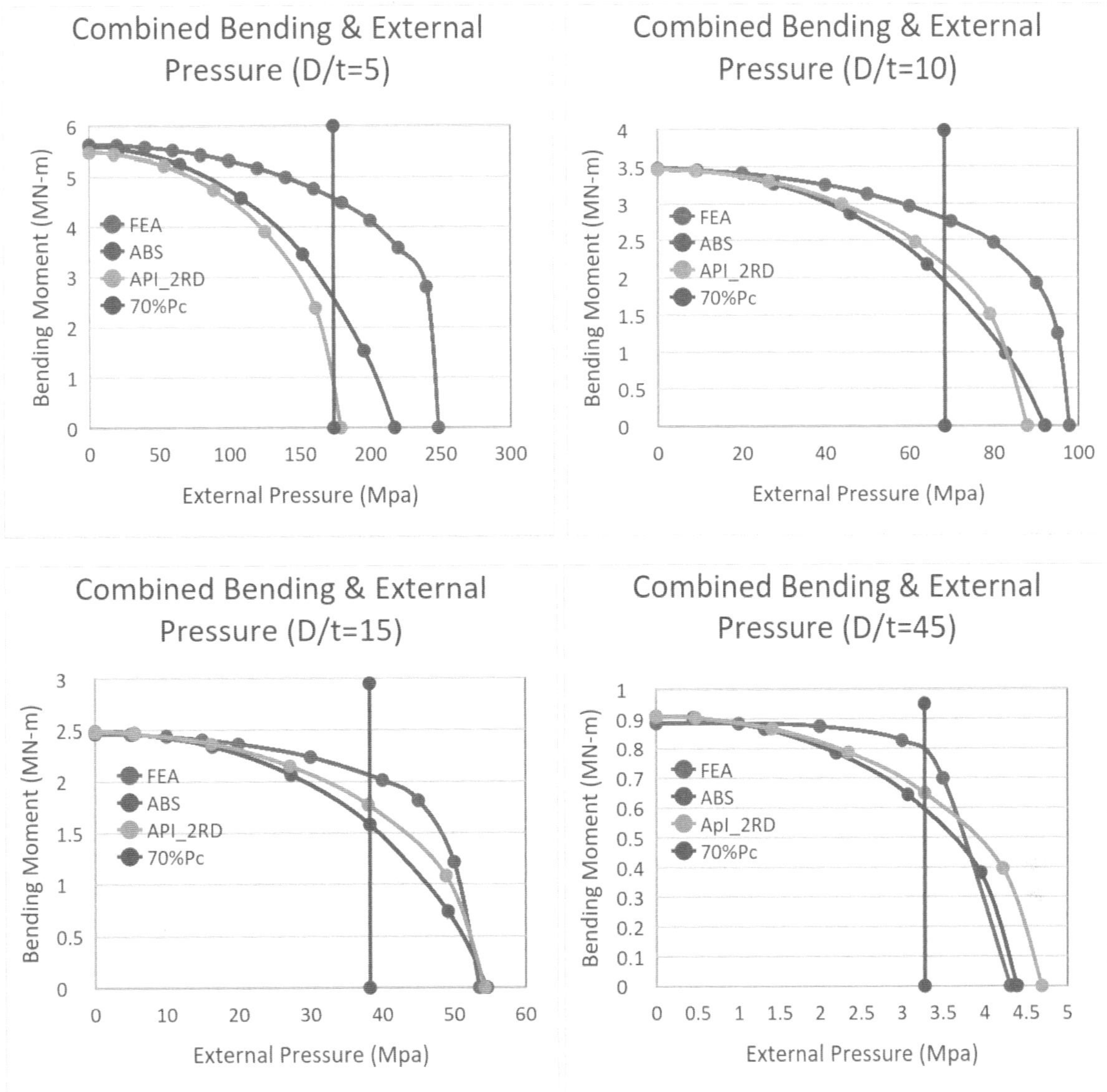

From Figure 8, it is observed that, for D/t ≤ 10, results calculated from API STD2RD and ABS Subsea Pipeline Guide are conservative for all cases comparing with the FEA results. For D/t ≥15, the FEA results become slightly less than those calculated from the equations at both relatively higher and lower pressure ranges. At D/t=45, the highest difference relative to the FEA results, at higher pressure range, is about 5% and 2% for API STD 2RD and ABS Subsea Pipeline Guide respectively, and is about 2.5% for both API STD 2RD and ABS Subsea Pipeline Guide at lower pressure range. Since the differences are small, both API STD 2RD and ABS Subsea Pipeline Guide are considered applicable for D/t ≥ 15.

In addition, the maximum allowable external pressure for normal operation is limited to be 70% of collapse pressure (Pc) (specified in API RP1111, API STD2RD and ABS Subsea Pipeline Guide); therefore, for external pressure ≥ 70% Pc (higher pressure range), the difference of the results relative to the FEA results is generally not a concern.

From the Figure 8 and the above discussion, the following can be concluded on ultimate strength assessment for pipes in intact condition due to combined bending and external pressure:

- Ultimate strength calculated from API STD2RD and ABS Subsea Pipeline Guide are applicable to all D/t ratios including D/t ≤ 10.

The FEA results for ultimate collapse pressure for pipe in a range of Beta (β) corrosion conditions (from 20° to 60°) for corrosion loss =0.25t and 0.5t are shown in Figure 9 and 10 respectively for different D/t ratios. In the figures, ultimate collapse pressures calculated from ABS Subsea Pipeline Guide (equation 11) for the selected D/t ratios are also included.

Figure 9: Collapse capacity due to external pressure for inner corroded pipe with corrosion loss=0.25t

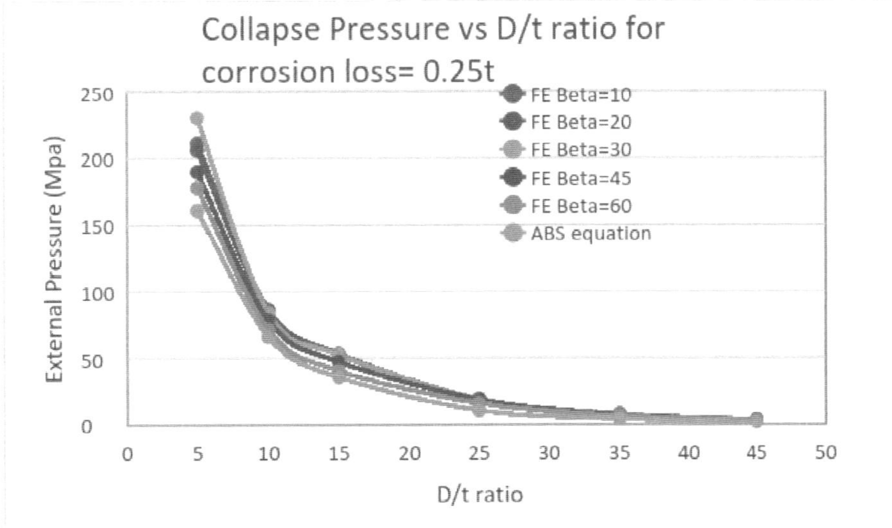

Figure 10: Collapse capacity due to external pressure for inner corroded pipe with corrosion loss = 0.5t

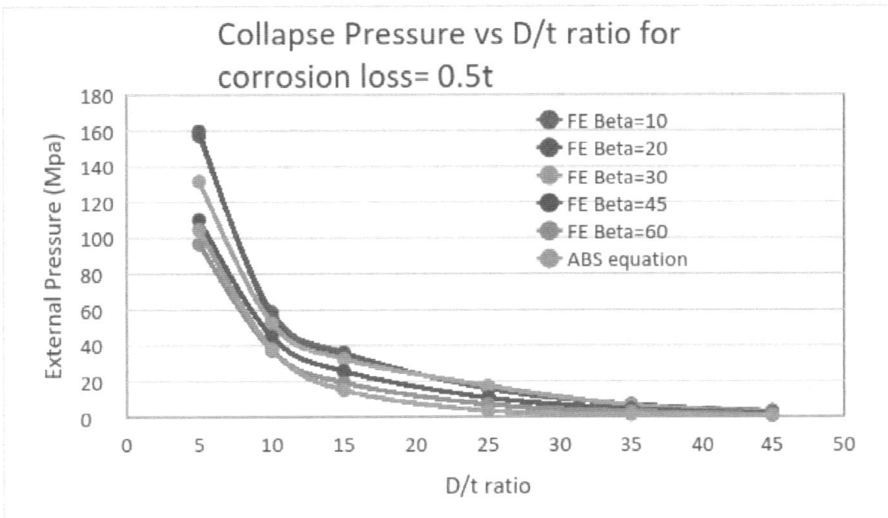

From Figure 9 (corrosion loss =0.25t), it is observed that, results from the ABS Subsea Pipeline Guide are conservative for all cases. From Figure 10 (corrosion loss =0.5t), results from ABS Subsea Pipeline Guide are conservative except for D/t ≤10 with Beta angle=60°. Therefore, the following can be concluded on ultimate collapse strength assessment for pipe in inner corroded condition due to external pressure:

- For small corrosion with corrosion loss = 0.25t, ultimate collapse pressure calculated from ABS Subsea Pipeline Guide (equation 11), are conservative relative to the FEA results for all Beta angles (≤ 60°) and all D/t ratios including D/t≤ 10.
- For large corrosion with corrosion loss = 0.5t, ultimate collapse pressures calculated from ABS Subsea Pipeline Guide (equation 11), are conservative relative to the FEA results except for D/t< 10 at Beta angles = 60°.
- These results show that ABS Subsea Pipeline Guide can be applicable to corrosion loss ≤ 0.5t for most of D/t ratios and Beta angles except for cases of corrosion > 0.5t with D/t < 10 and at Beta angel ≥60°.

The FEA results for ultimate capacity for combined bending and external pressure for pipe in a range of Beta (β) corrosion conditions (from 20° to 60°) for corrosion loss=0.25t and 0.5t are shown in figure 11 and 12 respectively for different D/t ratios. In the figures, ultimate strength results calculated from the ABS Subsea Pipeline Guide (equation 14) for the selected D/t ratios are also included.

Figure 11: Ultimate capacity for combined bending and external pressure for corroded pipe with corrosion loss=0.25t

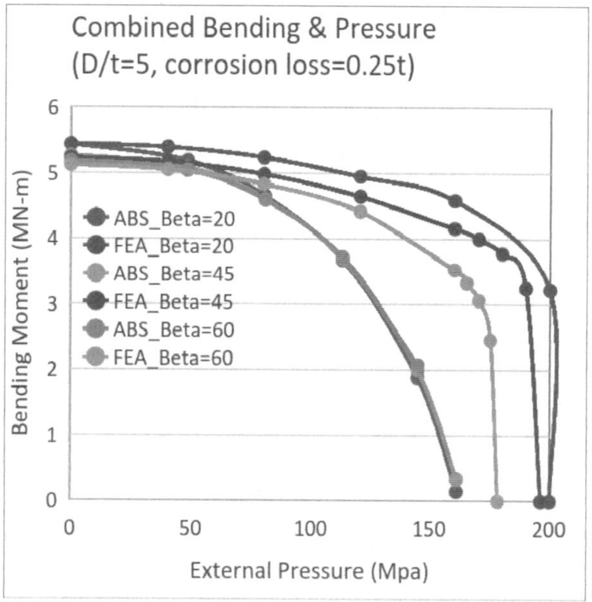

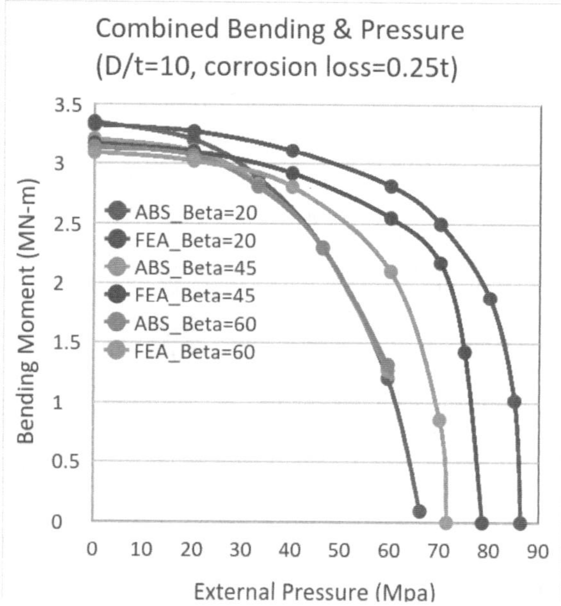

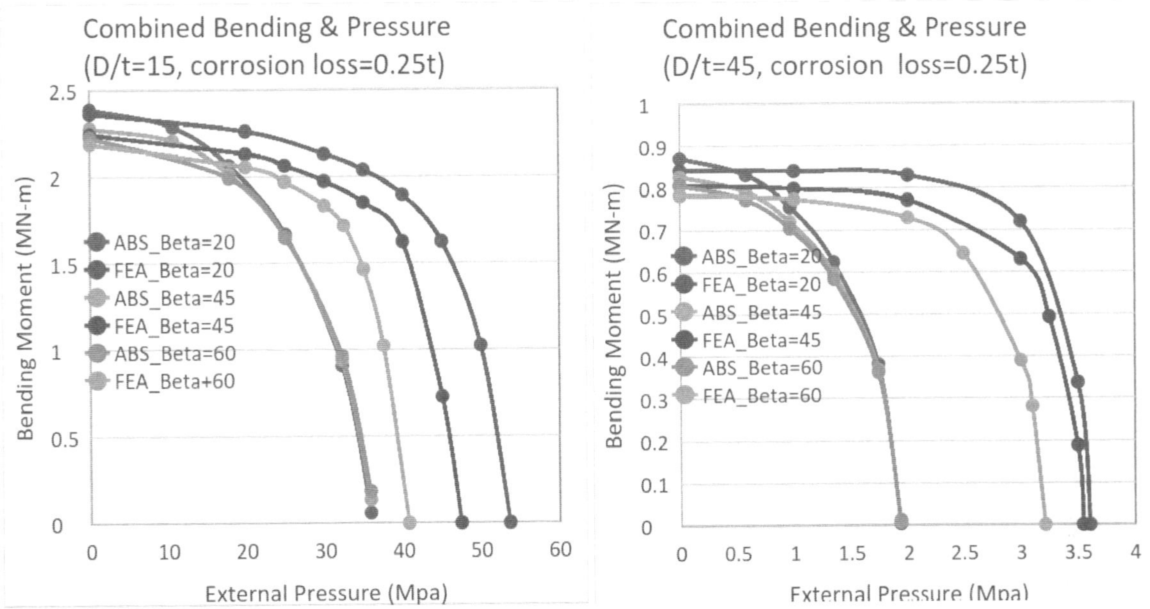

Figure 12: Ultimate capacity for combined bending and external pressure for corroded pipe with corrosion loss=0.5t

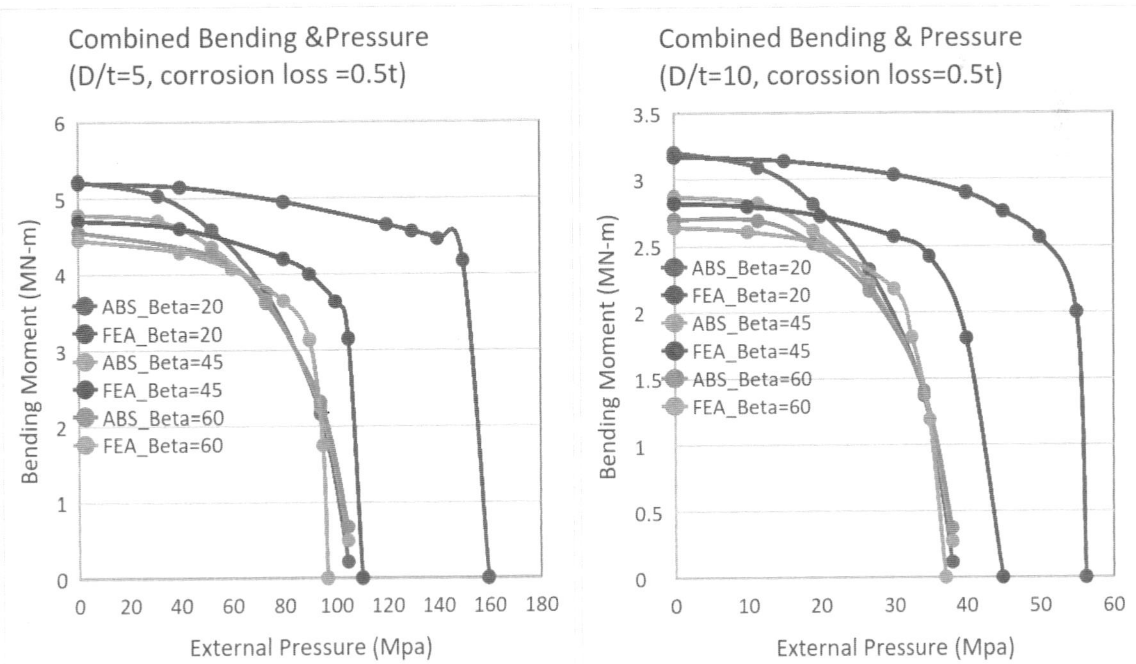

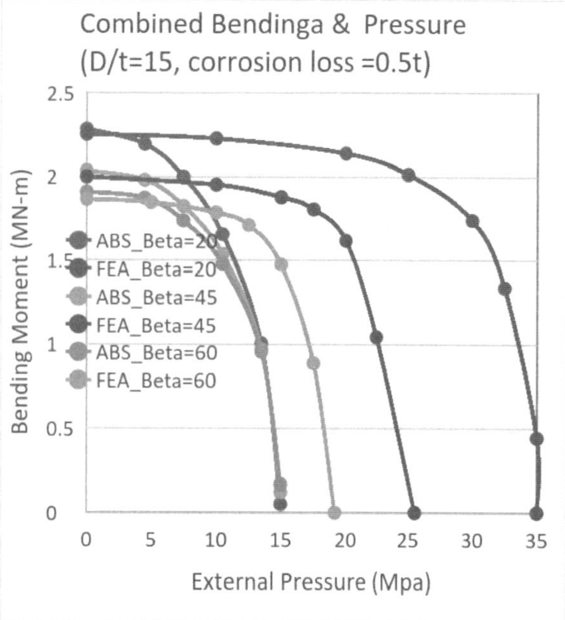

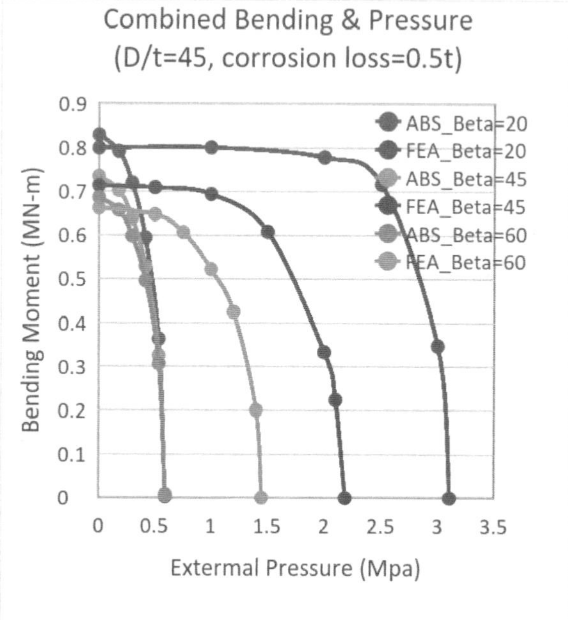

From Figure 11 (corrosion loss =0.25t), it is observed that, results from ABS Subsea Pipeline Guide are generally conservative relative to the FEA results except at relatively lower pressure range, where the FEA results are slightly less than those from the ABS Subsea Pipeline Guide. The highest difference is 3.25% at D/t=45, which is small; therefore, the ABS Subsea Pipeline Guide is considered applicable.

From Figure 12 (corrosion loss =0.5t), results from ABS Subsea Pipeline Guide are generally conservative relative to the FEA results except at relatively lower pressure range and at D/t ≤10 with Beta angle=60°. At relative lower pressure range, the FEA results are slightly less than those from the ABS Subsea Pipeline Guide, and the highest difference is 3.6% at D/t=45, which is small; therefore, the ABS Subsea Pipeline Guide is considered applicable. For D/t=10 at Beta angle=60°, the highest difference is 1.05%, which is also small. However, at D/t=5, the highest difference is 9.7%, which is not considered small; therefore, not considered applicable.

From Figures 11, 12 and the above discussion, the following can be concluded on the ultimate capacity assessment for combined bending and external pressure for pipe in inner corroded condition:

- For small corrosion with corrosion loss=0.25t, ultimate combined strength results calculated from ABS Subsea Pipeline Guide (equation 14) relative to the FEA results are applicable for all Beta angles (≤ 60°) and all D/t ratios including D/t≤ 10.
- For large corrosion with corrosion loss=0.5t, ultimate combined strength results calculated from ABS Subsea Pipeline Guide (equation 14) relative to the FEA results are applicable for most of D/t ratios and Beta angles except for cases of D/t< 10 and at Beta angles = 60°.
- These results show that ABS Subsea Pipeline Guide can be applicable to corrosion loss ≤ 0.5t for most of D/t ratios and Beta angles except for cases of corrosion loss=0.5t with D/t < 10 and at Beta angel ≥60°.

CONCLUSIONS

In this paper, the ultimate strength of inner corroded pipe was assessed for two limit states with a selected large range of D/t ratios and corrosion conditions. The two limit states are (i) collapse limit state resulting from external overpressure and buckling and (ii) collapsed limit state resulting from combined bending and external overpressure.

The assessments were carried out using Abaqus FEA software. From the FEA, ultimate capacities of pipes in both intact and corroded conditions were captured, and the results were compared with those calculated from equations of the two limit states as defined in the API RP1111, API STD2RD and the ABS Subsea Pipeline Guide. This was done to examine the applicability of these equations.

Since FEA is used to examine the applicability of the two limit state equations, the accuracy of the FEA method and results was first verified. The verification was carried out for pipe in both intact and corroded conditions, and very good alignment was demonstrated. Therefore, the same FEA method was applied to assess the ultimate strength of the two limit states for a selected range of corroded conditions and D/t ratios (5 ~ 45).

Comparing the FEA results with those from the API RP1111, API STD2RD and ABS Subsea Pipeline Guide calculations, the following conclusions were reached:

1. For pipe in intact condition, current API RP1111 and ABS Subsea Pipeline Guide on pure bending capacity assessment can be applied for all D/t ratios including D/t ≤ 10.

2. For pipe in intact condition, current API RP1111and ABS Subsea Pipeline Guide on ultimate collapse pressure relative to FEA results are applicable for all D/t ratios.

3. For pipe in intact condition, ultimate strength assessment due to combined bending and external pressure:

   - Ultimate strength calculated from API STD2RD and ABS Subsea Pipeline Guide are applicable to all D/t ratios including D/t ≤ 10.

4. For pipe in inner corroded condition, the ultimate collapse strength assessment due to external pressure:

   - For small corrosion with corrosion loss = 0.25t, ultimate collapse pressure calculated from ABS Subsea Pipeline Guide (equation 11), are conservative relative to the FEA results for all Beta angles (≤ 60o) and all D/t ratios including D/t≤ 10.
   - For large corrosion with corrosion loss = 0.5t, ultimate collapse pressures calculated from ABS Subsea Pipeline Guide (equation 11), are conservative relative to the FEA results except for D/t< 10 at Beta angles = 60°.
   - These results show that ABS Subsea Pipeline Guide can be applicable to corrosion loss ≤ 0.5t for most of D/t ratios and Beta angles except for cases of corrosion > 0.5t with D/t < 10 and at Beta angel ≥60°.

5. For pipe in inner corroded condition, the ultimate capacity assessment for combined bending and external pressure:

- For small corrosion with corrosion loss=0.25t, ultimate combined strength results calculated from ABS Subsea Pipeline Guide (equation 14) relative to the FEA results are applicable for all Beta angles (≤ 60°) and all D/t ratios including D/t≤ 10.
- For large corrosion with corrosion loss=0.5t, ultimate combined strength results calculated from ABS Subsea Pipeline Guide (equation 14) relative to the FEA results are applicable for most of D/t ratios and Beta angles except for cases of D/t< 10 and at Beta angles = 60°.
- These results show that ABS Subsea Pipeline Guide can apply to corrosion loss ≤ 0.5t for most of D/t ratios and Beta angles except for cases of corrosion loss=0.5t with D/t < 10 and at Beta angel ≥60°.

6. For ultimate capacity assessment of combined bending and external pressure, conditions of lower D/t ratios (e.g., lower than 10) and with high external pressure, direct calculation such as FE method presented in this paper, can be applied for both intact and corroded conditions.

In this paper, ultimate strength assessment for inner corroded condition was covered for two limit states. Other damaged conditions as well as other limit states involving longitudinal axial load, internal pressure, temperature, and large strain effect are not included, but will be considered in future works.

ACKNOWLEDGEMENTS

The authors would like to thank ABS for its support and permission in the preparation and publishing of this paper. The opinions expressed in this paper are those of the authors and do not necessarily represent those of ABS and affiliated companies.

REFERENCES

American Bureau of Shipping, "Guide for Building and Classing Subsea Pipeline Systems", 2014

American Petroleum Institute, "Recommended Practice for Design, Construction, Operation, and Maintenance of Offshore Hydrocarbon Pipelines (Limit State Design)", API RP 1111, 2015.

American Petroleum Institute, "Standard for Dynamic Risers for Floating Production Systems", API STD 2RD, 2013

Adilson C. Benjamin, and Divino J. S. Cunha, "Assessment of Hydrostatic Collapse of Submarine Pipelines: Historical Review of the Classic Methods", Proceedings of International Pipeline Conference (IPC), Calgary, Alberta, Canada. 2012.

Naoto Sakakibara, Stelios Kyriakides, Edmundo Corona, "Collapse of partially Corroded or Worn Pipe Under External Pressure" International Journal of Mechanical Sciences, 2008.

SHAPE MONITORING OF SUBSEA PIPELINES THROUGH OPTICAL FIBER SENSORS: S-LAY PROCESS CASE STUDY

FRANCOIS MICHELIN
INSTRUMENTATION ENGINEER, CEMENTYS, PARIS, FRANCE

PIERRE DE LACAZE
INSTRUMENTATION ENGINEER, CEMENTYS, HOUSTON, TEXAS

VINCENT LAMOUR
CHIEF TECHNOLOGY OFFICER, CEMENTYS, PARIS, FRANCE

Proceedings of the 22nd Offshore Symposium, February 2017, Houston, Texas
Texas Section of the Society of Naval Architects and Marine Engineers

ABSTRACT

Distributed strain monitoring of structures using optical fiber sensing is a new technique which has opened new possibilities in subsea pipeline installation survey. Spiecapag and Geostock, for the replacement project of a 1300m long immersed pipeline, asked Cementys to monitor the distortion of the pipeline. To certificate the pipeline, they needed to measure and save the strain conditions of the pipe-line during different steps of towing and laying. In addition, the required system had to offer a real-time control of the pipe curvature so that the marine operations could be driven without exceeding the threshold of the strain level and damaging the pipe. Cementys developed a sensor called SensoluxTM based on Raman and Brillouin Optical Time Domain Reflectometry. The sensor cable contains 4 optical fibers to measure the Brillouin (strain and temperature sensitive) and the Raman (temperature sensitive only) scatterings. The cable-sensing is sensitive at each point of its length, with a spatial resolution of 0.5m, a precision of ±5μdef and a reach of several dozens of kilometres. Three optical cables were glued in three grooves disposed at 120° in the concrete surrounding the metallic pipe. Within a few minutes, the distortion of the 3 optical fibers was measured and the pipe global shape calculated, with a one-meter spatial resolution. The small diameter of the sensor (2mm) made it easy to integrate on pipeline by gluing the optical cable in grooves carved on the concrete. The optical cable was thus protected, with a low level of intrusiveness.

Keywords: Optical Fiber, DTS, Raman, Brillouin, Asset Integrity Monitoring, Structural Health Monitoring, Subsea Pipe Line

INTRODUCTION

This article presents a method applied to calculate the flexion and the longitudinal strain on a subsea pipe line during operations of towing and laying. The method uses the distributed fiber optic sensors by Optical Time Domain Reflectometry measurements. Distributed measurements is well adapted for pipe-lines because it can be operated over long distances (several kilometers). Moreover, this sensors does not present any blind spot.
Distributed fiber optic sensing by Optical Time Domain Reflectometry is the only solution to measure so many points, with the possibility to retrieve the data in real time, over long distances by optical fiber.

TECHNOLOGY

What is an optical fiber?

In order to understand optical fiber sensors, one must first look at an optical fiber. An optical fiber is a flexible string made of extruded glass capable of transmitting light from one end to the other. It is extremely small (125µm, thinner than a hair) and totally passive: it does not transport energy other than light and does not need any power supply. The core fiber can be protected within a cable, making an optical fiber cable just as strong as a conventional electric cable.

In order to confine the light, an optical fiber is made out of different layers of silica, with different optical indices (see Figure 1). The difference of indices between the core and the cladding ensures that the light will stay within the core of the fiber. The size of the core can change, making the fiber either single-mode (core of typically 9µm) or multi-mode (typically 50 or 62.5µm). The fiber is then protected by a buffer and inserted inside a cable, within different thickness and protection depending on the application.

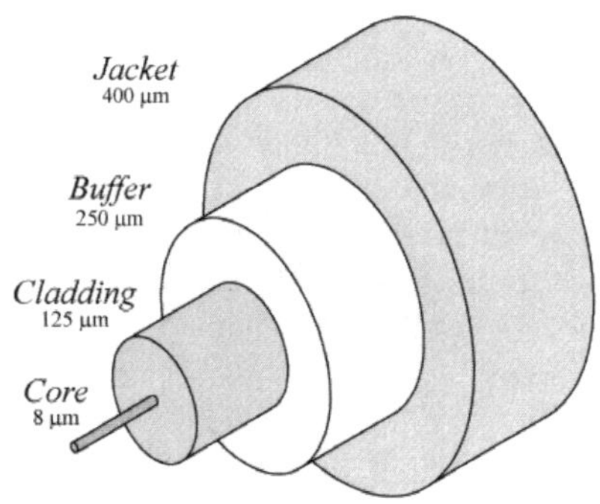

Figure 1 Cross section of an optical fiber cable

Distributed Fiber Optic Sensors by Optical Time Domain Reflectometry measurements

Distributed Sensing led a major breakthrough in the optical fiber sensing technologies: a single telecommunication fiber is able to provide measurements typically every foot for several miles ! No punctual sensors are necessary, the fiber itself reacts to its environment and sends the information back to both ends, it acts both as sensor and as a communication device (Figure 2).

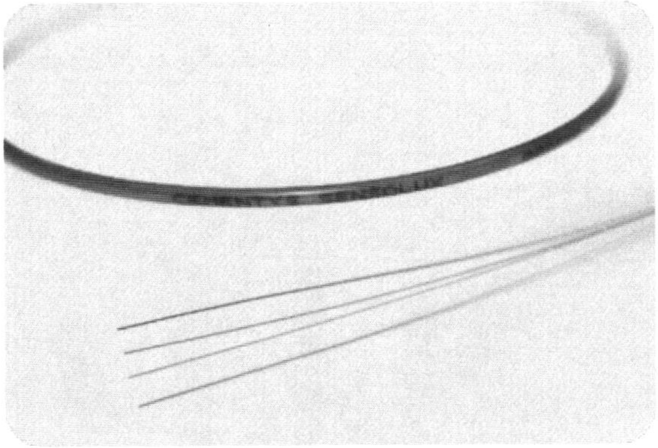

Figure 2 SensoLuxTM optical fiber sensor

This technology is based on Optical Time Domain Reflectometry, meaning that it analyses the light backscattered from a fiber (Figure 3). A laser pushes a short single frequency light pulse through the fiber, that pulse is backscattered in the fiber due to different physical phenomena (Rayleigh, Raman, Brillouin backscatterings), this backscattered light is then analysed. The location of the information is known thanks to the time of flight of the backscattered light (in the same way a radar computes distance).

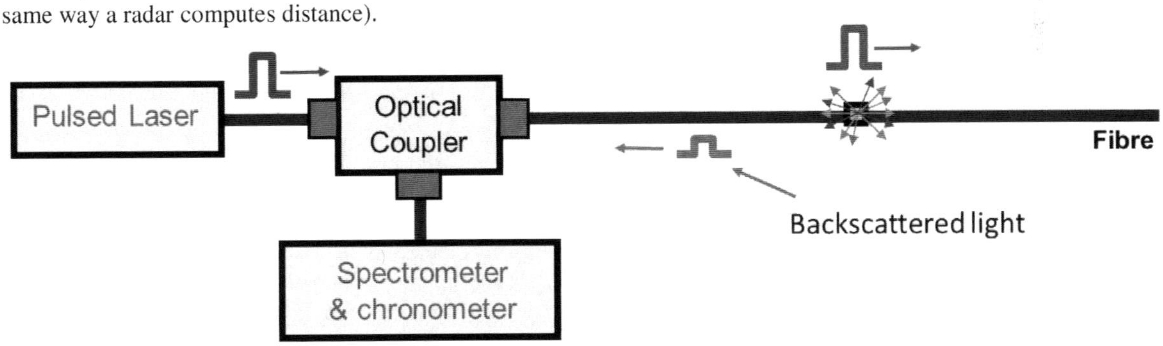

Figure 3: basic layout of a reflectometry measurement

The backscattered light is analysed in frequency thanks to an optical spectrometer. The received spectrum contains interaction peaks due to different physical phenomena (see Figure 4):

The Rayleigh peak has the same frequency as the laser pulse. Its intensity will vary with local vibrations and temperature

The Brillouin peaks will see their frequency shifted due to both local temperature and strain changes

The Raman peaks intensity will vary with temperature

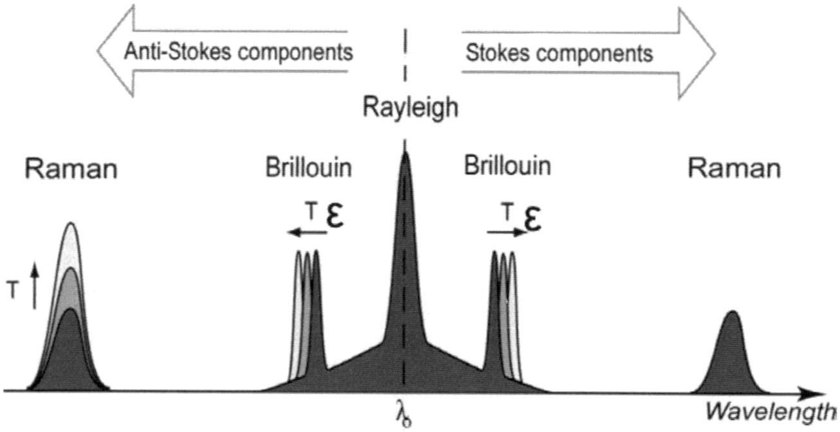

Figure 4 Typical spectrum of backscattered light

By analysing the different peaks, a single fiber can serve as a distributed sensor for temperature (DTS), strain (DSS), vibration (DVS) and acoustics (DAS).

APPLICATION

In the South of France, a 1300m long immersed pipeline has been replaced. The team chose a new technique to monitor the distortion of the pipeline. To certificate the pipeline, they needed to measure and save the strain conditions of the pipe-line during different steps of towing and laying. In addition, the system required to offer a real-time control of the pipe curvature so that the marine operations could be driven without exceeding the threshold of the strain level and damaging the pipe.

Figure 11: The instrumented pipe lines, before assembling, towing and S-Lay in shallow waters

Figure 4 The 1300m long offshore pipe line on the launching ramp

For that purpose, we used a sensor based on Raman and Brillouin Optical Time Domain Reflectometry: the SensoLuxTM. This sensor is a cable containing 4 optical fibers to measure the Brillouin (strain and temperature sensitive) and the Raman (temperature sensitive only) scatterings. Three optical cables were glued in three axial grooves disposed at 120° in the concrete surrounding the metallic pipe.

Figure 4 The sensor glued in a groove carved in the concrete weighing coating

The distortion of the 3 optical fibers could be measured to calculate the pipe global shape (flexion, traction or compression). The measurement lasts a few minutes to analyse the whole length of the pipe, with a fifty-centimetre spatial resolution. The small diameter of the sensor (2mm) made it easy to integrate on the pipeline by gluing the optical cable in grooves carved on the concrete. The optical cable was thus protected, integrated in the pipeline with a low level of intrusiveness.

The cable-sensing is sensitive at each point of its length, with a precision of ±5µdef. The data was retrieved continuously from a boat following the pipeline. We could save enough information to avoid radiographic inspections of the pipeline once laid down on the seabed.

RESULTS

The pipe line was monitored from a boat following the tugboat. One measure was realized every 10 minutes.
The following graph represents the strain of the upper sensor along the pipe at various times.
We can observe the progress of the "S" along the pipe line, from the left to the right.
Strains are precisely located on the pipe line. Positive strains represents a compression and negative strains represents a traction.

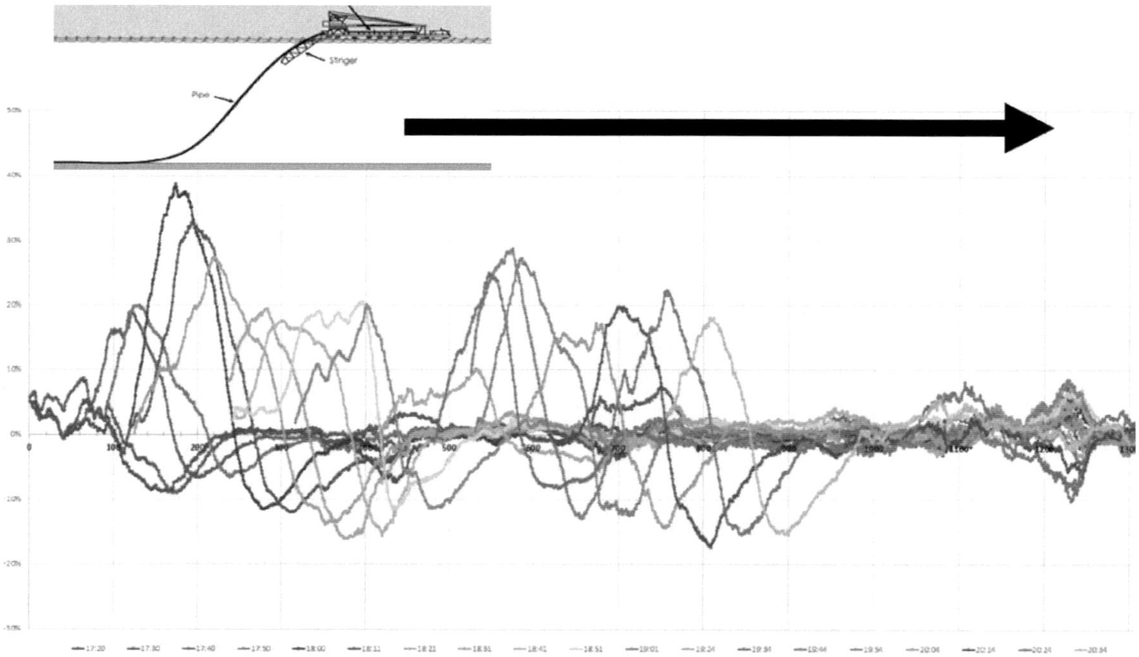

Figure 12: Strain distribution over the 1,3km pipeline during the S-Lay process (each line corresponds to an instantaneous distributed strain measurements, during the step-by-step S-Lay process)

CONCLUSION

A multitude of industrial sectors will benefit from the new optical fiber sensor technologies. From the oil well to the topside structure, optical fiber sensor can bring complementary and previously impossible data to improve safety and production of Oil&Gas assets. Such long term monitoring solutions can provide near-term and long-term cost-saving benefits such as:

- Supervised lifetime extension for existing assets, which are candidates for replacement
- Asset Integrity Monitoring during installation, commissioning and early life, reducing Vessel Time when inspection / expertise is needed.
- Risk Based Management and Condition Based Maintenance during service life

Although the optical fiber technology is fully developed and already in place in other industries (civil engineering, nuclear, health sciences…), installation challenges slowed the implementations of optical fiber sensors in the Offshore industry. Most of those challenges are getting resolved (wetmate connectors, optical pass-through assemblies, optical umbilicals and jumpers…), meaning that the offshore industry will soon benefit from this Asset Integrity Monitoring enabling technologies.

SNAME
THE INTERNATIONAL COMMUNITY
FOR MARITIME AND OCEAN PROFESSIONALS

DYNAMIC PERFORMANCE ANALYSIS OF A NEW SEMI TYPE FLOATING OFFSHORE WIND TURBINE PLATFORM

SUNG YOUN BOO[1)], STEFFEN ALLEN SHEEELY[2)], DAEJUN KIM [3)]
[1),2), 3)] VICE PRESIDENT, VL OFFSHORE, HOUSTON, USA

Proceedings of the 22nd Offshore Symposium, February 2017, Houston, Texas
Texas Section of the Society of Naval Architects and Marine Engineers

ABSTRACT

A new semi type floating offshore wind turbine platform called Y-Wind is developed to support a 5MW wind turbine, considering various design aspects from engineering through to installation, which can reduce the CAPEX. The platform is designed to achieve the integration of the tower and rotor to the platform at quayside in a US shipyard. Also a proper lightship draft is considered in the early platform design phase to enable wet-tow from the yard to the installation site with no dedicated extra equipment, as the majority of shipyards in US east and west coasts may not be able to accommodate floating wind platforms with deep lightship draft.

The Y-Wind platform with damping plates consists of four columns having the same diameter. No deck structures and braces are used. Center column supports a 5MW wind turbine and three other offset columns connected to the center column by respective pontoon are optimally placed such that the hydrostatic stability and minimal static heeling angle by wind turbine peak thrust are ensured. The platform is moored with three catenary lines with a sufficient length to avoid the uplift at the anchor.

Time domain coupled simulations under typical US offshore metocean conditions are conducted to assess the platform dynamic responses including motions, accelerations, and mooring tensions for the design load cases of the power production, 50-yr extreme and 500-yr survival conditions in accordance with ABS floating offshore wind turbine guide. The present design is checked against the ABS design criteria and confirmed to comply with the design requirements. Also significant reductions of motions are observed due to the damping plates, which also results in reducing the mooring line tensions. Preliminary global structural analysis was performed to assess the stress ratios of the primary structures.

Keywords: Y-Wind, Floating Offshore Wind Turbine, Semisubmersible, Damping Plate, Mooring Analysis, Coupled Analysis, Dynamic Response

INTRODUCTION

Several floating offshore wind prototypes, for instance Windfloat and Hywind, have been successfully installed world-wide and operated for several years to demonstrate the technologies. However, those prototypes are having difficulties in scaling up for the commercial utility for US offshore application due to higher levelized cost of energy (LCoE). It is therefore necessary to approach the offshore floating wind technology from the perspective of the platform with the lowest LCoE possible in order to compete with other energy resources. This can be achieved by fabricating the platform, assembling the tower and installing the integrated platform using existing local and regional facilities and equipment as inexpensively as possible. However, the current industry approach to floating wind production is to follow the development of land-based wind resources by continually increasing the size of the wind turbine in order to produce more power output. This following approach for offshore wind results in ever larger and more costly to build, install and maintain platforms

In the United States, due to the added cost associated with offshore floating wind, regulation and environmental issues, there has been relatively little focus on the offshore wind. Recently the first US offshore wind farm off Block Island was completed to produce 30MW power with five turbines of 6MW each with fixed foundations. In addition, a 1/8th scale semi-sub floating platform, a prototype of *VolturnUS,* was removed in late 2014 after about 18 moth testing operation (Viselli et.al, 2015A).

As the turbine sizes grows, the floating turbine platform (foundations) also get larger until the foundations are too large to fabricate in any but the largest offshore construction yards overseas which adds considerable transport costs to bring a platform from overseas into the US. Also, due to the deep drafts of these very large platforms, the platforms need to utilize very expensive offshore vessels to complete turbine integration and subsequent major maintenance activities at offshore. Thus, the VL Offshore (VLO) design approach of reducing the floating foundation costs significantly is to design a foundation that can be built and integrated locally in US shipyards. VLO has developed a semi-type floating wind platform called "Y-wind" by accommodating all the design factors stated above, Furthermore, the VLO Y-Wind platform can be disconnected from site and brought to quayside for major repairs of the turbine thereby also significantly reducing the operation and maintenance cost.

Through internal studies in terms of LCoE of the various size of the floating foundation, VLO have selected a floating foundation with 5~6MW power rate which could be a recommended size for US offshore installations. The present paper, thus, focuses on a concept design of a newly developed floating offshore wind turbine to support a 5MW wind turbine. The semi-type Y-wind platform has one center, three outer cylinder columns and three rectangular pontoons. Damping plate is attached to the platform to improve the hydrodynamic responses. In the following sections, sizing of Y-wind platform, metocean data, design criteria, platform motions, hydrostatic stability and preliminary structural analysis results are described. Several other design issues on floating wind platform design, for instance, tower base strength and fatigue, platform and mooring fatigue will be studied in the future.

METOCEAN DATA

Platform installation location is assumed at an offshore site in the US which may include the Gulf of Mexico, East and West coasts. Thus, the metocean conditions used in the present work was derived to cover the typical US offshore environmental conditions, by combining the available data listed in the documents (ABS, 2013; Weinstein, 2014; Viselli et. al, 2015A; Viselli et. al, 2015B) and summarized in Table 1. Water depth considered is 200m.

The Design Load Cases (DLCs) in Table 1 is based on ABS Guide for Building and Classing (ABS Guide, 2015). The operating condition is for the case that the turbine is in operation for power production which is in the speed ranges from cut-in to cut-out wind speeds while the turbine will be parked or idle under the 50-yr extreme and 500-yr survival conditions.

Table 1 Metocean Data

Platform Condition	Turbine Condition	ABS DLC	Wind	Wind Speed 10min@hub (m/s)	Wave (JOWNSWAP) Hs (m)	Tp (s)	Gamma -	Current (m/s)
Operating	Power Production	1.3	Rated, V_r	11.4	7.5	11.5	1.0	0.4
		1.6	Cut-out, V_o	25.0	8.5	12.7	2.2	0.6
Extreme	Parked	6.1	50-yr	40.0	12.5	14.2	3.3	0.8
Survival	Parked	SLC	500-yr	45.0	15.0	15.3	3.3	1.0

DESIGN BASIS

The offshore floating wind turbine platform moored at a water depth of 200m is designed to produce the power of 5MW for a service life of 25 year under the sea environments stated specified in Table 1. The platform with the tower integrated at a quayside has a sufficient buoyancy to support the platform at a wet-tow draft, which enables the tow-out of the platform to site with conventional tugs.

Floating offshore wind turbine platform design requirements in stationkeeping design and air gap are specified in ABS and API documents (ABS Guide, 2015; ABS FPI 2013; ABS Guidance Note, 2014; API RP 2SK, 2005) and presented in Table 2. The platform heel angle of 10 degrees and acceleration of 0.4g (3.92m/s^2) at the nacelle are selected according to the oil and gas offshore platform design practices and floating wind turbine design (Huijs et. al, 2014). The heel and acceleration requirements are applied to the turbine operating conditions to protect the components in the rotor assembly and wind turbine tower. The extreme conditions (DLC 6.1) will be used for mooring line safety check while survival condition for air gap estimate based on ABS guide (2015).

Table 2 Hydrodynamic and Mooring Design Criteria

Condition	Heel (deg)	Acceleration (g)	Mooring Line FoS	Air Gap (m)
Operating (DLC 1.3, 1.6)	≤ 10	≤ 0.4	-	-
Extreme (DLC 6.1)	-	-	≥ 2.0	≥ 1.5
Survival (SLC)	-	-	≥ 1.05	≥ 0.0

DESIGN STANRDS AND GUIDELINES

The present floating offshore wind turbine design and analysis follows the ABS and API requirements in hull, global performance, stationkeeping, stability and structure as summarized below.

- ABS Guide for Building and Classing Floating Offshore Wind Turbine Installations, 2015
- ABS Guidance Note on Global Performance Analysis for Floating Offshore Wind Turbine Installations, 2014
- ABS Rules for Building and Classing Mobile Offshore Drilling Units, 2008
- ABS FPI Guide for Building and Classing Floating Production Installations, 2013
- API RP 2SK Recommended Practice for Design and Analysis of Stationkeeping Systems for Floating Structures, 2nd edition, 2005

Y-WIND PLATFORM CONCEPT AND CONFIGURATION

The Y-Wind semi type wind turbine platform is designed to support the 5MW wind turbine on the top of the column. The platform consists of three outer (offset, outboard) columns and one center column. The outer columns are connected to the center column with corresponding rectangular pontoon. The platform has no decks and braces which can facilitate the efficient construction of the hull and remove the potential risk to the deck structure induced by the slamming loads. Also this scheme can allow us to reduce the column freeboard height which eventually provides additional benefit of structure weight reduction.

Water depth of shipyards in the US West and East coasts is not deep enough to accommodate a deep draft wind turbine platform. One of key drivers of the Y-Wind platform design is, thus, a proper lightship draft which enables to integrate the tower and rotor assembly at quayside and wet-tow out to site with no expensive and dedicated vessel. To achieve this goal, the pontoon and column sizes were determined accordingly with iterative manner. Also at the concept design stage, a static heel angle of the platform due to the peak rotor thrust at the rated wind speed was considered such that the static heel angle is maintained around 4 degrees. In addition, it was checked if the platform has a sufficient positive metacentric height specifically in wet-tow condition. In order to improve the motion performances, various damping structure options were introduced. All these design parameters were implemented to the in-house platform sizing program to configure the 5MW semi type Y-Wind platform.

The Y-Wind platform particulars with no damping plate (No DP) and with wide DP are summarized in Table 3. The total weight of the platform for the in-place conditions were estimated considering the hull structure, appurtenances, marine growth, ballast, turbines and marine system, where an appropriate contingency was applied based on the past experiences of the oil and gas platform design. Lightship draft with the tower assembled is about 5.3m which enables the quayside integration of tower and wet-tow out from the ports or 2nd-tier US shipyards. No active ballast is required due to the location of the turbine tower at the platform center.

Figure 1 represents the Y-Wind platform moored with three catenary mooring lines, where the damping plate is omitted.

Table 3 Y-Wind Semi-type 5MW Platform Particulars

Items	Units	No DP	Wide DP
Displacement	tonnes	7,748	7,770
Draft – Design	m	18.0	18.0
- Lightship (Wet-tow)	m	5.21	5.30
Outer Column Center Radius	m	35.0	35.0
Outer and Center Column OD	m	10.5	10.5
Outer Column Height	m	29.5	29.5
Center Column Height	m	28.0	28.0
Freeboard (Outer Column)	m	11.5	11.5
Pontoon Width × Height	m	4.5 X 4.0	4.5 X 4.0
Tower Base above SWL	m	10	10
Hub Height above SWL	m	90.0	90.0
CoG (above keel)	m	14.58	14.23

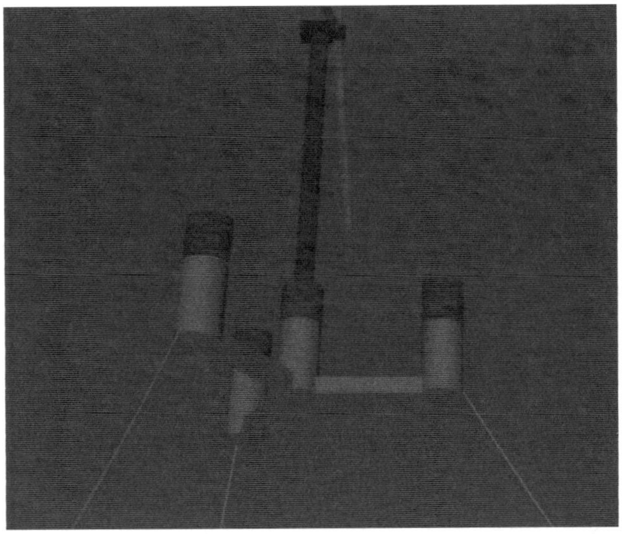

Figure 1. Y-Wind Semi Type Offshore Floating Wind Turbine Platform with Mooring Lines

The NREL 5MW reference wind turbine (Jonkman et. al, 2009; Robertson et. al, 2014) was selected and its properties are summarized in Table 4. The tower base and hub are located at 10m and 90m above SWL respectively. The rotor thrusts at the rated and cut-off wind speed presented in Table 4 were used to assess a static heel angle at the platform design stage due to the thrust, where the platform was designed to limit the heel angle about 4 degrees with no mooring induced restoring moment. The thrusts vary in time about the mean value due to wind speed change at the rotor but mean thrust was applied for the present analysis.

Table 4 NREL 5MW Wind Turbine Data

Power Rate	MW	5
Rotor Diameter	m	126.0
Number of Blades	ea	3.0
Tower Height	m	77.6
Tower Diameter Top / Base	m	3.87 / 6.5
Cut-in V_{in} / Rated V_r / Cut-out V_{out}	m/s	3.0 / 11.4 /25.0
RNA Weight	tonnes	350
Tower Weight	tonnes	250
Rotor Thrust - Rated Wind	KN	819
- Cutout Wind	KN	369

NUMERICAL MODELING

Figure 2 presents the Y-Wind platform coordinate systems used for the present analysis. The reference center is located at the platform center on the SWL. Wave direction (heading) is positive toward the x-direction and measured in counter-clock wise direction. Mooring line #1 to #3 with a 120 degree apart are also shown in Figure 2.

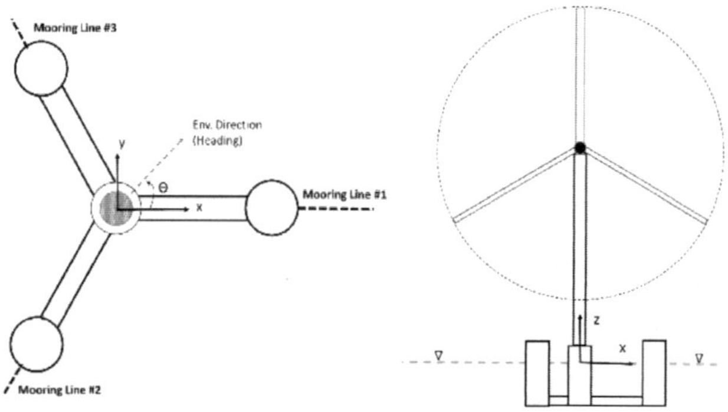

Figure 2 Platform Coordinate System Definition

The Y-Wind platform except the damping plate below the SLW was modelled with source panels for the radiation and diffraction analysis. The damping plate was discretized with the dipole panels assuming its thickness is very thin. The platform is moored with a total of three catenary R4 studless chain lines with 120mm diameter at a water depth of 200m. The one end of the 800m long line is connected to a fairlead located near the keel of the outer column and the other end is connected to an anchor foundation. Table 5 summarizes the mooring line properties. The uncorroded line is for the nominal chain with 120mm diameter whereas the diameter of the corroded line is reduced to 110mm due to corrosion over the service life of 25 years. In the strength factor of safety check, the MBL of the corroded chain is utilized.

Table 5 Mooring Line Properties

Item	Unit	Uncorroded	Corroded
Chain Type	-	R4 Studless	R4 Studless
Chain Diameter	mm	120	110
Dry Weight	kg/m	288	246
MBL	kN	13,573	11,856
Length	m	800	800

Wind forces on the hull above the SWL and wind turbine tower were estimated based on the reference (ABS FPI, 2013) and input to the numerical model as wind coefficients. The current loads on the submerged platform members and the mooring lines were implemented in the model with the drag coefficients given in Table 6.

Platform radiation damping is computed using the radiation-diffraction program but hull viscous damping can be presented in the model with the viscous drags with the drag coefficients of column, pontoon, damping plate and mooring lines in Table 6.

Table 6 Platform Member and Mooring Line Drag Coefficients

Column - Horizontal	0.65
- Vertical	1.5
Pontoon - Horizontal	1.9
- Vertical	2.2
Damping Plate - Vertical	4.5
Chain - Normal	2.4
Chain - Tangential	1.15

Vertical Cd of 1.5 of the column was based on the experimental data (Fisher, 1998). Drag coefficients on the pontoon also vary with the aspect ratio, AR (breadth/depth) and Kuelegan-Carpenter number. The horizontal and vertical Cds on the pontoons were read from the experimental data plot in reference (Venugopal, et al, 2009). According to DNV (2007), a recommended Cd of a square cylinder is about 2.2 for Reynold number$\approx 4.7\times10^4$. Damping plate drag coefficients on the vertical direction was selected to be 4.5, which is little lower than the values used in references (Robertson at al., 2014; Li et al, 2012).

The present time-domain numerical analysis using Orcaflex was based on a semi-coupled analysis (ABS Guidance Note, 2014), where the aero-elastic couplings are neglected. The tower and nacelle were modeled as part of the rigid body of the hull. The rotor thrust loads were modeled by applying the static loads in Table 4, at the top of the tower for the power production cases of DLC 1.3 and 1.6 in Table 1. This approach is very useful in practical sense to evaluate the platform global performances and mooring responses in the early design stages although it may not provide the dynamic responses of tower and accurate aero-hydro coupled effects to the platform.

NUMERICAL RESULTS AND DISCUSSIONS

Heave Free Decay Test and Damping Ratio

Figure 3 compares the free decay test results of the platforms without and with the damping plate. Logarithmic decrements are computed every cycle using an equation;

$$\delta = \ln (A_i / A_{i+1})$$

where A_i and A_{i+1} are two consecutive crest or trough values in the decay time histories. Damping ratio is then estimated as;

$$\zeta = \delta / 2\pi$$

Heave logarithmic decrements up to the first six cycles of the platform without and with the damping plate are compared in Figure 3. The decrements between crest and trough differ much in the first few cycles and converge as the cycle increases. Heave linear damping ratios were estimated by taking the average up to the first six cycles of the crest and trough and are summarized in Table 5. Heave natural period of the Y-wind platform is shifted from 11.25s for no damping plate to 15.92s for the damping plate, which is mainly due to increase of the added mass by the damping plate. The heave damping is mostly contributed by the hull and mooring line viscous drags with the associated drag coefficients in Table 6. The hull member drag coefficients will be correlated using the model test data when available.

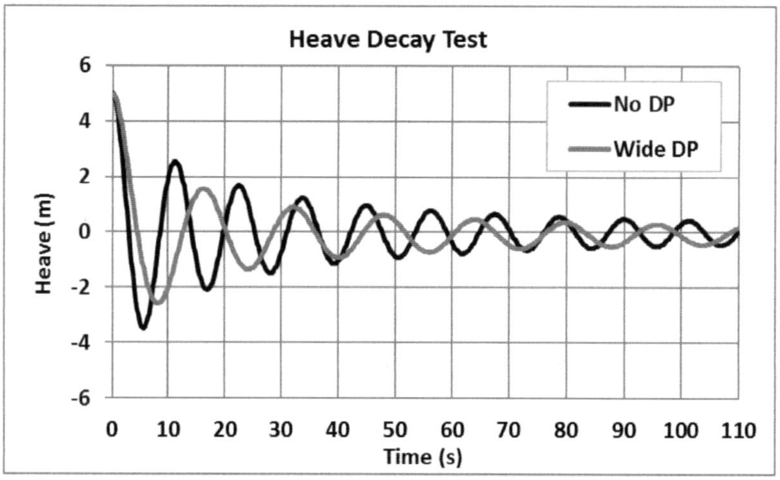

Figure 3 Time Histories of Free Decay Tests of Y-Wind Platforms

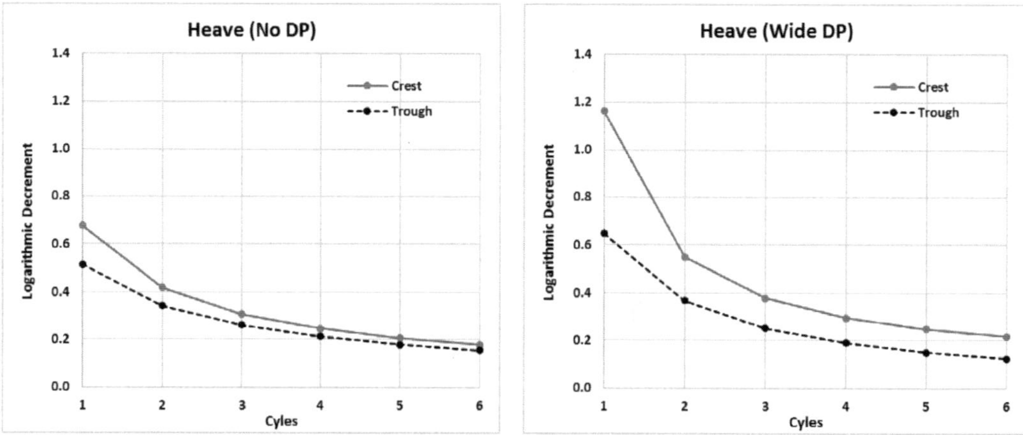

Figure 4 Comparison of Logarithmic Decrements per Cycle of Y-Wind Platforms

Table 5 Heave Damping Ratios and Natural Periods

Item	No DP	Wide DP
Damping Ratio (%)	4.9	6.1
Natural Period (s)	11.25	15.92

Platform Heave RAOs

Heave RAOs of Y-wind platform without and with the damping plate were estimated using a numerical white noise (WN) method where a wave heading of 180 degrees was considered and compared with RAOs derived from the potential-based radiation-diffraction program in Figure 5 (left-side plot). Significant reductions of the heave responses around the natural period by the viscous damping are observed. Also JONSWAP wave spectral densities of DLC 1.3 (operating), DLC 1.6 (extreme) and DLC SLC (survival) in Table 1 are plotted together with the RAOs from the white noise (right-side plot). The wave spectral peak of the survival condition is closely located to the heave natural period of the platform with the damping plate, which may affect the hydrodynamic responses to the platform. The effects are investigated in the following section.

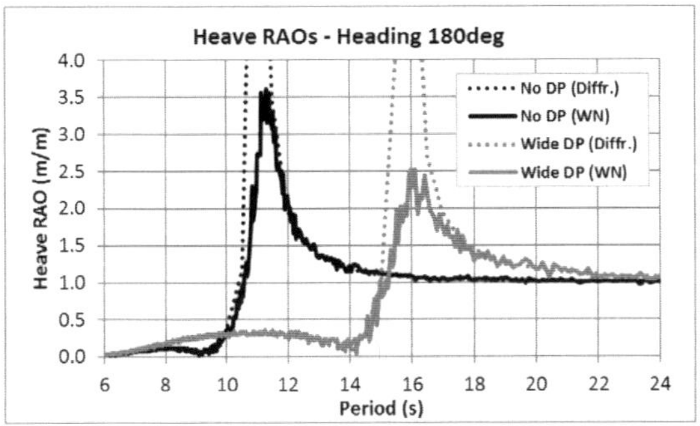

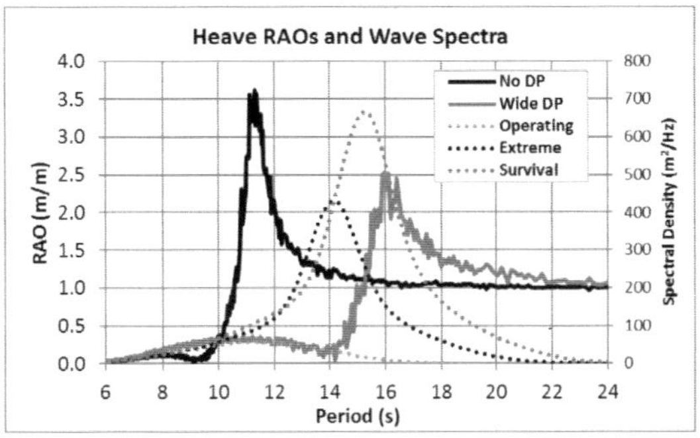

Figure 5 Heave RAOs of Y-Wind Platforms

Platform Hydrodynamic and Mooring Reponses

Numerical simulations were performed for a wave heading of 180 degrees which is identified to be a proper heading to investigate the platform responses in heave, pitch, acceleration and mooring line tension. Wind and currents were codirectional to the waves. Time domain simulations were run for three hours excluding initial ramp-time and post-processed to compute the mean, standard deviation, extreme values. Extreme values can be determined with Rayleigh 3-hour most probable extreme by,

$$Most\ probable\ Max = \mu + \sigma\,[2\,ln(n)]^{1/2}$$

or 3-hour extreme with 1% risk factor by

$$Max = \mu + \sigma\,[2\,ln(-n/ln(1-\alpha))]^{1/2}$$

where $n = T/T_z$ is the number of peaks and T, T_z, μ, σ and α are the storm duration, mean up-crossing period, mean, standard deviation and risk factor, respectively. All the results presented in this section are based on the most probable maximum except the mooring line tensions which used the 1% risk factor extreme.

The metocean conditions in Table 1 were considered to evaluate the corresponding responses. Figure 6 compares the heave motions of the Y-Wind platforms with and without the damping plate, showing that the heave motions increases with the higher sea states. It is seen that the damping plate contributes to the significant motion reductions for all the considered DLCs. There may be more improvement possible in heave motions with modification of the damping plate to swift the heave natural period away from the survival wave peak period. However, other design requirements in rotation and acceleration are met as stated in the following. The modification is, thus, recommended to be carried out to cause a minimal impact on the platform weight and associated cost.

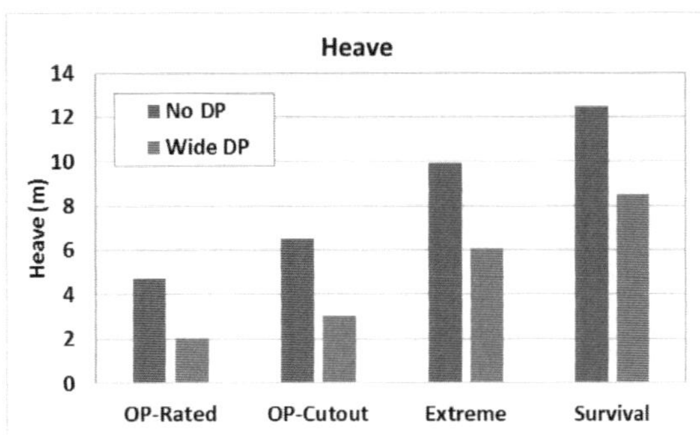

Figure 6 Comparison of Y-Wind Platform Heave Motions under the Various DLCs

Pitch motions of mean and maximum are presented in Figure 7. Mean pitch is about 3.3 degrees at the rated wind and 1.4 degrees at cut-out wind speeds. It is seen that the peak thrust by the rotor causes greater mean pitch than the values by other conditions. The mean pitch from the dynamic simulation is lower than the static heel angle of about 4 degrees considered in the platform sizing stage, which is likely due to the combined action of the mooring, current and wind induced moments. The pitch maximum for the power production (operating) condition is found to be lower than the allowable design maximum of 10 degrees, which complies with the turbine design requirement. The maximum pitch motions of the platform without the damping plate for the extreme and survival are very large. On the other hand, the pitch motions with the damping plate exceeds 10 degrees for the survival conditions, which may, however, be acceptable as the rotor is parked during such survival storm events. These results demonstrate that the damping plate is significant role in reducing the platform motions.

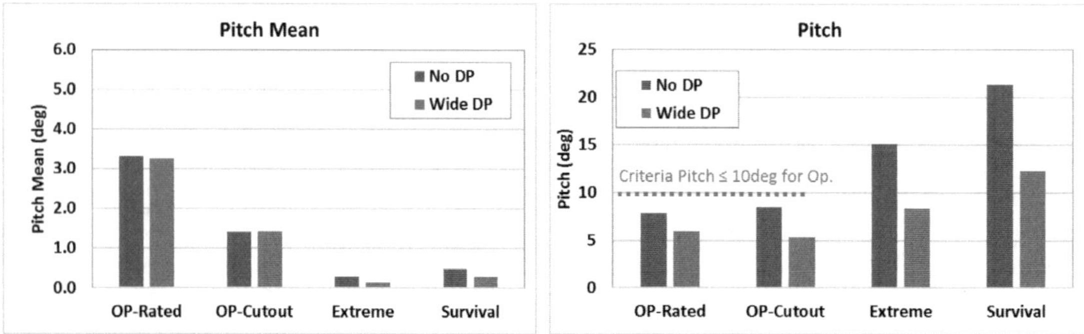

Figure 7 Comparison of Y-Wind Platform Pitch Motions under the Various DLCs

Vertical and horizontal accelerations at the nacelle are compared in Figure 8, showing that the accelerations during the power production is well below the allowable design maximum of 0.4g. Higher horizontal than vertical accelerations are observed, which is mainly induced by the pitch motions for the considered wave heading of 180deg under the extreme and survival conditions.

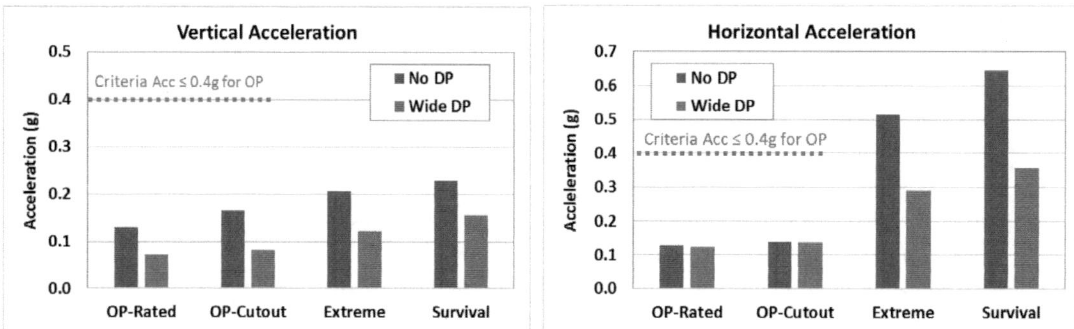

Figure 8 Comparison of Nacelle Accelerations under the Various DLCs

With the given wave heading of 180 degrees, mooring line #1 is the most loaded line so that the mooring line factor of safety was assessed on the mooring line #1 and its results for all the DLCs considered are presented in Figure 9. As described earlier, the mooring line FoS was based on the MBL of the corroded chain. The estimated mooring factors of safety of the platform with the damping plate are greater than the required minimum values for both extreme and survival conditions, which demonstrates the current mooring design satisfies the ABS mooring design requirements.

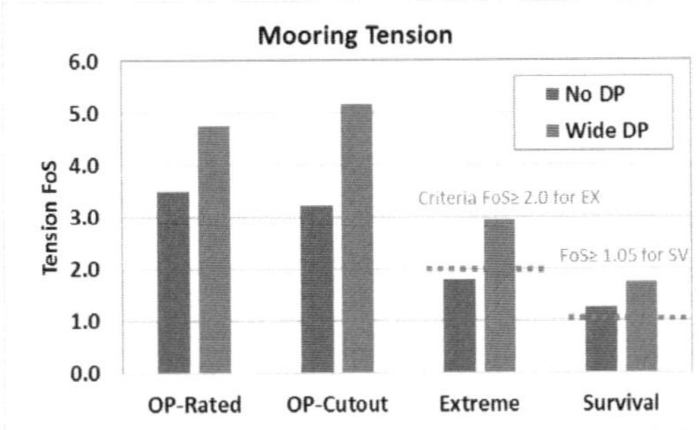

Figure 9 Mooring Line Factor of Safeties under the Various DLCs

STRUCTURAL DESIGN

The platform structural design consists of stiffened plates for both the pontoons and the center and outer columns. The design and scantlings of the structure is in accordance with ABS Guidelines (ABS Guide, 2015). The structural design is also developed with consideration of fabrication capabilities and efficiencies and in order to allow for steep and quick learning curves for series fabrication.

Longitudinal framing of pontoons is supported by transverse webs and bulkheads. Framing of columns includes webs and ring stiffeners (stringers), with intermittent watertight and not watertight decks. Pontoon to column connections are designed for simplicity and robustness. Material selected for design, including sizes and thicknesses, as well as material properties are within standard commercially available sizes and grades. Where possible, the structural design minimizes weldments. These features are incorporated in order to maximize simplicity and therefore minimize fabrication costs as much as possible.

A quasi-static, global frame analysis is used to verify the platform structure and to generate loads for local structural analysis. Of primary interest are the stresses in the pontoons connecting the columns. Loads applied are the maximum in two main platform design conditions: Operating (both rated, cut-out wind) and Extreme. The loads applied include turbine, mooring, inertial and ballast. Masses for secondary appurtenances and marine growth are also included in the analysis. In order to produce the most conservative results, the loads are applied in the most conservative manner possible (for example, the maximum turbine load is applied collinearly but opposite to the maximum mooring load).

The global structural analysis indicates that the design of the hull with no damping plate is sufficient, but near maximum allowable stresses in the pontoon (Figure 10). However, in the damping plate platform configuration, the platform motions and loads are reduced significantly and this results in a noticeable reduction in the calculated pontoon stresses (Figure 11). Table 6 compares the pontoon (east side), outer column and center column maximum calculated stresses.

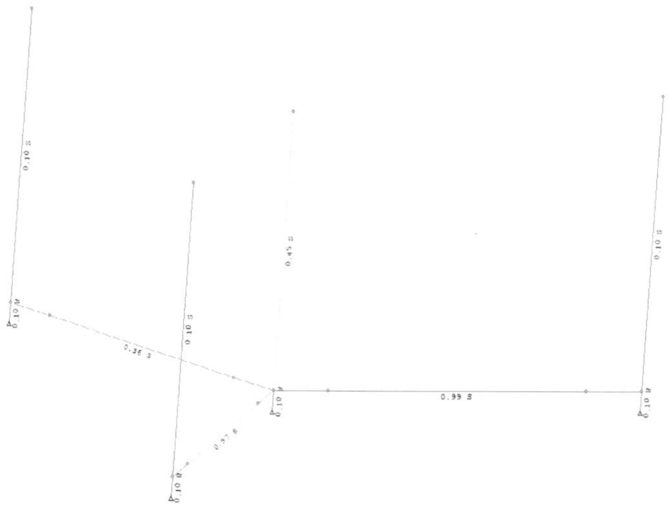

Figure 10. Maximum Calculated Stresses Y-Wind without Damping Plate

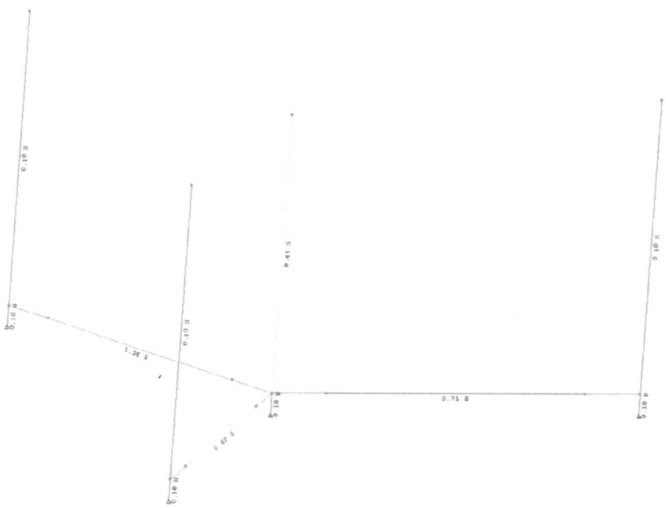

Figure 10. Maximum Calculated Stresses of Y-Wind with Damping Plate

Table 6. Y-Wind Maximum Calculated Stresses

Structural Component	Platform Configuration	
	No Damping Plate	With Damping Plate
Outer Column	0.10	0.10
Pontoon (East)	0.99	0.71
Center Column	0.45	0.44

The reduction in the platform motions due to the damping plate has a noticeable, positive effect on the pontoon stresses. While detailed local and fatigue analysis needs to be completed in order to confirm the local structural design of the pontoon, it is encouraging to note that the possibility may exists that the pontoon structural design could be improved and made a bit more efficient. Overall, the base structural design is considered sufficient to continue further development of the Y-Wind Platform design. According to preliminary screening, environment direction of 180 degree governs the platform structure strength design. Future works will consider other directions for comprehensive structural analyses.

STABILITY ANALYSIS

For the hydrostatic stability analysis, Y-wind platform without the damping plate was selected. Figure 10 shows the 3D model used for the analysis.

Figure 10 Y-Wind Platform Model for Stability Analysis

Various load cases were considered to check the intact and damage stabilities of the platform with no damping plate. Considered were two different drafts of the wet-tow and in-place conditions. For the power production condition, both rotor thrust and wind force acting on the hull were applied to give the combined heeling moment to the platform. In the parked case, the wind forces on the tower and hull were considered. Table 3 summarizes the load cases and associated wind speeds used for the stability analysis.

Table 2 Load Case and Wind Speed used for Stability Analysis

Hull Condition	Load cases	Draft (m)	Wind Speed (m/s)
Intact	Wet-tow	5.21	51.44 (100kts)
	In-Place Power Production: Rated	18.0	11.4 (V_r)
	In-Place Power Production: Cut-out	18.0	25 (V_o)
	In-Place Rotor Parked	18.0	51.44 (100kts)
Damaged	Wet-tow	5.21	25.72 (50kts)
	In-Place Rotor Parked	5.21	25.72 (50kts)

The stability analyses were performed for every 15 degrees of yaw (axis) angle so that the worst case can be identified. Figure 11 shows intact stability and heeling arm curves for in-place condition with rotor parked. Both intact and damage stabilities were checked as per the criteria in reference (ABS Guide, 2013). The platform has always positive GM at any intact and damaged conditions including two tank damage during wet-tow and one tank damage cases for in-place, which satisfies all stability requirements. It should be noted that Y-wind platform has maximum GM with highest margin at the wet tow so that the platform will be always stable during the ballasting at site. Due to this inherent stability feature, Y-wind platform will not require additional support means during the tow and ballasting operations, which can eventually provide the cost saving benefit.

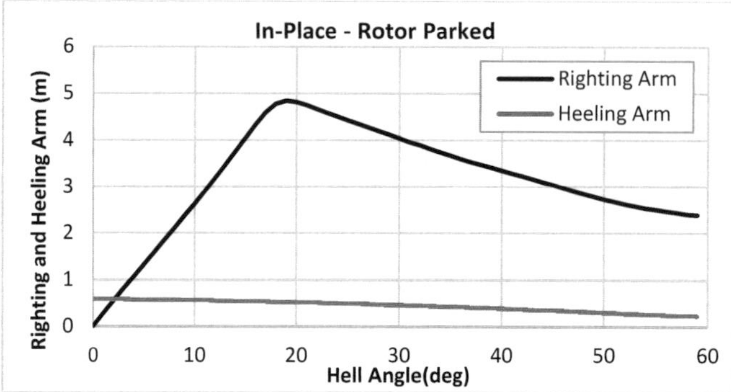

Figure 11 Intact Stability Curve for In-Place Rotor Parked Condition

CONCLUSIONS

A semi-type Y-Wind floating offshore wind turbine platform has been developed to support a 5MW wind turbine. The platform design is properly sized to be fabricated in US shipyards, integrated with existing equipment at quayside of the shipyards and wet-towed out to site with no dedicated expensive vessels, which all can reduce the LCoE significantly.

The platform consists of four columns having the same diameter. Center column supports the wind turbine and three other columns are connected to the center column by respective pontoon. The platform has VLO's proprietary damping plate attached to the platform

The platform moored at a water depth of 200m with three catenary chain legs is designed to comply with the requirements by ABS floating wind turbine guide, under the US environments in the power production, 50-year extreme and 500-year survival sea states.

Time domain numerical simulations were conducted to validate the present concept design in heave and pitch motions, accelerations, mooring and damping plate. The results with and without the damping plate are compared together and it is found that the Y-wind platform with damping plates satisfies all the design requirements. It is also demonstrated that the dynamic responses are reduced significantly by the proposed damping plate, which is promising in the design of structural members, wind turbine components and mooring system. In addition, it is confirmed that the hydrostatic stability and global structural strength comply with the ABS and industry standards. The stability analysis demonstrates that no dedicated equipment or vessels in wet-towing to installation operations are required,

The numerical viscous model in the damping plate will be correlated with the model test data when available. In addition, various damping plate configuration effects will be further studied.

ACKNOWLEDGEMENTS

Dr. Chankwon Jeong, Dockwise has helped the numerical modeling of the platform and his support is gratefully acknowledged.

REFERENCES

ABS, "Design Guideline for Stationkeeping Systems of Floating Offshore Wind Turbines", June 2013

ABS, "Guide for Building and Classing – Floating Offshore Wind Turbine Installations", January 2015

ABS, "Guidance Notes on Global Performance Analysis for Floating Offshore Wind Turbine Installations", February 2014

ABS FPI, "Guide for Building and Classing Floating Production Installations", 2013

DNV RP C205, "Environmental Conditions and Environmental Loads", April 2007

Fisher, F.J., Gopalkrishman, R., "Some Observations on the Heave Behavior of Spar Platforms", 17th Intl Conference on OMAE, 1998

Huijs, F., Bruijin, D.R., Savenije, F., "Concept Design Verification of a Semi-submersible Floating Wind Turbine Using Coupled Simulations", Energy Prodedia, 53, 2014

Jonkman, J., Butterfield, S., Musial, W., Scott, G., "Definition of a 5-MW Reference Wind Turbine for Offshore System Development", Technical Report, NREL/TP-500-38060, February 2009

Li, B., Huang, Z., Low, Y.M., Ou, J., "Experimental and numerical study of the effects of heave plate on the motion of a new deep draft multi-spar platform", Journal of Marine Science and Technology, June 2013

Robertson, A., Jonkman, J., Masciola, M., Song, H., Goupee, A., and Coulling, A., Luan, C., "Definition of the Semisubmersible Floating System for Phase II of OC4", NREL Technical Report, NREL/TP-5000-60601, September 2014

Venugopai, V., Varyani, K.S., Westlake, P.C., "Drag and Inertia Coefficients for Horizontally Submerged Rectangular Cylinders in Waves and Currents", Proc. Institution of Mechanical Engineers, Part M: J. Engineering for the Maritime Environment, Vol. 223, 2009

Viselli, A.M., Goupee, A.J., Dagher, H.J., "Model Test of 1:8-Scale Floating Wind Turbine Offshore in the Gulf of Maine", J. Offshore Mechanics and Artic Engineering, Vol. 137, 2015A

Viselli,A.M., Forristall, G.Z., Pearce, B.R., Dagher, H.J., "Estimation of extreme wave and wind design parameters for offshore wind turbines in the Gulf of Maine using a POT method", Ocean Engineering 104, 2015B

Weinstein, A., "2014 Wind Power Peer Review – Windfloat Pacific OSW Demo Project", US DoE, March 2014

SNAME
THE INTERNATIONAL COMMUNITY
FOR MARITIME AND OCEAN PROFESSIONALS

RANS/LIFTING LINE MODEL INTERACTION METHOD FOR THE DESIGN OF DUCTED PROPELLERS AND TIDAL TURBINES

WEIKANG DU
GRADUATE STUDENT, THE UNIVERSITY OF TEXAS AT AUSTIN

SPYROS A. KINNAS
PROFESSOR, THE UNIVERSITY OF TEXAS AT AUSTIN

ROBIN MARTINS MENDES
EXCHANGE STUDENT, ECOLE NAVALE, FRANCE

THOMAS LE QUERE
EXCHANGE STUDENT, ECOLE NAVALE, FRANCE

Proceedings of the 22nd Offshore Symposium, February 2017, Houston, Texas
Texas Section of the Society of Naval Architects and Marine Engineers

ABSTRACT

In this paper, a RANS/lifting line model interaction method is proposed to consider the duct geometry in the design of propellers and tidal turbines. In the lifting line model, the Lerbs-Wrench formulas are used for the wake alignment procedure. In the RANS solver, the blade is represented by a pressure jump profile. The blade loading is determined via a previously developed optimization algorithm which takes into consideration the effect of the duct via a simplified image model. An iterative procedure is developed in which the advance ratio (based on the ship speed) and the total thrust in the propeller case and the tip speed ratio (based on the inflow velocity far upstream) are kept constant. The procedure is tested for different cambers and thicknesses of the duct shape. The efficiency, inflow velocity and thrust on the blade and duct are obtained and analyzed. The influence of different factors, including the blade number, drag-to-lift ratio, advance ratio, thrust coefficient and tip speed ratio, are studied. This method is proved to be reliable and efficient in designing ducted propellers and tidal turbines.

Keywords: ducted tidal turbines and propellers, optimization, lifting line method, Reynolds Averaged Navier-Stokes (RANS) method, Lerbs-Wrench formulas

INTRODUCTION

The lifting line model is widely used in the first step of propeller and tidal turbine design to obtain the optimal loading on the blades in given conditions, and the results are then used in more accurate methods like the boundary element method (BEM) (Lee, 1987) and the vortex lattice method (VLM) (Kerwin et al. 1978). Xu (2010), Kinnas et al. (2012) and Menéndez (2013) followed this designing procedure and used a non-linear optimization code (CAVOPT-3D) and a data-base searching code (CAVOPT-BASE) to design the geometry of the propeller blades and tidal turbine blades, and compared the results with those from RANS solver. In the lifting line model, a simple way to consider the influence of the wake is by assuming a constant trailing pitch angle along the x-direction, known as the Betz condition (Kerwin et al., 2010). The induced velocities (both axial and tangential) can be evaluated by the formulas proposed by Lerbs et al. (1952) and Wrench et al. (1957). Menéndez et al. (2014) used an improved fully aligned wake model and had an assessment of the Betz condition.

In the lifting line model, the key blade is presented by discretized horseshoe vortices (Kerwin et al., 2010). The effect of the hub and duct can be considered by placing image vortexes for every vortex located on the blade. However, the image model for the duct is equivalent to the assumption that the length of the duct is infinite and the blade is placed in an infinite cylindrical tunnel. The shape of the duct is neglected, so this method is neither accurate nor plausible in the designing of the duct geometry.

In this paper, a Reynolds-Averaged Navier-Stokes (RANS)/lifting line model interaction method is proposed to consider the duct geometry in the design of ducted propellers and tidal turbines, and a numerical code called LLOPT2NS (lifting line optimization to Navier-Stokes) is developed. In this method, the Lerbs-Wrench formulas are used in the wake alignment procedure. In the propeller case, the duct is built by using a NACA a=0.8 camber line and a NACA 00 thickness distribution with the angle of attack fixed as 10 degree. In the turbine case, to make the ducted tidal turbine able to work with flow in both directions, a symmetric geometry is selected by using parabolic camber distribution and elliptic thickness distribution. The influence of the maximum camber and thickness on the efficiency, inflow velocity, KT and 10KQ are studied. Different factors on the results, including the advance ratio, the drag-to-lift coefficient, the number of blade, and the tip speed ratio, are compared with those obtained by running the propeller or tidal turbine without duct or by using the image model for the duct in the lifting line method without considering the geometry. The blade optimized loading can be used in designing the blade geometry and by using the VLM/RANS coupling method (Kinnas et al., 2013) or the BEM/RANS coupling method (Kinnas et al., 2016) on the ducted propeller and tidal turbines, this results from this paper can be tested.

METHODOLOGY

The actuator disk model

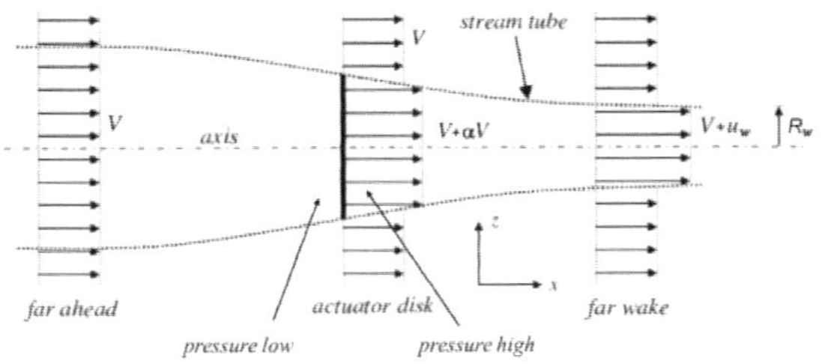

(a) Propeller case

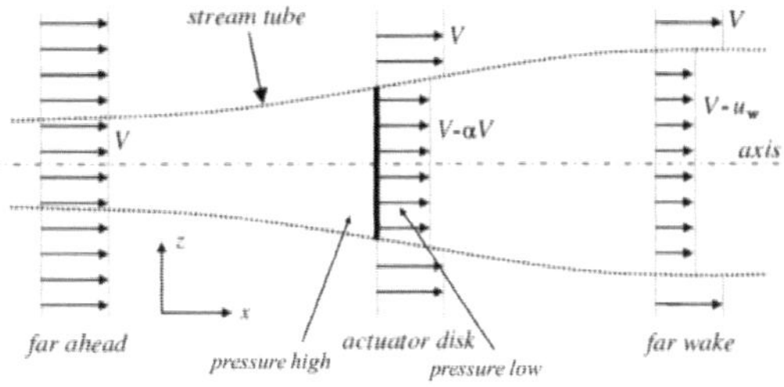

(b) Turbine case

Figure 1 Inflow velocity change and pressure in the actuator model

In the actuator disk model (Menéndez, 2013), the propeller and turbine blades are represented by a surface called an actuator disk. As shown in Figure 1, in the propeller case, the flow velocity increases because of the energy from the ship machinery, and from Bernoulli's equation the pressure decreases. There is a pressure jump on both sides of the actuator disk, and the low-pressure side is the left-hand side of the actuator disk for the inflow in the given direction. The integration of the pressure differences on both sides is the thrust acting on the propeller blade. In the turbine case, the velocity decreases and the low-pressure side is the right-hand side of the actuator disk.

The lifting line model with Lerbs-Wrench formulas

In the lifting line mode, the propeller and turbine blades are represented by discretized M horseshoe vortices, as shown in Figure 2 (Menéndez, 2013). The Lerbs-Wrench formula is given in Equation (1).

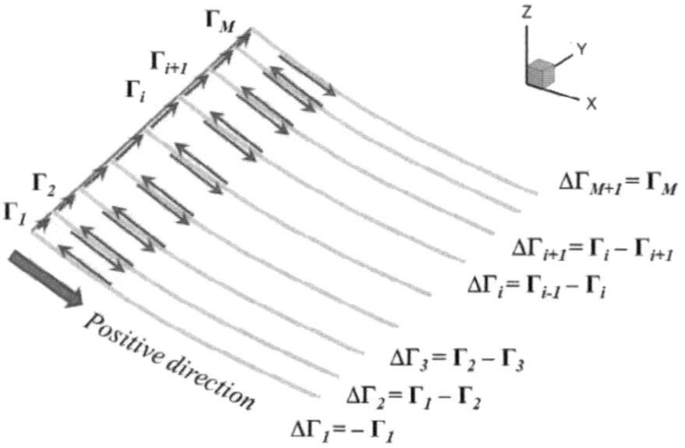

Figure 2 The lifting line model and the line vortex

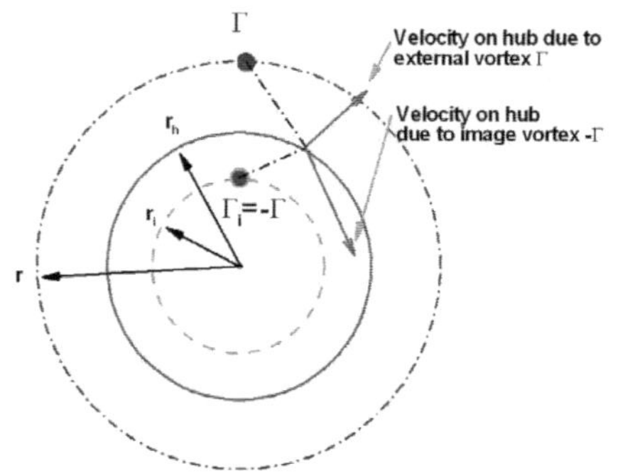

Figure 3 The image model for the hub

$$\frac{\tan(\beta)}{\tan(\beta_i)} = \gamma\sqrt{1 - w_x(r)} \tag{1}$$

where β is the undisturbed wake pitch angle, β_i is the wake pitch angle, $w_x(r)$ is the axial wake fraction distribution and γ is a constant.

The image model for the hub and duct

In the lifting line model, the influence of the hub and duct can be considered by in an image model, as shown in Figure 3 for the hub case. For each vortex located at radius r_v, an image vortex with the same strength but different sign is located at radius r_{image}, which is given in Equation (2). The length of the duct in the image model is infinite, so the blade is place in an infinite cylindrical tunnel. It should be noted that in this paper, the images of the image are not considered.

$$r_{image} = \frac{r_h^2}{r_v} \tag{2}$$

where r_h is the hub radius.

The RANS/lifting line model interaction method

In this paper, an iterative method is proposed to couple the RANS method for the duct with the lifting line method for the propellers and tidal turbines. In the propeller case, the far upstream inflow velocity is V_s (ship speed). The advance ratio J_s and the propeller rotational frequency n are kept constant through the iterations. The advance ratio J_s is defined in Equation (3) and the local advance ratio is defined in Equation (4).

$$J_s = \frac{V_s}{nD} \tag{3}$$

where the subscript *s* means *ship*, and *D* is the diameter of the propeller.

$$J_l = \frac{U_{in}}{nD} \tag{4}$$

where the subscript *l* means *local*, and U_{in} is the inflow for the lifting line model. U_{in} is different from the ship speed because the inflow velocity is changed due to the existence of the duct.

The total thrust coefficient C_{Ts} is non-dimensionalized by the ship speed as in Equation (5).

$$C_{Ts} = \frac{T_T}{\frac{1}{2}\rho\pi V_S^2 R^2} \quad (5)$$

where R is the radius of the propeller, and T_T is the total thrust from the duct and propeller blade.

The total thrust can be divided into two parts: the propeller-provided thrust T_P and the duct-provided thrust T_D. A non-dimensionalized factor τ is defined as

$$\tau = \frac{T_P}{T_P+T_D} = \frac{T_P}{T_T} \quad (6)$$

where $\tau > 1$ means there is drag on the duct and $\tau < 1$ means there is thrust on the duct.

The local thrust coefficient and the torque coefficient from the lifting line model is defined in Equation (7) and (8). The efficiency of the propeller is defined in Equation (9).

$$C_{Tl} = \frac{T_P}{\frac{1}{2}\rho\pi U_{in}^2 R^2} = \tau C_{Ts}\frac{V_S^2}{U_{in}^2} \quad (7)$$

$$C_Q = \frac{Q}{\frac{1}{2}\rho\pi V_S^2 R^3} \quad (8)$$

$$\eta = \frac{T_T V_S}{Q\omega} \quad (9)$$

where $\omega = 2n\pi$ is the propeller angular velocity.

Plug Equation (3~8) into Equation (9), the final expression for the propeller efficiency is shown in Equation (10).

$$\eta = \frac{J_S C_{Ts}}{\pi C_Q} \quad (10)$$

where J_S and C_{Ts} are both constant and the only changing variable during the iterations to in the efficiency equation is the torque coefficient C_Q.

In the RANS/lifting line model interaction method, the flow around the duct is solved in the RANS solver, and the inflow velocity for the lifting line model is modified considering the influence of the duct. The code for the lifting line model is called LLOPT (Lifting Line OPTimization). It should be noted that the swirl component of the induced velocity is not included in the RANS model. A flow chart of the coupling process is shown in Figure 4.

The viscous effect in the RANS/lifting line model interaction method is considered in two parts: the viscosity on the duct is considered as a non-slip boundary condition in the RANS model, and in the lifting line model a drag-to-lift ratio is use, as shown in Equation (11).

$$\kappa = \frac{C_D}{C_L} \quad (11)$$

where C_D is the sectional drag coefficient and C_L is the sectional lift coefficient.

In the RANS model, the actuator disk is represented as a *fan* boundary condition with a pressure jump profile as a function of the sectional radius r, calculated in Equation (12) and (13).

$$\Delta T = \rho Z\Gamma[(\omega r + u_t^*) - \kappa(U_{in} + u_a^*)]\Delta r \quad (12)$$

where Z is the number of blades.

$$\Delta p = \frac{\Delta T}{2\pi r \Delta r} \tag{13}$$

where u_t^* and u_a^* are the tangential and axial inducted velocities, respectively, and Γ is the optimized circulation on the blade, calculated from the lifting line model.

In the propeller case, the inflow velocity for the lifting line model is calculated by using Equation (14).

$$U_{in} = U_{RANS} - u_a^* \tag{14}$$

where U_{RANS} is the averaged velocity at the actuator disk from the RANS solver, and u_a^* is the averaged axial induced velocity from the previous iteration, calculated from the lifting line model.

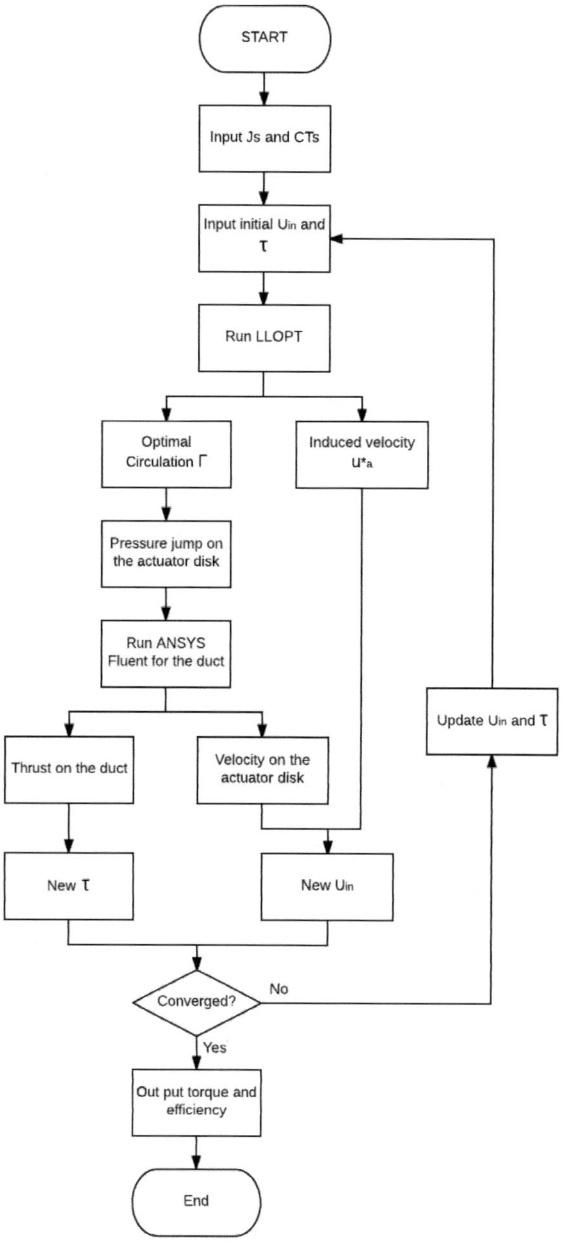

Figure 4 Flow chart of the RANS/lifting line model interaction method (propeller case)

In the tidal turbine case, the coupling scheme is the similar with the propeller case except the following details. First, in this case the thrust is not of interest so only the inflow velocity is calculated and updated during the iterations. Second, there is no advance ratio and thrust coefficient in this case. Instead, the far upstream tip speed ratio (TSR) is kept constant. Third, the inflow velocity is calculated in Equation (15) because the inflow decelerates as it approaches the actuator disk, as shown in Figure 1.

$$U_{in} = U_{RANS} + u_a^* \tag{15}$$

In the turbine case, the efficient is defined as the useful power from the blade over the total energy in this area, as shown in Equation (16).

$$\eta_1 = \frac{Q\omega}{\frac{1}{2}\pi\rho V_S^3 R^2} \tag{16}$$

where R is the radius of the turbine blade, consistent with the propeller case. However, the existence of the duct with a negative camber distribution brings more flow into the actuator disk, thus the total energy in Equation (16) is not appropriate and will end up with very high efficiency, as shown in later sections. To solve this problem, another efficiency in Equation (17).

$$\eta_2 = \frac{Q\omega}{\frac{1}{2}\pi\rho V_S^3 R_D^2} \tag{17}$$

where R_D is the maximum radius of the duct outer side. The relationship of R and R_D are shown in Figure 5.

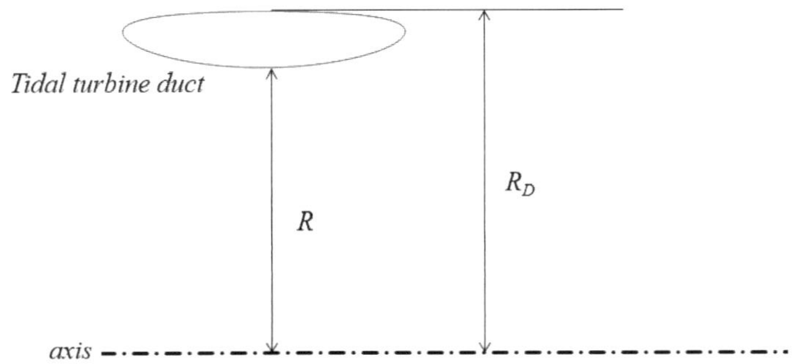

Figure 5 The turbine efficiency defined by two different radii

RESULTS AND DISCUSSION

The following runs for the propeller case have been performed in this paper:

- **Propeller_case1**: lifting line method for open propeller (no hub), without coupling with RANS (only the code LLOPT is used);
- **Propeller_case2**: lifting line method for propeller (no hub) inside a cylindrical tunnel, without coupling with RANS (only the code LLOPT is used);
- **Propeller_case3**: the RANS/lifting line model interaction method for propeller with real duct geometry (no hub) coupled with RANS (the code LLOPT2NS is used).

It should be noted that the current model can run a ducted propeller without a hub, but has not been reliable when a hub is included, so all the propeller cases in this paper are performed without a hub.

The following runs for the tidal turbine case have been performed in this paper:

- **Turbine_case1**: lifting line method for open turbine (with hub), without coupling with RANS (only the code LLOPT is used);
- **Turbine_case2**: lifting line method for turbine (with hub) inside a cylindrical tunnel, without coupling

with RANS (only the code LLOPT is used);

- **Turbine_case3**: the RANS/lifting line model interaction method for turbine with real duct geometry (with hub), coupled with RANS (the code LLOPT2NS is used).

1. The image model

The effects of the image from the lifting line model are shown in Figure 6. In the propeller case, the circulation is positive and in the turbine case, the circulation is negative. If there is no duct or hub, the circulation goes to zero at both the tip and the hub. With the image model, the gradient of the circulation goes to zero near the hub or duct, and the circulation profile is "flat". The full cosine spacing in the propeller case is applied for the case without hub or duct case, because it concentrates more panels near the blade tip and duct, where the gradient of the circulation is big. The half cosine spacing is applied when only one side of the image model is used, either for the hub or duct, and less panels are used where the gradient for the circulation profile is small. For the turbine case with both hub and duct, only the constant spacing is appropriate. The total number of panels is 70 for both cases. More discussion about the spacing and the panel numbers can be found in the appendix.

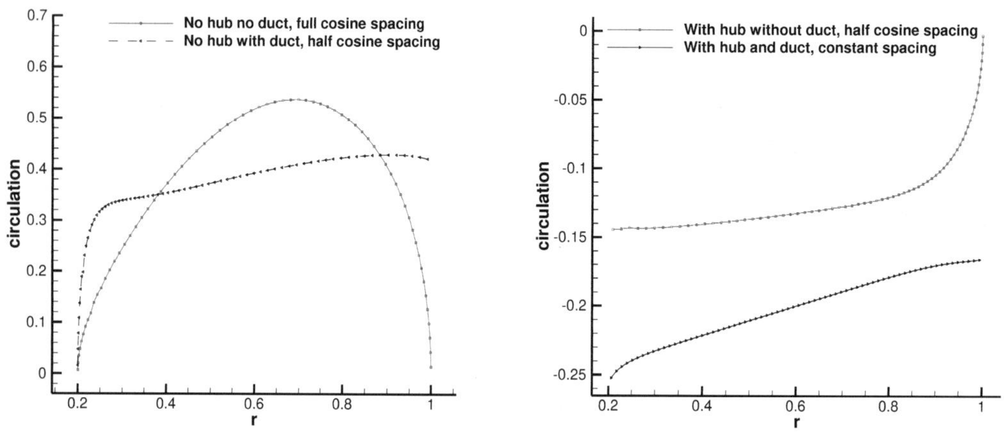

(a) Propeller case **(b) Turbine case**

Figure 6 Effect of the image on the circulation

2. The RANS model

The domain and mesh of the duct in the RANS model are shown in Figure 7. In the propeller case, the camber distribution is NACA a=0.8 and the thickness distribution is NACA 00. In the turbine case, the camber distribution is parabolic and the thickness distribution is elliptic. The geometries shown in the figure as samples are the cases where the camber equals -0.03 and the thickness equals 0.2. Negative camber makes more flow goes into the actuator disk plane and thus improve the performance of propellers and turbines. In both cases, the total grid number is about 50k. To make sure that the first layer of the mesh does not fall into the buffer layer, the y plus on the duct for the propeller case is over 50 and for the turbine case is over 40. The $k - \omega$ with SST turbulence model is used in the RANS solver and the Reynolds number is 10^6.

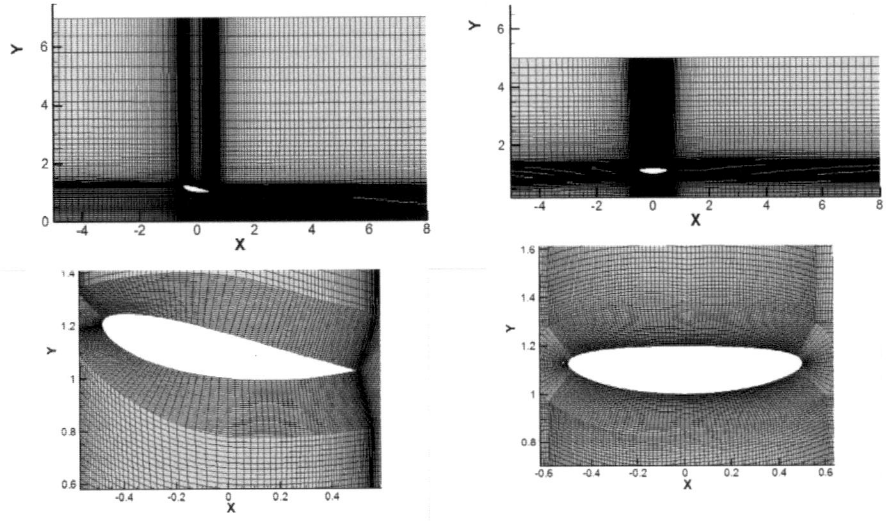

(a) Propeller case **(b) Turbine case**

Figure 7 Sample mesh in the RANS model (f_0=-0.03, t_0=0.2)

The pressure contour plots are shown in Figure 8. There is a discontinuity at the actuator disk because a pressure jump profile is used in the fan boundary condition. The high-pressure side and low-pressure side are consistent with Figure 1.

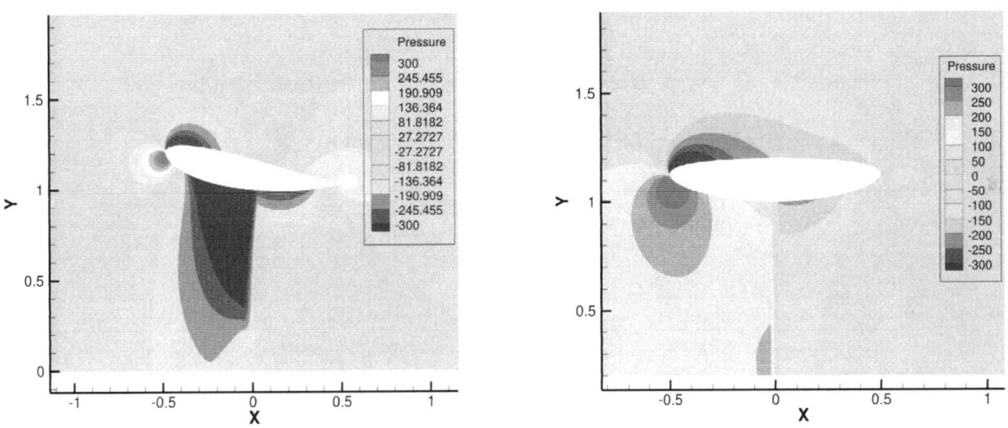

(a) Propeller case **(b) Turbine case**

Figure 8 Pressure contour in the RANS model, pressure jump is shown by the discontinuity at the actuator disk (Plotted with U_{in}=1m/s, $\rho = 1000kg/m^3$ and the reference location is at the far upstream)

3.1. Propeller case, J_s=0.5, C_{Ts}=1, influence of the duct geometry

In this case, the advance ratio and total thrust coefficient defined by the velocity at far upstream are fixed and different duct geometries are used in the RANS/lifting line model interaction method. For the camber, five maximum camber f_0 are used, including -0.01, -0.02, -0.03, -0.04 and -0.05. For each f_0, four t_0 are selected, including 0.15, 0.20, 0.25 and 0.30, so there are 20 different duct geometries in total. In each geometry, the duct is located from -0.5 to 0.5 in the x-direction, and in the r-direction, the middle point of the lower side is always at location (0,1), because the radius of the blade is 1. Each geometry is rotated by 10 degree in the clock-wise direction around point (0,1), giving a 10-degree angle of attack for the inflow coming from left to the right along x-direction. The four most "extreme" geometries are shown in Figure 9.

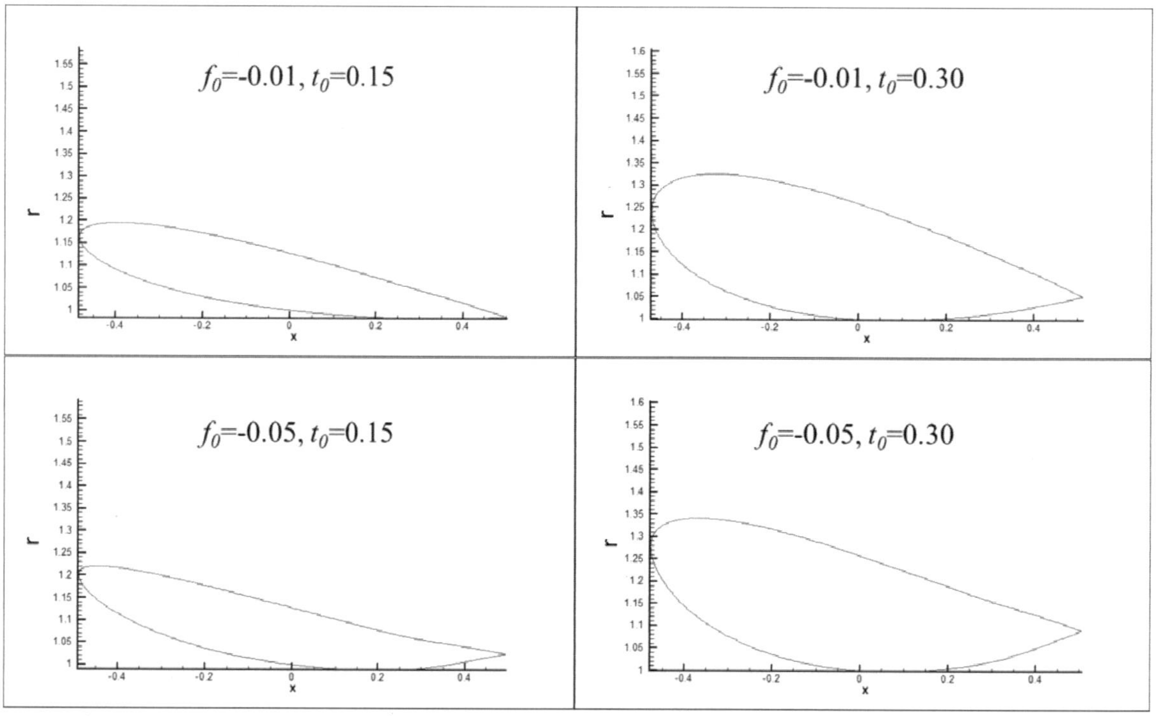

Figure 9 The most "extreme" duct geometries for the propeller case

Table 1 Input for LLOPT in the propeller case

Hub radius	0.2
Number of blades	3
Number of panels	70
κ	0.05
Image model	Duct only
Spacing	Half cosine with more panels near hub

Other parameters for the lifting line model is listed in Table 1 and used as input for LLOPT. The spacing is consistent with the image model. In the code LLOPT2NS, the tolerance for U_{in} and τ are 0.001, and each case converges within several iterations. For the case f_0 =-0.03 and t_0 =0.2, the convergence history for the pressure jump and circulation on the blade are shown in Figure 10, and the convergence history for U_{in} and τ are shown in Figure 11. From those figures, it is shown that the solutions from the first iteration to the second iteration change most comparing with other iterations.

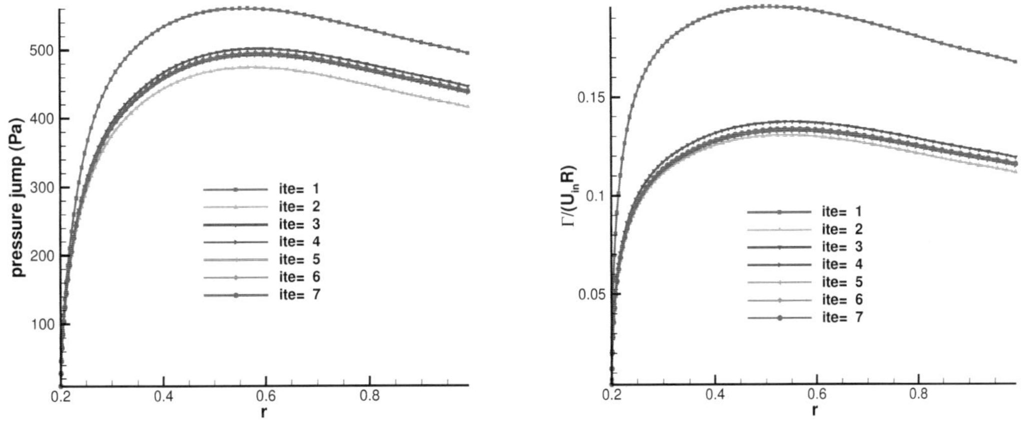

Figure 10 The convergence history of pressure jump and circulation

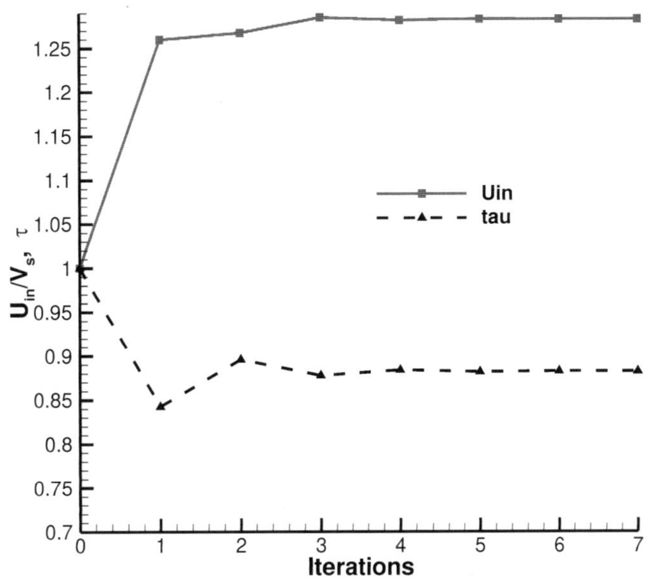

Figure 11 The convergence history of U_{in} (mean inflow) and τ

In Propeller_case3, the efficiencies of different duct geometries are shown in Figure 12, together with the efficiencies from Propeller_case1 and Propeller_case2 for the same advance ratio and thrust coefficient. The KT and 10KQ are plotted in Figure 13 and the inflow velocity and τ are shown in Figure 14. If the duct has bigger camber or thickness, the mean inflow for the actuator disk will be higher, and more thrust will be generated by the duct, so τ is smaller. If the duct geometry is not considered, all the thrust will be generated from the propeller blade, so for given total thrust, the KT will the same for Propeller_case1 and Propeller_case2, as shown in Figure 13. In Figure 14, the thrust contributed from the duct may vary from less than 5% to more than 20%, depending on different duct geometries for the given J_s and C_{Ts}. It is also shown that the increase of camber and thickness of the duct can both increase the efficiencies.

However, based on the following three reasons, the conclusion in this part must be evaluated more carefully and further tested in the future. First, in this paper the Lerbs-Wrench formulas are used for the wake alignment procedure, which is suitable in this iterative method due to its high computational efficiency, but may not be as reliable as the more complicated full wake alignment model (especially in the case of turbines). By using full wake alignment, the results of both the propeller case and turbine case might change. Second, the swirl component of the induced velocity is not considered in the RANS model, assuming that neglecting the swirl will not influence the thrust on the duct and the axial velocity in the inflow. This assumption needs to be better justified and the torque might be represented by the body force in the tangential direction, which will be addressed in future work. Third, as shown in Figure 14, the mean inflow velocity for some duct geometries can be 50% higher than the velocity at far-upstream, which might cause cavitation or separation and the high efficiency might not be achieved. The efficiency needs to be further verified after the blade geometry is designed based on the circulation, and then tested with other numerical tools, like by coupling of BEM or VLM method with the RANS solver. More details about these methods can be found in Kinnas et al. (2013) and Kinnas et al. (2016).

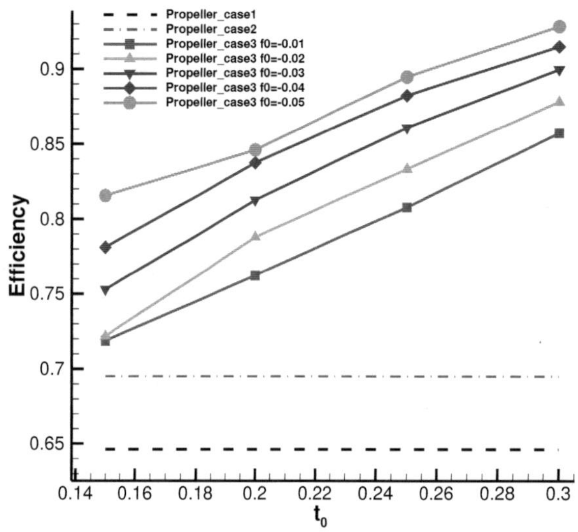

Figure 12 Efficiency for the propeller case for various duct geometries

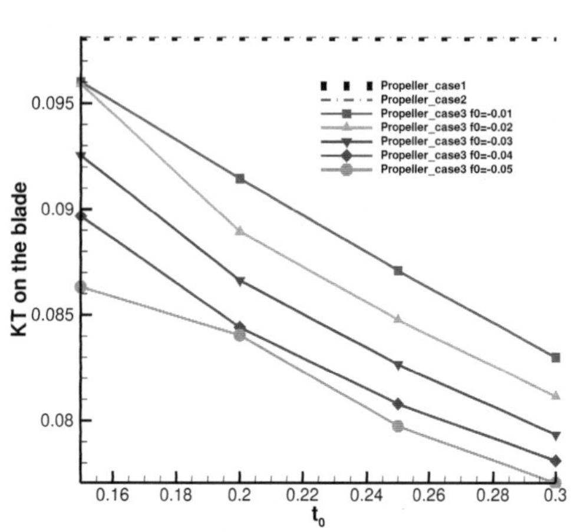

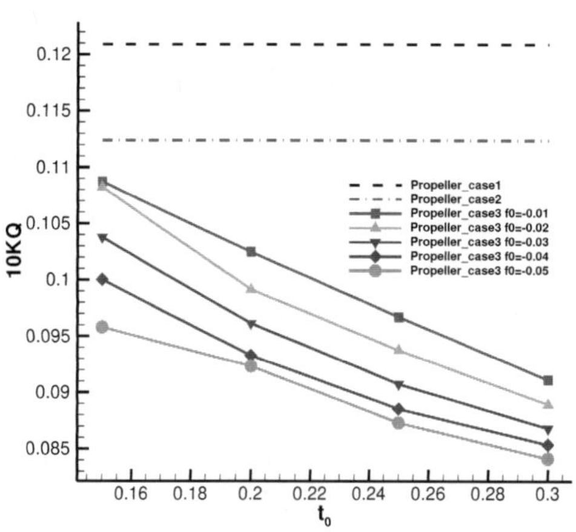

Figure 13 The KT and 10KQ on the blade for the propeller case for various duct geometries

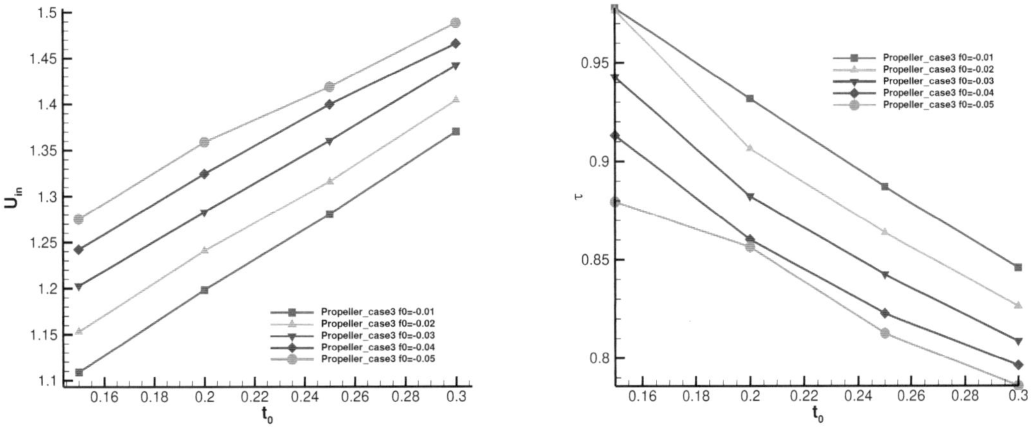

Figure 14 The inflow velocity and τ on the blade for the propeller case for various duct geometries

The inflow velocity and τ are shown in Figure 14. With higher camber and thickness, the duct brings more flow into the actuator disk, so the inflow velocity is higher. The ratio of thrust generated from the duct increases with the increase of camber and thickness.

3.2. Propeller case, f_0 = -0.03 and t_0 = 0.2, influence of J_s and C_{Ts}

In this case, the duct geometry is fixed with f_0 = -0.03 and t_0 = 0.2. The duct shape is shown in Figure 7. In Figure 15(a), the advance ratio changes from 0.5 to 1.0 for given C_{Ts} as 0.5, and in Figure 15(b) the total thrust coefficient is selected among 0.5, 1.0, 1.5, 2.0, 2.5 and 3.0 for given J_s as 0.5. Other parameters are the same as in Table 1. From the figure, in Propeller_case1, there is an advance ratio that the efficiency is highest for given C_{Ts}. In Propeller_case2 and Propeller_case3, the efficiency increase with the increase of advance ratio for given C_{Ts}. For fix J_s, the efficiency increase with the increase of C_{Ts} in Propeller_case3, because the duct provides more thrust.

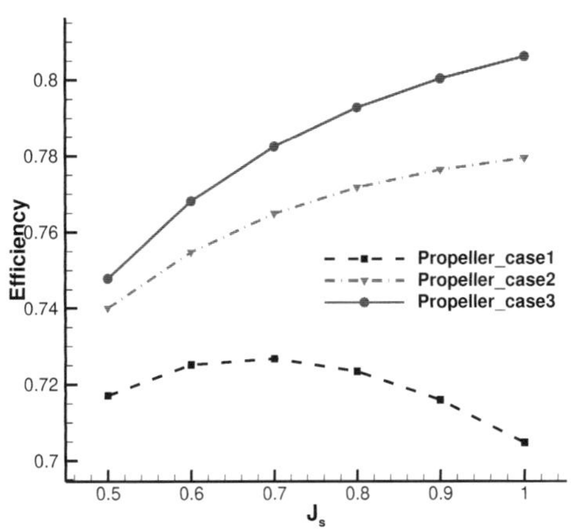

(a) Fixed C_{Ts} as 0.5

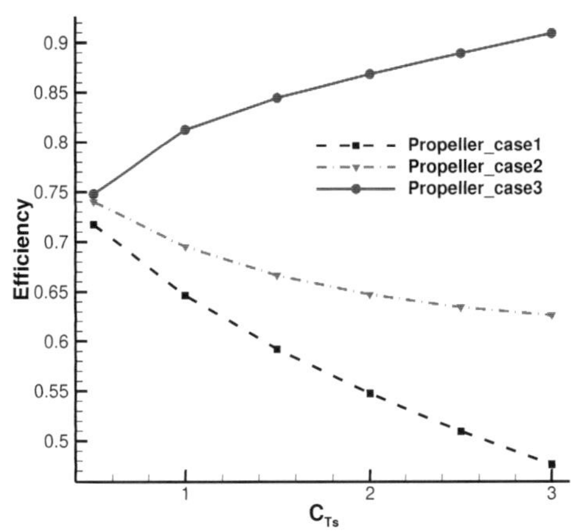

(b) Fixed J_s as 0.5

Figure 15 The efficiencies for a fixed duct geometry with changing J_s and C_{Ts}

The inflow velocity and τ are shown in Figure 16. With the increase of J_s and C_{Ts}, the inflow velocity and the thrust on the duct both increase.

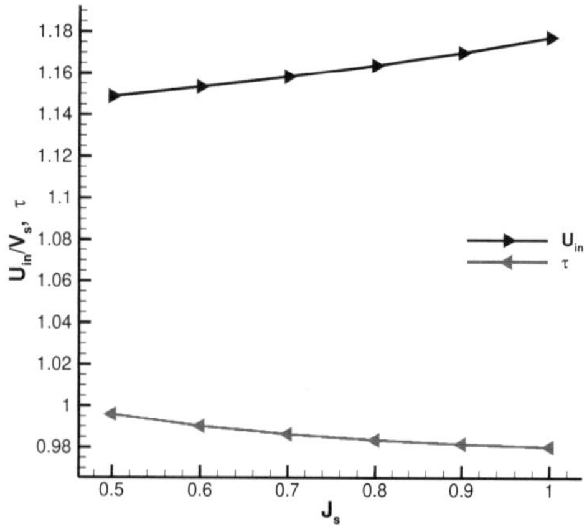

(a) Fixed C_{Ts} as 0.5

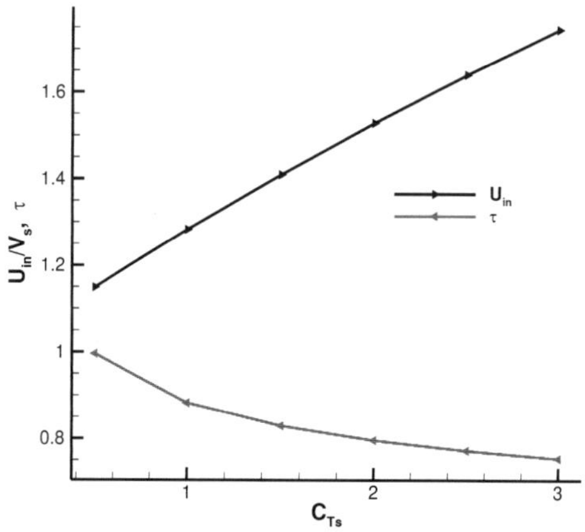

(b) Fixed J_s as 0.5

Figure 16 The inflow velocity and τ for a fixed duct geometry with changing J_s and C_{Ts}

3.3. Propeller case, influence of the blade number

In all the above cases, the number of blade is kept as 3. The influence of the blade number on the efficiency is studied in this section and results are shown in Figure 17. In Propeller_case3, the duct geometry is $f_0 = -0.03$ and $t_0 = 0.2$, as shown in Figure 7. The advance ratio is 0.5 and thrust coefficient is 0.5, and other parameters are the same as in Table 1.

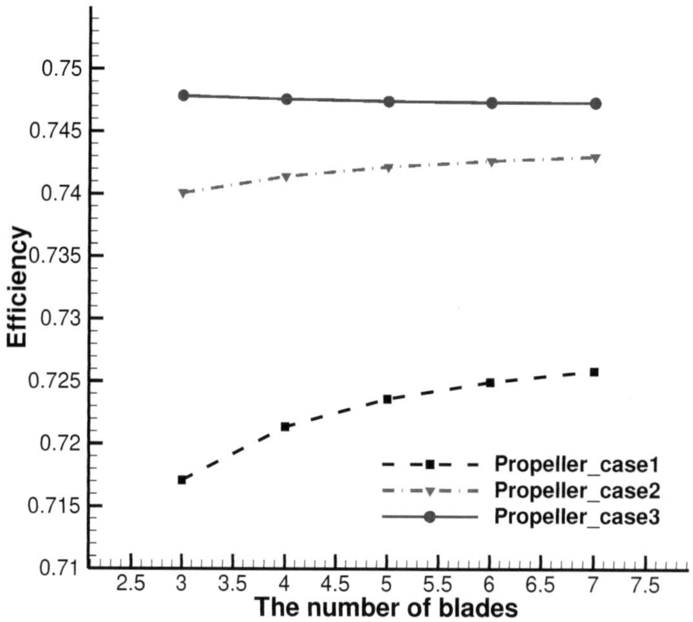

Figure 17 The influence of blade number on the efficiency, fixed duct shape, J_s and C_{Ts}

As shown in the figure above, as the blade number increases, the efficiencies from Propeller_case1 and Propeller_case2 increase but the efficiencies from Propeller_case3 decrease slightly.

3.4. Propeller case, influence of κ

In the lifting line model, the viscous effect on the blade is taken into consideration by adding a drag-to-lift ratio κ. In the above sections κ is kept constant as 0.05. In this section, κ varies from 0 (without any viscous effect on blade) to 0.05. The duct geometry is f_0= -0.03 and t_0= 0.2, as shown in Figure 7. The advance ratio is 0.5 and thrust coefficient is 0.5, and all the other parameters are the same as in Table 1.

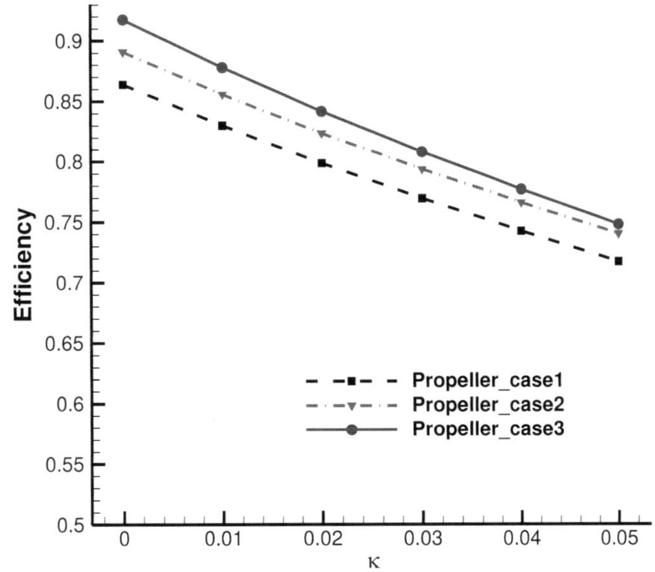

Figure 18 The influence of drag-to-lift ratio on the efficiency, fixed duct shape, J_s and C_{Ts}

As shown in Figure 18, as the drag-to-lift ratio increases, the efficiency decreases because more energy is consumed by friction.

4.1. Turbine case, TSR=10, influence of the duct geometry

In this case, the tip speed ratio, defined by the velocity at far upstream, is fixed as 10 and different duct geometries are used in the RANS/lifting line model interaction method. The parabolic camber distribution and elliptic thickness distribution are used to make the duct geometry symmetric. The camber and thickness selections are the same as in the propeller case (f_0= **-0.01, -0.02, -0.03, -0.04 and -0.05;** t_0= **0.15, 0.20, 0.25 and 0.30**), and the four most "extreme" geometries are shown in Figure 19.

Other parameters in the lifting line model is shown in Table 2. Since the image model is used near both hub and duct, the circulation is expected to have zero slope both near the tip and near the root, so only the constant spacing is appropriate in this case.

Table 2 Input for LLOPT in the turbine case

Hub radius	0.2
Number of blades	3
Number of panels	70
κ	0.05
Image model	Hub and duct
Spacing	Constant

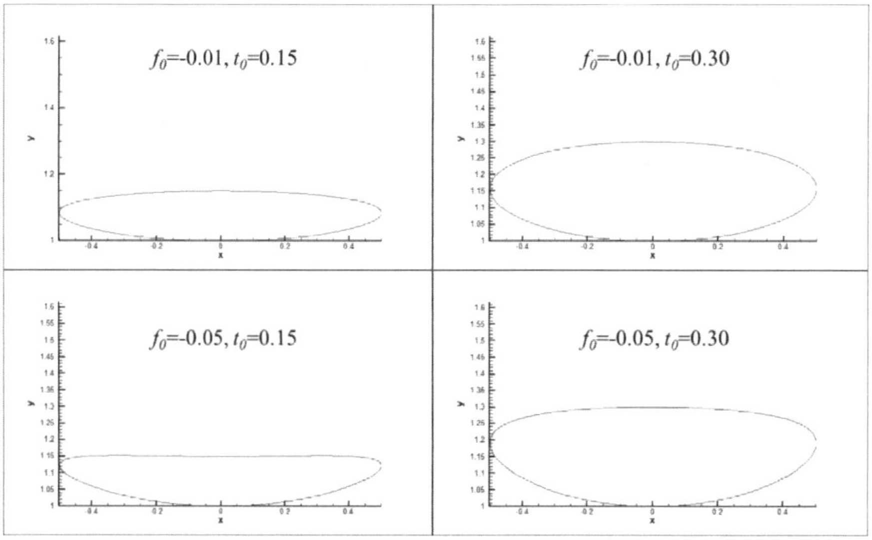

Figure 19 The most "extreme" duct geometries for the tidal turbine case

The convergence history for the pressure jump and velocity inflow are similar with that of the propeller case, so it is not shown here. It should be noted that in the RANS/lifting line model interaction method for the turbine case only the velocity inflow is updated during each iteration, and the pressure jump is negative, as shown in Figure 8. The efficiencies, defined in Equation (16), are shown in Figure 20.

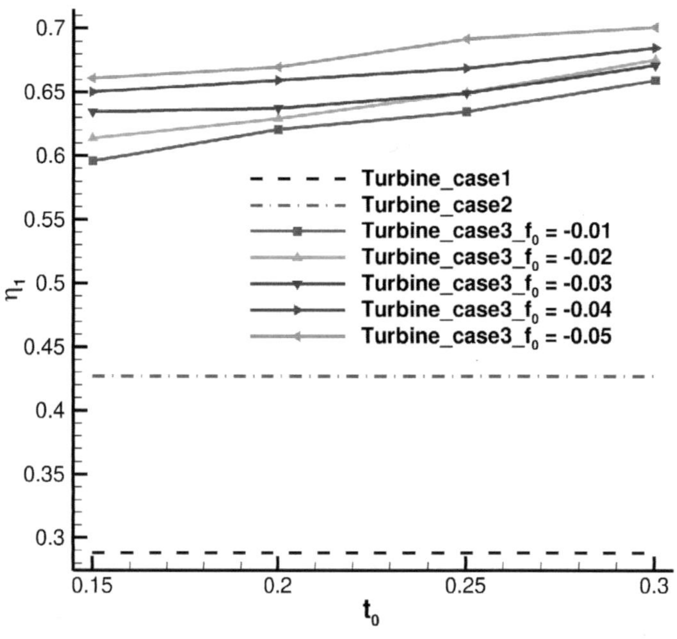

Figure 20 The efficiencies of different duct geometry, η_1 is based on the turbine radius, which is smaller than the outer radius of the duct (refer to Figure 5 and 21 as well)

As shown in Figure 19, the ducted tidal turbine has higher efficiency that the tidal turbine without duct in the same condition (tip speed ratio, hub radius, etc.). The efficiencies predicted by the LLOPT2NS method is much higher than the efficiency from the image model. For the same camber, the efficiency increases as the thickness and camber increase, for most points. However, as discussed in the introduction section, the turbine efficiency is defined as the useful power output over the total power in the actuator disk. The existence of the turbine duct increases the area and the total power. The other comparison for efficiency η_2, as defined in Equation (17), is shown in Figure 21.

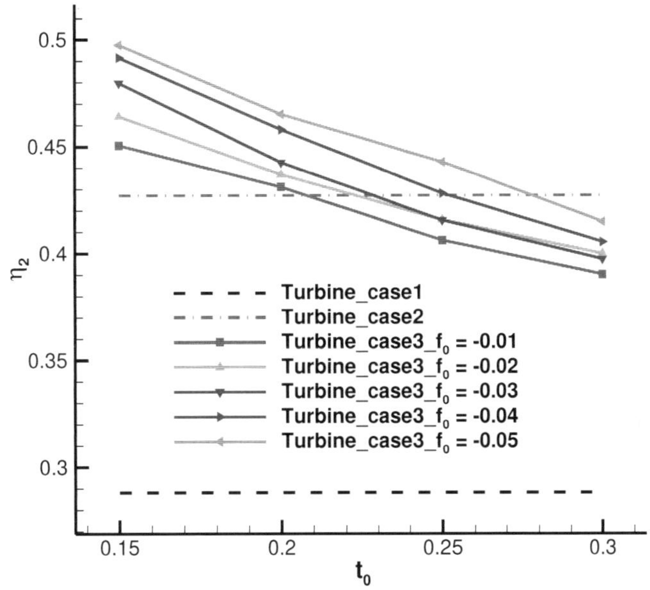

Figure 21 The efficiencies (η_2) of different duct geometry, η_2 is based on the largest duct radius

As shown in the above figure, under this definition, the efficiencies from LLOPT2NS decrease as the duck thickness increases because the area for evaluating the total power also increases. It should be noted that not all the duct geometries have higher efficiency than the result from only using the image model for the duct.

4.2. Turbine case, $f_0 = -0.03$ and $t_0 = 0.2$, influence of TSR

In this case the duct geometry is fixed with $f_0 = -0.03$ and $t_0 = 0.2$, as shown in Figure 7, and other parameters are the same as in Table 2. The tip speed ratio changes from 7 to 10 and the efficiencies are shown in Figure 22. With the increase of the tip speed ratio, the efficiencies decrease for all the three cases.

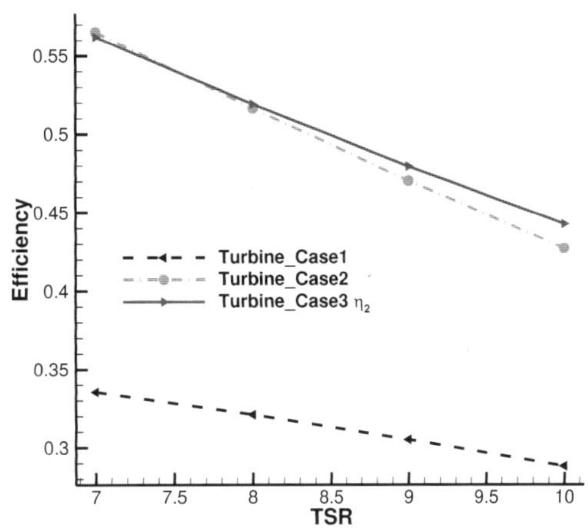

Figure 22 The efficiencies (η_2) of different tip speed ratio

CONCLUSIONS AND FUTURE WORK

In this paper, a RANS/lifting line model interaction method is proposed for the design of ducted propellers and tidal turbines, and a numerical code called LLOPT2NS is developed and tested for different cases. Results show that this method is efficient and reliable. In the lifting line model, the image effect on the circulation is considered for the hub and the duct, and the appropriate spacing is adopted. Lerbs-Wrench formulas are used for the wake alignment procedure. In the RANS model, flow around the duct with real geometry is solved and results are coupled with the lifting line model in an iterative way. The viscous effect is considered in the lifting line model by adopting a drag-to-lift ratio and in the RANS model by the non-slip wall boundary condition. In the propeller case, the duct is built with a NACA a=0.8 camber, a NACA 00 thickness distribution and a fixed angle of attack of 10 degree. In the tidal turbine case, a parabolic camber and an elliptic thickness distribution are used to have a symmetric geometry so that it can work with flow coming from both directions. In both cases the camber is negative so the duct can bring flow into the actuator disk.

In the lifting line model, by comparing the efficiencies with the open case and the duct case with image model, it is shown that the duct can increase the efficiency, but without coupling with the RANS method, the duct geometry is not considered.

In the RANS/lifting line model interaction method, with the increase of camber and thickness, the efficiency, local inflow velocity and thrust on the duct all increase while the KT and 10KQ on the blade decrease for the propeller case. For fixed duct geometry, it is shown that the efficiency and local inflow velocity increase with the increase of advance ratio and total thrust coefficients. As the blade number and drag-to-lift number increase, the efficiency for the ducted propeller decreases.

In the turbine case, a more appropriate efficiency definition is proposed. By using this definition, it is shown that to increase the efficiency, the camber should be increased while the thickness should be decreased. For fixed duct geometry, the efficiency decrease as the tip speed ratio increase.

In the future, the effect of the angle of attack can be studied for the propeller case. Fully aligned wake alignment model can replace the Lerbs-Wrench formula in both the propeller case and turbine case. The swirl component of the induced velocity can be added in the RANS model as body forces in the tangential direction. Furthermore, the blade geometry can be designed based on the optimal circulation from this method, and analyzed by couple the panel method or the vortex lattice method with the RANS solver to see if the circulation can be recovered.

ACKNOWLEDGEMENTS

We wish to thank the people who helped make this work possible: the members of Ocean Engineering Group (OEG), especially, graduate student Mr. Yiran Su and former graduate student Mr. Ye Tian. We also wish to thank Mr. Stanislas Barbraud, Mr. Martin Croué and Mr. Hagen Fritz, former students worked in OEG, for their attempts in developing the early stage of the work in this paper. This work was partly supported by the office of the Offshore Technology Research Center at UT Austin and Phase VII of the "Consortium on Cavitation Performance of High Speed Propulsors" with the following members: Kawasaki Heavy Industry Ltd., Rolls-Royce Marine AB, Rolls-Royce Marine AS, SSPA AB, Andritz Hydro GmbH, Wärtsilä Netherlands B.V., Wärtsilä Norway AS, Wärtsilä CME Zhenjiang Propeller Co. Ltd., and Steerprop Ltd.

REFERENCES

Kerwin, Justin E., and Jacques B. Hadler. "Principles of Naval Architecture Series: Propulsion." The Society of Naval Architects and Marine Engineers (SNAME) (2010).

Kerwin, Justin E., and Chang-Sup Lee. Prediction of steady and unsteady marine propeller performance by numerical lifting-surface theory. No. Paper No. 8. SNAME Transactions, 1978.

Kinnas, Spyros A., Chan-Hoo Jeon, Jay Purohit, and Ye Tian. "Prediction of the unsteady cavitating performance of ducted propellers subject to an inclined inflow." In International symposium on marine propulsors SMP, vol. 13. 2013.

Kinnas, Spyros A., Yiran Su, Weikang Du, and Seungnam Kim. "A viscous/inviscid interactive method applied to ducted propellers with ducts of sharp or blunt trailing edge." In 31st Symposium on Naval Hydrodynamics Monterey, 2016

Kinnas, Spyros A., Wei Xu, Yi-Hsiang Yu, and Lei He. "Computational methods for the design and prediction of performance of tidal turbines." Journal of Offshore Mechanics and Arctic Engineering 134, no. 1 (2012): 011101.

Lee, Jin-Tae. A Potential Based Panel Method for the Analysis of Marine Propellers in Steady Flow. No. OE-87-13. Massachusetts Institute of Technology, Department of Ocean Engineering, 1987.

Lerbs, Hermann W. Moderately loaded propellers with a finite number of blades and an arbitrary distribution of circulation. Society of Naval Architects and Marine Engineers, Transactions, 1952.

Menéndez Arán, David Hernán. "Hydrodynamic optimization and design of marine current turbines and propellers." Master's Thesis, Ocean Engineering Group, The University of Texas at Austin (2013).

Menéndez Arán, David H., and Spyros A. Kinnas. "On Fully Aligned Lifting Line Model for Propellers: An Assessment of Betz condition." Journal of Ship Research 58, no. 3 (2014): 130-145.

Wrench Jr, J. W. The Calculation of Propeller Induction Factors AML Problem 69-54. No. DTMB-1116. DAVID TAYLOR MODEL BASIN WASHINGTON DC, 1957.

Xu, Wei. "Numerical techniques for the design and prediction of performance of marine turbines and propellers." Master's Thesis, Ocean Engineering Group, The University of Texas at Austin. (2010).

APPENDIX

Propeller case and duct case validation: spacing, convergence study and different initial conditions

For different cases, different spacing should be used, and the corresponding circulations are shown in Figure 23. In the shown figure, both the advance ratio and the thrust coefficient for the propeller case are 1.0; in the turbine case, the tip speed ratio is 8. Other parameters are the same as in Table 1 and 2. In constant spacing with one ¼ inset and the half cosine spacing, more elements are concentrated where the gradient of the circulation is high. As shown in the figure, the half cosine or full cosine spacing agrees well with the constant spacing with one or two ¼ inset. In this paper, the propeller case has a duct but has no hub, so half cosine spacing is adopted. In the turbine case, the image models for both hub and duct are used, so constant spacing is applied.

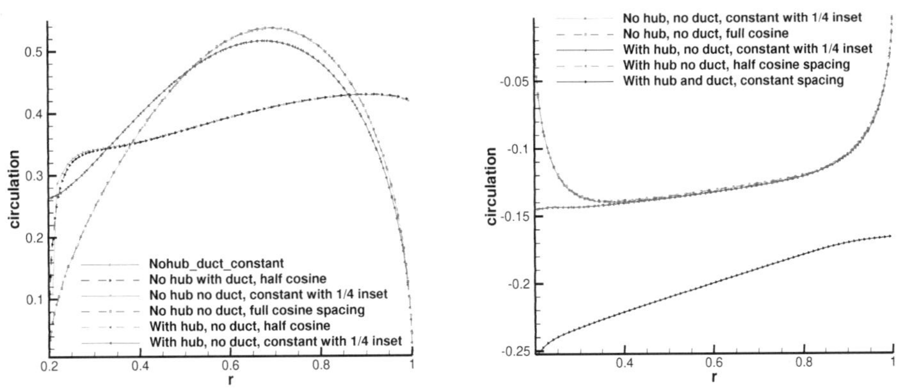

(a) Propeller case **(b) Turbine case**

Figure 23 The circulation from different spacing

In this paper, 70 panels are used in the lifting line model for both the propeller case and the turbine case. As shown in Figure 24, smaller panel number is enough for both cases. However, since the lifting line code can run very fast, 70 panels are used through this paper for both propeller case and turbine case.

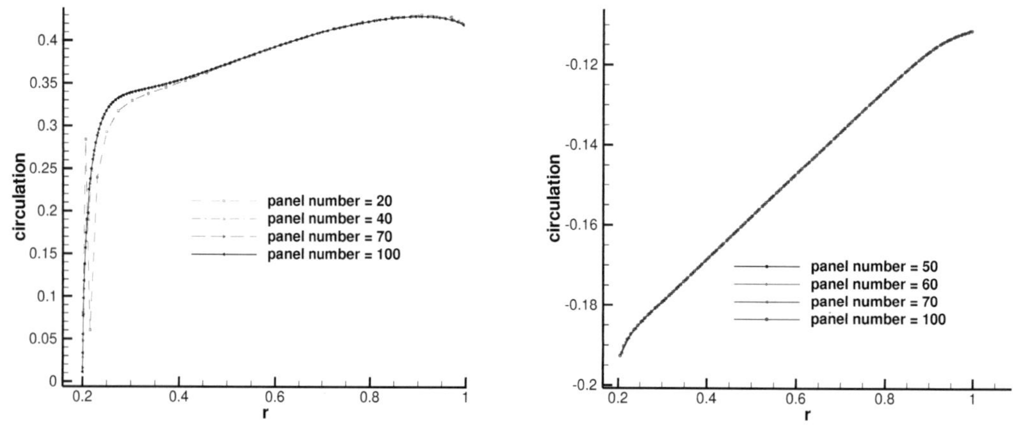

(a) Propeller case **(b) Turbine case**

Figure 24 Convergence study for the lifting line model

To show that the RANS/lifting line model interaction method does not depend on the selection of initial values, three sets of initial inflow velocity and τ are selected in the propeller case as: U_{in}=1.0, τ = 1.0; U_{in}=0.8, τ = 1.2; U_{in}=1.2, τ = 0.8. After running LLOPT2NS, all the three cases converge to the same final solution, as shown in Figure 25.

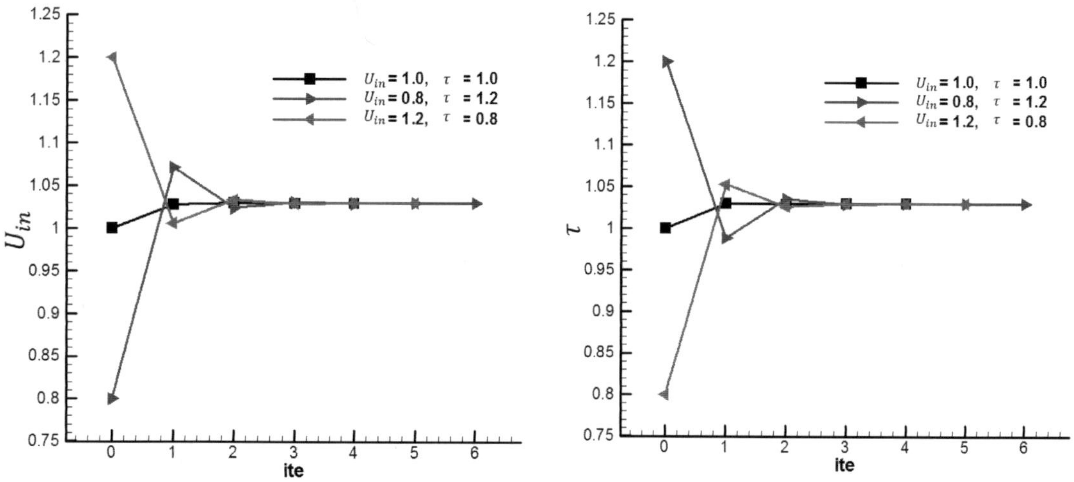

Figure 25 A propeller case by selecting different initial conditions

SNAME
THE INTERNATIONAL COMMUNITY
FOR MARITIME AND OCEAN PROFESSIONALS

A VIABLE DRY TREE SEMI-SUBMERSIBLE CONCEPT WITH TAPERED COLUMNS

WEI YE, ALEX RAN, JIM LI
OFFSHORETECH LLC, HOUSTON, TEXAS, USA

GUOJIE LI, XIUJUN TANG, ZHONGNIAN WANG, HONGWU WU
SANY MARINE HEAVY INDUSTRY CO., LTD, ZHUHAI, GUANGDONG, CHINA

Proceedings of the 22nd Offshore Symposium, February 2017, Houston, Texas
Texas Section of the Society of Naval Architects and Marine Engineers

ABSTRACT

A Tapered Column Deep Draft Semi-submersible (TCDD-Semi) concept has been developed for dry tree drilling and production applications. The new concept features enlarged lower hull geometry with gradually varying column cross-section, in contrast to conventional semi-submersible design with uniform columns. The enlarged lower columns, in combination with properly sized upper columns and pontoons, provide sufficient buoyancy and stability for topside integration at quayside and offshore pre-service operations such as submergence and wet-tow. Most importantly, this new TCDD-Semi concept with tapered columns optimizes the cancelation effect of wave force and is able to reduce the heave motion in harsh environmental condition for dry tree applications with top-tensioned risers.

This paper will present the topside and hull configuration, tank arrangement and ballasting system, stability, as well as mooring system, riser system and global performance design, platform global strength analysis and global fatigue analysis, hull vortex induced motion analysis. In addition, platform fabrication and topside integration methods will also be discussed.

The design analyses show that the proposed TCDD Semi-submersible platform has advantages in heave motion performance, topside integration flexibility, and transition stability during pontoon submergence. Its excellent heave motion performance enables the application of top-tensioned riser system with 35 feet (10.5 m) stroke limit. The new design also significantly reduces platform VIM compared with conventional deep draft semi-submersible.

Keywords: Floating Platform, FPS, Tapered Column, Deep Draft, Dry Tree, Semi-submersible, TTR stroke

INTRODUCTION

Compared with wet tree development, dry tree development with vertical access top tensioned risers (TTRs) allows direct monitoring, control, intervention of subsea wells, improves hydrocarbon recovery rates. Drilling, completion and work-over operation with a dry tree platform also provides benefits in operational costs and project schedules. Currently, only tension leg platforms (TLPs) and Spar platforms have been used as dry tree floaters because of their superior heave motion responses in waves. However, due to constraints in water depth limitation, payload and topside deck area, as well as costly and complex offshore installation operations of TLP and Spar platforms for deep and ultra-deep water drilling and operations, alternative dry tree concepts need to be developed. A dry tree semi-submersible can be a viable solution considering that it has no water depth limitation, allows quayside integration and commissioning, and provides larger topsides capacity.

The relatively large heave motions exhibited by many conventional semi-submersible platforms usually allow the use of steel catenary risers (SCRs) for wet tree service. The catenary shape of SCRs is able to accommodate and absorb the large heave motions and horizontal motions of the semi-submersible platform. However, due to their large motions, especially those in vertical direction, conventional semi-submersible platforms usually do not support top tensioned risers (TTR) with wellhead Christmas tree above waterline.

During recent years, dry tree semi-submersible concepts have been developed by either increasing the riser system compliancy, or improving the platform motion responses in waves. Halkyard et al. (2002) developed a deep draft semi-submersible system with a retractable heave plate to reduce heave motions. Mungall et al. (2004) proposed a semi-submersible FPS used as dry tree platform in Gulf of Mexico with Compliant Vertical Access Risers. Zen et al. (2014) presented solutions for a Deepwater Dry Tree Semi-submersible with TTRs and pointed out that the platform vertical motion is the critical challenge for a feasible and cost-effective concept. Muehlner and Banumurthy (2015) presented a new Offset-Pontoon Semi-submersible (OPS) featured pontoons that were offset to the outside of the columns which lent the vessel a low heave response. Das and Zou (2015) introduced the Paired-column Semi-submersible (PC-semi) concept and compared some characteristic responses with a traditional semi-submersible platform. It was proven that the motion characteristics of the PC-semi benefited from its unique hull form.

Generally speaking, increasing the draft of a semi-submersible offshore platform is one of the most effective ways to reduce platform motion in the vertical plane. A deep draft semi-submersible offshore platform with draft exceeding 45 m (~150 ft) usually has smaller vertical and rotational motions than those of a conventional shallower draft semi-submersible platform. In the meantime, tensioner technology is also progressing rapidly to address the continuous industry demands for longer stroke and higher capacity riser tensioning systems. Although tensioners currently installed on offshore floating units have stroke range of 8.5 m (28 ft) or less, new 10.5 m (35 ft) long stroke tensioner has already been designed and successfully tested (Koos 2013). Therefore, 10.5 m (35 ft) stoke tensioners are considered for the TCDD Semi-submersible design.

In terms of construction, topsides integration in the shipyard is the safest and most economical method. However, quayside water depths are usually about 9.0 – 10.5 m (30 – 35 ft), and thus, for quayside integration, the hull must provide sufficient buoyancy to support its own weight and topside weight while maintaining a draft less than 9.0 – 10.5 m (30 – 35 ft). It has been a challenge to maintain such a shallow draft at the quayside location with semi-submersible platforms designed for deep draft deployment. Alternatively, the semi-submersible hull and topsides integration may take place at a near shore site with deeper water depth, or at its operation site. In either case, the procedure is typically executed offshore, is performed in open seas, and is strongly dependent on weather conditions and the availability of a heavy lift barge.

After hull and topsides integration, a semi-submersible platform is usually transported to the operation site by wet tow or with a heavy transportation vessel. Both methods involve ballasting down the hull during pre-service operations. During the ballasting process, the stability of the floating structure typically decreases during the draft increases and the pontoons transition from being partially to fully submerged. This may be particularly problematic for deep draft semi-submersibles due to the increased length of the columns and the height of the topsides supported thereby. The pre-service stability issue will also be duly addressed during the TCDD Semi-submersible design.

Moreover, compared to conventional semi-submersibles, deep draft semi-submersible platforms have lengthened columns. Longer columns are more susceptible to a phenomenon known as "vortex-induced-motion" (VIM). On deep draft semi-submersible platforms with longer columns, VIM excitation forces are typically higher than those on conventional semi-submersibles with shorter columns, and hence, deep draft semi-submersible platforms are more likely to experience larger VIM motions and VIM lock-in. Xu (2011) and Kyoung et al (2014) introduced a new semi-submersible hull design with column blisters, and Zou et al. (2014) presented a Dry Tree Paired-Column Semi-submersible concept. Both design concepts were able to improve platform VIM response in strong currents. In this paper, the authors present a different approach featuring gradually varying lower column shape in order to reduce the TCDD Semi-submersible VIM responses.

In this paper, a Tapered Column Deep Draft Semi-submersible (TCDD-Semi) concept is presented for dry tree drilling and production application. The initial design and analysis have proven that this new concept TCDD Semi-submersible can effectively reduce heave motion responses, provide adequate buoyancy for quayside construction and integration, improve stability for pre-service operations, as well as minimize platform vortex induced motions.

DESIGN BASIS AND ENVIRONMENTAL CONDITIONS

The perspective field is located at an anonymous block in South China Sea with a production capacity of 5 billion m^3/yr (483.8 MMSCFD). The water depth in the vicinity of the platform location is approximately 1500 m. The TCDD Semi-submersible is designed to accommodate 12 dry tree risers (1 drilling riser, 11 production risers and/or water injection risers) that will be top-tensioned risers (TTR) and 10 steel catenary riser (SCR) tie-backs, and 2 export SCRs. The host facility will be anchored with a spread mooring system to the seabed. Gas and fluid condensate produced by the facility will be exported through one gas export SCR and one oil export SCR routed to the west of the platform. **Table 1** summarizes the design basis for the TCDD Semi-submersible platform.

Table 1. TCDD Semi-submersible Design Conditions Summary

Description	Value	Unit
Water Depth	1500	m
Gas Process Capacity	5×10^9	m^3/yr
Operating Life	20	year
Well Bay Slots	15	-
Drilling TTR	1	-
Production TTRs	11	-
Shut-in Tubing Pressure	34.47	MPa
SCR Tiebacks	12	-
Export Risers	1 Oil and 1 Gas	-

The environmental condition at the project site is characterized by predominant winds from northeast all year round. Consequently, the predominant wave direction is also from northwest most of the year, except for summer when strong wind and high wave due to typhoon may come from the southwest. The prevailing current direction is from west or west southwest all year around.

For the hull and topside structure design, the operating condition is 100-yr non-typhoon condition and the extreme condition is 100-yr typhoon condition, as defined in **Table 2**.

Table 2. Extreme and Operating Environmental Criteria

Condition	Dominant Environment	Wind (m/s)		Wave		Current (cm/s)
		1-hr	1-min	Hs(m)	Tp(s)	Surface
Typhoon	Wind	**40.9**	**51.5**	12.8	15.5	172
	Wave	37.0	46.6	**13.8**	**16.1**	174
	Current	34.8	43.8	11.7	15.1	**202**
Non-typhoon	Wind	**20.0**	**25.2**	7.3	12.5	57
	Wave	18.4	23.2	**7.5**	**12.7**	54
	Current	15.7	19.8	6.4	11.9	**63**

TOPSIDES AND HULL CONFIGURATION

The TCDD Semi-Submersible is sized to have adequate stability, strength, and column spacing as a permanently sited floating system with dry tree facilities. A ring pontoon arrangement is employed for this unit, which has four columns and four pontoons as shown in Figure 1.

Figure 1. 3D Model of TCDD Semi-Submersible (Patent Pending)

The TCDD Semi-submersible topside has two decks: lower deck and main deck. The total topside weight in operating condition will be 19,624 MT, including structure weight 6,884 MT and equipment weight 12,740 MT.

The column top has been set with a freeboard of 16.75 m. The lower deck has been set at a level of 21.354 m TOS and the main deck has a level of 31.354 m TOS. The estimated total topside deck area is 10,300 m2 (Table 3).

The platform hull has a deep draft of 48.75 m, and an operation displacement of around 65,000 MT. The pontoon is 55.50 m long and its cross section is 11.50 m in width and 9.00 m in height. The pontoons are to be filled with sea water (open to sea) after being installed and in in-place condition.

Table 3. Estimated Topside Deck Area and Elevation

Level	Deck Area	TOS Elevation
Lower Deck	6,400 m^2	21.354 m
Main Deck	3,900 m^2	31.354 m
Total Deck Area	**10,300 m^2**	-

The column height is 65.50 m, which is sufficient to support the topside deck with adequate clearance above the highest waves. The column cross section varies in different elevations. Lower column is the connection section between pontoon and column, which is also typically called Node. The lower column has a square cross section with side length of 25.00 m; the upper column has a square cross section with side length of 19.00 m. In the transition between lower and upper columns, the cross section gradually narrows from 25.00 m square to 19.00 m square. This hull geometry helps maintain adequate platform stability through the critical range of drafts. Inside each column, it includes an 8 m x 8 m square access shaft. The spacing between column centers is 69.50 m.

There are six (6) decks in each column and they locate from elevation 0.00 m (keel) to 65.50 m (Deck 6). Table 4 shows the main particulars of the TCDD Semi-submersible hull.

Table 4. Principal Dimensions of the Hull

Description	Value	Unit
Column Centerline Spacing	69.50	m
Column Height	65.50	m
Column Width from EL 0.00 m to 12.00 m	25.00	m
Column Width from EL 32.00 m to 65.50 m	19.00	m
Pontoon Width	11.50	m
Pontoon Length	50.50	m
Pontoon Height	9.00	m
Access shaft	8 x 8	m^2
Overall Length	100.50	m
Overall Width	100.50	m
Operating Draft	48.75	m

Operating Displacement	64,827	MT

The platform weights in operation condition are listed in Table 5. In stability analysis, the buoyancy and ballast weight of the flooded tanks (permanent ballast tanks) are not considered base on industry practice; while in global performance analysis, they both will be taken into account. Therefore, semi-submersible total weight is given in the table in two different ways: with and without permanent ballast weight.

Payloads

According to the weight information listed in Table 5, the payloads are calculated as follows:

Topside operating weight:	19,624 MT
Hull fluid:	926 MT
Vertical riser load:	4,044 MT
Total operating payloads:	24,594 MT

Table 5. Platform Weight in Operating Condition

Item	Weight (MT)
Topside	
Structure	6,884
Fixed-Weight Equipment (Dry Weight)	
Main Deck Equipment	2,107
Cellar Deck Equipment	4,099
Drilling Equipment (Moving)	2,228
Drilling equipment (Fixed)	955
Variable-Weight Equipment	
Main Deck Equipment	89
Cellar Deck Equipment	550
Drilling Equipment (Moving)	759
Drilling equipment (Fixed)	1,960
Total Topside Weight	**19,631**
Hull	
Hull Structure	14,249
Hull Structural Fittings	3,443
Corrosion Protection	461
Mooring Equipment	1,436
Piping and HVAC System	690
Top of Column Equipment & Steel Outfitting	55
Hull Fluids	926
Electrical	4
Marine Growth	330
Total Hull Weight	**21,594**
Ballast	
Permanent Ballast	52,638
Variable Ballast	15,857
Total Ballast	**68,495**
Mooring and Risers	

Mooring Vertical Tension	3,701
SCR Vertical Tension	1,517
TTR Vertical Tension	2,527
Total Mooring and Risers	**7,745**
Total Weight (with permanent ballast)	**117,465**
Total Weight (without permanent ballast)	**64,827**

TANK ARRANGEMENT AND BALLASTING SYSTEM

There are a total of sixteen (16) variable ballast tanks located between EL 12.00 m and EL 32.22 m, with total capacity of 33,839 MT. There are twenty-four (24) permanent ballasts tanks, 16 of which located between EL 0.00 m and EL 12.00 m in the columns, and another 8 tanks located inside the pontoons. The total capacity of permanent ballast is 52,638 MT. In addition, there are sixty-eight (68) void tanks with total volume of 43,412 m3. The compartments as modelled in MOSES software are shown in Figure 2.

During normal operations, bulkheads and watertight walls in the lower hull of a Deep Draft Semi-submersible platform generally need to withstand high hydrostatic pressure, and therefore require higher design standard to ensure structural integrity. The TCDD Semi-submersible in the current design employs pressurized air ballasting system (also known as air-over-water ballasting system) which greatly minimizes the pressure difference between inside the ballast tanks and external seawater pressure, and in turn reduces the hull structural design requirements as well as the structural steel weight.

Figure 2. Hull Compartment Model - Perspective View

The TCDD Semi-submersible hull consists of four columns and four pontoons, which are symmetrical in both directions. The ballast tanks adjusting the draft, position and stability of the platform are arranged in the lower part of the columns. Manifold for variable ballast piping, headers, and control valves are located on top of the columns. The U-shaped pipes for passage of pressurized air and sea water are arranged close to the wall of the access shafts, and penetrate into the variable ballast tanks through the lower part of the compartment wall, and then back up to near the roof of the tank. There are four compressed air ballast tanks in each column and totally 16 compressed air ballast tanks in the TCDD Semi-submersible platform hull. Schematics of the manifolds and U-shaped tube arrangement are shown in Figure 3.

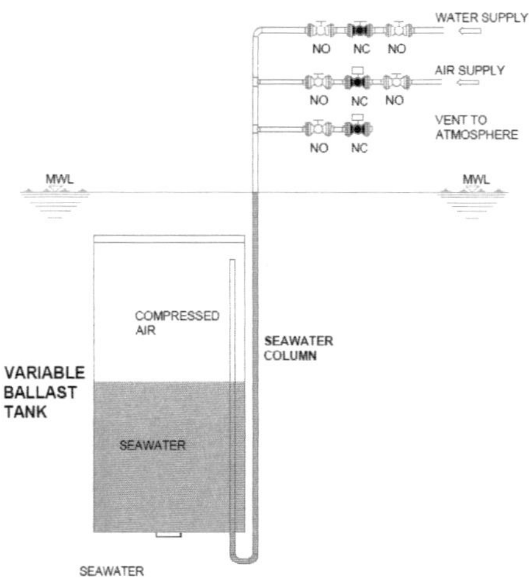

Figure 3. Schematic Illustration of Pressurized Air Ballast System

MOORING SYSTEM

Sixteen taught mooring lines grouped into 4 bundles are used to hold the platform in position. The mooring lines are arranged at four corners of the hull and the mooring system is a permanent system. Every mooring line has the same configuration in which a chain-polyester-chain pattern is employed. The arrangement of the mooring lines is illustrated in Figure 4. At each corner of hull, 4 mooring lines will be set at the EL +12 m. The mooring lines are arranged symmetrically with respect to the platform geometric center. Each mooring line consists of a short chain (platform chain), long polyester rope, and short chain (anchor chain). The make-up and properties of the mooring line are listed in Table 6.

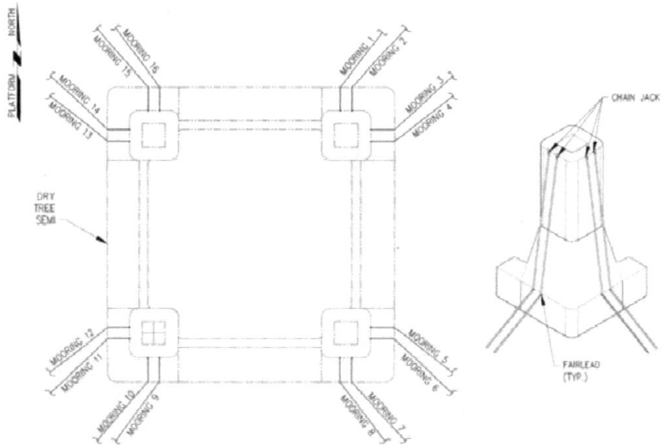

Figure 4. Mooring System Arrangement

Table 6. Mooring Line Configuration

Component	Diameter	Dry Weight	Length
	(mm)	(kg/m)	(m)
Platform Chain (Studless R4S)	142	402.7	100
Polyester Rope	239	36.1	2180
Anchor Chain (Studless R4S)	142	402.7	100

Mooring analyses were carried out using ANSYS AQWA. Under the design environments, the maximum tensions in mooring lines are 1,188 MT and 1,478 MT in intact condition and one line damage condition, respectively. The safety factors are summarized in Table 7. It can be seen that the designed mooring system satisfies both intact and damaged strength criteria.

Table 7. Safety Factors of Mooring Line Strength

Condition	Minimum Required Safety Factor	Calculated
Intact	1.67	1.73
One Line Damaged	1.25	1.39

RISER SYSTEM

The riser information can be found in Table 8. SCRs (Steel Catenary Riser) will be installed and connected onto semi pontoon. There are totally 12 SCRs, out of which, ten (10) production risers with diameter 8'', one (1) oil export riser with diameter 10'' and one (1) gas export riser with diameter 18''. Figure 5 shows the SCR layout.

Table 8. TCDD Semi-submersible Riser Information

Riser Type	Quantity
Top-tensioned production riser slots	15 (3×5)
Top-tensioned production risers (9.625" P-TTR)	11
Top-tensioned drilling risers (16.5" D-TTR)	1
Total number of production risers (8" SCR) -on north side	1
Total number of production risers (8" SCR) -on east side	4
Total number of production risers (8" SCR) -on south side	3
Total number of production risers (8" SCR) -on west side	2
Gas Export Risers (SCR) – 18", on west side	1
Oil Export Risers (SCR)- 10", on west side	1

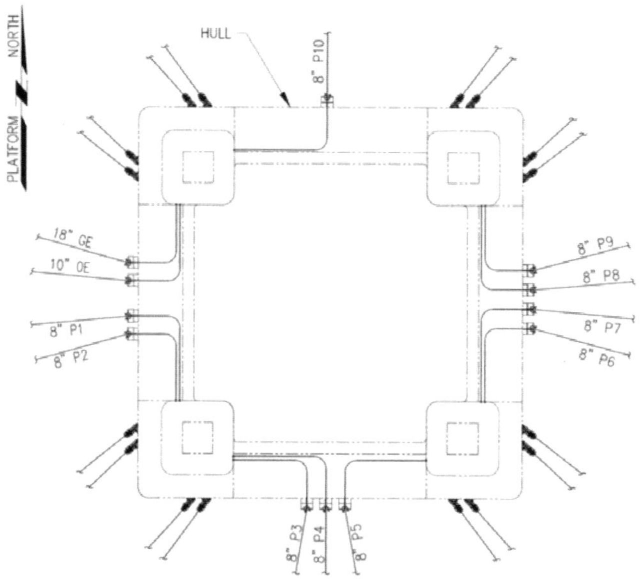

Figure 5. SCR Layout - Top View

The Production and drilling TTR properties including nominal OD, tensioner stiffness, top tension factor (TTF), and nominal tension are listed in Table 9.

Table 9. TTR and Tensioner Properties

TTR Type	OD (m)	Tensioner Stiffness (N/m)	Nominal TTF	Nominal Tension (MT)
Production	0.244	9.20E4	1.5	168.0
Drilling	0.419	2.47E5	1.4	679.0

The TTR well bay layout is shown in Figure 6. The well bay is arranged in a 3x5 rectangular pattern. The center slot is assigned to the drilling riser. Eleven (11) production TTRs will be arranged in the other 14 slots. There will be three (3) remaining empty slots which are reserved for future development. The nominal TTR spacing is 4.57 m (15 ft) in the well bay and 7.62 m (25 ft) on the seabed.

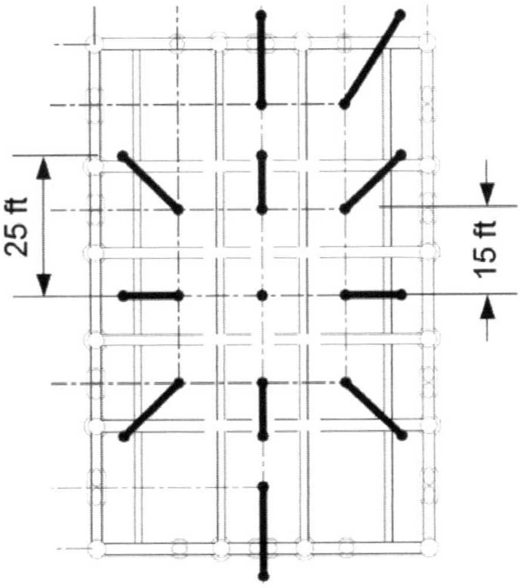

Figure 6. TTR Well Bay Layout - Top View

HYDRODYNAMICS AND GLOBAL PERFORMANCE

Platform hydrodynamic analysis was performed with ANSYS AQWA with mooring system and riser system stiffness included. Heave motion RAO curve of the TCDD Semi-submersible platform, compared with those for conventional uniform column shallower draft and deep draft semi-submersible platforms, are shown in Figure 7. The conventional shallow draft semi-submersible platform has a draft of 29 m and column width of 23 m. While the conventional uniform column deep draft semi-submersible platform has column size equal to the straight segment of the columns of the TCDD Semi-submersible. According to the figure, the TCDD Semi-submersible platform exhibits low heave responses for wave periods less than about 20 seconds. The first peak for the heave RAO is less than 0.2, and the heave natural period is greater than 20 seconds. This heave motion performance is much desirable for a platform for dry tree service. The comparison of heave RAOs indicate that the taper shaped columns, in combination with optimized pontoon size, enhances the wave force cancellation effects on the lower hull, and significantly reduces the platform wave frequency heave motion responses.

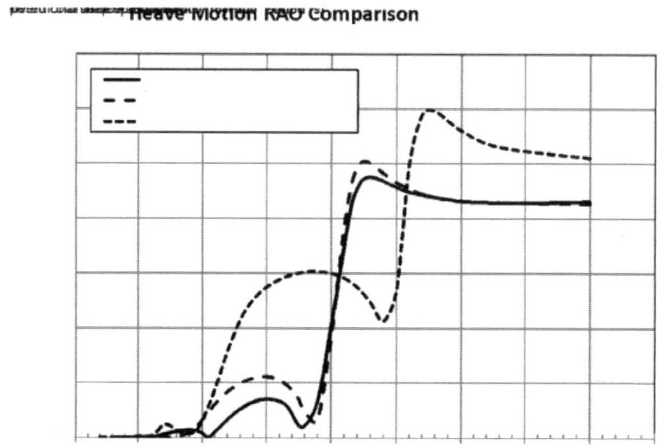

Figure 7. Comparison of Heave Motion RAOs

Global performance analysis was mainly focused on the calculation of platform global motions, minimum airgap between water surface and bottom cellar deck, and maximum TTR stroke range in in-place extreme and operating conditions.

In extreme environmental conditions, 100-year return period storms in typhoon season were applied as design environments to determine maximum motions of the platform. In operating condition, 100-year return period storms in non-typhoon season were applied. Global performance analyses were performed in time domain using ANSYS AQWA and the results are summarized below.

Extreme Motions

Table 10. Global Motion Summary in Extreme Condition

Global Motions		Max.	Min.
Surge	(m)	59.85	-60.41
Sway	(m)	60.79	-61.34
Heave	(m)	2.82	-3.10
Roll	(deg)	7.51	-7.83
Pitch	(deg)	7.74	-7.49
Yaw	(deg)	2.36	-2.44
Heave Range	(m)	5.92	-
Heel Angle	(deg)	8.19	-
Horizontal Offset	(m)	70.03	-
	(%WD)	4.67%	-
Horizontal Acceleration	(g)	0.179	-

Table 11 shows maximum global motions in operating conditions. The conclusions are:

The maximum heave range is 1.63 m
The maximum heel angle is 3.18 degrees
The maximum horizontal offset is 0.99% of water depth
The maximum horizontal acceleration at topside CG is 0.072 g.

Table 11. Global Motion Summary in Operating Condition

Global Motions		Max.	Min.
Surge	(m)	14.49	-14.47
Sway	(m)	14.71	-14.58
Heave	(m)	0.89	-0.74
Roll	(deg)	2.74	-3.08
Pitch	(deg)	3.03	-2.74
Yaw	(deg)	1.04	-1.03
Heave Range	(m)	1.63	-
Heel Angle	(deg)	3.18	-
Horizontal Offset	(m)	14.86	-
	(%WD)	0.99%	-
Horizontal Acceleration	(g)	0.072	-

Minimum Airgap

API RP 2FPS requires that the airgap should be greater than zero with the consideration of non-linear effects under design environments. In this study, wave crest asymmetry, wave enhancement near hull, platform dynamic responses to the environmental actions, nonlinear interaction between the hull and mooring/riser systems, etc., were accounted for. The analysis results indicate that the minimum airgap under design environments is 0.84 m (2.74 ft), which occurs for environmental heading of -112.5 degrees and under the edge of cellar deck on the weather side. Therefore, the platform design meets API minimum airgap requirements.

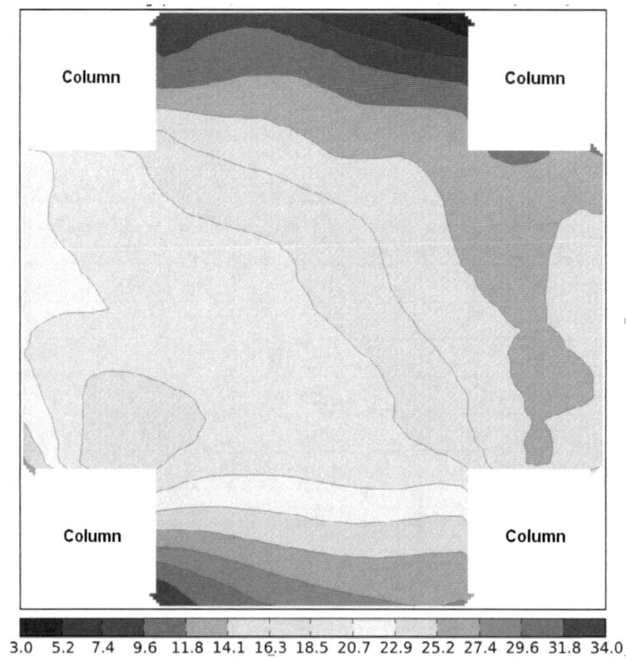

Figure 8. Minimum Airgap Contour for Most Critical Case (unit:ft)

Maximum TTR Stroke Range

Pull up style tensioners with the maximum stroke range of 10.67 m (35 ft) were assumed for the TTR tensioning system. The TTR stroke range was computed as summation of motion induced range and contributions from other effects such as seabed subsidence of 0.61 m (2 ft), tidal effect of 2.15 m (7.05 ft) and riser thermal expansion of 0.02 m (0.065 ft). The following four load cases listed in Table 12 were selected for the TTR stroke range prediction. Note the environmental conditions were defined in Table 2 above.

Table 12. Stroke Range Load Cases

Case ID	Environments	Mooring Condition	Hull Condition
1	Operating	Intact	Intact
2	Operating	Intact	One tank damaged
3	Extreme	Intact	Intact
4	Extreme	One line damaged	Intact

The final TTR stroke range calculation results are:

- In operating environments with intact hull, the calculated maximum TTR stroke range is 4.64 m (15.22 ft). Table 13 presents the detailed stroke calculation.

- In operating environments with one hull compartment damaged, the calculated maximum TTR stroke range is 4.64 m (15.22 ft). Note that upstroke and down stroke values are different from intact condition due to mean hull set down. The details are listed in Table 14.

- In extreme environments with intact mooring condition, the calculated maximum TTR stroke ranges are 9.74 m (31.95 ft). See Table 15 for details.

- In extreme environments with one line damage condition, the calculated maximum TTR stroke ranges are 10.09 m (33.10 ft). See Table 16 for details.

Table 13. Maximum TTR Stroke – Operating

Stroke Source	Up (m)	Down (m)	Range (m)
Tide	0.91	1.24	2.15
Subsidence	0.00	0.61	0.61
Thermal Expansion	0.02	0.00	0.02
Platform motions	0.85	1.01	1.86
Subtotal	**1.78**	**2.86**	**4.64**

Table 14. Maximum TTR Stroke – Operating with Tank Damage

Stroke Source	Up (m)	Down (m)	Range (m)
Tide	0.91	1.24	2.15
Subsidence	0.00	0.61	0.61
Thermal Expansion	0.02	0.00	0.02
Platform motions	2.95	-1.09	1.86
Subtotal	**3.88**	**0.76**	**4.64**

Table 15. Maximum TTR Stroke – Extreme Condition

Stroke Source	Up (m)	Down (m)	Range (m)
Tide	0.91	1.24	2.15
Subsidence	0.00	0.61	0.61
Thermal expansion	0.02	0.00	0.02
Platform motions	2.96	4.00	6.96
Subtotal	**3.89**	**5.85**	**9.74**

Table 16. Maximum TTR Stroke – Extreme with Mooring Damage

Stroke Source	Up (m)	Down (m)	Range (m)
Tide	0.91	1.24	2.15
Subsidence	0.00	0.61	0.61
Thermal expansion	0.02	0.00	0.02
Platform motions	2.56	4.75	7.31
Subtotal	**3.49**	**6.60**	**10.09**

These ranges for single events are all within the limiting criteria of 10.67 m (35 ft) required in Design Basis. All TTR stroke range calculations are also shown in Figure 9. The overall tensioner stroke range considering all operational scenarios is 10.49 m (34.4 ft) which still less than the specified stroke range limit.

According to the detailed calculations, platform motions, including horizontal offset, heave motion, and platform heeling, are the major contributors to TTR stroke range in design extreme conditions. Because of mild operating environments (100-yr non-typhoon conditions) compared to those in extreme conditions (100-yr typhoon), platform motions in operating conditions are also significantly smaller in all aspects compared to extreme platform motions. Consequently, motion induced TTR stroke range is much lower for intact operating conditions (1.87 m) than that in intact extreme conditions (6.96 m).

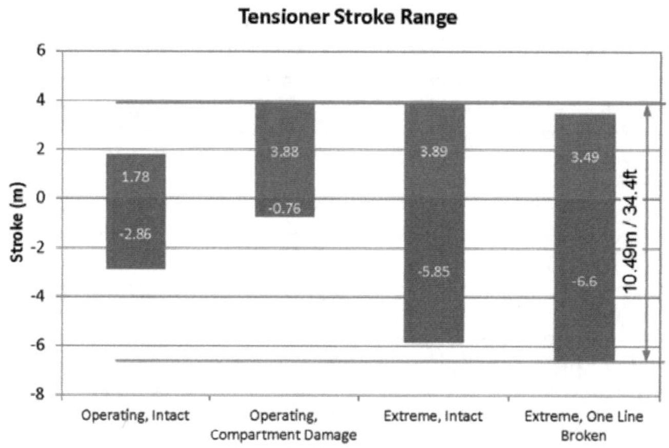

Figure 9. Maximum TTR Tensioner Stroke Ranges

PLATFORM GLOBAL STRENGTH

Global strength analyses were conducted for both operating and extreme conditions. The mean draft of the TCDD Semi-submersible unit is assumed 48.75m for both operating and extreme conditions.

Pure static strength analysis was conducted as the base case. The considered loads for this case includes the dead weight, buoyancy loads, air pressure in air-over-water ballast tanks, mooring line tension, and loads from riser lines. No environmental loads were involved for the pure static case.

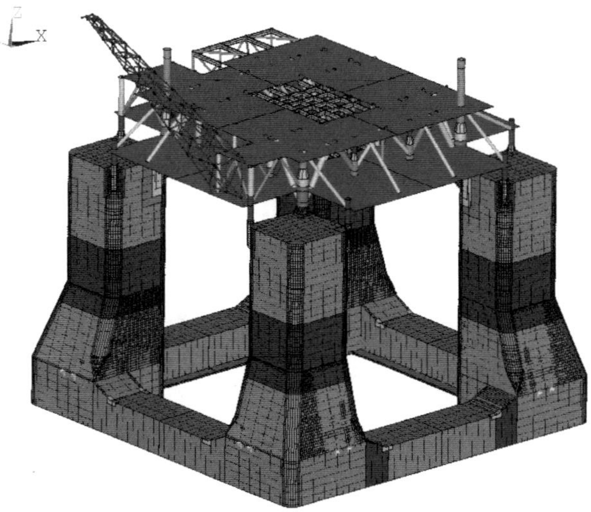

Figure 10. Global FE Model for Global Strength Analysis

For operating condition, operational (100-yr non-typhoon) environmental loads were combined with the pure static loads. No increase in the allowable stress (0.6Fy) was applied. For extreme condition, extreme environmental (100-yr typhoon) loads were added to the pure static condition. The allowable stress is 0.9Fy.

The buckling strength was screened using an in-house developed program in accordance with relevant ABS rules. Figure 10 shows the global FEA model.

The global FE analysis results indicate that for most areas of hull structure, the stress is within allowable range and the buckling usage factors are less than 1.0 for both operating and extreme conditions.

PLATFORM GLOBAL FATIGUE ANALYSIS

The main purpose of the global fatigue analysis is to assess the global fatigue strength adequacy, and identify the critical regions for further detail fatigue analysis.

The analysis was performed using industrial standard software programs AQWA and ANSYS. Only dynamic stresses induced by waves and associated low frequency motions were used in the calculation of principal stress ranges for global fatigue analysis.

The accumulated fatigue damage was determined by principal stress range with SCF and ABS S-N curves. The global fatigue screening were performed with SCF=1 and S-N curve F.

Based on the analysis results, fatigue lives for pontoon-to-column node connection, 44 m – 54 m column, top of column and nearby structure are low, thus detailed fatigue analysis for these areas are recommended. Otherwise, the major hull structure satisfies fatigue design criteria according to the screening. Figure 11 shows pontoon outer shell fatigue life results.

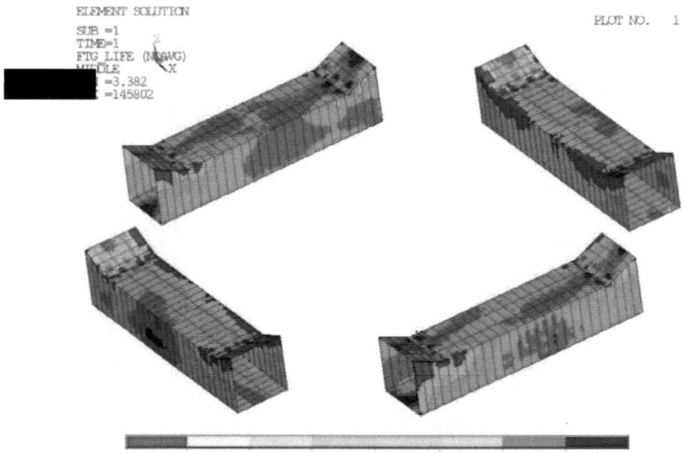

Figure 11. Pontoon Outer Shell Fatigue Life Plot

FABRICATION AND QUAYSIDE INTEGRATION

The TCDD Semi-submersible platform is divided into three principal parts, i.e. topside, column and pontoon. These three parts will be separately fabricated at different sites of the yard.

The topside and hull will be constructed on land at a shipyard in China. The columns will be constructed on two (2) skid tracks. The bottom part of the four columns (the nodes) will be built on the skid tracks directly. The pontoons will initially be built in blocks. The pontoon blocks will be fabricated at other sites and transported to the nodes for assembly on the skid tracks. Each skid track can provide 12,000 MT support capacity. The total dry weight of the hull is 21,600 MT, therefore two (2) skid tracks will have sufficient capacity to support the entire hull.

Two (2) other skid tracks will be used to construct the topside. The topside will be built by stick-built method for the decks, and the decks will be assembled from lower level to higher level. Two (2) rows of topside-to-hull posts will be supported on the skid tracks. The total lift weight of the topside is 13,000 MT before load-out, therefore two (2) skid tracks will have sufficient capacity to support the topside.

After the hull and topside are both fabricated on skid tracks, the integration of the hull and topside will follow the following steps:

Skid the hull onto a submersible barge in quay-side.
Tow the barge to the open sea, submerge the barge and have the hull float-off the semi-submersible barge.
Skid the topside onto a transportation barge, and move the barge to under the quayside sheer legs.
Lift the topside by sheer legs, and move the barge away.
Tow the hull under the sheer legs.
Lower the topside onto the hull and finish the integration and commissioning at quayside.
Tow the integrated platform out of the quay-side and submerge the semi.
The platform is ready for wet tow.

Figure 12 shows the schematics of the quayside topside integration using sheer legs.

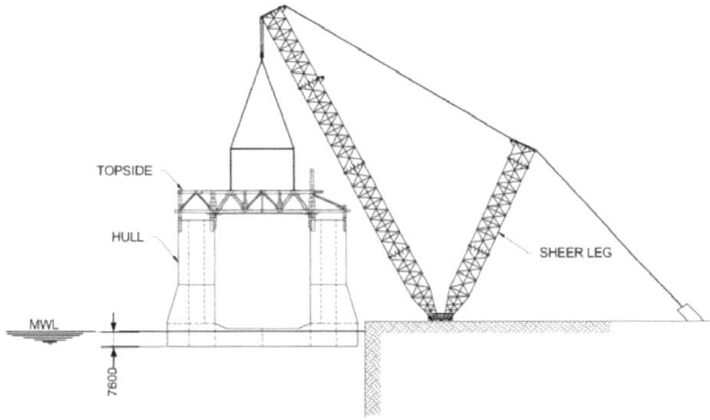

Figure 12. Quayside Topside Installation with Sheer Legs

It should be noted that quayside integration of the topside with sheer legs is only one of feasible installation options presented in this paper. In addition, the topside may also be integrated using super lift method or using offshore float-over method. In addition, the topside integration may also take place in dry dock facility using heavy lift gantry crane.

PRE-SERVICE STABILITY ANALYSIS

The pre-service analysis of TCDD Semi-submersible platform was performed for the following phases:

Hull Load-out
Hull Float-off
Quay Side Integration
Wet-tow
Submergence

After the hull is finished fabricating on the building berth, it will be load-out onto a barge. The total load-out weight of the hull is 20,102 MT. After the hull is loaded onto the barge, it will be float-off at a nearby site having enough water depth for float-off operating. The hull will then be towed to the quay side for integration of the topside. Considering 1.0 m under keel clearance, the required minimum water depth is 8.6 m. After the integration and outfitting at quay side, the platform will be tow to an offshore site which has enough water depth to submerge to towing draft of 48.75 m. Certain topside equipment may also be installed at the submergence site. The platform will finally be wet towed to the project site where the mooring and riser systems will be installed.

ABS and Noble Denton stability criteria were used to check the TCDD semi-submersible platform pre-service stability. The analysis results have confirmed that the platform has adequate stability (both intact and damaged) for all pre-service operation stages.

Figure 13 shows typical intact stability curves for wet tow.

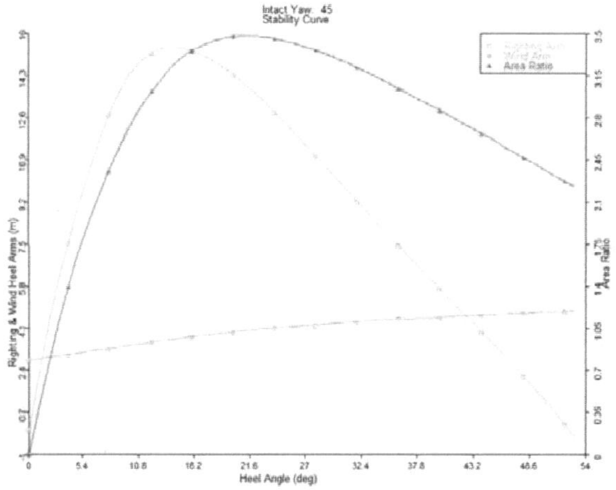

Figure 13. Intact Stability Curves for Wet Tow

VOTEX INDUCED HULL MOTIONS

When strong current flow passes a semi-submersible hull, vortices are usually created behind the columns. When the vortex shedding frequency is close to the natural frequency of the platform and mooring system, resonance or lock-in is likely to occur. The vortex shedding frequency is function of current speed, column size, and hull geometry. The tapered column shape helps alter the possible vortex shedding frequency along the column, and in turn reduces the possibility of lock-in occurrence, and eventually reduces the amplitude of VIM if resonance motion occurs at all.

Computational fluid dynamic (CFD) analyses method was utilized to investigate the TCDD Semi-submersible hull VIM motions in currents. The CFD analysis tool, STAR-CCM+ was used to perform a series of simulations for the platform.

The calculations were carried out at 1:50 scale with Froude scaling. The approximate mesh size was 1.5 – 2 million hexahedral cells and the exact number of cells depended on the platform orientation inside the computational domain. The computational domain size was 36m×36m ×18m (width × depth × length) at model scale. The time step was 0.1 s with 2nd order implicit time integration. The turbulence models were RANS and SST/k-ω Detached Eddy Simulation.

Figure 14 shows the TCDD Semi-submersible hull mesh and the free surface mesh for CFD analysis.

The VIM amplitude calculation results as well as recommended A/D envelope curves are shown in Figure 15. The results indicate that the lock-in A/D is 0.3 for diagonal currents and 0.1 for head-on currents. For consistency, the characteristic length D was taken as the minimum column width for both head-on and diagonal currents.

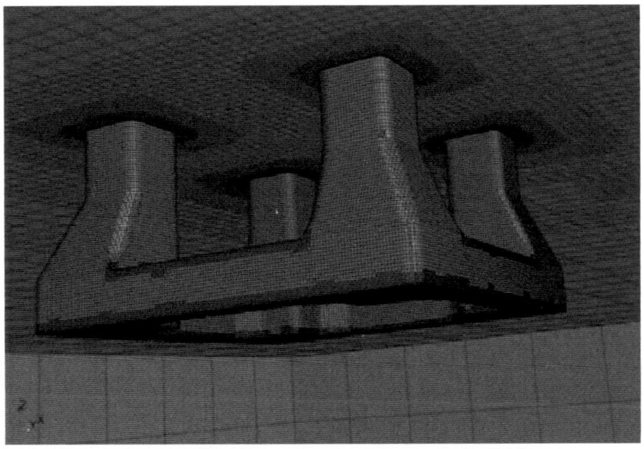

Figure 14. Semi-submersible Hull Surface Mesh

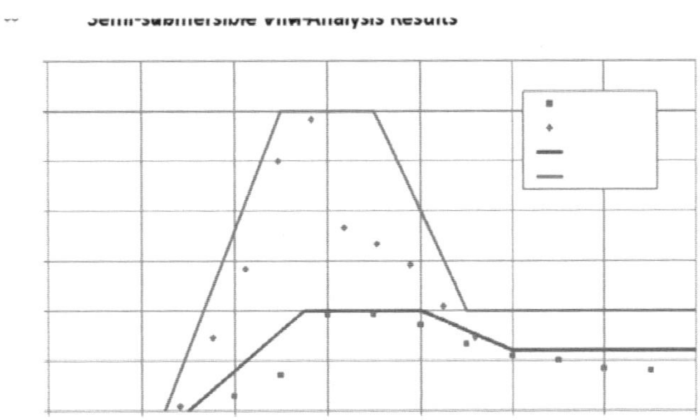

Figure 15. Semi-submersible VIM Analysis Results

CONCLUSIONS

The study has shown that the TCDD Semi-submersible concept is technically feasible for construction in China and operation in South China Sea. Among several benefits in the new concept design, the highlights are the superior heave motion responses enable dry tree applications. The maximum TTR stroke is within the target design value of 35 ft. The unique tapered column hull geometry provides easiness of fabrication and quayside integration. The hull configuration and compartmentation ensure the adequate stability and ballast capacity during the pre-service operation phases including hull load-out, float-off, topside integration, wet tow, and submergence.

REFERENCES

Das, S. and Zou, J. (2015). "Characteristic Responses of a Dry Tree Paired-column and Deep Draft Semisubmersible in Central Gulf of Mexico," Proceedings of the 20th Offshore Symposium, Texas Section of the Society of Naval Architects and Marine Engineers, Houston, Texas, USA.

Halkyard, J., Chao, J., Dagleish, J., Banon, H., and Thiagarajan, K. (2002). "A Deep Draft Semisubmersible with a Retractable Heave Plate," Offshore Technology Conference, Houston, Texas, USA, OTC 14304.

Koos, J. (2013). "Long Stroke Ram-Style Tensioner Technology," Offshore Technology Conference, Houston, Texas, USA, OTC 23963.

Kyoung, J., O'Sullivan, J., Kostas, L., Kim, J.W., and Ermon, J. (2014). "Low Motion Semisubmersible Hull for Dry Tree Application," Offshore Technology Conference, Houston, Texas, USA, OTC 25388.

Muehlner, E. and Banumurthy, S. (2015). "Low-Heave Semi-Submersible Enabling Dry-Tree Field Development," Offshore Technology Conference, Houston, Texas, USA, OTC 25666.

Mungall, C., Harverty, K., Bhat, S., Andersen, D., Sarker, I., and Wu, J. (2004). "Semisubmersible Based Dry Tree Platform with Compliant Vertical Access Risers," Offshore Technology Conference, Houston, Texas, USA, OTC 16199.

Xu, Q. (2011). " A New Semisubmersible Design for Improved Heave Motion, Vortex-Induced Motion and Quayside Stability," Proc 30th Int Conf on Ocean, Offshore and Arctic Engineering, Rotterdam, The Netherlands, OMAE 2011-49118.

Zeng, J., Smith, D., and Wu, W. (2014). "Challenges and Solutions of an Ultra-deepwater Dry Tree System for Drilling and Production in the Gulf of Mexico," Offshore Technology Conference, Houston, Texas, USA, OTC 25250.

Zou, J., Poll, P., Antony, A., Das,, S., Padmanabhan, R., Vinayan, V., and Parambath, A. (2014). "VIM Model Testing and VIM Induced Mooring Fatigue of a Dry Tree Paired-Column Semisubmersible Platform," Offshore Technology Conference, Houston, Texas, USA, OTC 25427.

MOONPOOL EFFECTS ON THE STRUCTURAL ASSESSMENT OF DRILLSHIPS

FLAVIA C. REZENDE
MANAGER TECHNOLOGY DEVELOPMENT, BUREAU VERITAS HOUSTON

DANIEL BARCAROLO
SENIOR HYDRODYNAMICS & PROJECT ENGINEER, HYDROCEAN

Proceedings of the 22nd Offshore Symposium, February 2017, Houston, Texas
Texas Section of the Society of Naval Architects and Marine Engineers

ABSTRACT

The purpose of the moonpool in a drillship is to provide a sheltered and protected access to the sea for the safe and easy deployment of equipment used for drilling, diving, etc. However, in certain wave conditions, this function is impaired due to a hydrodynamic resonance that can lead to significant amplification of the wave elevation inside the moonpool. Beside the impact on operability, this amplification can result in large pressures that must be accounted for in the design of the vessel. Usually for the direct computation of hydrodynamic pressures, potential flow theory is employed, which is based on the assumption of a perfect fluid. This approach is known to overpredict the wave elevations close to the moonpool resonance frequencies, as it neglects viscous effects that in reality play a major role in limiting the wave kinematics. Thus, applying potential theory, without special treatment, may induce undesired over conservatism in the scantling of the structural elements in the moonpool region. As an alternative to potential flow, modelling techniques based on the solution of Navier Stokes equations (CFD) can potentially provide accurate results. The main drawback of these tools is that they are costly solutions and their application should be practically limited to a number of selected critical cases. This paper will present the results obtained from the assessment of wave elevations / hydrodynamic pressures on the moonpool of a drillship using CFD computations in calm water for calibrating a potential flow model.

Keywords: Moonpool, resonance, pressures

INTRODUCTION

The purpose of the moonpool in a drillship is to provide a sheltered and protected access to the sea for the safe and easy deployment of equipment used for drilling, diving, etc. However, in certain wave conditions, this function is impaired due to amplification of wave kinematics inside the moonpool. This effect may lead to unexpected downtime as the wave conditions outside the vessel may be mild and within the operational limits, while wave conditions within the moonpool are outside of operating limits.. Beside the operability impact, the structural integrity of the vessel is of concern as this phenomenon may result in larger pressures than considered in the design base case either for extreme or for fatigue assessments.

The issue with entrapped water-induced resonance has been the subject of many publications as the problem has been identified for a broad range of practical applications. In Van't Veer and Tholen (2008) the moonpool effects on the resistance of drillships are discussed. In Chen *et al* (2007) the problem is presented for side-by-side offloading operations. Sphaier *et al* (2007) shows that the heave motions of a new platform concept, the Monocolunm Production, Storage and Offloading system (MPSO) [4], are significantly affected by the moonpool design.

As a result of this body of work, significant progress has been made in the comprehension of the hydrodynamic phenomenon. However, the problem remains difficult to be solved in practice with the classical potential flow models, which are based on the assumption of a perfect fluid. It is well known that, close to the resonance frequencies, potential theory over predicts the wave elevations inside the moonpool while in reality it is expected that the energy dissipation by viscous effects would play a major role in limiting the wave kinematics.

We may divide the problem into two steps, the first being the estimation of the resonance frequencies inside the moonpool, and the second being the estimation of the magnitude of the wave elevations around those frequencies. With respect to the resonant modes, we may distinguish two types: piston and sloshing. For the estimation of the frequencies of the natural modes, analytical formulations exist (see Molin, 1999 and 2000)). These formulations are based on simplified moonpool geometries (rectangular or circular) and on the assumption of an infinity body outside the moonpool. For an arbitrary geometry, potential flow solvers are theoretically able to present an accurate solution. However, it should be remarked that some numerical convergence issues may occur with the classical BEM (see Chen, 2010). The piston mode, which corresponds to the heave up and down of the water column inside the moonpool, generally occurs at lower frequencies than the sloshing modes for typical dimensions, having larger impact on the design loads.

The estimation of the resonant modes frequencies compared with the frequency range of incident waves will determine the probability of observing a resonant phenomenon and whether the designer should act to properly quantify the impact of that. Knowing that the impact is exaggerated with the classical potential theory approach, one should employ more sophisticated tools able to more accurately account for the viscous effects such as CFD or model tests. These tools however are both costly solutions and their use shall be limited to few selected design conditions, while the overall design procedure requires a long term assessment of the loads (e.g. for fatigue).

In this work we present the impact of a typical moonpool of a drillship on the dynamic loads applied to the vessel from the potential theory. CFD heave forced oscillation simulations have been performed in order to calibrate some damping coefficients for the piston mode to be used as input for potential flow. The loads after application of the damping coefficients have been compared to those obtained without any treatment.

THE USE OF POTENTIAL THEORY

In the structural design procedure, the assessment of the loads comes first. Therefore the accuracy of the dynamic loads is essential since it will significantly affect the scantlings. For ship-shaped vessels, Classification Societies generally prescribe the dynamic loads to be applied to the structure based on some design principles (ex. North Atlantic scatter diagram, etc). In the particular case of a drillship, however, the presence of the moonpool brings some special issues that may affect the loads in a way difficult to generalize as each moonpool may have different geometry and appendages to it (ex. damping devices). Thus, one may be interested in performing a direct assessment of the loads through a project specific seakeeping model. For that purpose, we normally rely on potential flow theory.

In order to illustrate the application of classical potential flow theory for the assessment of the dynamic loads acting on a drillship with rectangular moonpool we consider a typical BEM as illustrated in Figure 1.

Figure 1 – Hydrodynamic mesh of a typical drillship with rectangular moonpool

Figure 2 presents the relative wave elevation at a point close to the aft transverse bulkhead of the moonpool. We can identify a significant amplification at the frequency of 0.74rad/s and two other smaller amplifications at higher frequencies. The first peak corresponds to the piston mode while the two others correspond to sloshing modes. These high wave elevations around resonance frequencies will result in very large pressure values applied to the walls of the moonpool. If these pressures are taken without any treatment they will not only affect the local assessments but also the global behavior of the vessel (accelerations and hull girder loads) impacting the overall scantlings.

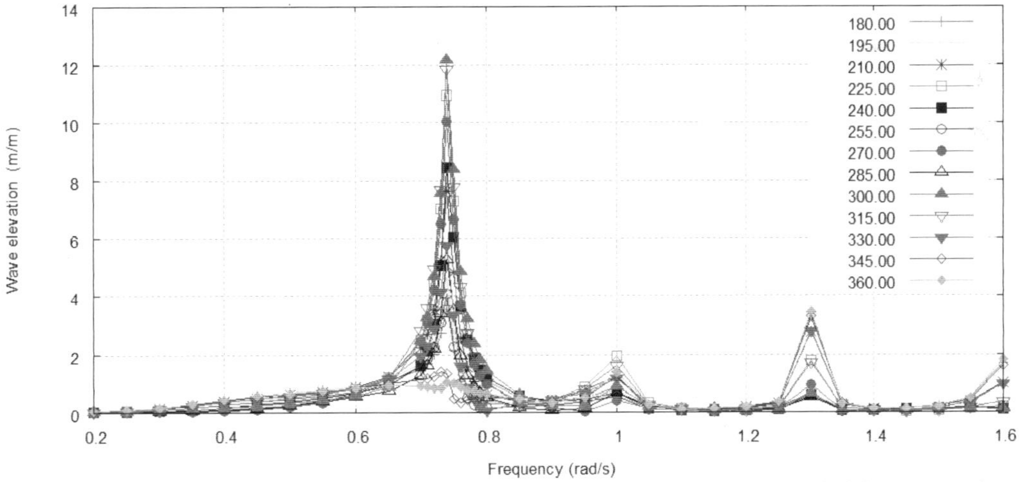

Figure 2 – Relative wave elevation at a point close to the aft transverse bulkhead of the moonpool

Figure 3 presents the heave acceleration at the center of gravity and the bending moment at the midship section. It is interesting to note that around the piston resonance frequency the value of bending moment is higher for quartering seas and beam seas than for head seas.

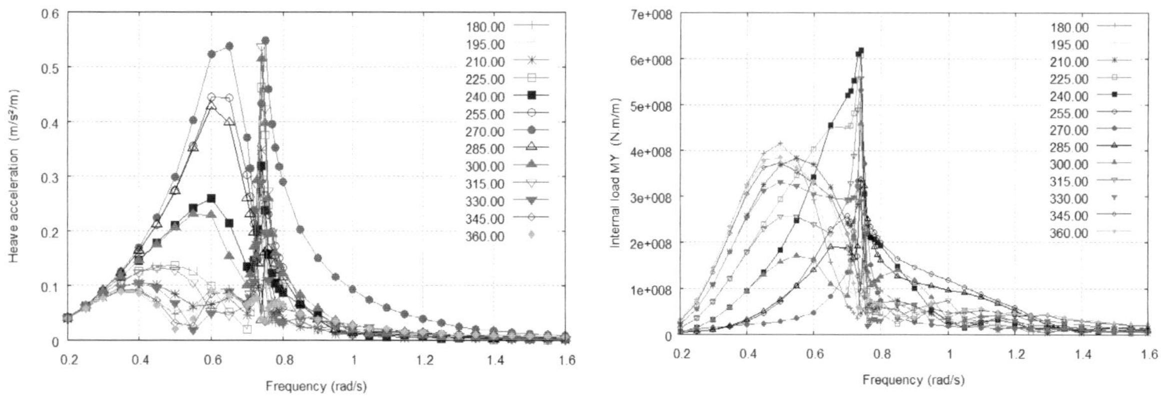

Figure 3 – RAOs of heave acceleration at the center of gravity (left) and bending moment amidships (right)

It is clear that the results presented above which were obtained from the classical potential theory, without any special treatment applied to the resonant wave kinematics in the moonpool, are exaggerated and the impact on the design could be very significant.

THE VISCOUS DISSIPATION INSIDE THE MOONPOOL

In order to assimilate the effects of viscous dissipation within the potential theory we may apply an internal force in the fluid, as described in Chen *et al* (2007), which is proportional to the magnitude of the fluid velocity but in the opposite direction:

$$f = -\mu\mathbf{V} \tag{1}$$

where **V** is the flow velocity which can be expressed as the gradient of a scalar potential Φ, for irrotational flow. The parameter μ is assumed positive and small.

This dissipative force has no physical relation to the energy dissipated by viscous effects, but aims at playing the same role without introducing any vorticity to the flow which is the condition for the existence of the potential. The fluid after including that fictitious force is called by Guével (1982) as "fairly perfect" fluid.

At first order, the dynamic pressure p is given by:

$$\frac{p}{\rho} = -\frac{\partial\Phi}{\partial t} - \mu\mathbf{\Phi} \tag{2}$$

As the dissipative force is fictitious, the parameter μ needs to be calibrated against models that better capture the physics of the problem, as CFD or model tests.

As shown previously, in the case of typical moonpool dimensions, we are mostly interested by the piston mode. So, heave forced oscillation simulations at the piston mode frequency have been performed in CFD and the wave elevation (or pressure) in the middle of the moonpool has been monitored. In potential flow, in the case of a heave forced oscillation in calm water, the pressure will only comprise the radiation potential due to the imposed heave motion. Then the dissipation parameter μ can be calibrated by matching the heave radiation pressure computed in potential flow code and the pressure obtained from the CFD computation.

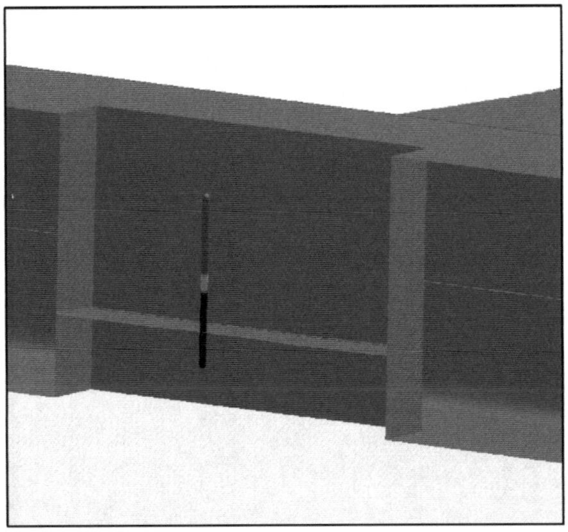

Figure 4 –Heave forced oscillation in CFD

Figure 4 presents a portion of the CFD model showing the amplification of the elevation inside the moonpool due to a unitary heave motion. Three forced oscillation simulations have been performed for heave amplitudes of 1m, 2m and 3m. Figure 5 shows the wave elevations computed at the middle of the moonpool for the three amplitudes of heave forced motions.

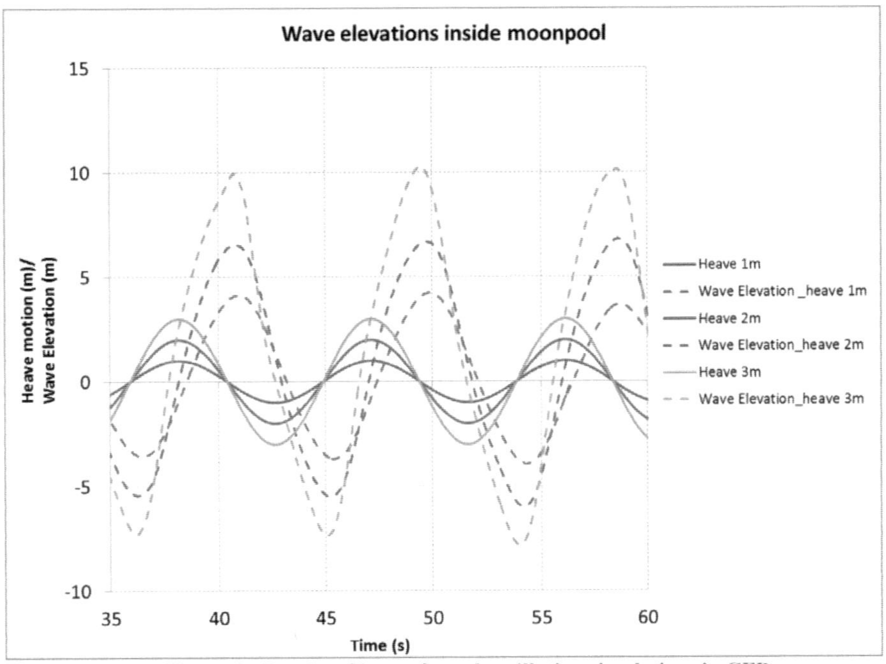

Figure 5 –Results of heave forced oscillation simulations in CFD

The dissipation parameter is set for the potential flow computation so that the radiation pressure due to heave reaches the same level as in the CFD computation. Figure 6 presents the comparison between the radiated wave elevation due to heave motion computed with potential flow and the CFD results before and after the calibration of the dissipation parameters. It is remarked that a different dissipation parameter has been calibrated for each amplitude of the forced heave motion which means that the dissipation is not linear as it would be expected.

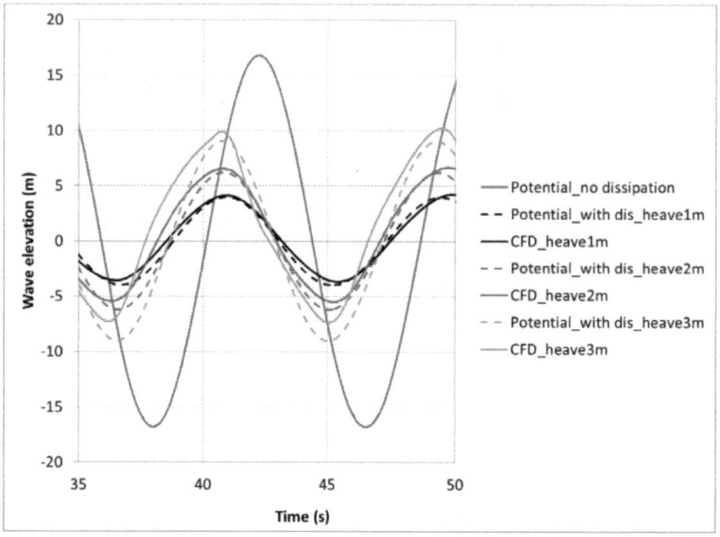

Figure 6 –Radiation pressure inside the moonpool due heave motion with 1m amplitude comparing with wave elevation during CFD forced oscillation simulations.

It is observed that applying a small dissipation parameter, the moonpool effects on the global behavior of the vessel become negligible as illustrated in Figure 7 for the heave acceleration and the amplification of the wave elevation inside the moonpool is much reduced as observed by comparing Figure 8 and Figure 2. It should be remarked that there is no unique dissipation parameter and its value should be specific for each wave condition assessed.

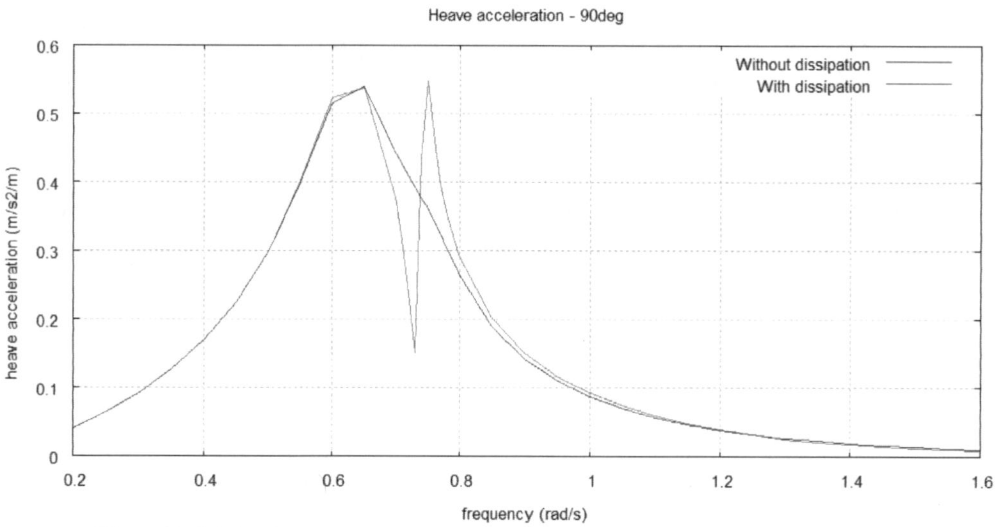

Figure 7 –RAO of heave acceleration before and after inclusion of dissipation parameter

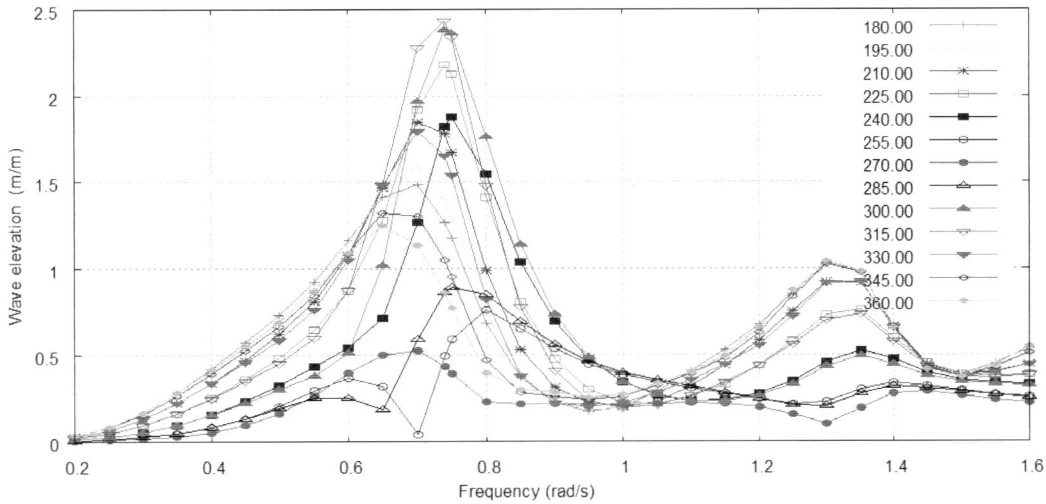

Figure 8 –RAO of wave elevation inside the moonpool after application of dissipation

PROPOSED METHODOLOGY

The proposed methodology to account for the piston mode inside a moonpool consists of the following steps:

1) Perform CFD forced oscillation simulations for different motion amplitudes;
2) Calibrate a dissipation parameter so that the motion radiation pressure at the free surface inside the moonpool resulting from the potential flow computation matches the result of CFD simulation;
3) Adjust a model that fits the evolution of the calibrated dissipation parameters with respect to the wave elevation inside the moonpool;
4) Pre-compute a set of RAOs for a range of dissipation parameters;
5) For a given sea state perform a spectral analysis assuming an initial dissipation parameter;
6) Obtain the response spectrum of wave elevation and the associated spectral moments;
7) Perform stochastic linearization of the dissipation parameter;
8) Iterate interpolating the set of RAOs until the linearized dissipation converges

Further work shall be performed to validate the methodology by comparing the resulting RAOs with either model tests or CFD simulations in irregular waves.

CONCLUSIONS

It is known that the resonance phenomenon inside moonpool of drillships may impair the operation of the vessels under certain wave conditions. Therefore, it is important to predict that type of resonant response at the design phase so that mitigation measures can be implemented avoiding unexpected downtime. Moreover, the amplification of pressures inside the moonpool, whenever they can happen, shall be properly accounted for on the structural assessments of the vessel (i.e. extreme and fatigue). The classical potential flow theory, which is usually employed in the assessment of the behavior of floating units, fails to compute the wave elevations inside the moonpool close to the resonant modes as it does not include viscous dissipation. It has been shown that, without any special treatment, the application of potential flow to vessels with large moonpools could result in over conservative designs as the wave loads (global and local) would be exaggerated. In order to provide an engineering solution to the problem, some potential flow software includes a fictitious linear dissipation parameter aiming at simulating the effects of viscosity under the assumptions of potential theory. As it is fictitious, such parameter needs to be calibrated against model tests or CFD simulations. Considering that in principle the dissipation is non-linear, it is expected that a different parameter would be needed for different wave conditions. In this work, we considered the piston mode, so that we applied forced heave CFD simulations to calibrate the dissipation parameter based on the decomposed heave radiation pressure obtained from potential flow software. Finally, a methodology has been proposed in which CFD could be employed to perform a series of forced oscillations enabling adjusting a model for the dissipation parameter with respect to the wave elevation inside the moonpool. After defining the model, it would be possible to apply a stochastic linearization approach to obtain a linearized dissipation parameter for each sea state of interest. Even though the methodology still requires further validation based on model tests or CFD simulations in irregular waves, such a method would provide an efficient solution to the long term assessment of vessels including moonpools.

REFERENCES

[1] I. de Vries, Y. Roux, M. Naciri, G. Bonnafoux, "Dynamics of entrapped water in large diameter turret moonpools", Proceedings of OMAE2013, Nantes, France, 2013

[2] R. Van't Veer and H. J. Tholen, "Added resistance of moonpools in calm water," Proceedings of OMAE2008, Estoril, Portugal, 2008

[3] X. B. Chen, F. Rezende, S. Malenica & J. R. Fournier, "Advanced Hydrodynamic Analysis of LNG Terminals", Proceedings of PRADS 2007, Houston, USA

[4] Sphaier S.H., Torres F.G.S., Masetti I.Q., Costa A.P. & Levi C., "Monocolumn behavior in waves: experimental analysis", Ocean Engineering, 34, 1724-33, 2007

[5] Gonçalves R.T., Matsumoto F.T., Malta E.B., Rosetti G.F., Fujarra A.C. & Nishimoto K., "Evolution of the MPSO (monocolumn production, storage and offloading system)", Marine System & Ocean Technology, 5(1), 45-53, 2010

[6] B. Molin, "On the piston mode in moonpools", Proceedings of the 14th IWWWFB, Port Huron, Michigan, USA, 1999

[7] B. Molin, "On the sloshing modes in moonpools, or the dispersion equation for progressive waves in a channel through the ice sheet", Proceedings of the 15th IWWWFB, Caesarea, Israel, 2000

[8] Martin P.A. & Risso F.J., " Boundary integral equations for bodies of small, but finite,thickness", Proceedings of the 8th IWWWFB, St John's, Canada

[9] X.B. Chen, "Multi-domain boundary element method", Proceedings of the 6th IWSH, Harbin, China, 2010

[10] H.X. Liu, W.Y. Duan and X.B. Chen, "Semi-analytical solutions to diffraction of cylindrical structures with moonpool of restricted entrances"

[11] F.C. Rezende, N.A.M. Neumann and M. Gundelach, "The dynamic effects of water column inside the turret of an FPSO", OTC Brasil, Rio de Janeiro, Brazil, 2013

[12] P. Guével, "Le problème de diffraction-radiation – Première Partie: Théorèmes fondamentaux", ENSM, University of Nantes, 1982

[13] X.B.Chen, "Hydrodynamics in offshore and naval applications – Part I" , 6th International Conference on Hydrodynamics, The University of Western Australia, Perth, Australia, 2004

OPTI® SERIES PRODUCTION SEMI-SUBMERSIBLE HULL SIZING PROCESS

ZHIGANG KEVIN TIAN
LEAD NAVAL ARCHITECT, EXMAR OFFSHORE COMPANY

BLAIR KIRKLAND
SENIOR NAVAL ARCHITECT, EXMAR OFFSHORE COMPANY

BRANKA RADANOVIC
NAVAL ARCHITECT MANAGER, EXMAR OFFSHORE COMPANY

OTTO DASILVA
VICE PRESIDENT OF ENGINEERING, EXMAR OFFSHORE COMPANY

Proceedings of the 22nd Offshore Symposium, February 2017, Houston, Texas

Texas Section of the Society of Naval Architects and Marine Engineers

ABSTRACT

Two production semi-submersibles that were based on Exmar Offshore Company's proprietary OPTI® Hull had been successfully constructed and installed in the Gulf of Mexico in the past five years. The OPTI® hull features pontoon lip extending outboard of five-sided columns. The pontoon lip improves heave motion and the five-sided column design has proven effective in Vortex-Induced-Motion (VIM) reduction.

This paper presents the methodology and main considerations in the OPTI® hull sizing process. With a set of limiting criteria, an in-house hull sizing program is used to generate thousands of hull configurations and to perform diffraction and radiation analysis. Subsequently, the hull steel weight and heave motion is examined to determine a design front on which a better compromise between hull steel and motion is achieved. A few candidate hulls may be selected with further considerations of constructability, mechanical and electrical equipment arrangement and structural design restrictions. Finally, Client's preference, riser design requirements, installation and operation preferences, and economics are fully considered.

Keywords: OPTI® Production Semi-Submersible, Hull Sizing

1 INTRODUCTION

Exmar Offshore Company (EOC) had successfully developed the OPTI® series hull design for semi-submersibles and had constructed and installed two Floating Production Units (FPU) based on the design in the Gulf of Mexico (GOM) in the past 10 years. The OPTI® series hull is composed of a ring pontoon and five-sided columns. The pontoon is uniquely configured such that a pontoon lip extends outboard of the columns and this configuration has proven hydrodynamic advantages in terms of heave motion reduction. The five-sided column design improves structural integrity and constructability, as the outer shells of the columns are natural extensions of the bulkheads and inboard shell of the nodes. Model tests also demonstrated that the OPTI® series hull is less subject to VIM, which is mainly attributed to the five-sided column design. A typical OPTI hull demonstrating these features is shown in Figure 1.

Based on the two abovementioned unique features of the OPTI® series hull, Exmar further developed the FAST™ riser installation technology, fully utilizing the declination angle formed between riser porches and deck edge (12 degrees as shown in Figure 1 for this example). Contrary to traditional steel catenary riser (SCR) installation method, the FAST™ technology essentially maneuvers the hull to catch the riser by using the ballast system and, hence, eliminates the usage of steering winch or divers. Field operations using the FAST™ technology had been successfully executed 10 times and greatly reduced the riser installation operation time and cost.

A typical design process of the OPTI® series semi-submersible starts with a given topsides or payload capacity requirement. First, an estimate of the required area to accommodate the equipment shall be determined, which produces an estimate of the deck truss dimension and weight. Second, a hull is sized to satisfy a set of criteria, including stability requirement, ballast and heave natural periods. Subsequently, a panel model is built for diffraction and radiation analysis, results of which are used in wave frequency and low frequency motion analysis for the free floating body. If the sized hull reaches the design target for motion, then a preliminary mooring size and air gap need to be assessed. As more accurate parameters used in the early phases become available in the later phases, iterations with these updated information need to be performed to identify a working solution. This is the so-called spiral design process, as demonstrated in Figure 2 (Chakrabarti, 2005; Ch.7).

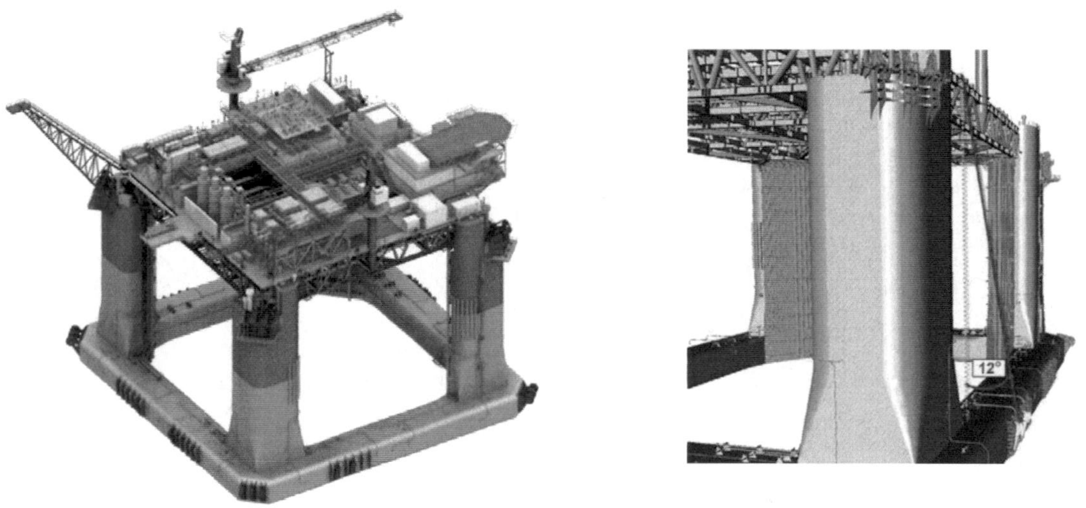

Figure 1 A demonstration of a typical OPTI® Hull and its features

Hull sizing is obviously an important process in the design spiral. Better estimates of parameters used in the sizing process, particularly weight estimate, can reduce the number of spiral passes needed to reach the working solution. The weight estimate in the early phase has to rely heavily on parent vessels and/or past project experience.

In practice, for a given topsides or payload capacity requirement, there exists an infinite number of hull configurations that can satisfy the design criteria. The selection of a superior hull configuration is also important in the sizing process. Depending on specific project and/or the Client's preference, a better hull can be one with better motion performance, better constructability and higher efficiency in terms hull steel weight. It is common to perform a hull sizing study in which serval hull particulars, such as draft, water plane area, and column center to center distance, are examined to evaluate their impact on hull steel and motion performance. It is expected that some sort of "optimized" hull can be identified through such a study, which may take weeks and even months to complete. Typically, this study can examine a limited number of hulls, at most several tens, due to time and budget limits.

To improve the efficiency and robustness of the hull sizing study, EOC has developed a program that can quickly generate a large number of hull configurations by systematically varying hull main particulars while satisfying given criteria. The program then automatically generates panel models and performs diffraction and radiation analysis and motion analysis. The produced candidate hull configurations can be further analyzed with additional constraints for the selection of the final hull configuration. The program significantly reduced the time to complete sizing and examination of a hull, from days to minutes; hence, it enables effective checking of thousands of hull configurations for a hull sizing study.

After this introduction, the next section discusses the main assumptions adopted in weight estimation. Section 3 presents the criteria imposed by rules, regulatory bodies and/or Client, practical considerations in hull selection, and the automated hull sizing subroutine. Section 4 presents a case study. The final section concludes the paper.

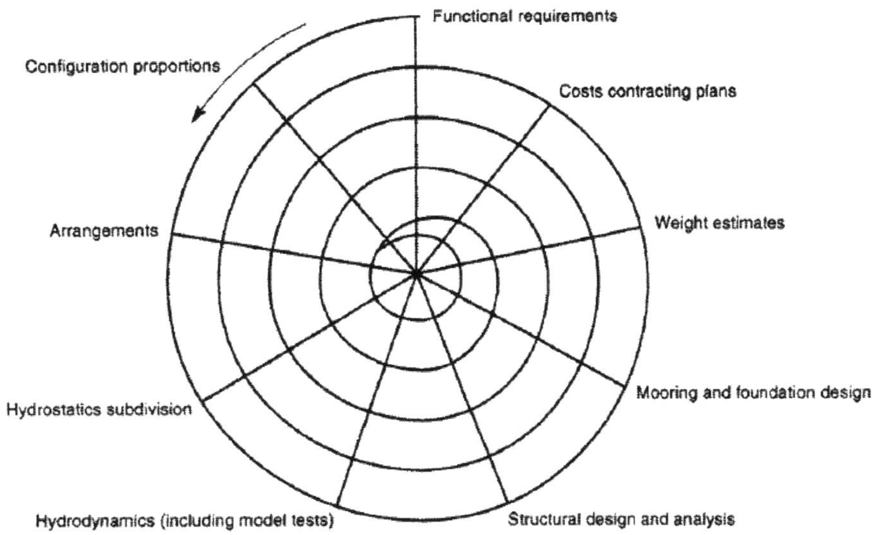

Figure 2 Illustration of the design spiral (Chakrabarti, 2005; Ch. 7)

2 WEIGHT ESTIMATE

For a production semi-submersible, the hull is required to provide enough buoyancy to support the weight of the hull, topsides, and equipment required, for the functions to be performed, and satisfy stability, station keeping and seakeeping criteria. The design of a production semi-submersible is a typical spiral in which weight, buoyancy, stability, motion performance, functionality and cost definition improves as the design passes around the spiral and, ideally, converges to a "better" working solution. Initial hull sizing has great impact on the spiral and can greatly affect the overall engineering time and cost. In this initial sizing, weight has to be estimated according to past experience as accurate weight information becomes only available in subsequent design phases. This section documents the methodology and main assumptions adopted in the initial sizing.

2.1 Topsides Equipment Weight

It is assumed that the topsides equipment weight and space required for hull sizing are known. This equipment weight for different loading conditions can vary as it is the sum of fixed ("dry") weight plus variable loads. The equipment weight can be provided by Client or based on results of previous phases of the project. For internal development projects for which specific field information is not available, the topsides equipment weight can be specified as a lump sum weight based on the target oil and gas production and processing capacity.

2.2 Deck Truss Structure

While the OPTI® series semi-submersible can use both deck box and truss, single level truss structure had been used in the past projects. The deck truss structure is considered as an integral part of the topsides; the sum of deck truss weight and the equipment weight is the overall topsides weight.

Initial deck truss length can be estimated by satisfying the Client specified space requirement, and the truss height can be determined using a reasonable deck length to height aspect ratio, approximately 8 to 10.

Accurate truss structural weight has to be obtained in a deck sizing study, fully considering the topsides equipment weight and its distribution, environmental loads, and vessel motions. The information only becomes available at a later stage. An initial deck truss weight estimate by scaling and using past project experiences can be done. EOC also performed internal studies to correlate the topsides weight and the deck truss weight, and the findings are consistent with the results of the empirical scaling methods. In general, the deck truss weight increases as the equipment weight increases.

Based on previous experiences, the vertical center of gravity (VCG) of the deck truss is taken at 60% of the truss height from truss bottom as heavier beams are present in the top chord.

2.3 Hull Steel Weight

A practical hull steel weight estimate method is using the enclosed hull volume and a "density factor", which is typically between 10 lb/ft^3 in the upper sections and 15 lb/ft^3 in the pontoons (Chakrabarti, 2005). The density factor shall be carefully benchmarked against parent vessels and constructed units. In case that the hull is much different in size than the parent vessels, the utilization of a density factor needs extra caution and a larger margin may be used for the estimate. The density factor can also depend on the hull subdivision that is typically controlled by stability requirements.

EOC uses hull steel density factors benchmarked against the Who Dat and the Delta House projects. Particularly, a density factor increase in the node, i.e. column to pontoon connection, is used to accommodate the strengthened structure members and the heavy equipment in the pump rooms.

2.4 Hull Outfitting Weight

With existing parent vessels, scaling can be applied to certain hull outfitting weight estimates. To perform this, one shall apply the appropriate scaling factor to the outfitting based on length, area, or volume. The following is a list of some of the hull outfitting equipment:

- Hull marine piping, equipment and ventilation.
- Stairs, ladders, elevators, tank anodes, ICCP, water tight doors, water line fenders etc.
- Electrical equipment.
- Chain fairleads, turndown sheaves, stoppers, and jacks.
- On hull production piping and adaptors for production and export lines.
- Pull tubes for umbilicals.
- Porches and pipe supports production and export risers.

2.5 Hull Variables Weight

Diesel, potable water, chemicals, and external hull piping liquids are variables stored in the hull. In addition, on-board chain and clump weights can be included in the hull variables.

For the purpose of a deep water production semi-submersible, it is assumed that 12 mooring lines will be used. The size of the mooring chain is unknown at the hull sizing stage; however, one may obtain an educated estimate of the chain size by investigating existing semi-submersibles and the associated mooring system. The length of on-board chain on board may be estimated as a function of column length. The on-board chain length at transit and operating condition are different, as additional chain may be pulled on-board after installation.

2.6 Risers and Mooring Tensions

For OPTI® series semi-submersible, hull sizing process is less sensitive to the risers and mooring tension, as the riser adaptor and fairleads are at the pontoon, where most of the ballast tanks are placed. The tension can be conveniently exchanged with ballast without affecting much of the overall vertical center of gravity (VCG) of the vessel.

At the same time, with a clear definition of water depth, number and size of the production and export risers in the Basis of Design, one can still make good estimate of the vertical riser tension. The estimate of the mooring vertical tension may be less accurate in the initial sizing but it can be compensated by swapping ballast.

2.7 Transit Load Case

It is assumed that the vessel will leave the shipyard under tow with 15% lighter payload than designed for in operating condition. For conservative consideration, in the calculation of the payload in wet tow condition, the future reserve and deck variables are also considered. The pontoons are sized to ensure a minimum of 0.5m freeboard during this operation and sufficient ballast is allowed for trimming.

3 HULL CONFIGURATION AND SELECTION

Another important task in hull sizing is to select a "better" hull configuration. The selection demands the availability of the following: 1) a pool of at least two hull configurations; 2) an evaluation criterion or definition of better.

As mentioned previously, an infinite amount of hull configurations can be found to support a given topsides payload. However, for practical reasons one may identify and examine only a finite amount of hull configurations produced by systematically varying the hull particulars. To further reduce the amount of work, a set of sizing criteria, such as setting a stability requirement (minimum GM), ballast capacity (minimum ballast), and maximum displacement allowed, can be implemented. Practical constraints due to space for equipment and inspection may be used to eliminate non-practical solutions.

The method of elimination discussed above may still result in a large number of candidates. It is therefore necessary to establish an evaluation criterion, i.e. what defines a better hull configuration. This evaluation criterion often involves hull motion performance, displacement, hull steel, equipment accommodation, constructability, and installation. Unfortunately, the parameters (e.g. motion vs displacement) are quite often in opposition, in which one is only improved at the expense of the other. From this perspective, it is favorable to determine such a "design front" for hull selection. This "design front" carries the characteristics of a "Pareto Frontier" often employed in optimization, i.e. one cannot make one parameter better without compromising the other parameter. It is also important to establish priorities among the parameters examined.

A flow diagram is illustrated in the figure below.

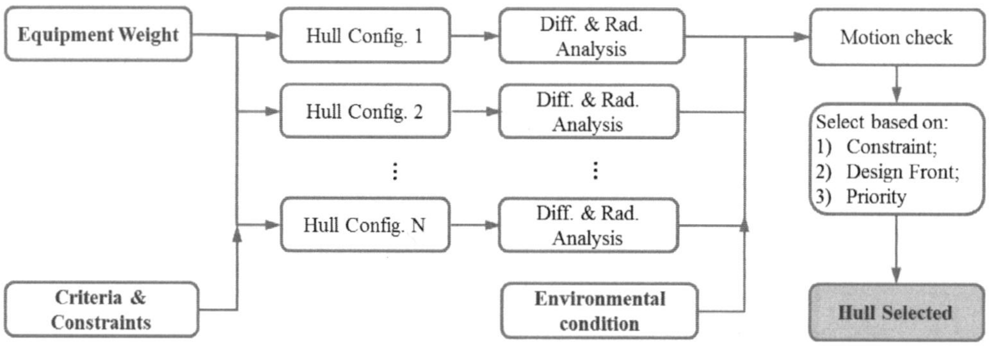

Figure 3 Hull sizing and selection flow diagram

3.1 Sizing Criteria

The main hull sizing criteria utilized in this study are related to stability as measured by GM, ballast available, heave natural period, and maximum unit displacement. The following criteria are provided as an example:

- GM at pontoon just submerged is no less than 1.0 m.
- GM at column transitional section top is no less than 1.0 m.
- GM at operating condition is no less than 3.0 m.
- Ballast in transit (wet tow) is no less than 400 mt.
- Ballast in operating condition is no less than 4,500 mt.
- Maximum unit displacement is no greater than 32,000 mt.
- Heave natural period is between 20 sec and 25 sec.

3.2 Practice Constraints on Hull Selection

There are practical considerations constraining the selection of a hull. These constraints include:

- The triangular portion of the column has to accommodate ventilation pipes and chemical tanks. Additionally, the riser connections and jumpers onboard the vessel will run through the triangular portion (inboard of columns) from pontoon top to the deck; therefore, enough space shall be available for the risers.
- The pontoon shall provide enough room/space to accommodate riser porches directly attached to it as well as access tunnels inside.
- Pontoon aspect ratio (width over height) can greatly affect current drag force and has to be reasonably configured.
- Fairleads are placed at the corners, where a cut to the pontoon is made. This cut length shall be enough for fairlead arrangements.

3.3 Automation Subroutine

EOC has developed an in-house subroutine, *AutoOptiHull*, that can perform a batch hull sizing task, with given payload capacity and specified criteria, using the weight estimate assumptions discussed in section 2. This subroutine reduces the engineering time to size one hull and calculate its motion response, from days to minutes. Figure 3 demonstrated the steps of the subroutine, which are also listed as follows:

- generate a large number of hull configurations by systematically varying the hull main particulars;
- examine the hulls against the criteria and make fine adjustments to comply;
- produce the panel model and perform diffraction and radiation analysis;
- complete frequency domain analysis with analysis results and environmental data;
- provide results to check hull steel weight against motion to facilitate hull selection.

The hull main particulars that can be varied include:

- still water airgap,
- draft,
- column center to center distance (usually the same as deck length),
- column width,
- pontoon size, and
- target operating metacentric height (GM).

4 CASE STUDY

This case study discusses a production semi-submersible that has the same production capacity as the OPTI-EX® unit. The production capacity is 60,000 BOPD, 150 MMSCFD, and 40,000 BWPD. The production equipment weight is 4,800 mt, including 600 mt deck variables and 500 mt for future reserve.

To reduce the total number of hull configurations, the still water airgap is fixed to 22 m and only two drafts are considered, i.e. 26 m and 28 m. Additionally, the number of variations of each of the rest of the parameters, i.e. the deck length, the column width, pontoon size, and target operating metacentric height, is limited to 4 to 6. With the five varying parameters mentioned above, the total permutation produces 1,400 hulls, which are examined using the criteria specified in section 3. A typical hull panel model is demonstrated in Figure 4.

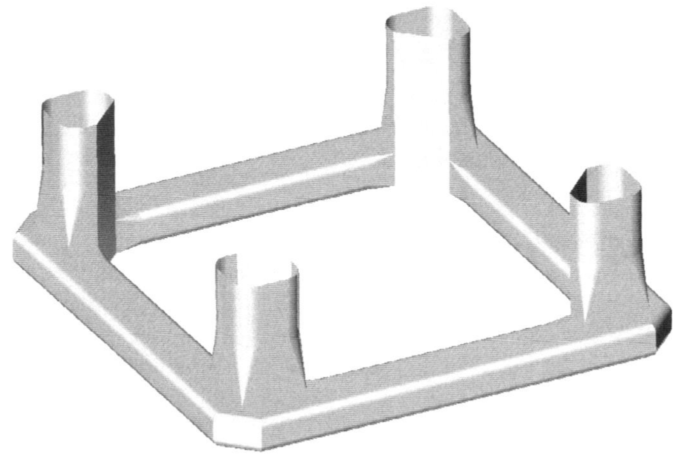

Figure 4 Illustration of the panel model of the submerged hull

Table 1 shows the environmental data represented by Hs and Tp of the JONSWAP wave spectrum. Combining the diffraction and radiation analysis results, the hull sizing subroutine can produce motion results of all the hulls. One may utilize the output of the hull sizing study in several ways to facilitate the final hull selection

In this study, hull motion characteristics, steel weight (cost) and accommodation of all topsides equipment (deck size) are assigned higher priorities. Figure 5 presents hull steel against the heave motion for all hulls examined. In the figure, each blue dot represents a sized hull. Also shown in the figure are a few candidate hulls selected and they will be discussed in detail later.

The importance of this output is that it presents a big picture as regards to each sized hull's performance relative to others. A visual inspection can easily reveal an approximate design front, represented by the blue dash line. This empirical front demonstrates similarities to the so-called Pareto frontier often used in optimization. In this example, the figure presents the constraint between the hull motion and steel weight along the design front.

The output of the hull sizing subroutine can be used to examine the impact of a particular parameter; for example, one may group the results according to column center to center distance (same as deck length) and plot the group results separately, as shown in Figure 6. In each of the four sub-figures, a design front may be estimated by visual inspection. These figures shows that, as the deck length increases from 57.6 m to 64.8 m, hulls with better constraint of motion and steel weight become more and more available.

The breakdown according to deck length may facilitate hull selection as demonstrated by the one or two selected hull candidates shown in the sub-figures and listed in Table 2. These selections were also subjected to practical constraints discussed in section 3.

As shown in the Table 2, for the hull candidates with the same draft (e.g. 26 m for HS-1 to 4), both the displacement and hull steel can be reduced significantly as the deck length increases while similar heave motion character can be maintained. However, due to the increase in the deck size, the truss steel increases. The overall effect is that the total hull steel decreases. For the hull candidates with the same deck size (e.g. 64.8 m for HS-4 and HS-6), the displacement and the hull steel increases as the draft increases; however, the heave motion improves.

For this case study, the deck area and motion takes higher priority and therefore, hull candidate HS-6 is selected as the initial hull configuration for subsequent spiral design. In the case that hull steel takes priority and motion is of less importance, hull candidate HS-4 could be a good selection. Other constraints such as water plane area, which depends on column size, could influence the selection if large chemical tanks or others hull storage is needed.

Table 1 Environments used in the motion calculation

Environment	Hs [m]	Tp [sec]	Gamma
100 year Winter Storm	8.2	11.7	2.0
100 year Hurricane	15.2	15.7	2.0
1000 year Hurricane	19.4	16.2	2.0

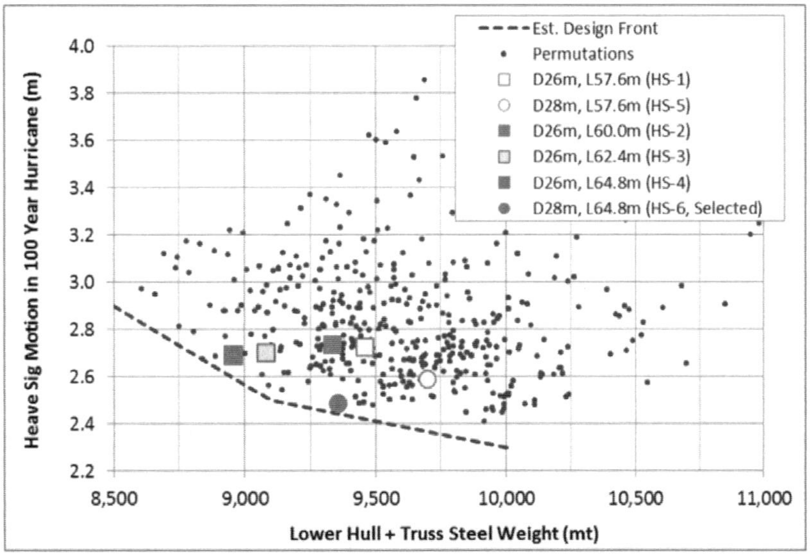

Figure 5 hull steel vs. significant heave motion in 100 year hurricane

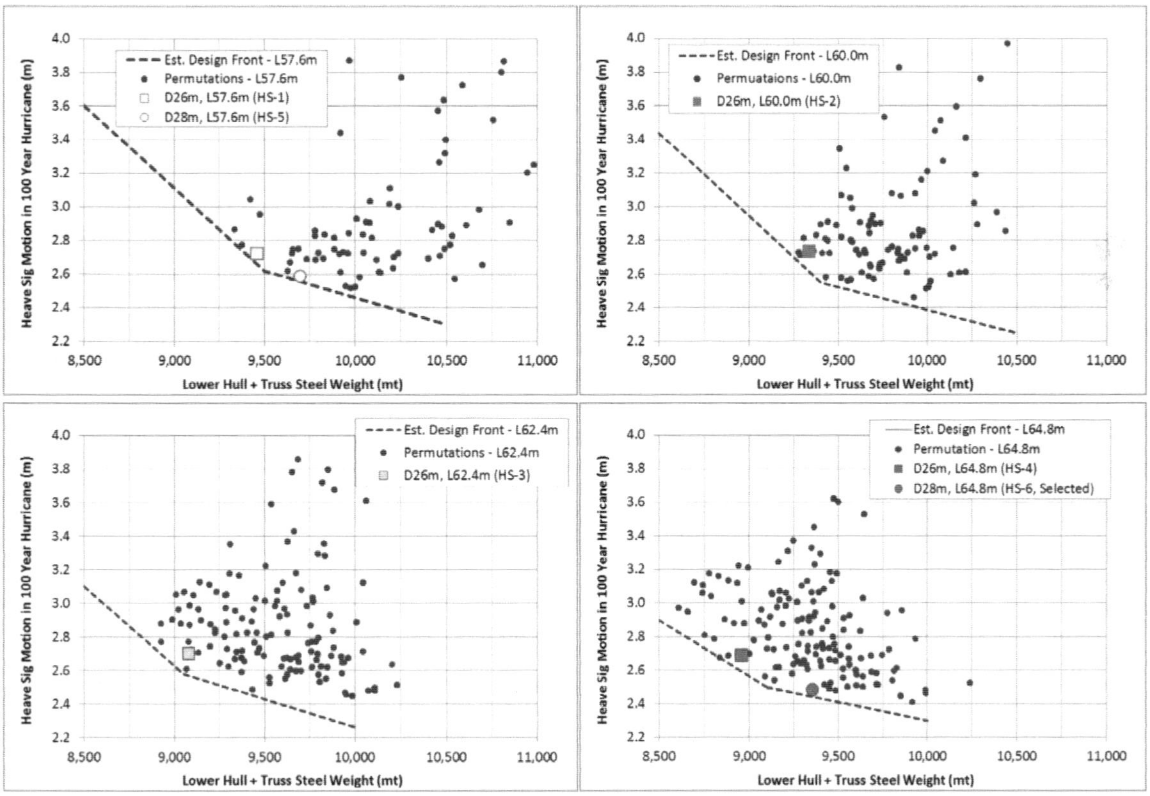

Figure 6 hull steel vs. significant heave motion in 100 year hurricane (grouped by deck length)

Table 2 Parameters of the identified hull candidates

Hull Selection ID	HS-1	HS-2	HS-3	HS-4	HS-5	HS-6 (Selected)
Description	D26m, L57.6m	D26m, L60.0m	D26m, L62.4m	D26m, L64.8m	D28m, L57.6m	D28m, L64.8m
Disp [mt]	29,573	28,806	27,382	26,681	30,865	28,922
Hull steel [mt]	7,551	7,348	6,939	6,740	7,790	7,211
Truss [mt]	1,906	1,986	2,065	2,145	1,906	2,145
Hull + Truss [mt]	9,457	9,334	9,079	8,957	9,696	9,356
Equipment + Truss [mt]	6,722	6,802	6,881	6,961	6,722	6,961
Draft [m]	26.0	26.0	26.0	26.0	28.0	28.0
Still Water Air Gap [m]	22.0	22.0	22.0	22.0	22.0	22.0
Deck Length [m]	57.6	60.0	62.4	64.8	57.6	64.8
Truss Height [m]	6.0	6.0	7.0	7.0	6.0	7.0
Column Length [m]	42.0	42.0	42.0	42.0	44.0	44.0
Water Plane Area [m2]	517.6	502.5	470.5	453.9	525.5	458.4
Pontoon_L [m]	80.4	81.6	82.8	84.0	80.4	85.2
Pontoon_W [m]	11.4	10.8	10.2	9.6	11.4	10.2
Pontoon_H [m]	6.0	6.0	6.0	6.0	6.0	6.0
VCG [m]	20.4	20.8	21.6	22.1	20.5	22.0
GM [m]	3.2	3.4	3.1	3.0	3.3	3.2
Heave Tn [sec]	22.1	21.9	21.9	21.8	22.2	22.8
Sig Heave in 100 yr WS [m]	0.83	0.87	0.92	0.92	0.84	0.81
Sig Heave in 100 yr Hur [m]	2.72	2.74	2.71	2.69	2.59	2.48
Sig Heave in 1000 yr Hur [m]	3.74	3.75	3.77	3.77	3.61	3.40

5 CONCLUSIONS

This study discusses the methodologies for weight estimate and the subroutine employed in a hull sizing for the OPTI® series production semi-submersibles.

Accurate weight estimate in initial sizing is of significant importance as it can reduce the number of spiral passes in the design, and hence, save engineering time and cost. The hull sizing of OPTI® series production semi-submersibles utilizes weight estimate methodologies that are based on EOC's extensive engineering practices and fully benchmarked against the Who Dat and Delta House projects.

The EOC developed *AutoOptiHull* subroutine is highly efficient; the time for sizing one hull configuration and examine its motion performance is reduced from days to minutes. The high efficiency enables one to generate and compare thousands of hull configurations, determine a design front, and set hull evaluation target, which greatly facilitates the hull selection process.

Implementation of the weight estimate and the sizing subroutine through a case study is presented. With the specified priorities, a better compromise between hull motion and steel weight is achieved.

REFERENCES

Chakrabarti, S. K., "Handbook of Offshore Engineering", Vol I, 2005.

SNAME
THE INTERNATIONAL COMMUNITY
FOR MARITIME AND OCEAN PROFESSIONALS

GSF GALAXY I JACKUP CASE STUDY FOR OPTIMIZING RIG MOVE PERFORMANCE IN NORTH SEA USING AN ADVANCED SIMULATION MODEL

DAMIEN CARRE, LUIS MCARTHUR
TOTAL E&P UK

ANDY SIMPSON, PHILIP ZHANG, PH.D.
TRANSOCEAN

JOSE H. VAZQUEZ, PH.D.
3DENT TECHNOLOGY

Proceedings of the 22nd Offshore Symposium, February 2017, Houston, Texas
Texas Section of the Society of Naval Architects and Marine Engineers

ABSTRACT

Through collaboration among TOTAL, TRANSOCEAN and 3DENT TECHNOLOGY, an initiative was launched to investigate the capabilities of the GSF Galaxy I jackup to perform a rig move in waves with periods greater than 6 seconds. The objective was to improve the rig move performance by reducing waiting on weather time.

GSF Galaxy I, is a Friede & Goldman-designed Mod VI harsh environment jackup. The generic Class-approved wave conditions for emplacement, preloading and leg extraction are simply stated in the operating manual as 6-foot wave height at 6-second period.

Consulting for TRANSOCEAN, 3DENT started the study by conducting so-called "Going on Location" studies in regular waves and random waves to establish permissible wave heights as a function of period. 3DENT later developed a methodology for analyzing the coming off location and preloading at large draft conditions. This paper discusses the various analyses, techniques and results of the studies conducted by 3DENT in the development of a user-friendly response assessment program to assist TRANSOCEAN rig personnel in the decision to carry out Going on Location, Preloading and Coming off Location operations with known information about the environmental conditions.

The response and therefore the limiting wave heights under which the various operations can safely be undertaken, requires a clear definition of the waves. Utilization of a wave radar, providing much-improved information on the waves when compared to either estimates by human observers or weather forecasts, was key to providing a verified and reliable input.

The response assessment program was successfully used while moving the GSF Galaxy I from the north to the south side of the TOTAL Elgin platform, saving 10 days of waiting on weather over a 15-day rig move. This application demonstrated it is possible to safely increase the environmental envelope for rig moving operations.

Keywords: Jackup, Going on Location, Coming off Location, Preloading at Large Draft.

INTRODUCTION AND BACKGROUND

Generally speaking, jackups can be thought of as being in one of three conditions or undergoing one of four operations, as listed below:

1. Afloat Condition (This can be in calm water, in waves, and/or while stationary or in transit)
2. Jacking Operations (Jacking up/down of legs or hull relative to one another)
3. Going on Location Operations (transitioning from afloat to soft-pinned condition)
4. Soft-Pinned Condition (Condition when soil supports the jackup with minimal penetration)
5. Preloading Operations (Pressing the legs to their final penetration)
6. Elevated Condition (At full airgap, after having preloaded the soil)
7. Coming off Location Operations (Transitioning from elevated to afloat conditions)

Jackups have class-approved Marine Operating Manuals (MOMs) which provide guidelines for their operations and list their design limits. The MOMs are usually prepared by the jackup designer and submitted to review by a Class Society. Each Class Society has its own set of requirements for classing a jackup and as of the writing of this paper, they do not include any specific criteria for Going on Location or Coming off Location. Therefore, if any specific guidance is included in the MOM, it has been solely to the discretion of the designer.

The *GSF Galaxy I* is a Friede & Goldman-designed Mod VI harsh environment jackup, built in 1991 and Classed by ABS. Its Class-approved MOM lists the following wave conditions for emplacement, preloading and leg extraction: *6-foot wave height at 6-second period.* Using a single limit seastate for the emplacement (going on location, or GoL) and extraction (coming off location, or CoL) implies that these operations can safely be carried out in those conditions, regardless of water depth, wave energy composition (seas dominated by swells, wind-driven waves or mixed), wave period(s), wave direction(s) or soil conditions. Furthermore, the MOM does not clarify whether the 6-ft wave height refers to maximum wave or significant wave, nor does it specify whether the 6-second period is T_p (peak spectral period), T_z (average zero-crossing period), T_s (significant period) or any other period.

In the summer of 2015, expecting to use the *GSF Galaxy I* on several of its fields in the North Sea, TOTAL requested TRANSOCEAN to investigate the possibility of expanding the conditions under which going on location, coming off location and preloading at large draft could safely be carried out. The objective for this request was to improve the rig move performance by reducing waiting on weather time. It is noted that the original scope of work included four locations with very comparable conditions (water depth and soil conditions). Furthermore, the prevalent conditions at these locations include combined seas with swells typically having periods in the 6-16 sec range and wind-driven waves having periods in the 3-8 sec range. Not only that, but the swells and wind-driven waves do not usually have the same direction. As such, it was important to provide a user-friendly program to allow onboard personnel to use the results of the study to make the Go/No Go decision.

This paper discusses the various analyses, techniques and results of the studies conducted by 3DENT in the development of a user-friendly response assessment program to assist TRANSOCEAN rig personnel in the decision to carry out Going on Location, Preloading and Coming off Location operations with known information about the environmental conditions. The paper also discusses the use of wave radar information and the use of the assessment program during a particular move from the north to the south side of the TOTAL Elgin platform which resulted in 10 days of waiting on weather saved over a 15-day rig move.

JACKUP DESCRIPTION AND MAIN PARTICULARS

The *GSF Galaxy I* unit has three levels of pinions at each of the chords for a total of 54 pinions. The unit is outfitted with rack chocks, but these are not engaged during the GoL, CoL or preloading operations. The pinion stiffness and capacity values are 3,920 kips/ft and 1,600 kips, respectively. Table 1 shows the principal characteristics of the *GSF Galaxy I* jackup, Ref 7).

Table 1 Principal Characteristics of *GSF Galaxy I* Jackup

Description	Value	
Length of Hull	244	ft
Breadth of Hull	250	ft
Depth of Hull (at Centerline)	36.0	ft
Overall Leg Length	560.0	ft
Longitudinal Leg Spacing	157.0	ft
Transverse leg Spacing	173.2	ft
Leg Chord Spacing (Pitch to Pitch)	50.0	ft
Loadline Draft	23.0	ft
Loadline Displacement	58,604.4	kips
Weight of Three Legs (excluding Spudcans)	11,911.7	kips
Weight of Three Spudcans (including permanent ballast)	10,674.6	kips

SITE INFORMATION

For the going on location analyses, the first soil immediately below the seabed is most relevant. For this reason, each of the penetration curves were reviewed to obtain apparent soil stiffness values and select properties for the going on location analyses based on the most conservative (i.e., stiffer) conditions. Table 2 shows a summary of the provided soil information and the apparent soil stiffness values. From these results, soil information from the South Face of Elgin A Platform Block 22/30, Bow leg was selected for the going on location analyses.

Table 2 – Soil Information at the Four Sites

Location	Water Depth (m)	Leg(s)	Penetration at Full Preload (m)	Penetration at 5,000 MT (m)	Apparent Soil Stiffness* (MT/m)
Elgin Standoff	93.0	Bow Leg	4.30	1.75	2,857
		Aft Legs	4.30	1.55	3,226
South Face of Elgin A Platform Block 22/30	92.7	Bow Leg	5.00	1.20	**4,167**
		Port Leg	5.00	1.75	2,857
		Stbd Leg	5.00	1.75	2,857
North Face of Elgin WHPA Platform Block 22/30	92.0	Bow Leg	5.00	1.65	3,030
		Aft Legs	5.00	1.25	4,000
West Franklin UKCS Block 29/5b	93.5	Bow Leg	14.20	1.85	2,703
		Aft Legs	13.70	1.85	2,703

* Apparent soil stiffness estimated as $K_a = F/\Delta z$, with F = 5,000 MT

ANALYSIS METHODOLOGY

As indicated in Reference 11, 3DENT personnel have developed and implemented a methodology for conducting GoL analyses that reflect some of the recent work on the subject, going beyond the previously used energy conservation approach. This methodology accounts for soil stiffness, spudcan shape, jacking speed, and compares the calculated loads with the structural capacity of the legs and jacking system to obtain permissible wave height curves. Permissible wave height curves for GoL are obtained by iteration.

For the *GSF Galaxy I* study, 3DENT developed a methodology for establishing permissible wave heights while the jackup is coming off location and/or preloading at large drafts (PLD). It is noted that both the CoL and PLD analyses are similar to the standard elevated analysis of a jackup, with the following exceptions:

1. In the CoL and PLD analyses, the significant loads are wave loads acting on the hull (as opposed to wind loads on the hull/legs and wave loads on the legs for the elevated condition)
2. In the CoL condition, the legs are in tension and there is a possibility of having hull inclination angles beyond the normally allowed 0.3-degree value.
3. In the CoL and PLD conditions, there is a larger degree of uncertainty on the effective soil stiffness properties than there is for the elevated condition after having completed preloading operations.

The CoL and PLD analysis methodology is made up of two types of analyses, as outlined below.

- *Frequency-domain diffraction analysis* (using *WAMIT*) to calculate the wave load as well as the added mass/damping coefficients of the jackup at the target depth.
- *Structural time-domain analysis* (*SAP2000*) to establish load response RAOs.

Given that the *GSF Galaxy I* jackup is outfitted with a fixation system (which is not used during these operations), pinion capacity proved to be the controlling mechanism in establishing permissible wave heights for the CoL or PLD operations. The methodology consists of isolating the contribution to the pinion loads due to waves and due to static (current, wind, gravity and buoyancy) loads. Then, analyses accounting for dynamic effects were carried out using waves with unit amplitude to establish pinion load response RAO curves.

The permissible wave height curves are obtained by combining the principle of superposition with the known relationships between wave spectra, RAOs and Maximum Most Probable Extreme values. As such, the load levels on all possible critical members are determined and saved prior to carrying out time-domain simulations for waves of unit amplitude. From the time-domain simulations, response RAO curves for each of the possible critical members are determined for use in the calculation of the permissible wave height curves.

MODELS, RESPONSE, PERMISSIBLE WAVE HEIGHT CURVES AND THE NEED FOR A PROGRAM

Figures 1 and 2 show the diffraction model for the GoL analyses and the free-floating heave/pitch motion RAOs of the *GSF Galaxy I* in head seas. Figures 3 and 4 show the diffraction and structural models for the CoL analyses and the excitation horizontal load RAOs of the *GSF Galaxy I* at maximum overdraft.

While the free-floating motions are not used in the determination of GoL permissible wave heights, they can be used in conjunction with wave heights to determine whether or not the conditions are deemed safe to carry out the GoL operations. As can be seen in Figure 4, wave loading (and response) is sensitive to direction as well as period. For this reason, permissible wave height curves are usually prepared for waves with different directions, covering a range of periods in the 4-20sec range. One issue to keep in mind is that the motion RAOs are for a specific wave train (i.e., a single regular wave with a uniform wave height, period and direction), but it is seldom the case that the conditions are described by a single wave.

One way to use the permissible wave height curves on sites with combined swells and wind-driven waves (often having different direction and periods) is to focus on maximum wave height, and check the curves for the most severe direction and most onerous period (whether the wind-drive T_z period or the swell period). Doing so can be expected to produce conservative results, but it also requires the onboard user to go through multiple graphs to select the appropriate one before making the GO/NO-GO decision. A much better approach is to use a simple program to do the wave combination and needed interpolations, based on the permissible wave height curves.

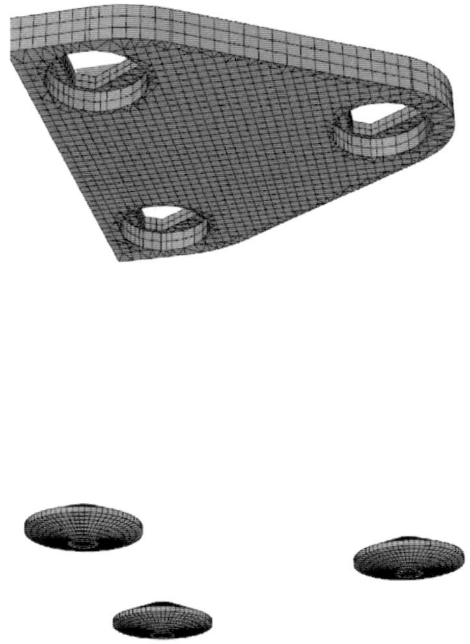

Figure 1 GoL Diffraction Model of *GSF Galaxy I*

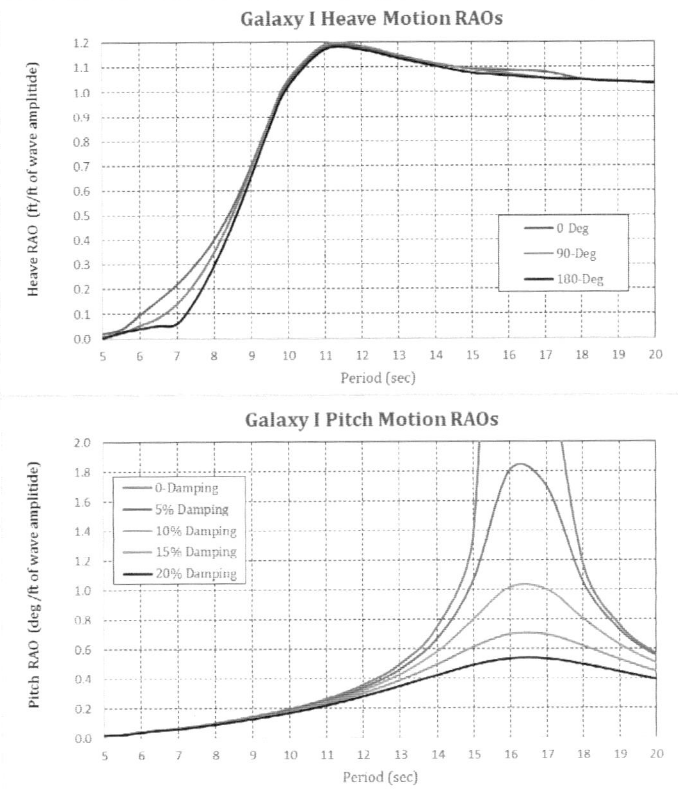

Figure 2 Free-Floating Heave/Pitch Motion RAOs of *GSF Galaxy I*

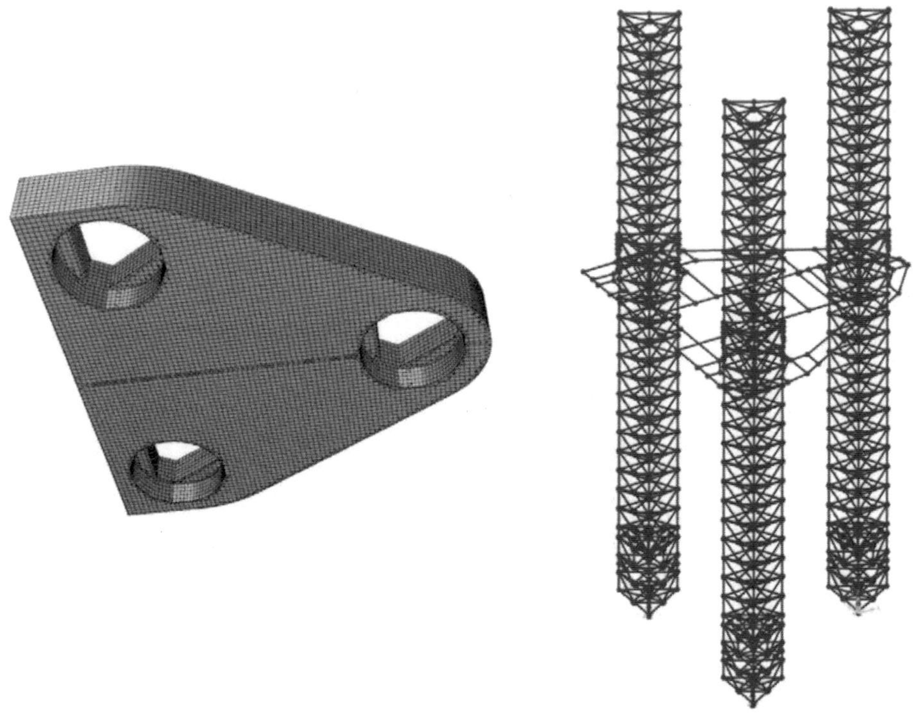

Figure 3 CoL Diffraction/Structural Models of GSF Galaxy I

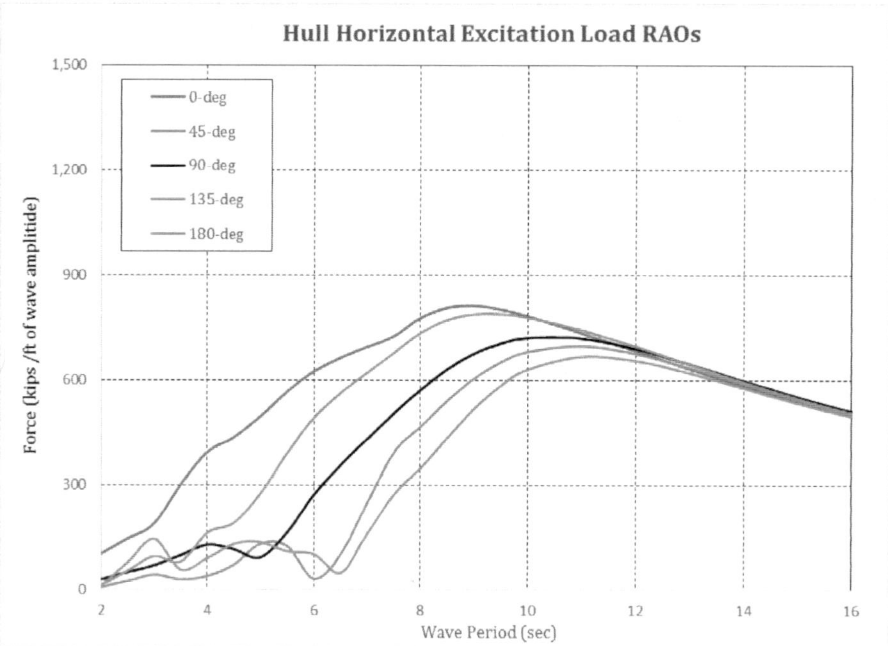

Figure 4 Horizontal Excitation Load RAOs for CoL Analysis of *GSF Galaxy I*

GOL/COL/PLD EXCEL-BASED PROGRAM

As is usually the case, conditions at a given site are such that the seastate is a combination of wind-driven waves (represented by random waves) and swells (represented by regular waves), with different periods and directions. Therefore, while the permissible wave height curves are useful, in order to avoid being overly conservative, a program that estimates the response of the jackup in combined seas can facilitate the decision-making process.

The figures below show sample screen shots of the Excel-based program developed for use onboard the GSF Galaxy I jackup to provide utilization factors in any combination of wind-driven waves and swells. The program screen shots shown focus on the CoL operations, but the program is suitable for GoL, CoL and PLD operations.

Since this program was developed for a particular set of sites with known water depth and soil conditions, the input only for the CoL assessment consists of the following parameters: a) draft, b) rig orientation, c) wind-driven wave parameters (direction, period and wave height) and d) swell parameters (direction, period and wave height). Based on the user-provided input, the program displays a graphical representation of the rig orientation and the direction of the waves. The program also shows the calculated utilization ratio for the unit under the provided wave/swell conditions.

The program is essentially a database populated with utilization ratios for combinations of wind-driven waves and swells having different periods and directions. The cases analyzed and included in the Excel-based program cover the expected wave heights and periods for the site, with combination headings at 22.5-deg intervals to cover the standard cardinal directions as well as the intercardinal and secondary intercardinal directions (N, NNE, NE,... NNW). Once a database is populated, the program's main function is for lookup and interpolation. Figure 5 shows a screen shot of the program for one of the conditions encountered during the move.

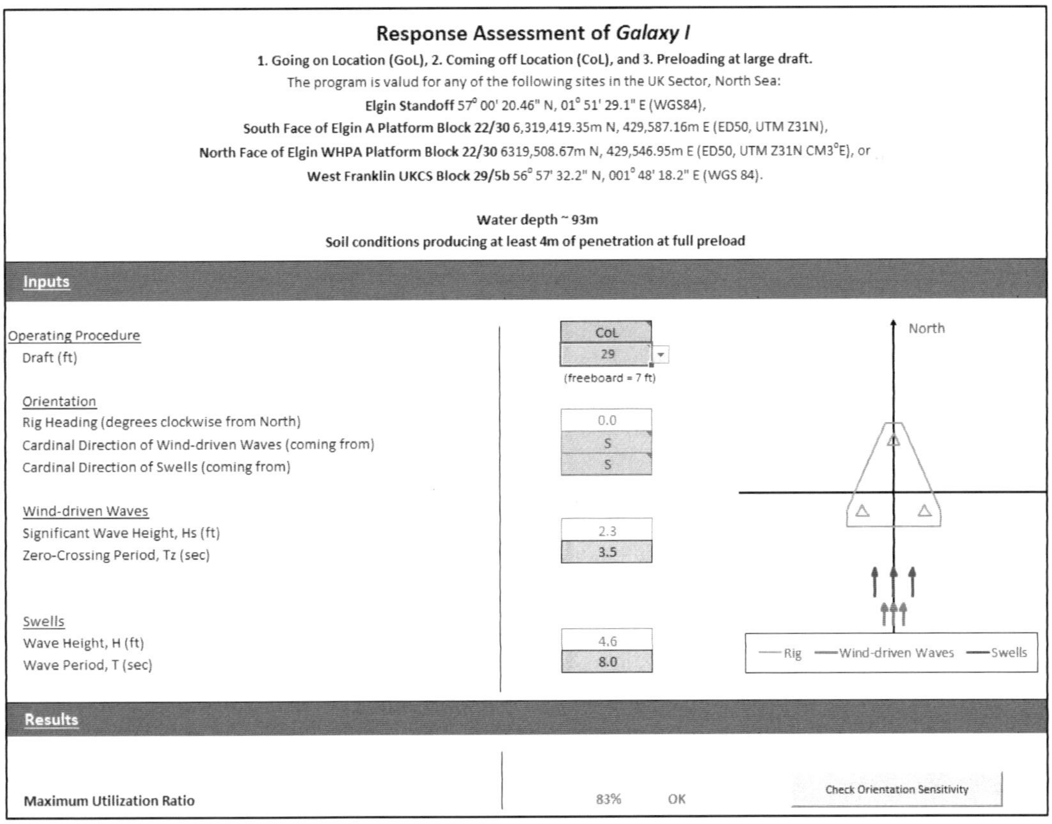

Figure 5 Sample Screen Shot of Program

Given the fact that the analyses were conducted for waves in 22.5-deg intervals, the results are not perfectly continuous. As such, the possibility exists that a small change in rig orientation would produce large differences in utilization ratios. For that reason, the program has an option to generate UC results for the requested wave directions and rig orientations from 0- to 360-degrees. This is depicted in Figure 6.

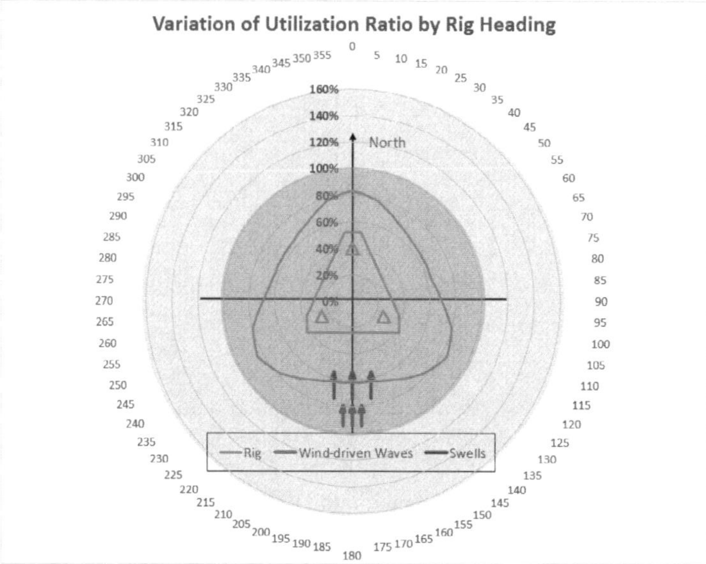

Figure 6 Sample Screen Shot Depicting Variation of UC by Rig Heading

WAVE RADAR INFORMATION AND RIG MOVE

As can be seen in Figure 5, accurate real-time metocean inputs are required for the program in order to produce meaningful output. There are three primary options for obtaining the metocean conditions:

- Manual Observations from personnel on the Rig
- Weather Forecasts
- Metocean instrumentation observations

Clearly, manual observations by personnel are subjective and prone to error and are not preferred.

Weather forecasts are scientifically based but are still not necessarily reflective of local conditions at the time, being subject to temporal, spatial and amplitudinal error and so are not ideal inputs for the program. Weather forecasts are of course a valid input into the tool if it is being used predictively, to ascertain for example if the predicted weather in 24 or 48-hours will still be suitable for the operation to continue. It is also important to have full confidence in the forecast being utilized. In this case specific studies and comparisons were conducted in advance of the operation to ensure the veracity of the forecast and to give the crew executing the operation full confidence.

The most accurate inputs for the program being used in real-time can be gained using accurate Metocean instrumentation, particularly from the static Platform rather than the dynamic Rig.

The TOTAL Elgin Platform has a suite of Metocean sensors as a part of the ICSS (Integrated Control & Safety System) which are integrated specifically into a RigNet NIMOS (Nessco Integrated Metocean System). The systems which specifically gave the inputs required for the program are:

- SAAB REX D Sea Level Microwave Wave Radar
- Two (2x) Gill Ultrasonic IS WindObserver Anemometers.

Proven instrumentation accuracy and reliability are required for metocean inputs and adoption by those carrying out the rig move. On the TOTAL Elgin Platform this is assured through documented installation and initial calibration, routine documented service visits for ongoing accuracy, and monthly independent Quality Control of the raw and processed output data from the systems in conjunction with comparison checks against the five closest comparables systems on nearby Platforms as depicted in Figure 7 in addition to correlation against forecast data. It should further be noted as a recommendation that all weather forecasts supplied during the operation include access to the metocean instrumentation observations and that modeling takes into account the readings as constraints.

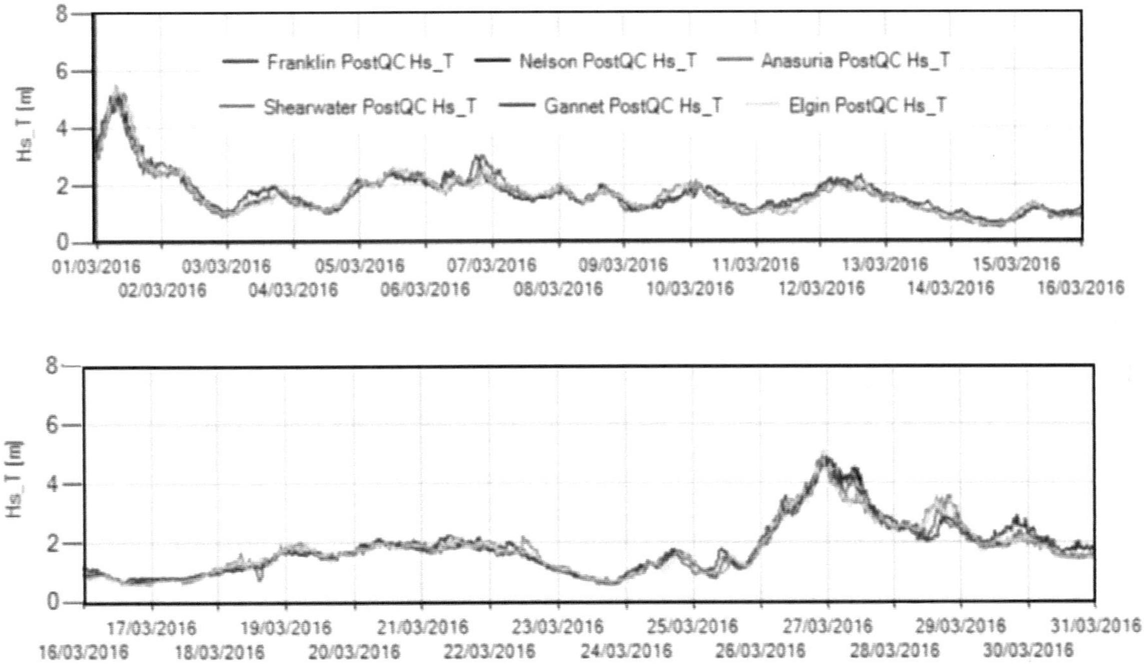

Figure 7 Sample QC Graph of Hs(m) from Total Elgin Platform and Nearby Systems

The real-time metocean observations for the TOTAL Elgin Platform were made available from the Platform to the Rig during the operation through the RigNET MET-REACH web service. Both instantaneous values and time-series data for all key parameters are made available through this service increasing the usability as depicted in Figure 8.

The following Metocean parameters were required for input, in two categories: a) wind-driven wave parameters (direction, period and wave height) and b) swell parameters (direction, period and wave height). A minor software modification was required to split the wave specta return into swell and wind/wave components by applying frequency bands and then carrying out spectral processing on the reduced spectrum to obtain height and period information for these two components of the wave. Even so, it was necessary to substitute T2 as the spectral equivalent of T_Z. The anemometers are used to give wind driven wave direction. Finally, the swell wave direction was taken from the weather forecast as the wave radar available on the Platform is a Sea Level wave radar. A clear recommendation for future subsequent optimization of the program is to ensure the availability of wave component direction information from a directional wave radar on the Platform or on a Buoy.

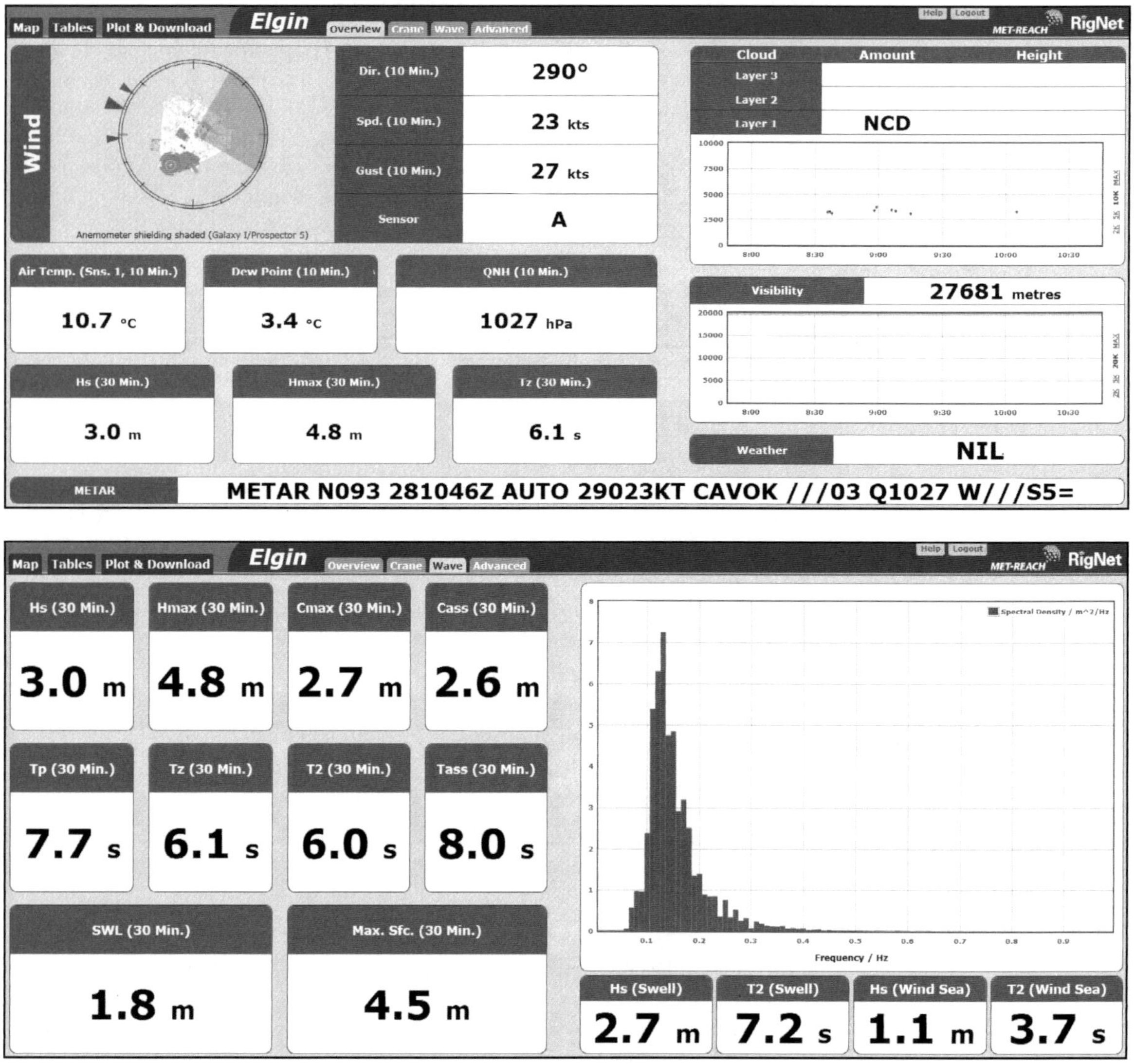

Figure 8 Typical Output from TOTAL Elgin Platform NIMOS

Waiting on Weather during 2015 Rig Moves

On the 17th of March 2015, rig moving operations commenced to move the GSF Galaxy I (GX1) from Total's Elgin Platform South face to the North face. The Rowan Gorilla V (RGV), which has a larger foot print area and exerts a greater preload pressure than GX1 had been on the North face previously.

The soil conditions and existing RGV footprints made this move challenging, inducing high Rack Phase Difference (RPD). There were extended periods of RPD manipulation, leg extraction and single leg preloading; including preloading with one leg off bottom while the hull was maintained level at draft.

In the central North Sea during March, possible weather windows for jackup rig move operations are limited. Including moves to intermediate stand off and standby locations, effectively 5 rig moves were completed, requiring 627 jacking system start/stops. 93 of a total of 122 days were spent waiting on weather.

When established on the North face, leg penetrations ranged from 6.4m to 6.9m (21ft to 23ft).

Like Transocean's other North Sea jack ups, GX1 has operated since build using a generic limiting sea state of 1.8m H_s x 6 secs T_z (6ft x 6 Seconds). Following the 2015 rig move experience, Total requested that this limiting sea state be investigated to determine if the rig move environmental criteria could be increased for the soft sea bed soils at Elgin.

Rig moving from North Face to South Face 10-18th May 2016 using GoL/CoL/PLD Program

On the basis of the 3Dent study, specific for this location, in certain combinations of wind and swell waves, relative to rig heading, the rig can safely be moved in greater than 1.8m H_s x 6 secs T_z (6ft x 6 Seconds), i.e. in sea states which would have previously been considered unworkable.

A rig move onto an open water location or standby location, where heading is not restricted, could be safely completed on one heading but not on another. This could increase moving opportunities when staged moves are being considered.

The rig move from the North to the South face of the Elgin platform was indeed requiring moving the rig between 5 different locations and stand-off locations in order to connect and disconnect anchors and to tow the rig between stand-off locations 2 and 3 as per Figure 9. Such sequence made the rig move highly dependent on at least 3 good weather windows (as opposed to a normal rig move, where a single weather window is required to arrive or leave a single location).

As a result, 3 waiting on weather periods for a total of 27.5 days had been budgeted: to leave the north face location #1, to move from the north stand-off location # 2 to the south stand-off location #3, to move to the south face locations #4 and #5.

Post rig move, an analysis of the weather conditions determined the improvement (or savings) from the utilization of the response assessment program.

Based on the weather forecast from May the 16th, the final approach could not be done. The Marine Operating Manual's 6ftx6sec criteria was only forecasted for 24 hours (Figure 10) when a 72 hour window was required. Using the newly developed criteria and the Excel-based program did allow a sufficiently large window for performing the final approach; which was successfully executed. Without the program, the rig would have been waiting on weather for 10 days up to May the 27th, as shown by Figure 11.

As a result, from the 27.5 days budgeted for waiting on weather:

- 3.7 days were actually spent waiting on weather
- 14 days were saved due to good weather
- <u>10 days were saved thanks to the successful utilization of the program</u>

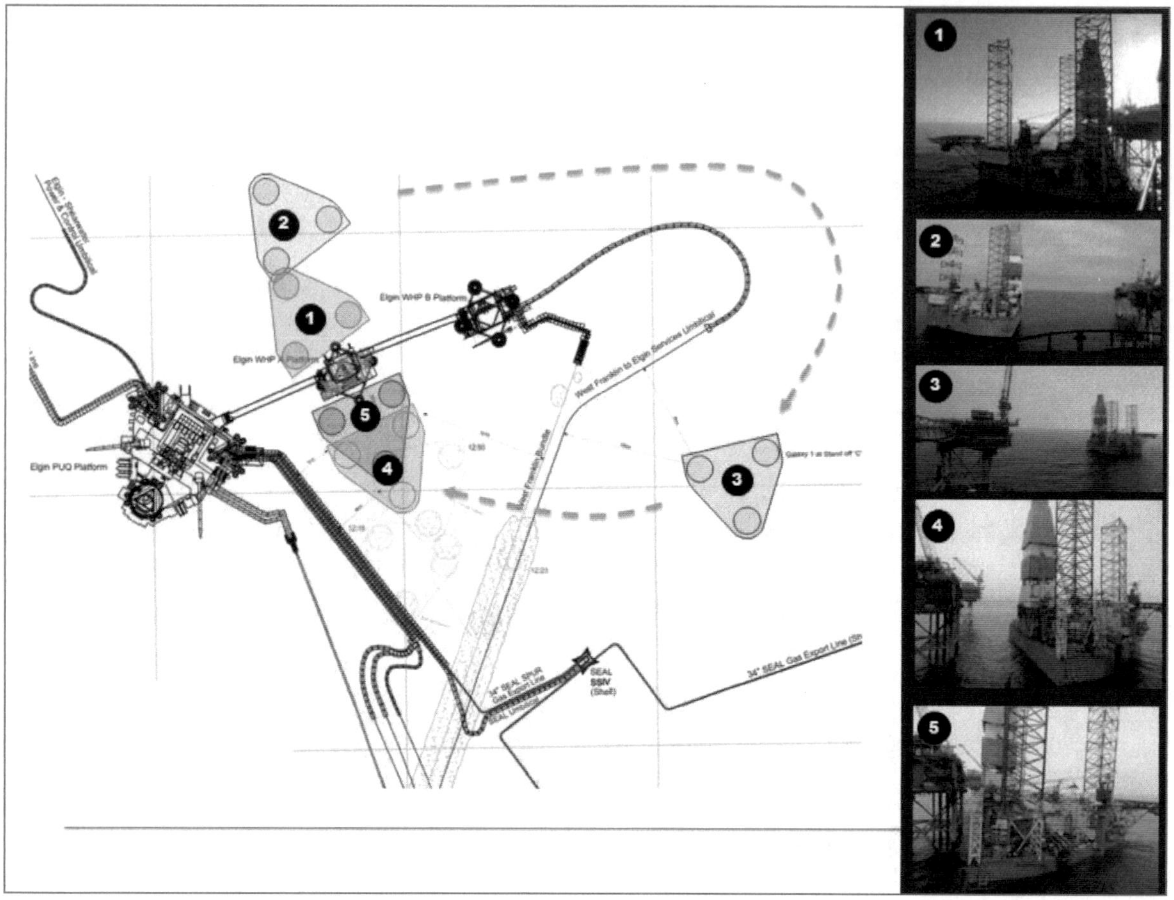

Figure 9 Rig move from North to South Face of the Elgin A platform

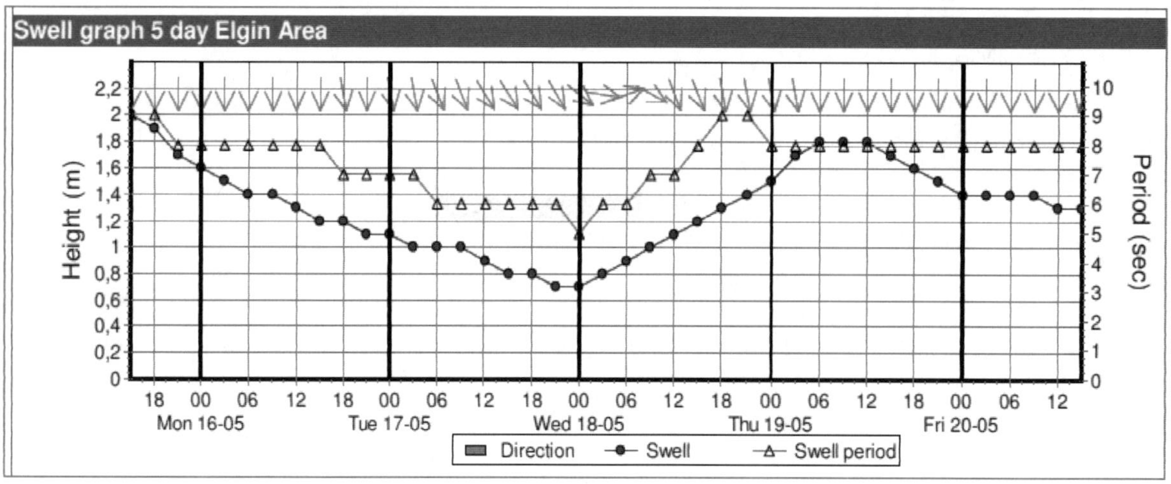

Figure 10 May 16th 2016 Weather Forecast Showing a Window Limited to 24 Hours.

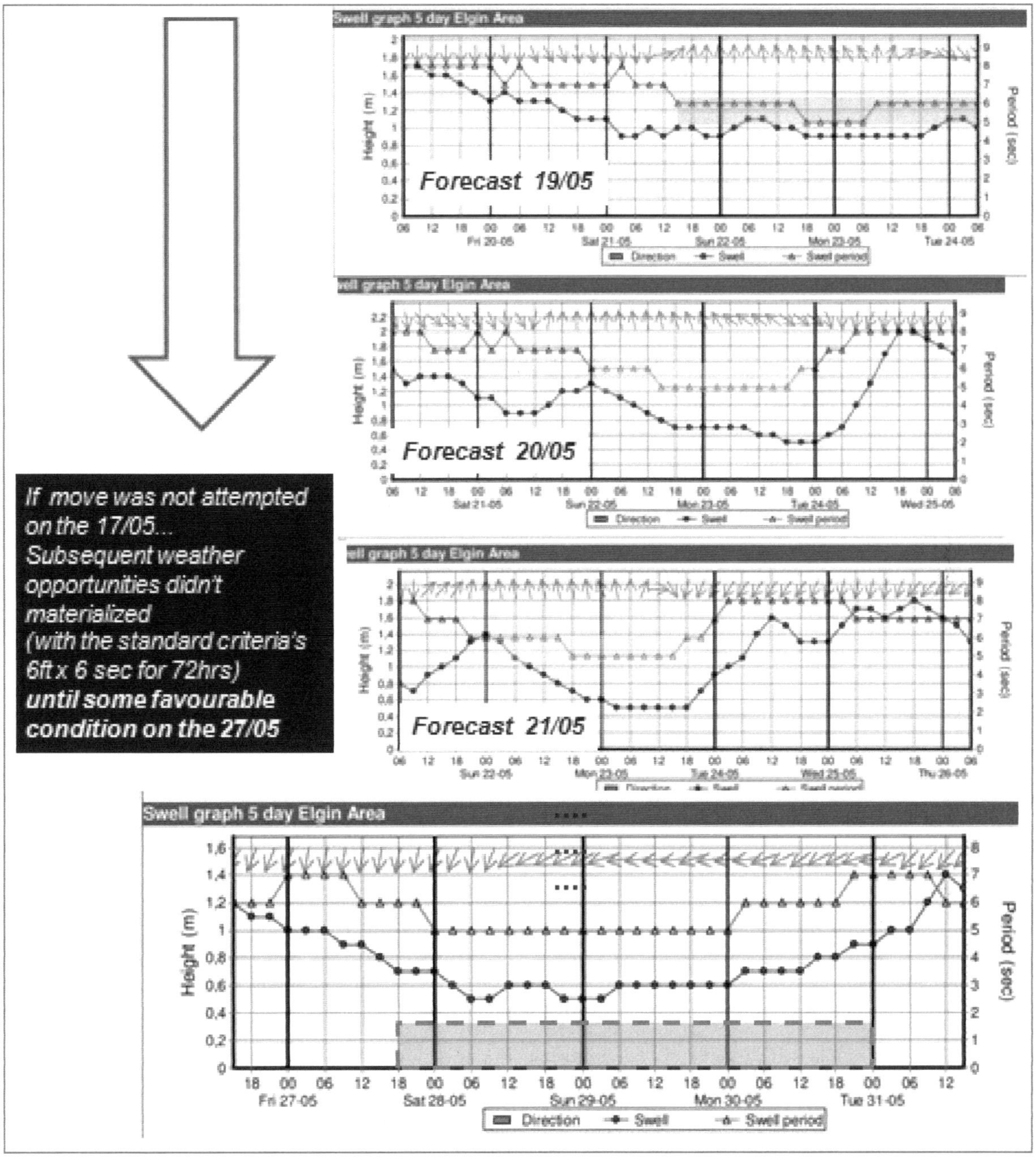

Figure 11 Absence of Favorable Weather Window Until May the 27th

SUMMARY AND CONCLUSIONS

By carrying out a series of detailed GoL, CoL and PLD analyses, the inherent conservatism in the standard 6ftx6sec limit for rig moves in the Marine Operating Manual of the GSF Galaxy I was modified to account for the site's water depth, soil type, rig direction and wave/swell direction. Given the large possible combinations of period and directions for wind-driven waves and swells, the results of the analyses were incorporated into an Excel-based program for ease of use. The Excel-based program proved to be a valuable tool to help the rig team in the decision making during the move.

In practice more work is required to look at what happens when extracting legs (i.e., when one leg comes free, the rig motions change instantly). The main benefit of the Excel-based program is that it provides the user a simple interface for establishing the degree of structural utilization of the rig under combined wind-driven waves and swells, with different periods and in different directions, specific to the rig orientation and for the CoL and PLD cases, at specific drafts. Without the program, the user would have to consult a multitude of graphs to identify the one that provides conservative results.

Some cautions

The leg guides may be subjected to high loads increasing wear and damage to the leg guides/wear plates.

Even small changes to rig heading, and/or wind wave and swell wave direction can rapidly change utilisations, so this must be constantly monitored by the rig movers.

Conclusion

The rig Operating Manual quotes generic sea state limit for rig moves of 1.8m H_s x 6 secs T_z (6ft x 6 Seconds). This may be derived from the most onerous seabed engagement onto hard sand and/or leg extraction draft over pull from deep penetrations.

By completing a location specific study, 3Dent proved it is possible to safely expand the sea state limit for some phases of the rig move and for some sea beds.

ACKNOWLEDGEMENTS

The authors would like to thank the management of our respective companies for their support in the publication of this work.

Additional thanks are expressed to Wei Wang, Marcus Gamino and Bart Grasso of 3Dent Technology for their contributions to the work contained in this paper.

REFERENCES

1. ABS, "Rules for Building and Classing Mobile Offshore Drilling Units," Part 3 Chapter 2 Appendix 2, 2014.
2. DNV "Classification Notes NO. 31.5 – Strength Analysis of Main Structures of Self-Elevating Units," Section 5.8, 1992.
3. GALAXY I AT ELGIN STANDBY, BRITISH SECTOR, NORTH SEA LOCATION, CERTIFICATE OF APPROVAL. Dated March 9, 2014.
4. GALAXY I AT NORTH SIDE, ELGIN WHP PLATFORM, BRITISH SECTOR, NORTH SEA LOCATION, CERTIFICATE OF APPROVAL. Dated December 15, 2014.
5. GALAXY I AT SOUTH SIDE, ELGIN PLATFORM, BRITISH SECTOR, NORTH SEA LOCATION, CERTIFICATE OF APPROVAL. Dated May 9, 2014.
6. GALAXY I AT WEST FRANKLIN WHP, BRITISH SECTOR, NORTH SEA LOCATION, CERTIFICATE OF APPROVAL. Dated February, 2015.
7. GSF GALAXY I, a Self-Elevating Mobile Offshore Drilling Unit Operating Manual.
8. ORCAFLEX User Manual, by ORCINA.
9. SACS User Manual, by Bentley.
10. Tan P, Chen X, Yu Q, Perry M, Mu H, Chang T, Wong M, Chen, D. "Jackup Going on Location Analysis." In: Proceedings, The 15th International Conference on the Jack-up Platform Design, Construction & Operation, London, UK, 2015.
11. Vazquez J, Grasso B, Gamino M, Wang, W "Jackups Going on Location - Understanding Energy Principles on Leg Impact Loads." Proceedings of the 21st Offshore Symposium, February 2016, Houston, Texas
12. WAMIT User Manual, by WAMIT INC.
13. 100-AP-001 Rev 01 Analysis Procedure for Simplified GoL Analysis by 3DENT, dated October 5, 2015.
14. 100-AP-002 Rev 0 Coming off Location of Jackups – Analysis Method by 3DENT, dated August 10, 2016.

SNAME
THE INTERNATIONAL COMMUNITY
FOR MARITIME AND OCEAN PROFESSIONALS

RELIABLE AND ACCURATE DETERMINATION OF LIFE EXTENSION FOR OFFSHORE UNITS

JONATHAN BOUTROT
OFFSHORE BUSINESS DEVELOPER AT BUREAU VERITAS

YANN GIORGIUTTI
ASSET INTEGRITY SERVICES SPECIALIST AT BUREAU VERITAS

FLAVIA REZENDE
TECHNICAL DEVELOPMENT MANAGER AT BUREAU VERITAS

STEVEN BARRAS
MANAGING DIRECTOR AT BUREAU VERITAS

Proceedings of the 22nd Offshore Symposium, February 2017, Houston, Texas
Texas Section of the Society of Naval Architects and Marine Engineers

ABSTRACT

As offshore units near the end of their design life, Operators are often faced with the requirement of extending the operating life of the facility, either to continue producing from an existing and still productive field, or to pursue new opportunities and/or defer new CAPEX as is often the case with MODUs, construction and support vessels and leased production units.

Life extension presents a unique engineering challenge from the perspective of risk management; requiring a complete structural re-assessment of the facility, from its current, as-is condition in order to determine the remaining life of the unit, along with any repairs required in order to achieve the desired additional service life.

This paper describes a methodology, developed by Bureau Veritas, for the engineering reassessment of aging offshore units that will help operators with decision making relating to life extension. It will focus on the two main degradation mechanisms affecting the structure: corrosion and fatigue.

Resulting directly from Bureau Veritas's digital transformation strategy, the methodology is built around the concept of a 3-D "digital twin" model of the asset, which is updated periodically and automatically with critical inspection data collected by connected surveyors. This facilitates continuous updating of the model to ensure a "real-time" equivalence to the "as-is" physical asset.

Furthermore, interfaces between the "digital twin" and conventional structural and hydrodynamics analysis software enable accurate condition assessment analyses to be performed, at any time, on the current state of the asset; providing critical information to be used in the assessment of feasibility of life extension and in determining actual expected remaining life.

In the event of extended life operations, the output of these condition assessment calculations can also be used to develop a Risk-Based Inspection (RBI) programme, based on the criticality and condition of each structural element, to monitor the critical degradation mechanisms, and measure and predict on-going degradation and calculate any consequences to the actual remaining life, using the continuously updated "digital twin". This approach maximizes the useful life of any facility while minimizing the overall cost of inspection, without compromising safety.

While the focus of this paper is life extension of existing assets, this methodology is equally applicable and valid for application on new-build assets, allowing risk-based and just-in-time management of maintenance and repair operations as well as real-time determination of existing asset life from the earliest stages of an assets deployment.

Keywords: Life Extension, Asset Integrity, Digitalization

ABBREVIATIONS

- CAPEX Capital Expenditure
- OPEX Operational Expenditure
- FPS Floating Production System
- SIM Structural Integrity Management
- AIM Asset Integrity Management
- IMR Inspection, Maintenance & Repair
- RBI Risk-Based Inspection
- UTM Ultrasonic Thickness Measurement
- FEM Finite Element Model
- SCF Stress Concentration Factor

Contents

LIFE EXTENSION – MARKET SIZING

Oil companies and leasing contractors have applied different investments and development strategies for their FPS fleet. Most of the oil companies have invested mainly in newbuild FPS for mid to long-term developments. Leasing contractor's strategy has generally been oriented toward fast track conversion for short to mid-term operating periods. Between each leasing contract, leasing contractors are then able to perform necessary modifications to extend the life of their asset and to reconfigure the unit to match the new field requirements. Average hull age is then generally higher for units under a leasing contract than it is for units directly owned by an oil company.

The life extension market can be broadly defined as the number of FPSO units within a 5-year time frame of reaching their design life. Indeed, this period of 5 years is selected to match the last cycle related to the renewal of the class certificate, hence representing an appropriate timeframe in which to establish the life extension strategy for the hull and to plan all relevant inspections. As of Q4-2016, there are about 300 Floating Production Systems (FPS) units in-service, available or ready to be redeployed worldwide, which comprises FPSO, FPU, Semi, SPAR and TLP units. Focusing on FPSO for which more public data is available, average hull age is close to 27 years as almost 65% of them were built from converted tankers that were trading for 10 to 20 years prior to refitting. Based on the FPS database published by Energy Maritime Associates, at the time of this report and considering an average design life of 25 years, it is anticipated that almost 40% of the current FPSO units will reach their initial design life within the next 5 years, which represents more than 60 units.

As a direct consequence of the current oil price situation it is now more and more common to see FPSO owners investigating the possibility of extending the life of their operating units in order to avoid additional costs associated with decommissioning and/ or newbuilding. As these assets reach the end of their initial design lives, operators are generally faced with the following four options:

- Life Extension
- Preservation / Lay up
- Redeployment
- Scrapping

Life extension on a same site is often performed to maximize production of economic reserves. Reservoir performance as well as conditions of the units are the key inputs to a life extension choice. Leasing contractors can have other specific reasons to extend the life of an asset especially when a redeployment is possible and if their units would fit easily with the new field characteristics. Based on the FPS database published by Energy Maritime Associates, in the next 5 years 30 additional units will be about to reach the end of their design life and will then enter in a likely life extension period.

Lay-up is generally not a choice. It is imposed by the market (supply & demand) when there is no room for an additional FPS or when the FPS facilities are not easily reconfigurable to accommodate the necessary requirements for an available field. On average, it is estimated that an average of 5 FPSOs will be in a layup mode each year.

Redeployment is intrinsically linked to the leasing contractor's business model. Leased installations can be redeployed 2 or 3 times during their life and it is generally possible through relatively small processing capabilities easily adaptable. It is estimated that one or two FPSO per year over the next 5 years will be redeployed which is in line with the past trend observed.

Scrapping is the decision of dismantling a unit when it becomes obsolete or too costly to maintain or redeploy. The current trend is for 1 or 2 FPSOs scrapped on average every year.

Offshore assets increasingly work in more and more hostile environments further and further from the shore and in deeper and deeper waters. Together the tighter regulatory framework and raised pubic expectations combined with harsher environments mean that aging structures may not only have to be life extended to the original as-built criteria, they may have to be actually upgraded, to perform better than originally design in order to reduce the likelihood of an incident or failure in the future.

Extend life or replace? It is clear that Asset Integrity Management and Life Extension are key challenges for operators. The following sections of this paper will help operators answer this question through the implementation of an innovative solution based on the concept of a digital twin.

3D DIGITAL TWIN CONCEPT

A common trend observed in several offshore developments has been the loss of information between CAPEX and OPEX project phases, as well as throughout the asset life. This has been one of the main targets for improvement through Digital Transformation of the Marine & Offshore industry. For Bureau Veritas, this Digital Transformation relies on the creation and maintenance of a 3D digital twin in which the condition of the actual asset can be tracked through time.

The value of a 3D digital twin is dependent on methodology used to implement it. BV have elected to apply the Structural Integrity Management (SIM) cycle as defined in reference (ISO19902:2007) as the methodology basis. This is presented in Figure 1 with slight modifications which have been developed in order to cope specifically with Asset Integrity Management (AIM) challenges.

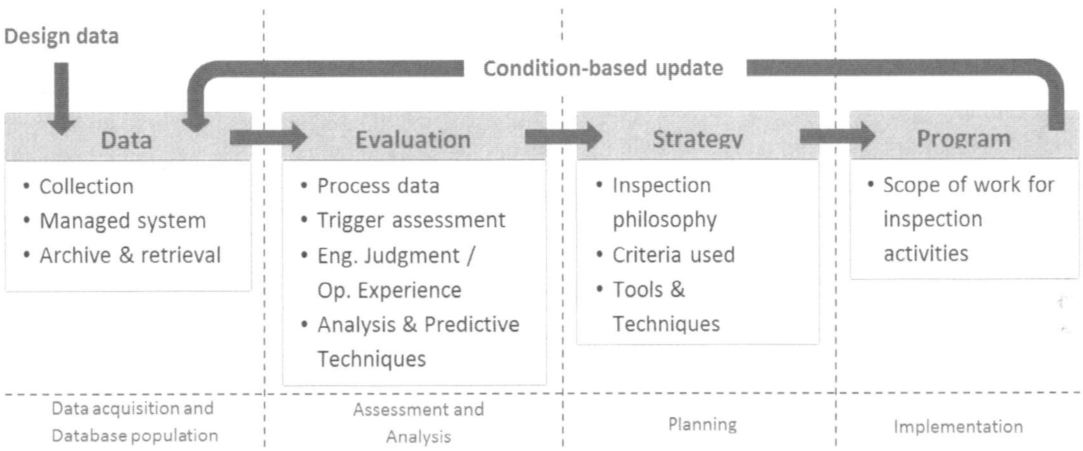

Figure 1 – Asset Integrity Management cycle

This diagram shows how the lifecycle management process can feed itself in order to provide a reliable Asset Integrity Management solution by collecting and evaluating the information, elaborating and performing inspection & maintenance strategies and then feeding itself again with results coming from the execution of these strategies. In consequence an Asset Integrity Management platform should mirror all these steps, providing a data management and storage element, capacity to exchange with sophisticated analyses, track I&M planning, support execution of the I&M activities and ultimately help with the reporting and Decision Making. All this is illustrated in Figure 2 below.

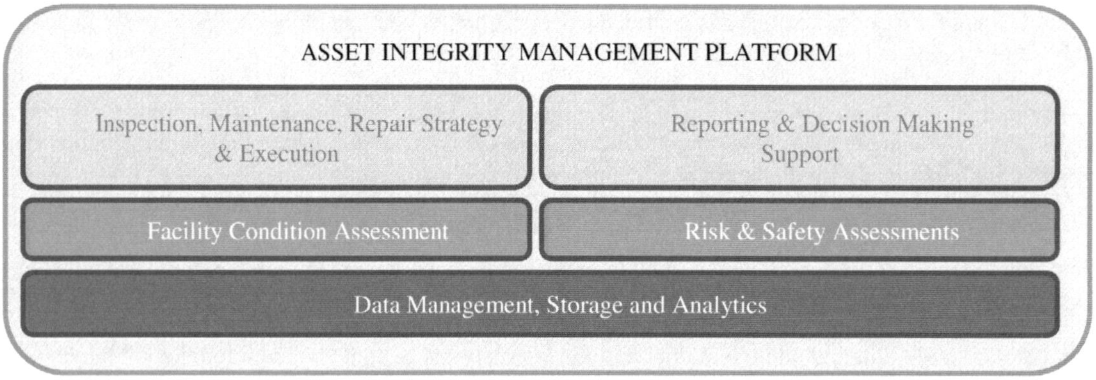

Figure 2 – Holistic AIM platform

The cornerstone of such platform is the 3D Digital Twin concept which is defined by a highly representative 3D model of the asset onto which metadata coming from the design, the operation and external computations are attached.

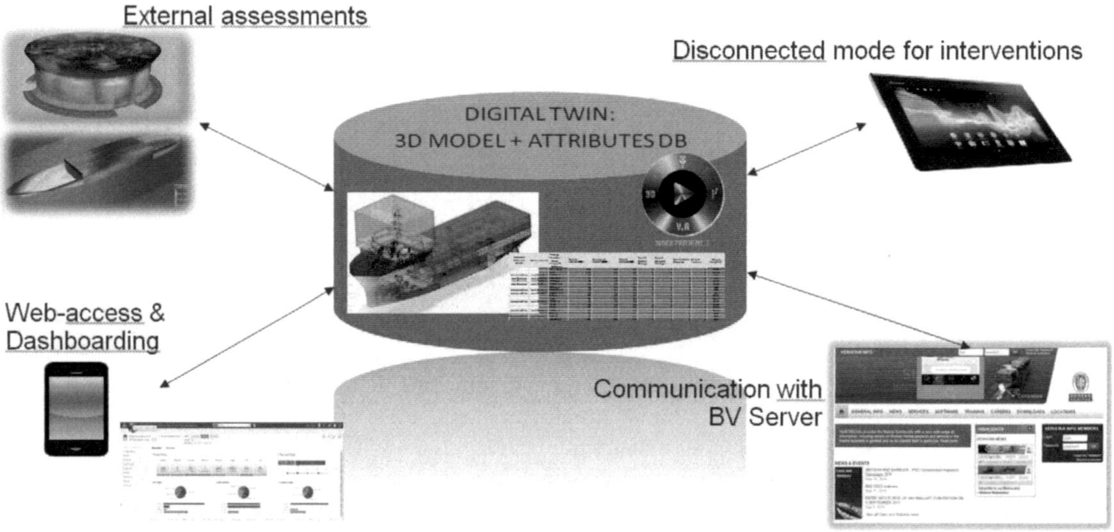

Figure 3 – Digital Twin concept

The intention of the AIM platform is to change the way of working by integrating all elements required for optimizing efficiency and reducing costs related to Asset Lifecycle Management. Indeed, it is now possible to work directly in 3D and to generate drawings on demand if needed. The metadata regarding inspection results or thickness measurement campaigns included in the platform can be issued as reports at any time.

Figure 4 – The 4 main capabilities of the AIM platform

As illustrated on Figure 4, the platform should have the following four main capabilities:

- A comprehensive Asset Data Management system:
  Comprehensive data storage, management and analytics. Using a detailed 3D Model as underlying support, metadata is attached to the individual 3D objects representing actual structural and equipment items present in the physical asset.
  Such metadata shall include:
  - Design factors: Yielding and buckling ratios, fatigue damage factors, as-built thickness, owner addition thickness, pressure, temperature...
  - IMR strategy and execution data: Next inspection due, last inspection done.
  - Criticality ratings or RBI based factors: Safety Criticality Rating value, Risk rating (pair consequence, probability)
  - Condition reporting data: UTMs taken and their value, coating condition, findings reported (buckling, cracks, pitting, corrosion...)

- A dedicated Integrity Management module, for criticality and risk assessment, supported by the 3D model in order to help visualizing where and what needs attention:
  This includes data analytics, such as steel renewal forecasts, defect criticality assessments, based on the metadata stored and attached to each structural and equipment item of the Digital Twin. It is therefore possible to present 3D views of the asset highlighting those items requiring immediate attention.

- A dedicated Integrity Monitoring module, for customized role-based and user-based dashboarding capabilities:
  Getting the right information in concise and clear format to the right person is key. The AIM system must have separate dashboards dedicated to different roles and each user can, in turn, configure their own dashboard to show only the information they need. It is expected that different types of users may have different needs for information and interaction with the AIMS, as follows:
  - Fleet or Country Management: Status report through mobile website
  - OIM / Operations Manager / Superintendent: Status report through mobile website and PC Dashboard
  - QHSE/Compliance Manager: Status report through mobile website and PC Dashboard plus Actions transmitted to Calendar application
  - Integrity Management team: Direct access through PC Console and actions transmitted to Calendar application

- Allowing users to work in disconnected mode using mobile devices or accessing remotely:

Access to the Digital Twin data shall be possible through web access using any device. Queries and dashboards should be available to users as long as they are connected to the Internet. The AIMS allow the users to extract partial models with relevant metadata to work in offline mode in order to carry out the interventions (on a laptop or mobile device) and to upload the reported new data (UTMs, findings) back into the AIMS database when the intervention is finished.

The platform should be collaborative and make any new information immediately available to all having access to the same asset. Finally, analysis modules together with field data allow predicting asset condition and then minimizing the reactive mode of work.

In the next part, examples will be shown on how the metadata included in the platform can be used in order to support condition assessment for life extension; the aim being to predict the asset condition based on realistic data and dedicated methodologies.

STRUCTURAL REASSESSMENT / CORROSION FORECAST SIMULATION

In the case of the life extension of a given asset, the objective of the structural assessment is to evaluate and to forecast the steel renewal requirements needed to allow the asset to operate safely for an additional time from its current life expectancy. As input it uses the inspection results in term of coating condition, thickness measurements, anode depletion, the asset topology and the modifications along its life. The output is the detailed scope of work to be performed according to life extension expectation and to help assess the corresponding cost.

The first steps of the basic methodology proposed are:

- To retrieve the results of last inspection campaigns;
- To build a 3D model of the asset;
- To report the thickness measurements on the 3D model.

In the case that a maintained Digital Twin model already exists, all of these and time-consuming steps are no longer needed as the up-to-date Digital Twin already embeds a 3D model and the associated metadata related to last surveys results.

The next steps of the methodology comprise forecasting the estimated amount of steel (i.e. the sum of individual weight of all structural elements) that is forecast to have a thickness loss above acceptance criteria prior to a given point in time. The acceptance criteria can be chosen according to the owner or the operator. By default the Bureau Veritas standard criteria for thickness diminution is proposed.

The parameters of the degradation mechanism have also to be defined. If sufficient inspection results are available, the coating life expectancy, the corrosion rate and the coating damage effect on corrosion can be extrapolated. Otherwise some hypotheses must be considered. Bureau Veritas proposes default parameters in case that the available inspection results are not sufficient.

Coating life expectancies	
PSPC level	15 years
Repair yard quality equivalent	5 years
Touch-up	2 years

Table 1 – Proposed default Coating life expectancies

Table 1 : Corrosion additions t_C and corrosion rate C_R, for each exposed side

Compartment type		General (1)		Upper zone (2)	
		C_R mm/yr	t_c max mm	C_R mm/yr	t_c max mm
Ballast tank (3)		0,050	1,00	0,063	1,25
Cargo oil tank and fuel oil tank	Plating of horizontal surfaces	0,038	0,75	0,050	1,00
	Plating of non-horizontal surfaces	0,025	0,50	0,050	1,00
	Ordinary stiffeners and primary supporting members	0,038	0,75	0,050	1,00
Accommodation space		0,000	0,00	0,000	0,00
Compartments other than those mentioned above Outside sea and air		0,025	0,50	0,025	0,50

(1) General: corrosion additions t_c are applicable to all members of the considered item with possible exceptions given for upper and lower zones.
(2) Upper zone: area within 1,5 m below the top of the tank or the hold. This is not to be applied to tanks in the double bottom.
(3) Ballast tank: does not include cargo oil tanks which may carry ballast according to Regulation 3 of MARPOL 73/78.

Table 2 – Proposed default average corrosion rates (mm/yr) as per Bureau Veritas Guidance Note NI 593

From the data collection and/or the hypotheses made, thickness loss distributions and trends can be deduced. The S-curves in Figure 5 show the distribution of thickness loss per compartment for one particular vessel. These trends are translated into corrosion models to allow assessment of the thickness reduction at any point in time for a particular location according to different degradation mechanism parameters reflecting the reality or based on agreed standards.

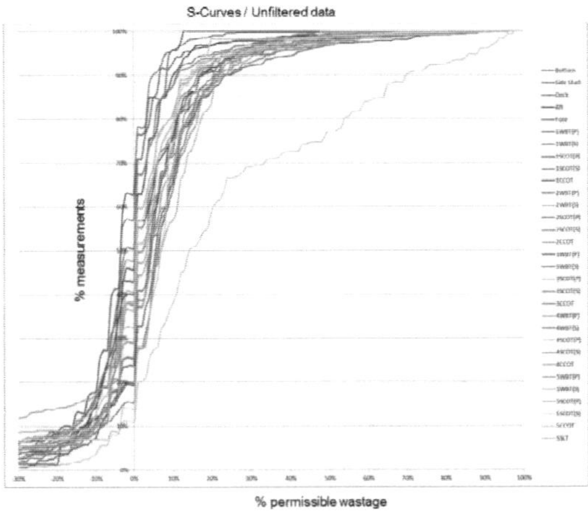

Figure 5 – S-curves by compartment for a ship

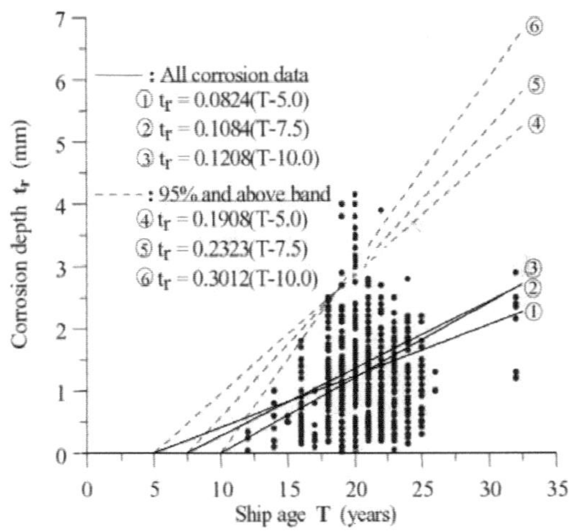

Figure 6 – Example of comparison between time-dependent corrosion wastage models and measured corrosion data for given structure members (source Ship-Shaped Offshore Installations: Design, Building, and Operation, Jeom Kee Paik,Anil Kumar Thayamballi)

At this step, the methodology permits to obtain the thickness reduction for structural elements which have been part of a thickness measurement campaign. As shown in Figure 7, even after several cycles of inspections, an important part of the structure remains unmeasured.

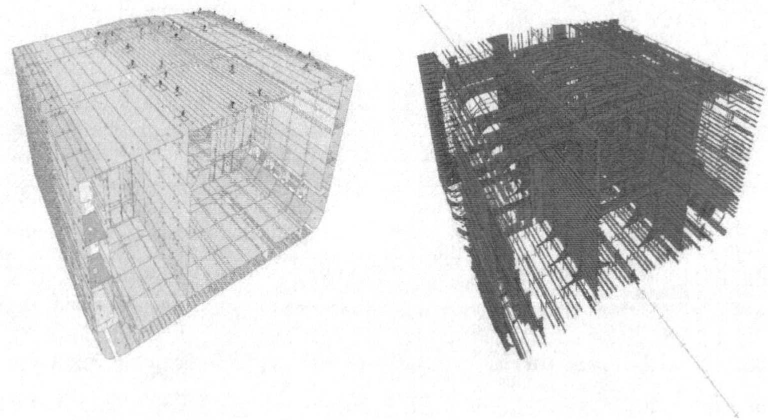

Figure 7 – Measured vs Unmeasured structure elements

For these unmeasured elements and according to their locations, the methodology proposes to apply 'randomly' diminutions and corrosion trends using Monte-Carlo simulations for the life extension period, based on thickness loss distribution and corrosion models determined by inspection results and the definition of degradation mechanism. Refurbishment scenarios can eventually be considered. The results of these simulations are then statistically post-treated in order to issue the renewal steel weight.

The methodology application shows the added value to have such a holistic Asset Integrity Management platform with a Digital Twin up-to-date. In the next part, the advantage brought by this platform will be shown for forecasting the effect of the other main degradation mechanism: fatigue.

STRUCTURAL REASSESSMENT / FATIGUE

Fatigue and corrosion have been identified as the main sources of time-driven degradation when assessing hull structural integrity failure. The aim of assessing structural failure is to provide an As-Is condition assessment of the asset's hull which will identify points of weakness in the structure, estimate the remaining fatigue life of the hull, gauge steel renewal and repair needs, and prepare forward inspection and maintenance plans.

As explained in the previous chapter, in the case that a maintained Digital Twin model already exists, the up-to-date Digital Twin already embeds a 3D model and the associated metadata related to last surveys results such as corrosion losses. The 3D model can easily be converted to a Finite Element Model (FEM) and the associated metadata imported into the FEM database.

This means an FE model can automatically be generated at any time with net thickness accounting for the output of the corrosion forecast, revealing the state of the asset at the end of the life extension period.

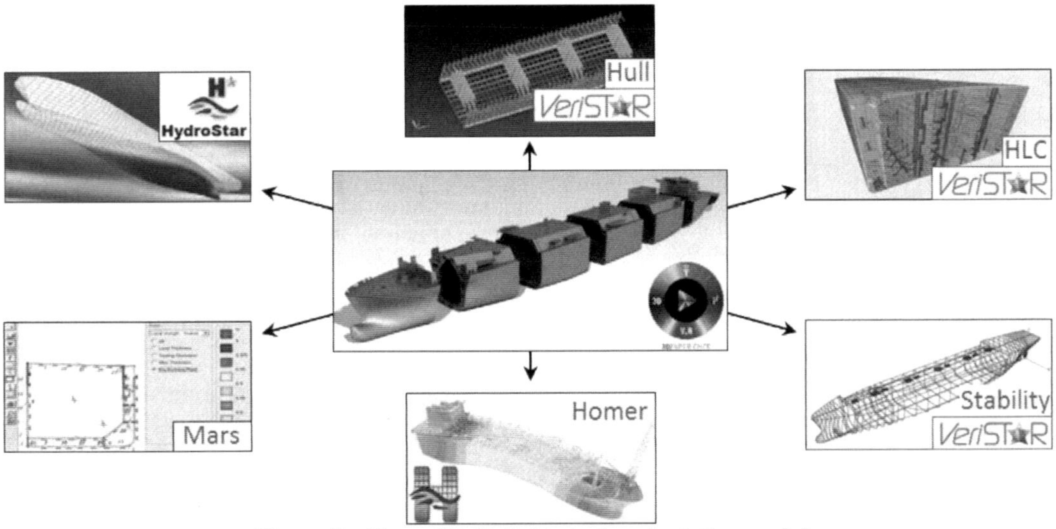

Figure 8 – From 3D digital twin to calculation models

Finite Element Analysis of the As-Is structure examines, in detail, yielding, buckling and spectral or deterministic fatigue using partial or full-length 3D models. A simpler 2D analysis can be used for hull girder analysis.

Fatigue analysis is most effective when employed together with a preliminary fatigue screening analysis to scan critical areas of fatigue weaknesses and assess typical longitudinal connections, supplemented by detailed assessment of typical locations determined by experience and the output of the fatigue scan. A full analysis of previously detected onboard cracks must be made, and special attention paid to areas such as the turret interface in the case of turret-moored FPSOs and topside mountings. The following chapters further describe the fatigue screening and analysis methodologies.

Fatigue loads

The fatigue assessment is based on the current environmental loads and latest rules & regulations, by adopting the most advanced computation methods for offshore structures. The reason for using the latest rules is that the new rules are updated with the most advanced-practice of the industry.

For a life extension assessment, the past & current periods must be evaluated to calculate the cumulated fatigue damage: this includes the current in-service phase but also trading tanker history and transit from yard to site in the case of conversions.

Environmental loads are evaluated by hydrodynamic analysis, taking into account the current site conditions and tow/transit phases.

Long-term and short-term environmental data should be considered. Site loading conditions must cover the Max/Min load conditions, intermediate conditions, inspection conditions, etc…

The parameters and their distribution obtained by hydrodynamic analysis comprise:

- Wave induced vertical bending moment
- Wave induced vertical shear force
- Motions and accelerations in the three direction including gravity
- Relative wave elevation

The wave frequencies should cover the anticipated sea states and spectra typically 0.1 rad/s to 2 rad/s with interval less than 0.05 rad/s. Refinements should be performed around the natural periods of the unit, in particular roll. Wave headings from 0 degree to 360 degree with steps of not less than 5 degrees should be considered. A heading analysis should also be carried out if a Single-Point-Mooring system is adopted in order to determine the proportion of time spent at each vessel heading.

Recognized software is to be used, such as HydroSTAR. The whole calculation should follow the Class rules or industry guidelines.

The environmental loads derived from hydrodynamic analysis will then be used for the 2D sections analysis and 3D FEM calculations.

Fatigue screening methodology

Two different types of screening shall be performed depending on the type of structural details in the dominant loads inducing fatigue:

Longitudinal connections

The fatigue screening of longitudinal stiffeners' end connections to transverse frames & bulkhead uses estimations of the components of longitudinal stress ranges due to waves: hull girder effects and local effects – internal pressures and external pressures.

It should be performed based on the following:

- All 2D sections analyzed within the cargo block area
- Damage arising from tanker history (if any)
- Damage coming from waves for the FPSO in-service prior to life extension period for 4 relevant loading conditions from min draft to maximum operating draft and as-is corrosion state (retrieved from the digital twin database)
- Damage coming from waves for the FPSO in-service at the end of the targeted life extension period for 4 relevant loading conditions from min draft to maximum operating draft and forecast corrosion state (retrieved from the digital twin database after corrosion analysis)
- Damage coming from Loading/Unloading sequences (FPSO on site connected)
- Relative deflection coefficients for connections in way of the most adjacent section to the bulkhead

This simplified and quick 2D spectral fatigue assessment reveals efficiently the most loaded longitudinal connections to be further refined by a full 3D FEM spectral fatigue analysis.

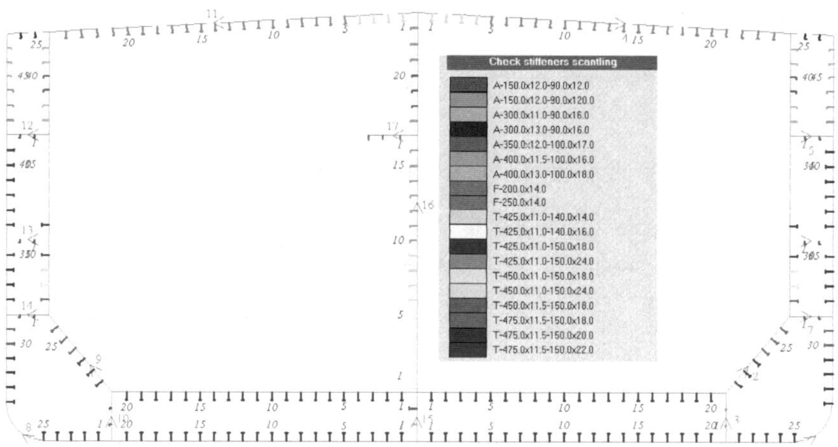

Figure 9 – 2D MARS section for 2D SFA

Transverse details

The selection of coarse mesh elements to represent fatigue details is a challenging exercise depending on good knowledge of principal stresses direction in typical fatigue sensitive details. The coarse mesh elements and stresses correlated to the fine mesh principal stresses generating fatigue will be the ones selected for screening. The fatigue screening of transverse repetitive details as illustrated in Figure 10, such as transverse bracket toes, openings in web frames, transverse lower knuckles, etc..., should be done following a simplified 3D FEM spectral fatigue approach based on the following:

- 3D FEM full hull length model
- Structural details within the cargo block area analyzed
- Minimum of 3 FPSO operating loading conditions from min to max drafts
- Fatigue screening performed on coarse mesh elements and Stress Concentration Factors: The typical size of elements is equal to the stiffeners spacing size
- Damage coming from Loading/Unloading sequences (FPSO on site connected)

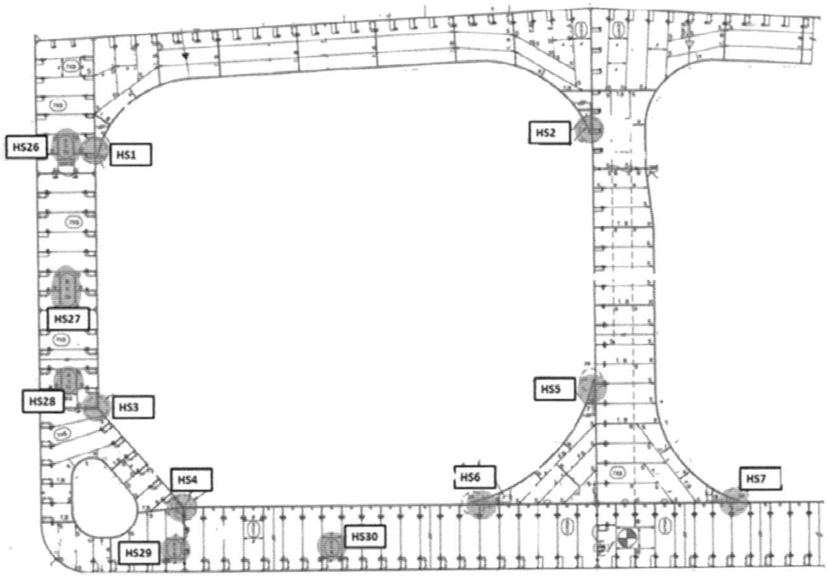

Figure 10 – Typical transverse structural details for fatigue screening

For each type of detail, a Stress Concentration Factor (SCF) shall be calculated based on a comparison of stresses obtained in coarse mesh and fine mesh models under unitary loads. SCFs aim to calculate meaningful fatigue lives, since it is not possible to capture the effect of geometric discontinuities by extracting the stresses directly on a coarse mesh.

Description	SCF
Bracket toe	2.6
Hopper knuckle	1.9
Opening	3.0

Table 3 – Stress Concentration Factors

The output of this fatigue screening is a numerical fatigue life for each analyzed detail. However, the absolute accuracy challenges associated with fatigue life calculations requires that results should be assessed qualitatively rather than quantitatively. For instance, the estimated fatigue life for a given structural detail can be plotted per tank, with transverse frames within each tank shown in abscissa: see next figures.

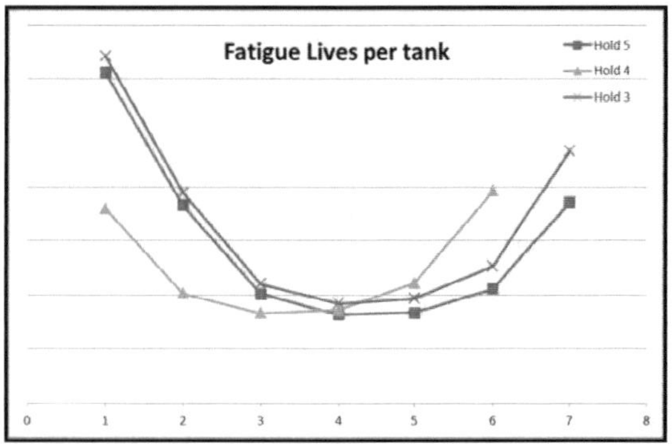

Figure 11 – Fatigue life at each longitudinal location for a given spot

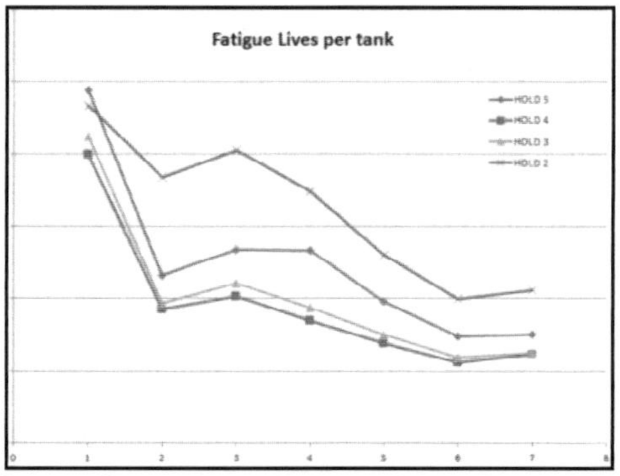

Figure 12 – Fatigue life at each longitudinal location for a given spot

Figure 11

This type of plot shows that the lowest fatigue lives for this particular structural detail (the outboard deck transverse bracket toe – HS1) are obtained at mid-tank (frame #4 on the plot. Frame #0 being the aft bulkhead and frame # 8 is the fore bulkhead of each tank).

It can also be deduced that the hold considered within the cargo block area (hold #3, 4 or 5) has very little influence on the fatigue performance of the detail analyzed.

As a conclusion of this screening for this spot, it is advised to further analyze this structural detail in full 3D spectral fatigue at a frame located at mid-tank, whatever the tank.

Figure 12

This next example, obtained for the upper hopper knuckle, shows that:

- Within 1 given tank, the detail adjacent the foremost bulkhead gives the lowest fatigue performance
- Tanks located amidships (hold #2 and #3) are more prone to fatigue than tanks located at the extremities of the cargo region.

Consequently, this spot should be further investigating by means of full spectral fatigue at the afterward connection with a transverse bulkhead around midship.

Ideally, an Asset Integrity Management system should allow tagging or enriching the 3D virtual representation of the structure with the conclusions from the structural analyses, see Figure 13. This type of software is useful for processing and for visualization, but it is dependent for its effectiveness on a good inspection regime which collects relevant, unambiguous data. A properly executed inspection plan is crucial. It can make use of Risk-Based techniques to define what to inspect, and how and when to inspect it. And it is only useful if a common and simple system has been set up for recording, transmitting and storing the inspection data. Put simply, it needs an overall management tool.

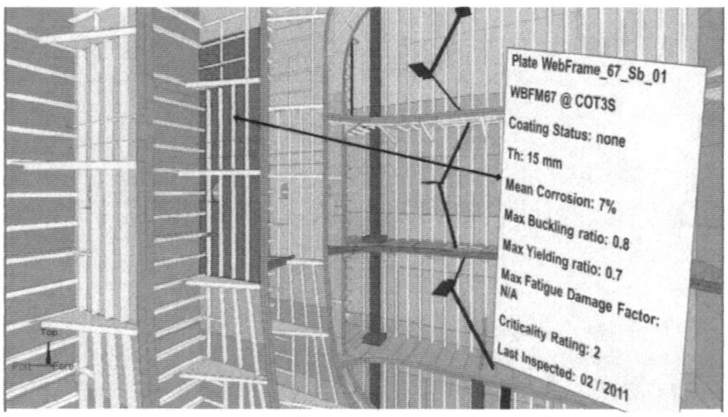

Figure 13 – Population of fatigue results onto the 3D digital twin

CONCLUSIONS

Life extension is increasingly the norm for offshore assets. There is an increase in the number of ageing assets needing life extension decisions within the next few years. Of these assets, the most challenging are FPSOs, few of which were designed, converted or built with life extension or an Asset Maintenance Integrity plan in mind. Structural integrity is one of the main challenges in the marine environment. A thorough Condition Assessment Program (CAP) can provide a platform for informed decisions on life extension in the absence of a functioning long serving AIMS plan, but following the decision to extend life an AIM and SIM system must be put in place. The CAP will include, in addition to inspection, Finite Element Analysis to provide answers to the actual status of the hull structure and how to assess its remaining life. Having an efficient AIM system in place and getting all parties to collaborate on the SIM process is a key factor to successful life extension. Innovative software solutions providing complete support to the life extension decision processes are available, but they are only as good as the people who use them and the methodologies that they are applied under.

REFERENCES

EMA Energy Maritime Associates – Floating Production Systems Quarterly Report Q4 2016

ISO 19902:2007 "Petroleum and natural gas industries – Fixed steel offshore structures"

American Petroleum Institute, "Recommended Practice for Structural Integrity Management of Fixed Offshore Structures", API RP 2SIM

UK HSE Books 2006 – Plant Ageing, prepared by TWI Ltd, ABB Engineering Services, SCS Ltd and Allianz Cornhill Engineering

Bureau Veritas. Safety Studies Workshops Methodology Statement. Bureau Veritas Internal Guide-line. 2010

Health and Safety Executive (HSE). Reducing Risks, Protecting People. HSEs Decision-making Process. HMSO, London. 2001b

International Maritime Organization (IMO) risk matrix – IMO recommendations. In Guidelines for Formal Safety Assessment (FSA) (MEPC/Circ.392 MSC/Circ.1023). International Maritime Organization (IMO). 2002

TNO. Guideline for Quantitative Risk Assessment (The Purple Book). Netherlands: Directorate General of Labour.

Renard, Biasotto and Lanquetin (2006), "Specific Tools and Services for Integrity of F(P)SOs – Case Studies for Girassol FPSO and a Concrete Unit," OTC 18142

Lanquetin (2006), "Experienced Gained From Floating-Units Integrity Management," OTC 18146

Bhattacharjee, Angevine and Sandstrom (2009), "Effective Fleet Management Strategy for Floating Assets – An Operator's Perspective," OTC 20231

Wisch and McMaster (2009), "Floating System Integrity Management: Developing a Process, OTC 20184

Kemp (2013), "A Risk-based Approach to Managing the Integrity of Aging Floating Production Storage and Offloading Topsides," OTC 24151

SNAME
THE INTERNATIONAL COMMUNITY
FOR MARITIME AND OCEAN PROFESSIONALS

BIG DATA - THE TRANSFORMATION OF THE OFFSHORE OIL AND GAS AND SHIPPING SECTORS THROUGH BIG DATA AND ADVANCED ANALYTICS

COLIN S. FROST
FOUNDER, ENERGECTIVE

Proceedings of the 22nd Offshore Symposium, February 2017, Houston, Texas
Texas Section of the Society of Naval Architects and Marine Engineers

ABSTRACT

Objectives/Scope: *To explain and illustrate a data-driven enterprise model, which enables significant performance improvements to oil and gas operators, oilfield services companies and the shipping industry that serves it..*

Methods, Procedures, Process*: Two foundational operating models will be explained, which together enable advanced data analytics based on well managed, trusted data. Both are applicable in business and technical domains. The 7 key factors to succeed in Data Analytics and the 10 critical factors in Data Management will be presented.*

Results*, Insights will be provided into how different aspects of the operating model were implemented, illustrated through case studies obtained through working with small, mid-size and large oil and gas operators; small, medium and large oilfield services companies, shipping brokers, owners and Class Societies*

Technical insights*: Insights into some novel technical aspects will be shared.*

Keywords: *Big Data, Data Analytics, Data Management, IoT, Data Lake, Data Science*

INTRODUCTION

The Marine and Offshore Sectors have been slower than other sectors to transform their performance through advanced data analytics techniques. Their 'Data-driven' strategies are still emerging. This paper serves to examine the phenomenon of the 'Data-driven Enterprise' and describes two operating models, one on Data Analytics and the other on Data Management, the likes of which the Marine and Offshore sectors must adopt - at pace - to keep up with other sectors.

Since the inception of the computer, the rate of generation of data has been exponential. To put it into perspective, the 2014 MacBook Air could store 2 million times as much data as the entire data set that was used to transport man from Earth to the Moon in 1969. Looking at the volume of electronic data that exists in the world today, there was approximately 4.4 zetabytes (10 to the power of 21 bytes) in 2014. Incredibly, 90% of that was created in the prior two years. 99.5% of that data has not even been accessed. Furthermore, by 2020, it is anticipated that the world will have generated ten times the data that exists today. (Business Insider, 2015).

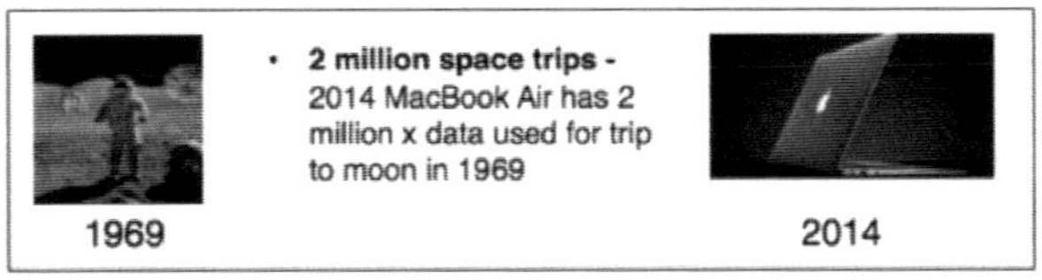

Figure 1 Exponential Growth of Data Capture

The huge potential of Data Analytics has become apparent because of the 'Big Data' phenomenon. Applying incisive analytics can deliver significant safety and performance improvements in all areas of a business. Early adoption has been driven by the recent and rapid decrease in the cost of advanced sensors, the expansion of wide area communication networks, the availability of high data storage capacity, and the ever-increasing computer processing power.

The Marine industry contributed to the birth of Big Data and Data Analytics through the foresight and ingenuity of Commander Matthew Fontaine Maury of the U.S. Navy in the early 19th century (Mayer-Schönberger & Cukier, 2014). He passionately gathered information about trade routes, winds and currents from thousands of ship voyages to help identify the most efficient routes from point A to B (See Fig. 2). He began by analyzing old ships' log books and then augmenting insight through feedback from ships' Captains when they returned from each of their voyages. As the industry began to reap the benefits of his insights, optimizing their routes and becoming more efficient, they started to actively participate in the continuous updating of the data to improve those insights.

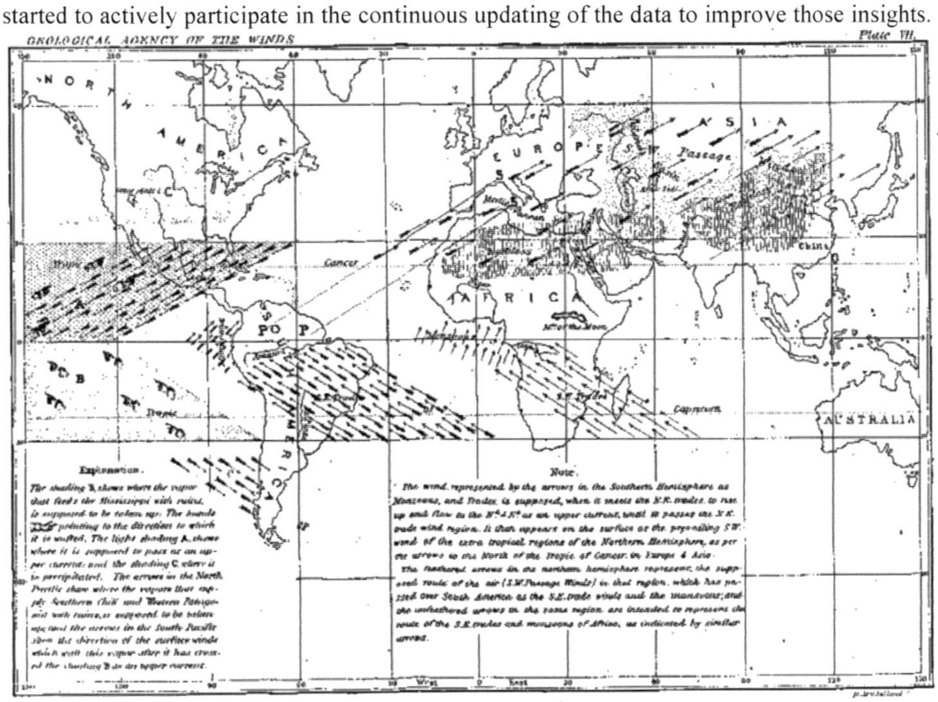

Figure 2 Commander Maury Conceived one of the Earliest Examples of Big Data in the 19th Century

Why Now?

Why is Data Analytics getting the attention of the Marine and Offshore industries today? The answer lies in a number of contemporaneous trends in sensor, communication, storage and processing capabilities. There has been a proliferation of small, intelligent sensors, which measure changes in physical attributes, and transmit the resulting data through extensive, easily accessible, and fast wide area communication networks. The data can be stored in massive data centers, and subsequently processed and analyzed by extremely powerful computers and processors to deliver critical insights. Those insights can then be acted upon and facilitated by computer-aided interventions. A key point is that today all this can be done at a relatively low cost.

One of the main reasons for the rapid adoption of low cost sensors has been the inclusion of accelerometers, digital cameras, GPS receivers, motion sensors, etc. into the mobile smart phone; this has had the effect of commoditizing sensors (Wazir Zada Khan, 2013).

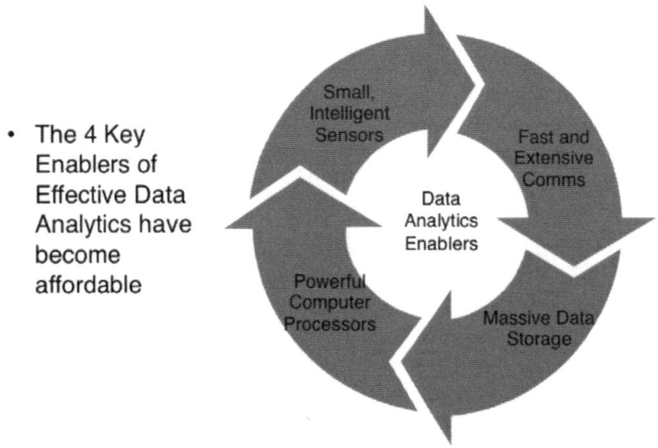

Figure 3 - Key Enablers of Effective Data Analytics

Trying to make sense of such large-scale volumes of heterogeneous and disparate data is challenging. In 2009 the term 'Big Data' started to gain popularity, describing an approach to accessing and analyzing data to deliver insights never before revealed from large but historically unconnected data sets. Instead of restricting oneself to traditional SQL (Structured Query Language) queries run on a tightly defined, limited scope, structured database associated with one particular application, Big Data was now combining all the data sources together into one large virtual "data lake" and revealing previously unknown inter-relationships. For example, it became possible to analyze how regional purchasing behavior changed depending on the how a sports team performed in a high profile tournament such as the World Cup, or an impending weather event such as a hurricane.

There are many problems and opportunities in the Marine and Offshore industries that can benefit from Data Analytics. A first point to emphasize is that Data Analytics may or may not include 'Big Data'. It should be performed on the 'right data', appropriate to a particular problem or opportunity. Organizations can start by applying Data Analytics to current business processes based on currently available data and realize meaningful value. Further significant value can be created by iteratively adding more of the 'right data' to improve analytical processes and programs and evolve towards 'Big Data'. The key point is 'Big Data' is not a prerequisite.

However, it is a requisite to turn the right data into actionable knowledge or insight. Typically, an enterprise will organize its data across many different systems and applications. Data Analytics is a key technique for combining different data sources, sometimes significant in data volume, and analyzing them in innovative ways to create new insights and knowledge, as

Data Management is a key enabler of not only Data Analytics, but also of all critical decision making in all business processes. Data is a valuable asset to most businesses, but is often not recognized or managed as one. There is plenty of scope for improvements in its management, which can deliver significant enhancements to safety, environmental protection, and operational efficiency with the associated economic benefit.

Broadly speaking, the generic opportunities presented to the Marine and Offshore industries, are represented in Figure 4. The color scheme, which ranges from red to green as defined in Figure 6, is not a linear, but a logarithmic scale. Each color transition, for the purposes of appreciating relative value, should be thought of as an order of magnitude greater than the previous level.

Companies in the Marine and Offshore industries can realize benefits of significant value relatively quickly (less than a year) at the component, system, and even vessel/asset levels. Such initiatives can be conceived and run by asset and vessel managers. Foresighted executive management in companies like GE (McKinsey, 2015), have shown the way to maximizing value across the enterprise - by creating the 'data-driven enterprise'. They set a clear vision and are driving the whole company towards it. Furthermore, they acknowledge this is a relatively new area, with much to learn, and nurture a culture of trying new, carefully qualified ideas, and seeing how well they work. They believe the most successful companies will be those set up to continuously discover and respond to strategic insights in order to drive improvements in business performance.

To give an idea of where the industry is today, Figure 7 shows a distribution of Data Analytics maturity within the shipping, offshore oil and gas, and aviation sectors. This is based on observations of applications against the matrix shown in Figure 4 and is subjective. It is provided here for illustration purposes and may serve as a tool for future surveys on behalf of the industry on a regular basis to monitor the advancement of the industry in Data Analytics and to help guide it forward. As the industry matures over time, bell curves can be expected to move to the right and eventually settle their zenith firmly in the top right of the plot (i.e. there will be a high percentage of companies of high maturity).

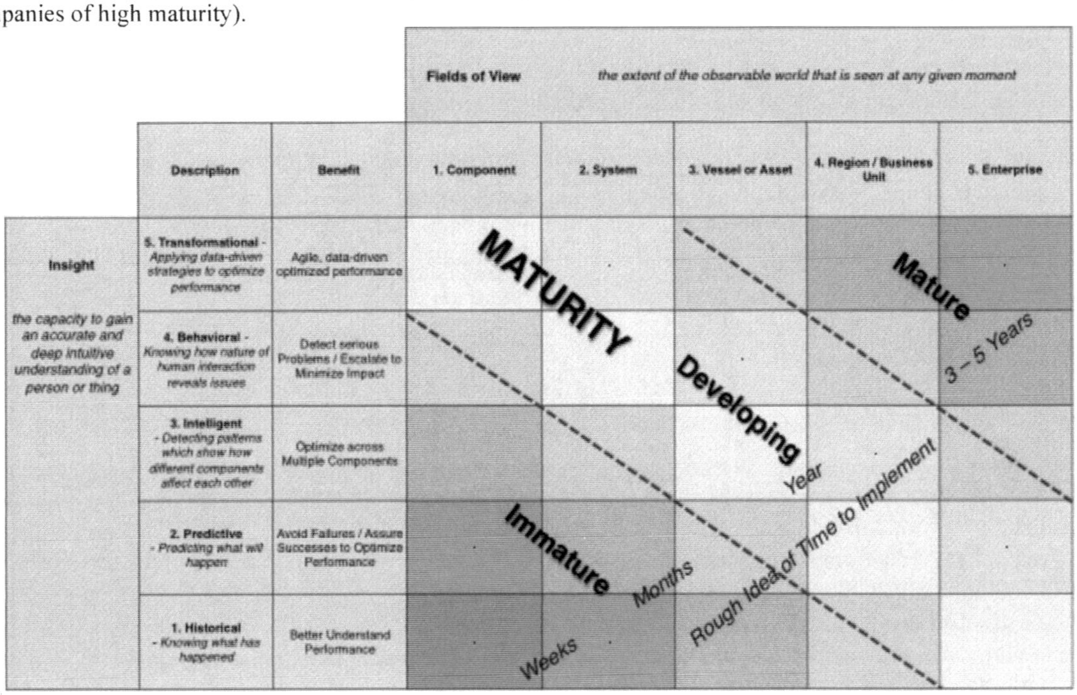

Figure 4 - Data Analytics Maturity Model

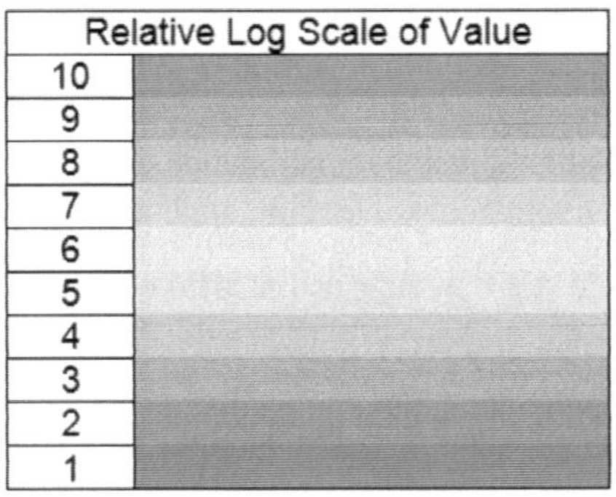

Figure 5 – Logarithmic Scale of Value

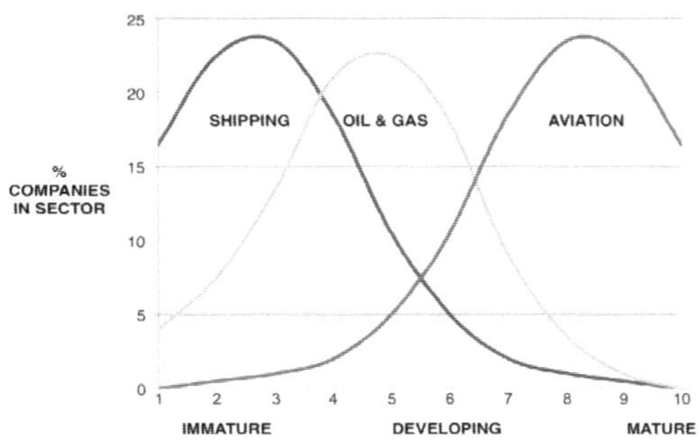

Figure 6 – Distribution of Data Analytics Maturity

Data Analytics Success Factors

There are 7 critical steps to successful Data Analytics, which are shown in Figure 7. The first two components are concerned with establishing the vision and capability for Data Analytics in the enterprise and putting in place a strategic planning process. The next four components concern the execution of individual Data Analytics projects. They focus on assessing the best Data Analytics approach to solving a problem, gathering the appropriate data, analyzing it, and executing the appropriate response to the insights found. The final component is continuous improvement, a critical process that will help ensure the ongoing success of Data Analytics within an organization.

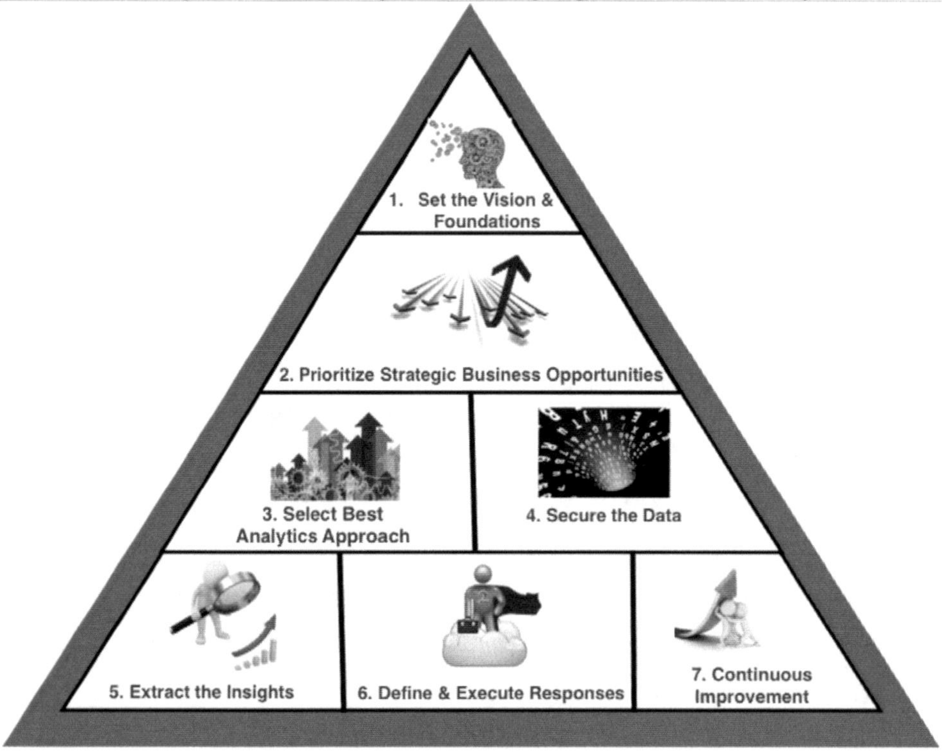

Figure 7 - The 7 Steps to Successful Data Analytics

1. Set the Vision and Foundations for a Data-Driven Business

Clear Vision and Strategic Directives

Setting a clear vision for Data Analytics and demonstrating executive support is key. The Vision describes what the business expects from Data Analytics. The associated strategic directives describe how it is going to achieve the Vision. Both set the tone and make clear that all are expected to play their role, as appropriate, to realize the strategic potential.

In defining the Vision, it is important to make clear that Data Analytics is not to be done for Data Analytics' sake. To ensure value creation, the first consideration has to be about the strategic priorities and objectives of the business. The role Data Analytics has to play to serve those priorities can be decided through consultation with the business, understanding strategic priorities, and determining where the opportunity or challenge areas are. An appropriate investment budget can then be set, which will further underline the business' commitment. All this should be done as part of the normal business governance and strategic planning processes.

Cultural and Organizational Readiness

Value will only be realized from the insights that Data Analytics will bring if there is willingness to accept and respond to them. A culture and organization has to be nurtured to allow this to happen, which recognizes the valuable knowledge and experience of employees and acknowledges how they can be helped in their roles and objectives through Data Analytics. Achieving this is non-trivial, and demands changes to the management system and operating model. Such a program must be led from the top of the organization and may be supported by management consulting resources experienced at transforming companies to become data-driven enterprises.

Correlation Vs. Causation

In Data Analytics, two common terms are 'correlation' and 'causation'. Correlation discerns patterns and trends between different types of data that are not necessarily connected but appear to trend with each other, and so one can be used as an indicator of the other (e.g. increased same-transaction sales of diapers and beer on a Friday afternoons, because male partners are being sent out to replenish emergency supplies of the former and topping up the latter while there (Rao, 1998)). Causation identifies root causes, which can then be addressed to prevent unwanted outcomes (e.g. not using appropriate lubricating oil leads to premature maintenance requirements). Understanding causation may have more impact and therefore more value than correlation.

Precisely because big data is based on correlations, it is not a great tool to help determine causality. However, correlations can be found far faster and cheaper than causation, and therefore are often preferable. Applying equations and science that model the physical world to Data Analytics can be very powerful in this respect and can sometimes deliver more value. For example, the relationship between the chemical composition of different grades of steel and their comparative strength after being subjected to repeated stress and strain can be modeled scientifically. Such a model can process received data about the operating conditions of different grades of steel on ships and vessels and potentially predict failure conditions. Huge, statistically significant data sets are not required to provide those insights.

Skills and Competencies

Undoubtedly, the skills and competencies required to perform Data Analytics successfully include deep domain expertise and experience in the Marine and Offshore industry itself. Invaluable, are the people who are able to walk into the engine room of a ship, for example, and know from the sounds that they hear, and perhaps the odors that they smell, whether or not there is a problem. Augmenting this traditional expertise with new complementary talents and skill sets will bring out the best value from Data Analytics. These skills include mathematics, statistics, computer programming, and visualization of insights. Such skills are in high demand and the people with the required knowledge and talent - often referred to as 'Data Scientists' - are scarce. Though the term Data Scientist is fluid and not universally defined, it is a gaining popularity and prevalence due to the high demand for those with related skills.

The Marine and Offshore industry can benefit from working with academia and professional societies to help develop their people and capabilities in these areas. They could be young undergraduates with particular industry-oriented Data Analytics courses or they could be current employees who are intrigued - or even passionate about - this relatively new and potentially game-changing area and wish to be given the opportunity to take their careers in a new direction.

Awareness and Optimization of Technology Enablers

Awareness of which technologies are available to perform Data Analytics is important. Broadly speaking the technology in question fits into the overall technical architecture, which comprises three broad categories:

- **Applications**
  The portfolio of software used by an enterprise
- **Platform**
  Data Lake, Databases, Middleware, Master Data Management platform and Integration for sharing data between applications and across business workflows
- **Infrastructure**
  Storage, computing, sensors and communication resources – the hardware.

When defining the best technical architecture for Data Analytics for a particular enterprise, a strategic approach should be taken which is cost-effective, efficient, flexible and secure. For a particular individual Data Analytics solution, the technical architecture may need to be fine-tuned to ensure data flows are workable and viable, particularly if high bandwidth data streams are involved.

Cloud

Not all this technology has to be run 'in-house'. It is becoming increasingly common to outsource or put the technology on the so-called 'Cloud' or 'Infrastructure as a Service (IAAS)' which is easily and highly scalable and therefore potentially cheaper than 'on premises' alternatives. Leading providers include Amazon Web Services, Microsoft Azure and Google Cloud Platform. (Gartner, 2016)

Fog / Edge Computing

A new trend based on old ideas is also gaining popularity – the so-called 'Fog' or 'Edge' computing – where the model is not to bring all data into a central storage and processing location, but to allow the processing to be performed where the data is being generated, e.g. with storage and processing capability built into a sensor. This local 'intelligence' can identify trends and insights and determine when it is appropriate to communicate those back to a central monitoring location. This saves on communication costs and delays. In essence, Fog computing permits intelligent alerts which spreads the load on storage and computing resources and help to optimize communication channels. (Cisco Systems, 2004)

Internet of Things (IoT)

To date, the Internet has primarily served communications between humans, particularly in its application in the World Wide Web, Email, Social Media etc. The Internet-of-Things (IoT) simply refers to internet-enabled communications between non-human entities, such as devices and equipment, with data stores, applications and computers. The idea is that items as diverse as temperature sensors and washing machines can easily connect to IP networks and communicate data about their own condition or what they are measuring and also respond to external requests for data and actuation of certain actions. The IoT revolution has primarily been driven by the convergence of three factors. One of them is the proliferation of GPS technology. The second is the rapid increase of high-speed wireless Internet connectivity. The third is rapid development of software, tools and frameworks to apply IoT. Its market size is estimated to grow from USD 157 Billion in 2016 to USD 662 Billion by 2021, at a Compound Annual Growth Rate (CAGR) of 33.3% from 2016 to 2021. (Markets and Markets Inc., 2016).

IoT-enabled devices will be able to 'wake up' when certain unexpected conditions arise and send alerts and associated data about emerging conditions, which require a response. For example, one idea might be that a thruster in a dynamic positioning system (DPS) may notice significantly different excursions on a semi-submersible drilling platform compared to the other thrusters. The anomalies would initially be detected by communications and self-diagnostic analysis between the thrusters. When deviations beyond certain acceptable tolerances are noticed, an alert can be issued to a central control center for human oversight and intervention. Associated diagnostic data can also be provided, for further analysis to be taken and the root cause of the problem confirmed and/or resolved.

The Marine and Offshore industries appear to have fallen behind other sectors, such as general transportation and logistics, in IoT adoption (Ericsson, 2016). One reason could be the relative difficulty and cost of collecting and transmitting data from vessels and offshore units. Satellite bandwidth has been very limited and expensive in comparison to terrestrial wireless networks. However, an exponential increase in satellite bandwidth will soon be available, with new services being launched or planned by Ericsson, Inmarsat, Iridium, and Intelsat and poised to come into service over the next three years (Rhodes & Soccoli, 2015). So the means are becoming available for the Marine and Offshore industries to emulate the achievements of other transport sectors, and to concoct their own transformational solutions.

2. Prioritize Strategic Business Opportunities

Opportunity and Problem Definition

A scope of work along the lines of 'analyze several data sets and see what patterns can be found' is too loose, and yet seems to be how many Data Analytics projects begin. To ensure focused application of Data Analytics that serves the business, scope and objectives should be clearly defined in terms of business priorities and associated KPIs. For example, "Reduce down time of top drives on offshore drilling units by 5%" or "Reduce non-productive time of drilling rigs by 3%." Not only does this approach encourage focus on a specific positive outcome, it also gives a clear indication of when that objective has been achieved.

Prioritizing and Planning Data Analytics Initiatives

The business value and cost of each candidate Data Analytics project should be estimated so they can be prioritized and planned accordingly. A focused program of high impact initiatives will result.

3. Select the Best Data Analytics Approach

Insight Characterization

Having defined the opportunity or problem, the next step is to consider new 'insights' that Data Analytics may provide to help solve them. By *insight* is meant a deep understanding of a topic – a very general term. To extract value from analytics, it is important to clearly define the type of insight being sought.

Note it is imperative not to be overly prescriptive about target insights at this stage. There may well be insights from data that will be discovered and which cannot intuitively be anticipated. However, this is an important step to characterize the 'type' of insights required, which in turn helps to target the right data.

For example, reducing down time of top drives requires insights into historical reliability, maintenance work history, and root causes of downtime. To reduce the non-productive time of drilling rigs, an analysis of the consistency - or lack thereof - of the work processes between more efficient and less efficient rigs is foundational. An understanding of their crews' relative experience could also be instructive.

Data Identification and Selection

Having considered the problem and/or opportunity sets and the initial insights that would be useful, it becomes possible to consider which data sets should be accessed and/or combined for the analysis. The mode of thinking should be 'what would be useful, in an ideal world', and not be constrained by what data is known to exist. In this way, new data sources, not currently in place or available, could be conceived and implemented, e.g. by installing new or additional sensors.

To further illustrate this point, in the case of condition monitoring of the propulsion system of a ship, vibrational data is clearly useful. If it is already being generated, then the source of the data must be identified and its integrity and validity for Data Analytics confirmed. If not, new sensors should be considered, of which there are many robust, lost cost, intelligent examples available today.

Depending on the nature of the problem or opportunity being addressed, the number of candidate data sources for analysis may be many, and prioritization will be required. In addition to selecting the data sources, the subsets of data from different selected sources should be clearly scoped out. This is where the subtleties of 'Big Data' become apparent. It is not always necessary to look at all data, although in some cases it will be useful to do that, particularly when on a discovery mission to identify new patterns or correlations across large disparate datasets. Also, some applications will require analysis of real-time data, where all data generated is analyzed from a continuous stream. However, in many cases, the data required can be honed in on, and 'right-sized', whether it be according to a particular time period, a particular equipment type, or a geographic location, etc.

At the end of this exercise, a list of prioritized and well-defined data sources will have been created. Details of source systems or providers and data locations will have been gathered, as will relevant organizations and personnel from whom to get the data relevant to the objective. In addition, a list of gaps in data will have been generated and prioritized, and then addressed as appropriate, for example, through generation of additional data via new sensors.

Businesses which take a strategic approach to Data Analytics, across high priority challenges or opportunity sets for a particular business, will make investments across the enterprise in new data sources. A strategic approach to investment in new data sources will help advance the maturity of the enterprise with respect to data-driven performance management. Both the *Insight Definition* step and the *Data Identification and Selection* step can be iterative, because considering the data available to solve a problem can give rise to ideas about additional, and perhaps better, insights not originally considered, which may in turn inform on the data sources required.

Method Selection

Data Analytics is largely about identifying trends and patterns in data and between data to create insight. Many different methods can be applied and it is essential to select those that are most appropriate to realize a particular opportunity or solve a specific problem.

There are four main types of Data Analytics methods:

1. **Classification**
   Classifying a new observation, on the basis of historical observations which have already been categorized e.g. assigning a given email into "spam" or "non-spam" classes
2. **Association Rules**
   Discovering interesting relations between variables in large datasets. e.g. if a large plane is leaving London Heathrow Airport in the afternoon and is heading East it is likely to be destined for the Asia Pacific region
3. **Prediction**
   Exploiting patterns found in historical and transactional data to identify risks and opportunities associated with current data, e.g. gradually increasing vibrations of a certain frequency in a particular type of pump may show that it is about to fail
4. **Clustering**
   Grouping a set of objects in such a way that objects in the same group are more similar (in some sense or another) to each other than to those in other groups (e.g. all ball valves with mean time between failures greater than 5 years which have never been operated in the tropics)

4. Secure the Data

Data Generation

Data Generation is the act of creating and storing data from the identified sources. When it comes to asset intensive industries like Marine and Offshore, this may include access to data from in situ SCADA or control systems for equipment or plant on board the vessel.

Role of Equipment Vendors

Equipment and Control System vendors have the opportunity to play their part in providing access to the data they generate and internalize to share with vessel owners, operators, and other industry stakeholders. That data can be combined with other data sets to potentially reveal performance inter-dependencies between different vendors' equipment and/or configurations.

Sharing of manufacturer's data should improve the industry safety record as well. One of the recommendations from the National Academy of Science (NAS) investigation of the Deepwater Horizon incident was the reinforcement of a recommendation from a 2004 study by West Engineering Services, Inc., which called on industry (and specifically the manufacturers) to develop better predictive methods and to establish a database that can be shared by all (NAS, 2012, p. 54).

Data Qualification

Due attention should be applied to qualifying data for analysis, including data quality assessments. If the quality of the collected data, particularly from older, legacy data stores is suspect then data cleansing may be required. [Note: data quality and qualification is addressed further in the Data Management section of this paper.] This step may not be required, particularly if there is already high confidence in the provenance of the data and especially if the data analysis will apply statistical methods to remove 'noise' caused by exceptional or outlying values in the data. Data qualification will recognize faulty data, apply clever techniques to close gaps or correct data (such as linear regression) or eliminate it from the sample data set, and make clear its absence.

Data Integration

Data Integration is the act of securely integrating data from the source databases or data channels, into a specially designated data store on which to perform the analytics. Over the last few years, new approaches to creating and managing large data sets have been developed.

Some companies will already have invested in a Data Warehouse (DW), which is a database that collects and stores integrated sets of transactional data from multiple operational systems. A DW is intended to enhance the flexibility of an organization to analyze and discover new information important to operations of the organization. It is designed for the purpose of query and analysis rather than transaction processing or shared data storage. However, a Data Warehouse tends to target specific data prescribed from specific applications in order to drive pre-selected insights (e.g. I want to compare the productivity of my crews in my North Atlantic operations versus Indian Ocean operations and therefore I need data from these systems to drive the analysis).

Becoming more popular these days is the concept of the 'Data Lake'. The philosophy of the Data Lake is a key differentiating factor –to gather all data (as much data from as many systems as possible) in their native forms into a single repository. The result is a massive repository of data of different types (structured, unstructured), sourced from different systems (enterprise business applications, social media, free text, emails, sensor readings etc). Having got it there, the business can apply it to address any problem or opportunity they so wish. Alternatively, they can choose to analyze it in its entirety through applying advanced Data Analytics and Data Science techniques and reveal new insights and patterns, which can help address business problems and realize unforeseen opportunities.

The very nature of a Data Lake calls for huge, highly scalable, high performance storage platform, which is typically created in a 'HADOOP' environment, which enables data to be efficiently stored across huge clusters of commodity storage servers, either on the Cloud or on-premise and easily retrieved for analysis.

Secure Environments

Due diligence must be applied to ensuring sensitive data is secure - both the source data and the data generated by data analytics to derive insights. The data should be categorized and segregated according to the sensitivity of the source or of the potential ensuing decisions and appropriately robust data security mechanisms put in place for each.

If operational data is being integrated from Equipment and Control Systems then particular attention will need to be made to avoid compromising the integrity of those operating environments when the data is being accessed and integrated. Typically, the systems environments of SCADA and control systems environments are to be kept highly secure, to avoid potentially catastrophic hacking. Recent examples of hacks into the control systems of operating automobiles demonstrate the importance of this security.

Furthermore, who can see what data must be carefully controlled to remain compliant with export trade controls. In particular, data about advanced technologies with potential military applications must be carefully controlled and must not be 'exported' to embargoed countries. This means that an individual's nationality, role and work location all come into play when carefully controlling what data they can and cannot see.

5. Extract the Insights

Data Analysis

Having obtained the data and selected the analytical methods, the next step is to apply the selected methods to the gathered and prepared data sets and execute the analysis of the data. The analysis could take the form of a one-off exercise on a huge historical data set (e.g. a modern day equivalent to what Commander Maury undertook to identify optimal routes for shipping), or a continuous, real time analysis on data as it is generated (for example, regularly monitoring vibrations on a ship's hull to detect structural issues if and when they arise).

Insight Delivery and Visualization

The analysis will deliver insights in the form of trends, patterns, anomalies, and forecasts. The notional 'half-life' of a response, which indicates how long an insight will be valid, useful or indeed valuable, must also be considered. Commander Maury's insights had a long half-life and much time could be spent deliberating how to respond to them. Conversely, the pressure and temperature readings from a safety critical valve, for instance, have a much shorter half-life and require a faster response.

In general, the longer the half-life of insight, the more strategic its nature will be; conversely the shorter, the more tactical.

Insights can be delivered in the form of reports, graphs and alerts. The way in which an insight is delivered will depend on the nature of the insight and the overall objective. If the objective is to illustrate a long-term trend - for example the gradually decreasing reliability of an aging fleet - then a report with supporting graphs will be appropriate. This is a long half-life insight example, which can afford discussion and deliberation amongst executives to decide if and how to respond.

If the objective is to ensure safety in response to a perilous corrosion trend in a riser, then the insight will be an alert requiring an immediate response, with a graphical representation of supporting data to show the trend that triggered the alert.

Data delivery: Provisioning of methods for delivery of both raw and summary data for internal analytics users, and for external customers, as required.

6. Define and Execute the Responses

Response Triggers

Some insights will be informational but others will invoke actions and are referred to as Response Triggers. For short half-life responses, time is of the essence and an appropriate response will need to be invoked and executed quickly, potentially in seconds (e.g. to activate an emergency shutdown valve if a dangerous high pressure transient is detected and an imminent breakthrough beyond acceptable norms is predicted). Automatic responses are appropriate in cases such as these and need to be 'triggered'. Relevant experts must carefully prescribe the criteria for the triggered action or actions.

Response Workflow Design and Implementation

In some cases, human intervention will be required as a response. The workflow for intervention needs to be sensibly thought out. It may be as simple as asking a crew-member to confirm the weight-on-bit of a drillstring or the oil pressure of a combustion engine. It may be more involved, e.g. to adjust the choke setting on a well-head or increase mud-weight by 20% during an offshore drilling operation. Or it may be critical, e.g. to initiate evacuation procedures. Whatever the response, the people who execute it must be appropriately trained and ready for invocation at all times.

Outcome Review

Performance improvement driven by Data Analytics is not an exact science, particularly as much of it is driven through statistics, which by definition is not exact. The actual results and consequences of completed actions need to be compared with the expected results and any differences noted, so that the algorithms applied and or the response actions invoked can be reviewed and refined, as appropriate. This will provide a feedback loop to the overall analytical process.

7. Continuous Improvement

Data Analytics is an enabler of 'Continuous Improvement' in an organization and, crucially, Continuous Improvement is critical to a highly effective and sustainable application of Data Analytics.

Continuous Improvement should be applied to Data Analytics at two levels, firstly to each Data Analytics project and secondly to the way Data Analytics is being applied across the enterprise.

The approach to any Data Analytics initiative should be iterative, aiming to realize the key objectives according to the specified KPIs (see Step 2 above). At the end of each cycle a detailed review should be undertaken. The following questions are examples of what might be considered in the review:

- What data was used?; Was it the right data?
- How was the data analyzed?; Can the analytics methods be improved?
- What types of insights were obtained?; Were the right criteria used to trigger a response?
- How effective was the response?; Was its execution efficient?
- What were the outcomes?; How well did they meet the objective?
- How can the whole cycle be improved?; What are the next iterations?

Regularly sharing information and insights about Data Analytics initiatives across business "silos", and using collective feedback and reviews of what worked well and what could be done better, will maximize the value of Data Analytics to an enterprise and ensure that value is sustained. It will also allow successes to be recognized, celebrated and shared.

Data Management

Introduction to Data Management

Data Analytics offers a wealth of benefits and is built upon data management. Data management provides some structure to the data and facilitates access to the data for analysis. However, before an organization can benefit from Data Analytics, it must have the right data accessible. A solid Data Management regime must be in place before Data Analytics can be performed reliably and effectively.

While the benefits of Data Analytics may alone justify the cost of Data Management, Data Management brings value to the organization in and of itself. Every decision made in any business is based upon data. This data can take many forms, from unstructured documents and engineering drawings to structured data, which provides measurements and factual information. Typically, the better informed a decision, the better the decision and the better the data, the faster and the more accurate the decision. Therefore, better Data Management leads to better business performance.

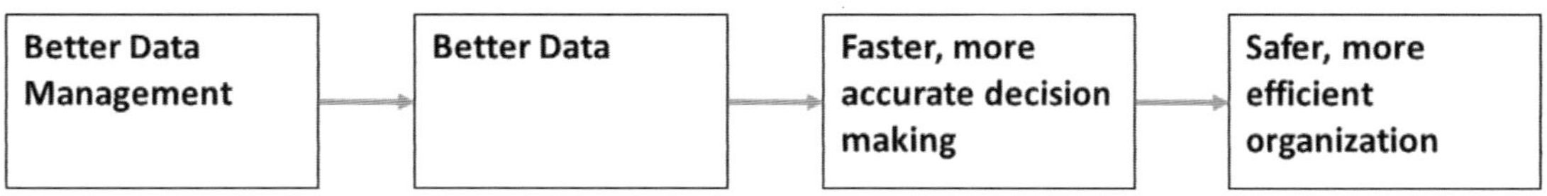

Figure 8- Data Management Impacts the Organization

In the past, Data Management was simply a discipline tasked with ensuring the organization could access their data, or had simply enough disk space to store it. As the value of data has increased for organizations, so too has the importance of the role of Data Management.

To understand the value of Data Management, the significant value of data must first be understood and fully appreciated. Consider the Data Asset schematic below, which gives some insight into the huge relative value of data to a business.

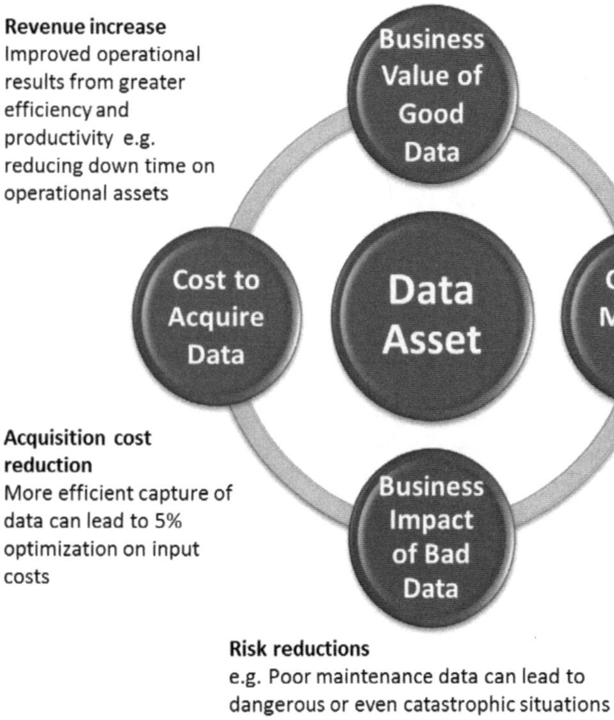

Figure 9 - Data Value Model

Figure 9 above depicts a simple value model for data. It shows that there are four main ways to value an organization's data. Applying that model to the upstream oil and gas industry, as an example, the value of data considerations for a typical supermajor would approximate to the following:

- **Cost of Acquisition**
  A typical integrated oil company invests $2 to $3 billion per year in acquiring data (from seismic to real time operations data and much more). Applying a straight-line depreciation of the value of that data over 10 years, gives an asset which is worth $10 to $15 billion. A physical asset worth the same amount would not be neglected. The data asset should be treated with the same level of care and investment.
- **Cost of Managing Data**
  A supermajor oil company may spend well over $1 million per annum centrally managing data. If regional costs are added, then annual operating costs for data management can easily exceed $10 million per year. Improving the way data is managed and avoiding re-work can reduce associated costs by at least 2%.
- **Value of Good Data**
  Critical decision makers typically spend 50 – 75% of their time looking for the data they need and confirming its validity and provenance before actually using it. Increasing trust in the data will make decisions makers much more efficient, focusing on activities which deliver true value rather than just gathering data, and the quality of their analysis work and decision-making will improve significantly as a result.
- **Risk of Bad Data**
  Making decisions on bad data exposes operators to significant risk and potentially negative impacts running into billions of dollars; e.g. bidding too much (or too little) in a competitive bid for a license block, pulling an anchor of a supply boat through a pipeline, mismanaging safety critical drilling operations, misjudging requirements for midstream capacity related to new production, etc.

The principles underneath these upstream oil and gas examples apply equally well to the marine industry and the offshore and subsea subsectors of upstream oil and gas. However the value of an organization's data is measured, it should be viewed - and treated - as an asset.

Data Management maintains and increases the value of data through three primary tasks; ensuring data is *secure*, *accessible*, and of sufficient *quality*.

Security

Just as an organization wouldn't leave a vessel unguarded, so too must they protect their data from unwarranted access. As hackers have gotten more sophisticated, so too have their traditional, data heavy targets. As a result, hackers have been turning to targets such as those in the Marine and Offshore industry. Even without hackers, uncontrolled data can easily be exposed. Confidential information is frequently shared too broadly through internal reports distributed outside of their intended audience, emails forwarded outside of an organization, or even in annual reports. Uncontrolled data is insecure data. For data security, data must be controlled while in processing, in transit, and at rest. In addition, new security threats are emerging, including altering GPS data, modifying cargo data, or simply shutting down vessels. Strong Data Management is needed for strong cybersecurity.

Accessibility

Data can be a useful benefit to an organization but inaccessible data is unused data. The challenge is not just to share the data while ensuring it remains secure, but also to make sure the data can be found and available when needed. This means that data needs to be where it is expected to be, organized and searchable. An interesting side effect of accessible data is that it can discourage data duplication and potentially data errors, by accessing related data already captured.

Quality

Finally, the quality of the data needs to be sufficient for an organization to make use of it. Data that isn't trusted won't be used at best and could cause poor decision making at worst. To make the data valuable to an organization, it must be admissible, have required confidence levels, and be relevant to how it is being applied, as depicted in Figure 10.

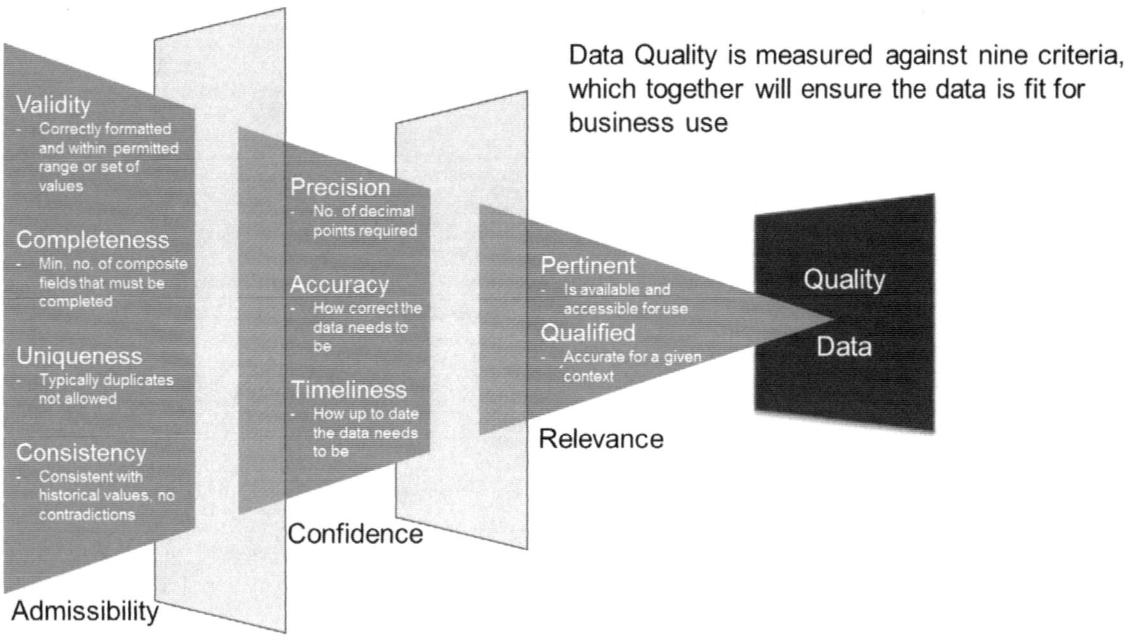

Figure 10 - Attributes of Data Quality

Data Management in the Marine, Offshore, and Subsea Industries

Data Management has its roots in some data intensive industries such as the Technology sector, but managing logs, drawings, equipment records, and even subsea geological data has been important in the Marine, Offshore, and Subsea industries for decades. However, the amount of data in these industries, and the speed at which it is generated is increasing exponentially due to new technology and practices from other industries. The way Data Management was done in the past is being rapidly outdated by the sheer volume of modern data and the demands being put upon it. The longer an organization waits to establish a strong Data Management program, the greater the backlog and the longer it will take to inform business decisions.

Data Management Success Factors

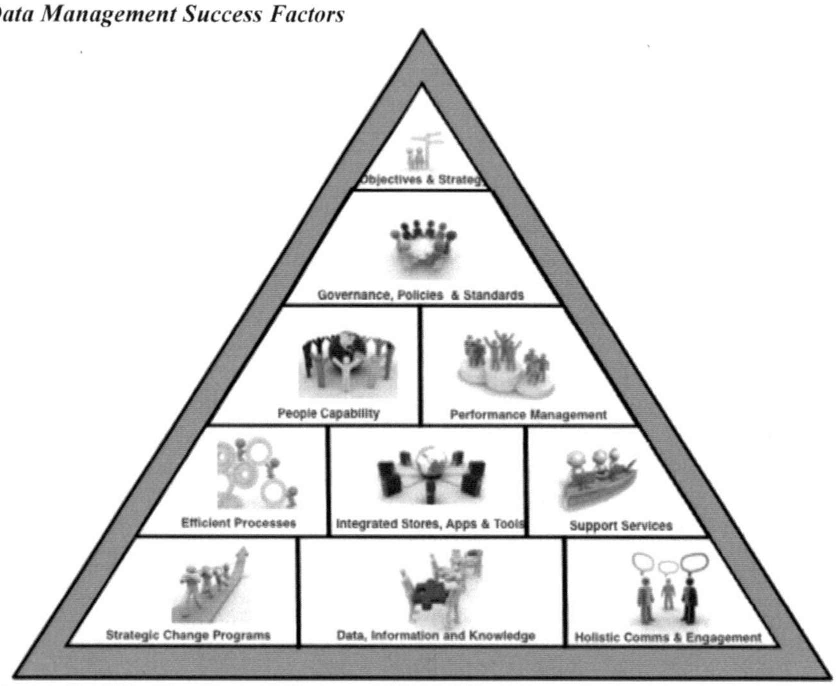

Figure 11 - The Ten Success Factors of Data Management

A holistic approach to Data Management includes ten success factors, as depicted in **Figure 11**:

1. ***Objectives and Strategy*** - *Strategic Alignment of Data with Business Priorities*

Formalizing approach to align Data Management activity planning (e.g. 3 – 5 year plan) to business priorities (safety, environmental protection, reliability, optimized asset operations etc.)

2. ***Governance, Policies and Standards*** - *Business Ownership of Data and Data Governance*

Formalized business executive champion, governance, clear accountabilities, business ownership of data – both structured and unstructured, defining policies and standards and addressing security and access controls

3. **Efficient Processes** - *Defining the role of Data in Key Work Processes and the Processes for Managing it*

Documented interplay and interdependency between critical work processes and data and clear definitions of the process requirements of Data. Putting in place standard definitions of key processes such as Data Cataloguing, Data Migration, Data Security, Master Data Management, Data Quality Management, etc. and ensuring compliance with those processes.

4. **People Capabilities** - *Data Management Organization and People*

Clearly defined roles, accountabilities and responsibilities, formally defined skills and competency requirements, career development paths etc.

5. **Performance Management** – *Proactively monitoring and managing data performance*
Putting in place performance management KPIs, metrics and associated processes and systems specifically for monitoring and managing performance in Data Management (e.g. around data quality or OPEX spend on DM)

6. **Integrated Apps, Stores and Tools** - *Data Management Systems and Infrastructure*
Ensuring there is a formalized approach to promoting standardized and optimized key DM systems and tools e.g., for storing source data, master data, migrating data between systems, cleaning data to standards, archiving to secure storage, ensuring easy access, etc.

7. **Support Services** – *Implementing services to support the business to sustain trusted data*
Services such as helpdesks and offshored operations to support the creation, updating and deletion / archiving of business data and to maintain its quality to conform to business needs.

8. **Strategic Change Programs** – *Ensuring continuous change and improvement to drive strategic objectives*
Recognizing the diversity and abundance of stakeholders who have a dependence on quality data, understanding the impact of changing the way data is managed and ensuring a proactive approach to stakeholder management and change management to secure support and minimize disruption.

9. **Data, Information, Knowledge** – *Proactively managing the data value chain*
Ensuring a well thought out and executed management framework is in place to optimize all value adding processes which convert data to information and actionable insight and then into prevailing corporate knowledge.

10. **Holistic Communications and Engagement** – *Ensuring transparency and alignment across the enterprise*
Recognizing the diversity and abundance of stakeholders who have a dependence on quality data or a key role in delivering and maintaining it and ensuring proactive and holistic engagement across all of them to ensure alignment and consistency in approach.

Energective has published another paper on the Data Management operating model, which is available on request.

CONCLUSIONS

Advanced and affordable sensor, communications, storage and processing technology is enabling advanced analysis of large disparate data sets, sometimes in real time, to increase knowledge, extract new insights, and improve decision-making. Data Analytics is already delivering significant value in several industry sectors, both tactically and strategically. The Marine and Offshore industries appear to be lagging other industry sectors, such as Aviation, General Transportation and Logistics. However, extrapolating from what other industries have already achieved and from reports of early adopter initiatives in Marine and Offshore, the realizable benefits are certainly worthy of attention, arguably running into hundreds of millions of dollars per year for larger enterprises.

Like its physical counterparts, Data should be recognized as a valuable business asset and managed as such. It is the raw material for Data Analytics and for all critical business decisions and processes. The better managed the data is, the more trust can be put in it. Faster, higher quality analysis and decision-making will result, making many business processes more efficient. Better data management will deliver significant benefits in its own right by ensuring the right fit-for-purpose data is securely delivered to where it is needed at the right time.

Companies in the Marine and Offshore industries can start realizing tactical benefits within a few months through focused initiatives on individual vessels and assets. Examples include applying low cost sensors to equipment and systems to enhance the knowledge and insights provided to existing maintenance procedures, reducing downtime and repair costs to increase operational efficiency.

The more significant benefits are strategic in nature and require C-suite leaders to set the vision and drive the organization forward. The opportunities can be transformational and may require carefully managed change to aspects of a company's operating model, to enable data-driven decision making and business processes across the enterprise. Two critical success factors of a Data-driven enterprise are implementing operating models for i) Data Analytics and ii) Data Management. The latter provides and sustains trusted and complete data to the former, and the former derives trusted, performance improvement insights from the latter.

The benefits of better Data Management and advanced Data Analytics to the Marine and Offshore industries are real and significant. Together, robust Data Analytics and Data Management will transform the way the Marine and Offshore industries think and the way they work.

ACKNOWLEDGEMENTS

Dr. Henrique Paula and colleagues at the American Bureau of Shipping (ABS) for the opportunity to share insights and explore potential of Data Analytics and Data Management to the Marine and Offshore sectors.

REFERENCES

Browning, L. (Director). (2015). *The Mind Blowing Growth and Power of Big Data* [Motion Picture].

Cisco Systems. (2004, 01 01). *Fog Computing and Its Role in the Internet of Things*. Retrieved 12 10, 2016, from Amazon Web Services: http://s3.amazonaws.com/academia.edu.documents/46135664/SIGCOMM-MMC-Fog.pdf?AWSAccessKeyId=AKIAIWOWYYGZ2Y53UL3A&Expires=1485108351&Signature=vTZd6DdvZTgH1M1hBe71rTLkMy4%3D&response-content-disposition=inline%3B%20filename%3DFog_Computing_and_Its_Role_in_the_Intern.pdf

Ericsson. (2016, 01 06). *Maritime ICT Cloud enables ships to join the Networked Society* . Retrieved 01 22, 2017, from Ericsson: https://www.ericsson.com/news/1884299

Gartner. (2016, 08 03). *Magic Quadrant for Cloud Infrastructure as a Service, Worldwide*. Retrieved from Gartner: https://www.gartner.com/doc/reprints?id=1-2G2O5FC&ct=150519

Markets and Markets Inc. (2016, April 11). *Internet of Things (IoT) Market by Software Solution (Real-Time Streaming Analytics, Security, Data Management, Remote Monitoring, & Network Bandwidth Management), Platform, Service, Application Domain, and Region - Global Forecast to 2021*. Retrieved November 25, 2016, from Markets and Markets: http://www.marketsandmarkets.com/Market-Reports/internet-of-things-market-573.html?gclid=COrAibL71dECFQ6paQoddEUF6Q

Maury, M. F. (1855). *The Physical Geography of the Sea*. New York, NY, USA: Harper & Brothers.

Mayer-Schönberger, V., & Cukier, K. (2014). *Big Data: A Revolution That Will Transform The Way We Live, Work and Think* (First Mariner Books edition 2014 ed.). Boston New York, USA: Mariner Books / Houghton Mifflin Harcourt.

McKinsey. (2015, October 1). *GE's Jeff Immelt on digitizing in the industrial space*. Retrieved December 10, 2015, from McKinsey: http://www.mckinsey.com/insights/organization/ges_jeff_immelt_on_digitizing_in_the_industrial_space

NAS. (2012). *Macondo Well Deepwater Horizon Blowout: Lessons for Improving Offshore Drilling Safety*. National Academy of Sciences. Washington, DC: The National Academies Press.

Rao, S. S. (1998, April 06). *Diaper-beer syndrome*. Retrieved November 25, 2015, from Forbes: http://www.forbes.com/forbes/1998/0406/6107128a.html

Rhodes, J., & Soccoli, F. (2015, July 01). Big Data: Big Value? Big Risk? Both? *Maritime Reporter & Engineering News* .

Wazir Zada Khan, Y. X. (2013). *Mobile Phone Sensing Systems: A Survey*. IEEE. New York: IEEE.

THE INTERNATIONAL COMMUNITY
FOR MARITIME AND OCEAN PROFESSIONALS

SOLVING THE DRONE DATA DILEMMA

ROBERT A ALBRIGHT JR
PRODUCT ENGINEERING MANAGER, HUVRDATA LLC

Proceedings of the 22nd Offshore Symposium, February 2017, Houston, Texas
Texas Section of the Society of Naval Architects and Marine Engineers

ABSTRACT

Technological advances in several key areas have led to a global proliferation in small unmanned aerial systems (sUAS, UAV, or "Drones"). In addition to the increasing availability of hardware systems, recent changes in the regulatory framework governing the commercial use of these systems has dramatically increased the number of companies offering services in this field. Numerous uses exist for this new technology, from 2D surveying and mapping, to asset inspection, fugitive gas detection, post incident investigation, routine infrastructure inspection, and many others.

Because of the potential benefits enabled by this technology, many energy companies are evaluating how to best make use of UAV ("Drone") technology into their monitoring and inspection toolkit. One aspect of implementing an sUAS inspection plan that is often underestimated is the process of data transfer, ingestion, storage and access. This process is referred to in this paper as the "Drone Data Dilemma".

A single UAV performing inspection or mapping activities can easily generate more than 30 Gigabytes of data per day, typically in the form of high resolution images, or video. This data must pass through several important workflow processing steps before it is ultimately useable to the client. A set of recommendations and best practices has been developed based upon over 500 asset inspections and over a dozen mapping missions. The recommendations are comprehensive and generic, and applicable regardless of inspection company or UAV inspection company. Several case studies will document these practices in action.

Keywords: SNAME, Offshore, Symposium

INTRODUCTION

Technological advances in several key areas have led to a global proliferation in small unmanned aerial systems (sUAS, UAV, or "Drones"). Recent changes in the regulatory framework governing the commercial use of these systems have increased the number of companies offering services in this field. The two primary categories of UAV service products are mapping and inspection.

Mapping and surveying products utilize photogrammetry and photogrammetric software to stitch together aerial images to create a 2D maps (Orthomosiac), and 3D models.

Photogrammetry is defined as:

"the art, science, and technology of obtaining reliable information about physical objects and the environment, through processes of recording, measuring, and interpreting images and patterns of electromagnetic radiant energy and other phenomena." – *source: American Society for Photogrammetry and Remote Sensing (ASPRA)*

From these photogrammetric properties, valuable information can be extracted, such as measurements, locations, volumes of stockpiles and associated information. The agriculture industry utilizes near infrared cameras to assess plant health, and accurately plan treatment of crops. This is known as "precision Ag". The mining and construction industries utilize mapping technology to calculate stock pile and fill volumes in addition to assessing elevation and site drainage. Commercially available software such as Agisoft Photoscan, and Pix4D offer straightforward processing tools, enabling UAV operators to quickly process and produce these products. Current commercially available technology offers Global Positioning System (GPS) accuracy. The accuracy for uncorrected GPS is defined by 15 meters horizontal by 25 meters vertical. Numerous methods are available to improve this base uncorrected accuracy. Supplemental navigation systems such as "Real Time Kinematics" (RTK) offer centimeter level accuracy. RTK is a satellite navigation system used to enhance the precision of position data derived from global navigation system satellite systems (GNSS), such as GPS.

Regarding inspection activities, the UAV offers an unprecedented ability to safely and economically inspect assets that would hitherto require access aids, such as man lifts, rope access, or installation of scaffolding. In addition to the expense, these existing methods expose human technicians to fall-associated risks. The deliverables from a UAV inspection are typically images from high resolution cameras. Video cameras, thermal cameras, optical gas detection cameras, and other sensors are also available currently. Currently application for inspection includes offshore asset integrity inspection for MODU's and production facilities, pipeline construction and monitoring, flare stack inspections, bridge inspections, wind turbine inspections, cell phone tower inspections, to name a few. Figure 1 shows the recently commissioned Block Island wind farm, located off the Rhode Island coast. This wind farm is the first offshore wind farm in the United States, and consists of five 6-megawatt Wind Turbine Generators (WTG). Numerous offshore wind farms exist in Europe. Figure 4 shows a typical offshore production facility.

Figure 1 - GE Wind Turbines, Block Island Wind Farm, Rhode Island, (image courtesy Washington Post)

Figures 2 and 3 show typical imagery from a UAV equipped with an infrared (thermal) imager. The thermal imager is a FLIR Tau core with a resolution of 640 by 512 pixels. Figure 2 is an aerial view of two flares. In this image the bottom flare only has the pilot lit. The top flare is actively operating. Figure 3 is an aerial view of a tank battery, showing the fluid levels in the individual tanks.

Figure 2 - Infrared image of two flares, acquired via UAV (Image courtesy HUVR Data LLC)

Figure 3 - Infrared image of production oil tanks, acquired via UAV (Image courtesy HUVR Data LLC)

Figure 4 - Offshore production platform (image courtesy BSSE)

Both mapping and inspection activities create very large data sets. This is due to the high resolution of current digital camera sensors and hence file size. The trend for increasing image sensor resolution can be expected to continue into the foreseeable future. Figure 5 shows a typical small Orthomosiac utilized for civil planning.

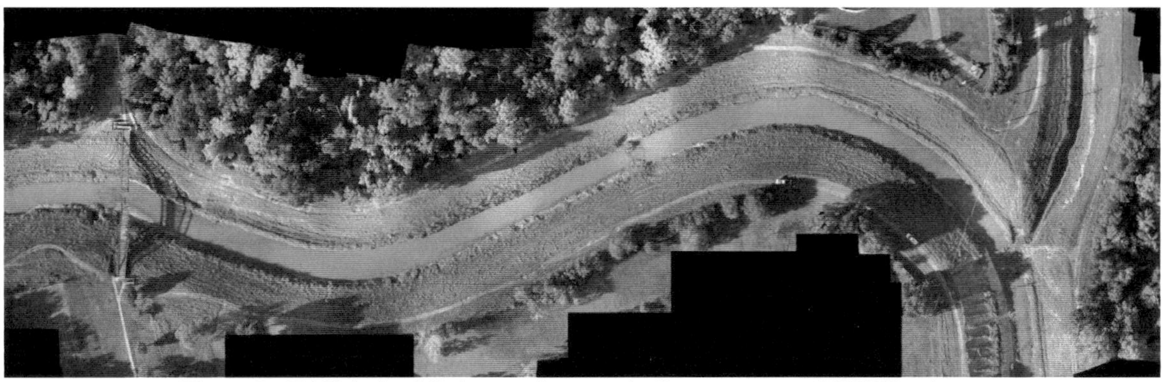

Figure 5 - UAV derived Orthomosaic example (Image courtesy HUVRData)

The current state of the art camera for UAV inspections is the Sony™ α7R, which utilizes a 36 Megapixel sensor. At this resolution, a single UAV performing inspection or mapping activities can easily generate more than 30 Gigabytes of data per day. With normal compression JPEG images, the file size for a typical image is around 15 Megabytes for a 36 Megapixel sensor. A UAV with this camera is shown in Figure 7. Note that at the time of writing, these exists a commercially available Sony™ α7R II, which has a 42 Megapixel sensor. A chart showing pixel resolution from 1994 – 2013 is shown in figure 6.

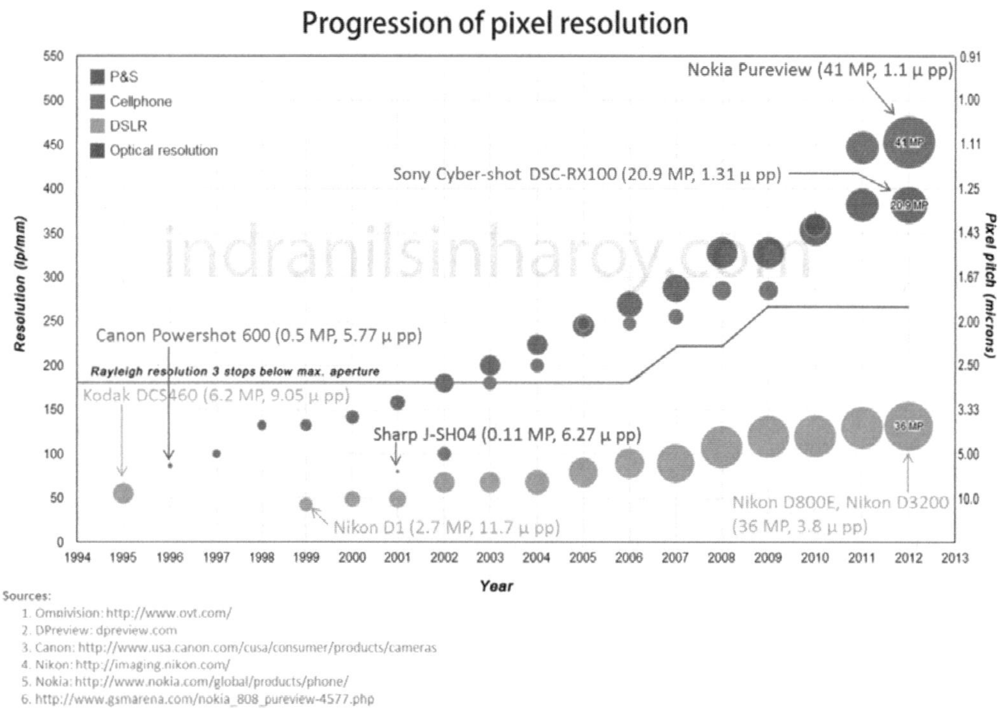

Figure 6 - Pixel resolute over time (Courtesy *Indranil Sinharoy* © *2012* licensed under Creative Commons Attribution-NonCommercial-NoDerivs 3.0 Unported License)

Figure 7 - Intel™ (AscTec) Falcon-8 UAV with 36 Megapixel Sony™ α7R DSLR (Image courtesy HUVRData)

Because of the numerous potential benefits enabled by this technology, many companies and institutions are evaluating how to best make use of UAV technology into their monitoring and inspection toolkit. One aspect of implementing an sUAS inspection plan that is often underestimated is the process of data transfer, ingestion, storage and access. The basic dilemma for both UAV service providers as well as their clients is to devise a systematic methodology to store and manage the data, in addition to critical task of parsing and evaluating the data for relevant asset condition information.

METHODOLOGY

In developing a comprehensive solution, the problem must be broken down into functional steps. In general, these functional steps are the same regardless of the asset being inspected. It should be noted that some steps are more problematic given the specific asset being inspected. For example, existing communication infrastructure strongly affects the method as well as the timing of data transfer. Components of a comprehensive solution are as follows (shown in Figure 8):

1. Raw data management (Get Data off the drone)
2. Backup and Storage of raw data
3. Organize Data
4. Process Data (Knowledge)
5. Produce / present issues (Actionable knowledge)

Figure 8 - Data work flow steps (Courtesy HUVRData)

The initial step is to obtain the data. This occurs in the field. The result of this first process is a large amount of data, hundreds to thousands of images typically. The next step of the process is to organize the data, also occurring in the field. Each image is assigned to the client, asset, and asset component. The result of this is an organized collection of images, assigned to appropriate client, asset, and component. Once this step is complete, multiple independent backup copies are created. A modified "3-2-1" compliant backup strategy is utilized. In brief, a "3-2-1" backup strategy is to always retain 3 copies of the data, on 2 independent devices, 1 of which is off-site. The concept is modified to account for field collection of data and the transmission of the data.

The third step of the process is to parse the organized data to detect visible flaws. There are two options to this step of the process. Machine learning and artificial intelligence hold the promise and appeal of automation. Currently, the automated processing takes far longer than having an experienced human parse the data. As the technology for pattern recognition evolves, this will likely change in the future. The specific flaws, methodology and the evaluation of those flaws is a client or industry driven standard.

The final step is to organize the data from the third step into "Actionable Knowledge", a format that can readily be utilized by clients and key decision makers to make informed business and operational decisions regarding the operation of their assets. The presentation of the "Actionable Knowledge" is via a web-based portal customized to present a client-specified dashboard. This final distillation step provides the value to the client, allowing them to make asset management decisions based upon the vast data collected.

Wind Farm	Turbine	Status	Side	Weld	Link
Farm One	U3	red	0	t15	U3-red-[0][t15]-at-127.196-seconds.png
Farm One	S6	red	180	b7	S6-red-[180][b7]-at-451.666-seconds.png
Farm One	S6	red	90	m13	S6-red-[90][m13]-at-290.018-seconds.png
Farm One	T6	red	0	t15	T6 red [0][t15] at 191.683 seconds.png
Farm One	u2	red	0	m13	u2-red-[0][m13]-at-67.277-seconds.png
Farm One	u2	red	0	t1	u2-red-[0][t1]-at-67.277-seconds.png
Farm One	Q14	red	90	t1	Q14-red-[90][t1]-at-212.805-seconds.png
Farm One	T4	red	180	b7	T4-red-[180][b7]-at-411.153-seconds.png
Farm One	T4	red	270	b7	T4-red-[270][b7]-at-719.599-seconds.png
Farm One	S7	red	90	b7	S7 red [90][b7] at 317.565 seconds.png
Farm One	S3	red	90	t15	S3-red-[90][t15]-at-144.76-seconds.png
Farm One	R7	red	90	t1	R7-red-[90][t1]-at-286.415-seconds.png
Farm One	R8	red	180	t1	R8-red-[180][t1]-at-449.473-seconds.png

Figure 9 - Analytics dashboard (Courtesy HUVRData LLC)

RESULTS (CASE STUDY)

During the latter half of 2016, HUVR was awarded contacts to inspect over 500 assets. The specific assets inspected were Wind Turbine Generators (WTG), located in the continental United States. Several clients were involved, and the assets were in 6 states ranging from Washington, Oregon, Michigan, Pennsylvania, Indiana and Texas. The raw data from this inspection campaign exceeded 3 Terabytes, at an average of 15 Megabytes per image, approximately 500 images per asset. Field collection of the data initiated in September and concluded at the end of December, 2016. Table 1 shows the key results of this inspection campaign.

Table 1

Location	State	Number assets	Time to Data Delivery*	Total number of images	Total data file size
Wind Farm 1	Oregon	76	10 days		
Wind Farm 2	Washington	116	5 - 7 days		
Wind Farm 3	Michigan	62	5 - 7 days		
Wind Farm 4	Pennsylvania	44	5 - 7 days		
Wind Farm 5, 6	Indiana	133	5 - 7 days		
Wind Farm 7, 8	Texas	90	2 days		
Total		521		387,491	3.66 TB

* Defined as Number of business days from the completion of inspection until data loaded on web portal.

Figure 10 - Typical inspection finding (Image courtesy HUVRdata)

CONCLUSIONS & RECOMMENDATIONS

It is possible and practical to conduct asset inspections utilizing UAV's. One primary key to conducting a successful inspection program is to plan for the data that will be generated. The recommendations for managing the acquisition of large amounts of UAV –sourced data are broken down into several functional areas.

Regulatory Compliance

FAA regulations govern the utilization of UAV / UAS systems in the United States. A thorough understanding of these regulations is required to assure compliance. The reader is encouraged to visit the FAA web site for specifics on the 2016 part 107 ruling. The FAA regulates the US airspace and for safety, qualification, and training purposes. Privacy regulations are governed by the individual states.

The National Unmanned Aircraft Credentialing Program provides creditable, independent third party for vetting of UAV companies and service providers. Companies implementing a UAV inspection program or assessing UAV contractors can utilize this program as a part of or in place of their vendor approval process.

Data categorization

One of the key concepts required for data analytics is to create a systematic and consistent methodology and approach to categorizing the flows and conditions on the assets that are to be inspected. Industry standards from ASNT, ASTM, etc. should be adopted if available. In the absence of global industry standards, close coordination with clients to develop universal criteria for flaw identification and categorization should be implemented.

Physical safety

Most physical safety recommendations regarding the use of UAV's for commercial purposes are explicitly defined by FAA regulations. Additional safety precautions and requirements are often dictated by client safety programs. Several additional safety practices relating to charging of Lithium Polymer (LiPo) batteries are recommended:

- Avoid Charging LiPo batteries inside vehicles
- Avoid Charging batteries in adverse weather (rain)
- Adhere to manufacturers recommended practices when charging, storing, and discharging LiPo batteries.
- Utilize proper safety guidelines if a generator is used in charging batteries

Data security

The camera memory cards are the initial link in the data acquisition process. The following guidelines are adapted from Kim Olson Photography.

- Format cards instead of erasing the files on them
- Format in the camera being utilized, not in the computer
- Format new cards before using
- Format cards before using in other cameras
- Use quality name-brand memory cards with appropriate speed rating
- Use appropriate-capacity cards
- Plan the mission to not exceed the card's capacity
- Don't remove the card while it is writing / reading
- Turn off the camera when removing a card (& wait until card 'writes' are complete)
- Don't reuse 'problem' cards
- Monitor the camera's battery level

Once the images are copied off the cards, implement a data backup plan that is compliant to the "3-2-1" at a minimum. Modify for field operations if required.

Data Storage

Plan the storage needs per the inspection plan and requirements. Include IT professionals early in the process as they usually are responsible for the server capacity. Note that storage of raw data on a client's IT infrastructure may not be a desirable solution. If raw data is stored with the UAV service provider, ensure that their storage capacity, security and backup systems and procedures comply with your organizational standards.

ACKNOWLEDGEMENTS

The author would like to acknowledge the hard work and efforts of the HUVRData team, specifically, Bob Baughman, Ben Williams, Anthony Jones, and Mark Cotton, in addition the dedication of pilots William Wigfield, Jordan Giangiulio, and Travis Bradfute. Without their efforts, none of this would be possible.

REFERENCES

The nation's first offshore wind farm is ready to go, despite critics' blowback. (n.d.). Retrieved January 05, 2017, from https://www.washingtonpost.com/national/health-science/the-nations-first-offshore-wind-farm-is-ready-to-go-despite-critics-blow-back/2016/08/27/7a43c6d6-693f-11e6-99bf-f0cf3a6449a6_story.html?utm_term=.9f20a91739e5

Indranil's world. (n.d.). Retrieved January 05, 2017, from https://indranilsinharoy.com/tag/sensor-resolution/

Olson, K. (n.d.). Memory Card Best Practices: 15 Things You Should Already Be Doing. Retrieved January 05, 2017, from http://tbexcon.com/us/2012/10/24/memory-card-best-practices-15-things-you-should-already-be-doing/

*** Formatting Issue - Best Available Paper/Graphic ***

SNAME
THE INTERNATIONAL COMMUNITY
FOR MARITIME AND OCEAN PROFESSIONALS

KEY TECHNICAL ALIGNMENTS IN THE HARMONIZATION OF FIXED OFFSHORE PLATFORM DESIGN CODES

SHIH-HSIANG LIU*

JIUN-YIH CHEN*

ALBERT KU*
*ENERGO ENGINEERING, A KBR COMPANY

Proceedings of the 22nd Offshore Symposium, February 2017, Houston, Texas
Texas Section of the Society of Naval Architects and Marine Engineers

ABSTRACT

The new API (American Petroleum Institute) RP 2A (Recommended Practice for Planning, Designing and Constructing Fixed Offshore Platforms) will have two co-existing versions: API RP 2A-WSD (Working Stress Design) 22nd edition and API RP 2A-LRFD (Load Resistance Factor Design) 2nd edition. API RP 2A-LRFD 2nd edition will adopt most of the technical provisions of ISO 19902:2007 1st edition, which was originally derived from API RP 2A-LRFD 1st edition. In parallel, ISO 19902 2nd edition is being prepared for publication. The objective of this paper is to present key technical alignments in the harmonization of the latest editions of API RP 2A-LRFD and ISO 19902 in the following aspects: 1) unity check (UC) of tubular members with hydrostatic pressure, 2) UC of conical transitions, and 3) minimum capacity requirements of tubular joints.

Based on findings from the AKADEME project (Energo Engineering, 2016), API RP 2A-WSD 21st edition and ISO 19902:2007 1st edition have reasonably good alignments of UC ratios for tubular members without the hydrostatic pressure checks. When hydrostatic pressure checks are included, UC ratios vary depending on whether members are in axial tension or compression. API RP 2A-WSD 21st edition and ISO 19902:2007 1st edition have identical member code check equations when members are in axial compression with hydrostatic pressure. For members in axial tension with hydrostatic pressure, API RP 2A-WSD 21st edition code check equations provide a reasonably conservative fit to existing test data. For conical transitions, API RP 2A-WSD 21st edition computes UC ratios using the tensile strength instead of yield strength. ISO 19902:2007 1st edition checks the combined longitudinal, hoop and local bending stresses, resulting in more conservative UC ratios. The Norwegian NORSOK standard checks the combined longitudinal and hoop stresses without considering local bending stress, and checks local bending stress separately. Review of a study performed by Maersk/DNVGL shows that ISO 19902:2007 1st edition produces the most conservative results while NORSOK produces the most reasonable (least conservative) results comparing with Finite Element Method (FEM) analyses. It is recommended to consider adopting the NORSOK conical transition code check equations for future revisions of API RP 2A-LRFD and ISO 19902. The minimum capacity requirements for tubular joints in API RP 2A-WSD 21st edition and ISO 19902:2007 1st edition are different. It is well known in the offshore structure community that the ISO 19902:2007 1st edition minimum joint strength requirement is overly conservative. As

*** Formatting Issue - Best Available Paper/Graphic ***

such, adopting the API RP 2A-WSD 50% effective strength requirement is recommended.

Keywords: Unity Check, Hydrostatic Pressure, Conical Transition, Minimum Joint Capacity

INTRODUCTION

In 1993, the 1st edition of the Load and Resistance Factor Design (LRFD) version of API (American Petroleum Institute) RP 2A (Recommended Practice for Planning, Designing and Constructing Fixed Offshore Platforms) was published (API RP 2A-LRFD). The LRFD version of RP 2A was chosen as the base document of ISO 19902 (Fixed Steel Structures) published in 2007 (ISO 19902). Despite its worldwide recognition, API RP 2A-LRFD was not widely used in the United States.

The new API RP 2A will have two co-existing versions: API RP 2A-WSD (Working Stress Design) 22nd edition and API RP 2A-LRFD 2nd edition. API RP 2A-LRFD 2nd edition will adopt most of the technical provisions of ISO 19902:2007 1st edition.

The AKADEME project, commissioned by API to prepare the 2nd edition of API RP 2A-LRFD and to assess the consistency of member and joint unity checks (UCs) based on API RP 2A-WSD 21st edition and ISO 19902:2007 1st edition, was carried out by a consortium of consultants led by Experia and Energo Engineering (Energo Engineering, 2016). The case studies performed in the AKADEME project demonstrated overall consistency in the member and joint unity check (UC) procedures between API RP 2A-WSD 21st edition and ISO 19902:2007 1st edition with two key differences, that is, member unity checks with hydrostatic pressure and minimum joint capacity requirements.

Additional studies for technical alignments in the harmonization of API RP 2A and ISO 19902 on member unity checks with hydrostatic pressure and conical transition unity checks were conducted. This paper presents the key technical alignments in the harmonization of the latest editions of API RP 2A-LRFD and ISO 19902 in the following areas: 1) UC of tubular members with hydrostatic pressure, 2) UC of conical transitions, and 3) minimum capacity requirements of tubular joints.

STUDY ON HYDROSTATIC PRESSURE IN HARMONIZATION OF API RP 2A AND ISO 19902

API RP 2A-WSD 21st edition and ISO 19902:2007 1st edition have reasonably good alignments of UC ratios of tubular members without the hydrostatic pressure checks. When hydrostatic pressure checks are included, UC ratios vary depending on whether members are in axial tension or compression. This section looks into the detailed hydrostatic code check equations in API RP 2A-WSD 21st edition and ISO 19902:2007 1st edition, with a comparison to existing experimental data used to develop the equations, and presents a parametric study of the hydrostatic code check equations to recommend the modifications of the equations in API RP 2A-LRFD.

Hoop Buckling Stress

For tubular members satisfying the out-of-roundness tolerances, the hoop stress due to hydrostatic pressure should be less than the critical hoop buckling stress,

(Table 1).

The elastic hoop buckling stress,

, and the critical hoop buckling stress,

, are determined using equations shown in Tables 2 and 3, respectively. The elastic hoop buckling stress defined in API RP 2A-WSD 21st edition is slightly different from that in ISO 19902:2007 1st edition when $< 0.825-$

(Figure 1b). For critical hoop buckling stress, API RP 2A-WSD 21st edition is more conservative than ISO 19902:2007 1st edition (Figure 2). However, ISO 19902:2007 1st edition equations are still on the conservative side

Proceedings of the 22nd Offshore Symposium, February 2017, Houston, Texas
Texas Section of the Society of Naval Architects and Marine Engineers

compared to the test data.

Table 1 Hoop Stress in Hydrostatic Pressure Unity Check

API RP 2A-WSD 21st edition	ISO 19902:2007 1st edition
$\leq$ —— where $=$ $=$	$\leq$ —— , where $=$ factored , $=$

Table 2 Elastic Hoop Buckling Stress

API RP 2A-WSD 21st edition		ISO 19902:2007 1st edition	
$= 2 \quad -$			
where	for	where	for
$= 0.44-$	$\geq 1.6-$	$= 0.44-$	$\geq 1.6-$
$= 0.44- + \frac{0.21(-)^3}{4}$	$0.825- \leq \quad < 1.6-$	$= 0.44- + \frac{0.21(-)^3}{4}$	$0.825- \leq \quad < 1.6-$
$= \frac{0.736}{(\ - 0.636)}$	$3.5 \leq \quad < 0.825-$	$= \frac{0.737}{(\ - 0.579)}$	$1.5 \leq \quad < 0.825-$
$= \frac{0.755}{(\ - 0.559)}$	$1.5 \leq \quad < 3.5$		
$= 0.8$	< 1.5	$= 0.8$	< 1.5
$= -(\frac{2}{-})^{0.5}$			

Proceedings of the 22nd Offshore Symposium, February 2017, Houston, Texas
Texas Section of the Society of Naval Architects and Marine Engineers

where = , ,

Table 3 Critical Hoop Buckling Stress

API RP 2A-WSD 21st edition	ISO 19902:2007 1st edition
= for ≤ 0.55	= for ≤ 0.55
$= 0.45 \quad + 0.18$ $0.55 < \quad \leq 1.6$	$= 0.7(—)^{0.4}$ $0.55 < \quad \leq 2.44$
$= \frac{1.31}{1.15 + (—)}$ $1.6 < \quad \leq 6.2$	
= > 6.2	= > 2.44

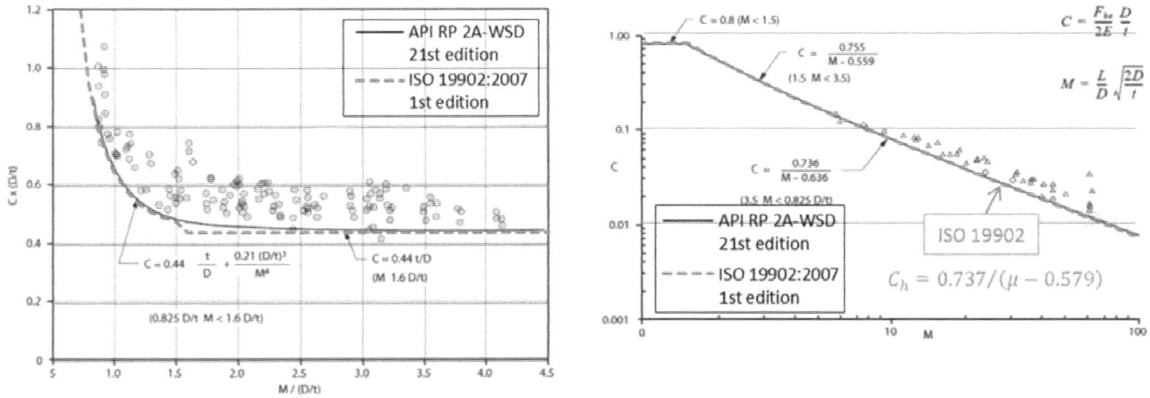

(a) M > 0.825 D/t (b) M < 0.825 D/t

Figure 1 Comparison of Elastic Hoop Buckling Code Check Equations between API RP 2A-WSD 21st Edition and ISO 19902:2007 1st Edition (API Curves and Test Data Adapted from API RP 2A-WSD 21st Edition Commentary)

***** Formatting Issue - Best Available Paper/Graphic *****

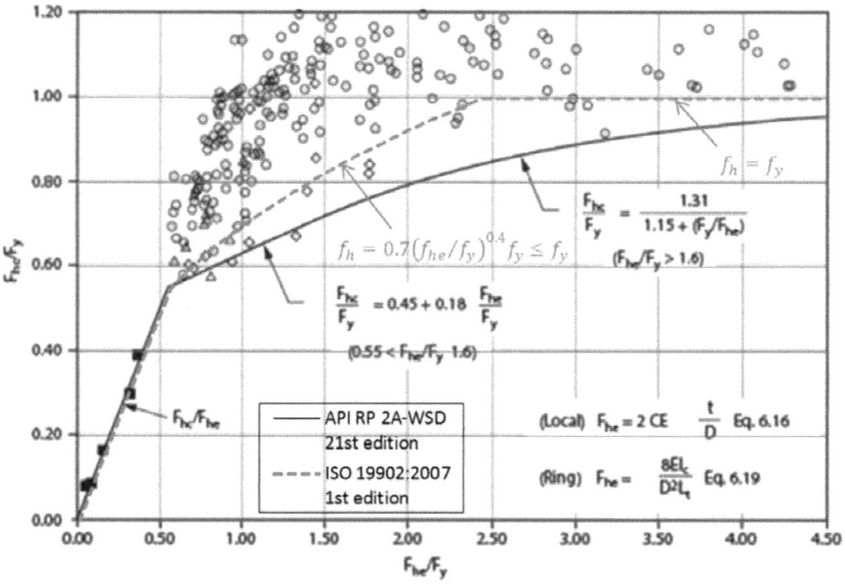

Figure 2 Comparison of Critical Hoop Buckling Code Check Equations between API RP 2A-WSD 21st Edition and ISO 19902:2007 1st Edition (API Curves and Test Data Adapted from API RP 2A-WSD 21st Edition Commentary)

Axial Compression with Hydrostatic Pressure

The code check equations when tubular members are subjected to axial compression with hydrostatic pressure are summarized in Table 4. For API RP 2A-WSD 21st edition, the code check equations for combined axial compression and bending, whether with or without the presence of hydrostatic pressure, need to be satisfied [Equations API WSD (1) and (2) in Table 4]. As summarized in Table 4, API RP 2A-WSD 21st edition and ISO 19902:2007 1st edition have identical code check equations when members are in axial compression with hydrostatic pressure. The only difference is in the safety factors for API RP 2A-WSD and partial factors for ISO 19902. A comparison of interaction equations for various stress conditions are shown in Figure 3. Note that the comparison excludes the safety factors in API RP 2A-WSD equations and the partial resistance factors in ISO 19902 equations.

Table 4 Unity Check Equations for Tubular Member Subjected to Axial Compression with Hydrostatic Pressure

API RP 2A-WSD 21st edition	ISO 19902:2007 1st edition
Axial Compression and Bending $\frac{}{0.6} + \frac{\sqrt{\ ^2 + \ ^2}}{}$ API WSD (1)	Hoop Buckling — — , ISO (1)
$- + \frac{\sqrt{[\frac{}{1 - \frac{}{,}}]^2 + [\frac{}{1 - \frac{}{,}}]^2}}{}$	Axial Compression and Hydrostatic Pressure $\frac{,\ \ ,}{} + \frac{,\sqrt{,\ ^2 + \ ,\ ^2}}{,}$ ISO (2)

*** Formatting Issue - Best Available Paper/Graphic ***

API WSD (2)	, = + 0.5 for Marine Buoyancy Method $\frac{,}{,} + \frac{,}{,} [(\frac{, \quad ,}{1 - \frac{}{,}})^2 + (\frac{, \quad ,}{1 - \frac{}{,}})^2]^{0.5}$ ISO (3) $\frac{- 0.5 \frac{}{,}}{} + (\frac{,}{})^2$ $, \; - 0.5 \frac{}{,}$ ISO (4)
Axial Compression and Hydrostatic Pressure $\frac{+ (0.5 \quad)}{} \times \quad + \frac{}{} \times$ API WSD (3) $\times \frac{}{}$ API WSD (4) > 0.5 $\frac{- 0.5}{- 0.5} + (\frac{}{})^2$ API WSD (5)	
where = = = = $= \frac{}{}$ $= \frac{}{}$ =	where = , , = , = = , = , =

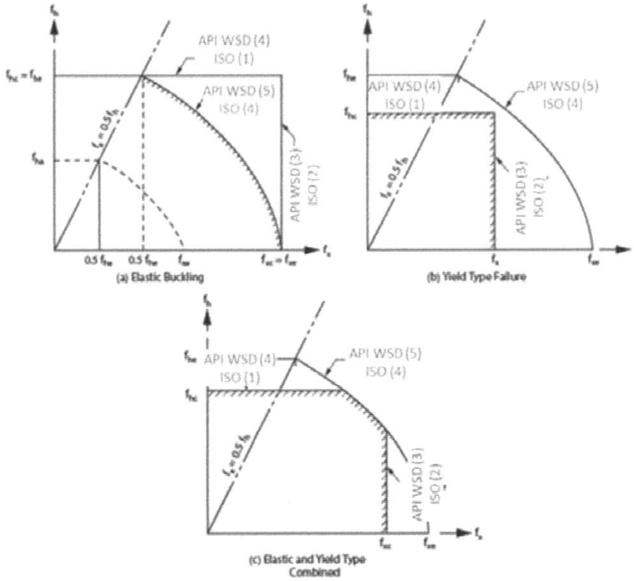

Figure 3 Comparison of Interaction Equations between API RP 2A-WSD 21st Edition and ISO 19902:2007 1st Edition for Various Stress Conditions (Adapted from API RP 2A-WSD 21st Edition Commentary)

<u>Axial Tension with Hydrostatic Pressure</u>

The code check equations when tubular members are subjected to axial tension with hydrostatic pressure in API RP 2A-WSD 21st edition are different from those in ISO 19902:2007 1st edition, as summarized in Table 5. Two typical tubular member diameter-to-thickness ratios were used to create data points where unity ratios are equal to 1.0 based on ISO 19902:2007 1st edition code check equations. By comparing these created data points with API RP 2A-WSD 21st edition curve and existing test data in the Commentary, it was found that API RP 2A-WSD 21st edition code check equations provide a reasonably conservative fit to the test data (Figure 4). Note that the comparison excludes the safety factors in API RP 2A-WSD equations and the partial resistance factors in ISO 19902 equations.

Table 5 Unity Check Equations for Tubular Member Subjected to Axial Tension with Hydrostatic Pressure

API RP 2A-WSD 21st edition	ISO 19902:2007 1st edition
<u>Axial Tension and Bending</u> $\frac{\quad}{0.6} + \frac{\sqrt{\;^{2} + \;^{2}}}{\quad}$	<u>Hoop Buckling</u> — — ,
	<u>Axial Tension and Hydrostatic Pressure</u>

***** Formatting Issue - Best Available Paper/Graphic *****

Axial Tension and Hydrostatic Pressure $^2 + \ ^2 + 2 \ \| \ \|$ $= \frac{+ \quad - (0.5 \quad)}{} \times$ $= \frac{}{} \times$	$\frac{, \quad ,}{} + \frac{, \sqrt{\ ,^2 + \ ,^2}}{}$ $, = \quad - 0.5$ for Marine Buoyancy Method $, = \quad (\sqrt{1 + 0.09 \ ^2 - \ ^2} - 0.3 \ )$ $, = \quad (\sqrt{1 + 0.09 \ ^2 - \ ^2} - 0.3 \ )$ $= \frac{,}{}$ $= 5 - 4 \frac{}{}$
where $=$ $=$ $=$	where $=$ $=$ $, =$ $=$ $=$

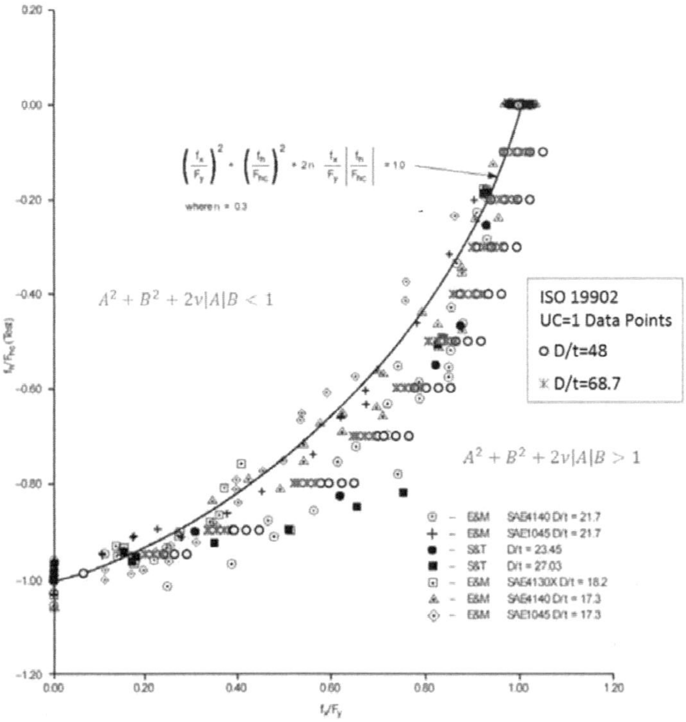

Figure 4 Comparison of Interaction Equation for Axial Tension with Hydrostatic Pressure between API RP 2A-WSD 21st Edition and ISO 19902:2007 1st Edition (API Curve and Test Data Adapted from API RP 2A-WSD 21st Edition Commentary)

A Parametric Study to Compare Code Check Equations

A parametric study using various combinations of axial (
), bending (
) and hoop (
) stress utilizations to compare the member UCs based on API RP 2A-WSD 21st edition and ISO 19902:2007 1st edition code check equations was conducted. Two typical tubular member diameter-to-thickness ratios were considered. The utilization levels were assumed to be 10%, 30% and 50% of the corresponding allowable stresses based on API RP 2A-WSD 21st edition (3 different stress components and each with 3 different utilization levels resulting in a total of 27 cases). These stresses were used to calculate the UCs of the tubular member under combined axial, bending and hydrostatic pressure. The loading condition was assumed to be the extreme (storm) loading condition. As such, for API RP 2A-WSD, the allowable stress was increased by one-third. For ISO 19902, a partial action factor of 1.35 for axial and bending stresses and a partial action factor of 1.1 for hoop stress were assumed.

Figures 5 and 6 compare the UCs for tubular members under various combinations of axial, bending and hoop stress utilizations (27 cases) with diameter-to-thickness ratios equal to 48 and 68.7, respectively. API RP 2A-WSD 21st edition and ISO 19902:2007 1st edition have identical code check equations for members under axial compression with hydrostatic pressure. ISO 19902:2007 1st edition yielded generally higher UCs than API RP 2A-WSD 21st edition due to the assumption of partial action factors and allowable stress increase under the extreme (storm) loading condition (Figures 5a and 6a). For members under axial tension with hydrostatic pressure, API RP 2A-WSD 21st edition and ISO 19902:2007 1st edition resulted in a reasonably good agreement of member UCs

*** Formatting Issue - Best Available Paper/Graphic ***

(Figures 5b and 6b), even though the code check equations for axial tension with hydrostatic pressure are different between API RP 2A-WSD 21st edition and ISO 19902:2007 1st edition.

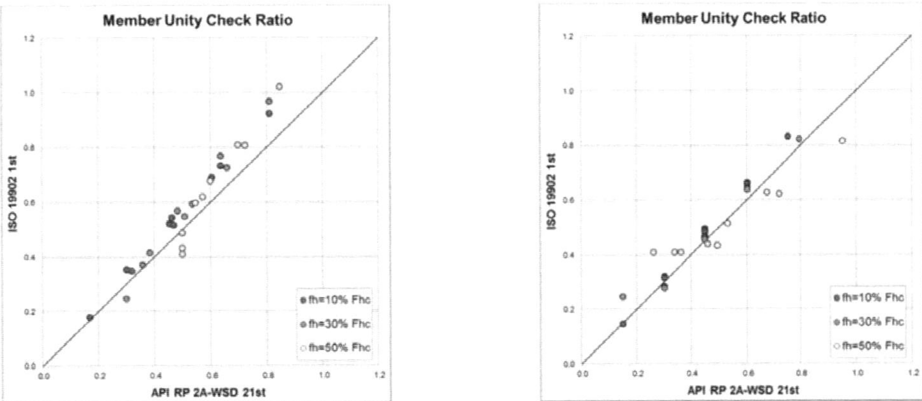

(a) Axial Compression, Bending and Hydrostatic (b) Axial Tension, Bending and Hydrostatic

Figure 5 Comparison of UCs for Tubular Member (D/t = 48) under Various Combinations of Axial, Bending and Hoop Stress Utilizations

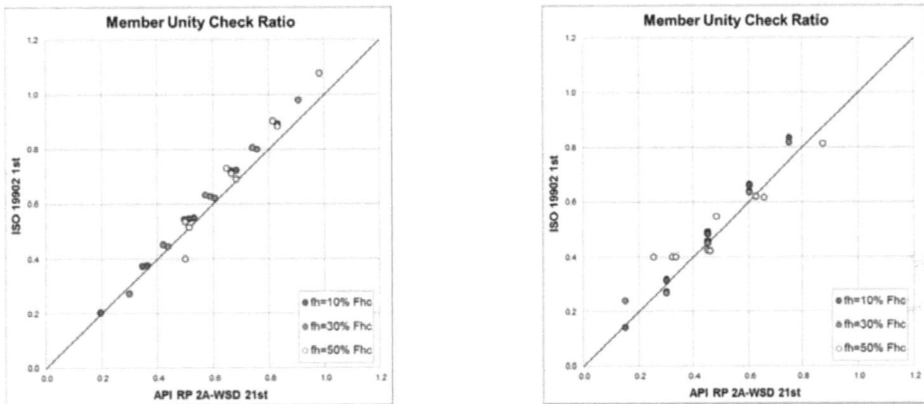

(a) Axial Compression, Bending and Hydrostatic (b) Axial Tension, Bending and Hydrostatic

Figure 6 Comparison of UCs for Tubular Member (D/t = 68.7) under Various Combinations of Axial, Bending and Hoop Stress Utilizations

STUDY ON CONICAL TRANSITION IN HARMONIZATION OF API RP 2A AND ISO 19902

The conical transition code check equations in API RP 2A-WSD 21st edition are distinctly different from those in ISO 19902:2007 1st edition. This section compares the conical transition code check equations of API RP 2A-WSD 21st edition, ISO 19902:2007 1st edition and NORSOK standard and presents the UC results of the conical transition member of Platform 3 based on these codes and standards. Additionally, this section reviews a finite element method (FEM) study of conical transition members performed by Maersk and DNV GL in an attempt to identify the best set of equations to use for the harmonization of these offshore codes and standards (Maersk and DNV GL). In summary, the Norwegian NORSOK standard has similar code check equations as ISO 19902:2007 1st

*** **Formatting Issue - Best Available Paper/Graphic** ***

edition for conical transition and produces the least conservative and most reasonable results as compared to the FEM analyses.

API RP 2A-WSD 21st Edition Conical Transition Code Check

Table 6 summarizes the strength requirements for conical transitions in API RP 2A-WSD 21st edition. The code check equations for combined axial and bending stresses need to be satisfied. Note that API RP 2A-WSD 21st edition computes UCs based on tensile strength instead of yield strength for the total stress check. The total stress is defined as the summation of the axial and resultant bending stresses in the cylindrical section at the junction, and the longitudinal stress (i.e., localized bending stress). In addition, the summation of axial and resultant bending stresses must be limited to the allowable stress. The hoop stress at the conical transition needs to be checked as well.

Table 6 Strength Requirements for Conical Transition in API RP 2A-WSD 21st Edition

API RP 2A-WSD 21st edition Conical Transition Requirements
Combined Stresses for Tubular Member $\frac{\;}{0.6} + \frac{\sqrt{\;^2 + \;^2}}{\;}$ $\frac{\;}{\;} + \frac{\sqrt{[\frac{\;}{1-\frac{\;}{\;}}]^2 + [\frac{\;}{1-\frac{\;}{\;}}]^2}}{\;}$
Total Stress Limited to Tensile Strength $\frac{(\;+\;+\;')}{\;}$ where $' = \frac{0.6\sqrt{(\;+\;)}}{2}(\;+\;)$ $=$ $=$ $=$ $=$ $=$
Allowable Stress $(\;+\;) \leq 0.6$ ≤ 0.8
Hoop Stress

$$' = 0.45\sqrt{-(\quad + \quad)} \qquad \leq 0.6$$

$$\leq 0.5$$

where

$$' =$$

ISO 19902:2007 1st Edition Conical Transition Code Check

ISO 19902:2007 1st edition requires local buckling, junction yielding, and junction buckling at the conical transition to be checked. Table 7 summarizes the strength requirements for conical transition without hydrostatic pressure in ISO 19902:2007 1st edition. For junction yielding and junction buckling, ISO 19902:2007 1st edition checks the combined longitudinal, hoop and local bending stresses, resulting in more conservative UCs.

Table 7 Strength Requirements for Conical Transition in ISO 19902:2007 1st Edition

ISO 19902:2007 1st edition Conical Transition Requirements	
Local Buckling	
$\frac{\quad , \quad ,}{}$	where $, = \frac{\quad , + \quad ,}{}$ $, = \frac{}{(\quad - \quad)}$ $=$ $\qquad -$ $, = \frac{4}{(\quad - \quad)^2}$ $=$ $-$

*** Formatting Issue - Best Available Paper/Graphic ***

$=$ $-$ $=$ $-$
<u>Junction Yielding</u> (when hoop stress is tensile) $\sqrt{\ ^2 + (\ )^2 -}$ $\sqrt{\ ^2 + (\ )^2 + \lvert\ \rvert}$ where $= \ , + \ , + \ ,$ $= \frac{, + ,}{} + \ ,$ $= \frac{0.6 \sqrt{(\ +\ )}}{2} (\ ,$ $+ \ ,)$ $= \frac{0.6 \sqrt{(\ +\ )}}{2} (\ ,$ $+ \ ,)$
<u>Junction Buckling</u> (when hoop stress is compressive) $^2 + \ ^2 + 2 \ \times$ where $= \frac{,}{}$ $= \frac{,}{}$

*** Formatting Issue - Best Available Paper/Graphic ***

$$= \; , = 0.45\sqrt{-}(\; , + \; ,)$$

$$= \; , = 0.45\sqrt{--}(\; , + \; ,)$$

NORSOK Standard Conical Transition Code Check

The Norwegian NORSOK standard also checks local buckling, junction yielding, and junction buckling. The code check equations for conical transitions without hydrostatic pressure are similar to those in ISO 19902:2007 1st edition. However, the NORSOK standard checks the combined longitudinal and hoop stresses without considering local bending stress, and checks local bending stress separately (Table 8).

Table 8 Strength Requirements for Conical Transition in NORSOK Standard

NORSOK Conical Transition Requirements
Local Buckling where $=$ $= \frac{, + ,}{}$ $, = \frac{}{(-)}$ $=$ $, = \frac{4}{(-)^2}$ $=$

***** Formatting Issue - Best Available Paper/Graphic *****

= —   = —
Junction Yielding (when hoop stress is tensile) $\sqrt{\ ^2 + (\ )^2 -\ }$ $\sqrt{\ ^2 + (\ )^2 + \lvert\ \rvert}$ where = , + ,   = , + ,   *Local Bending Stress* $, = 0.85\sqrt{—}(\ , + \ ,) \quad \leq 1.5—(1 - \frac{2}{2})$
Junction Buckling (when hoop stress is compressive) $^2 + \ ^2 + 2 \times$ where = ——   = ——   ——   —— $= \ , = 0.45\sqrt{—}(\ , +\ ,)$ —   — $= \ , = 0.45\sqrt{—}—(\ , + \ ,)$

*** Formatting Issue - Best Available Paper/Graphic ***

Local Bending Stress

$$_{,} = 0.85\sqrt{—}\left(\ _{,} + \ _{,}\right) \qquad \le 1.5—\left(1 - \frac{2}{2}\right)$$

Comparison of API RP 2A-WSD 21st Edition, ISO 19902:2007 1st Edition and NORSOK UCs for Conical Transition Member of AKADEME Platform 3

AKADEME Case Study Platform 3 is a free-standing caisson structure. Figure 7 shows a 3-D rendering and key characteristics of Platform 3. The conical transition member of Platform 3 was used as an example to compare the code check results using API RP 2A-WSD 21st edition, ISO 19902:2007 1st edition and NORSOK standard. The material factor in the NORSOK code check was assumed to be 1.15. The calculated UCs using API RP 2A-WSD, ISO 19902 and NORSOK standard are summarized in Tables 9, 10 and 11, respectively. Note that the governing UC ratios are highlighted in red. In this example, ISO 19902:2007 1st edition produces the most conservative result, and API RP 2A-WSD 21st edition and NORSOK standard produce comparable results. However, API RP 2A-WSD 21st edition computes UC ratios using tensile strength instead of yield strength, which is not typical for unity checks.

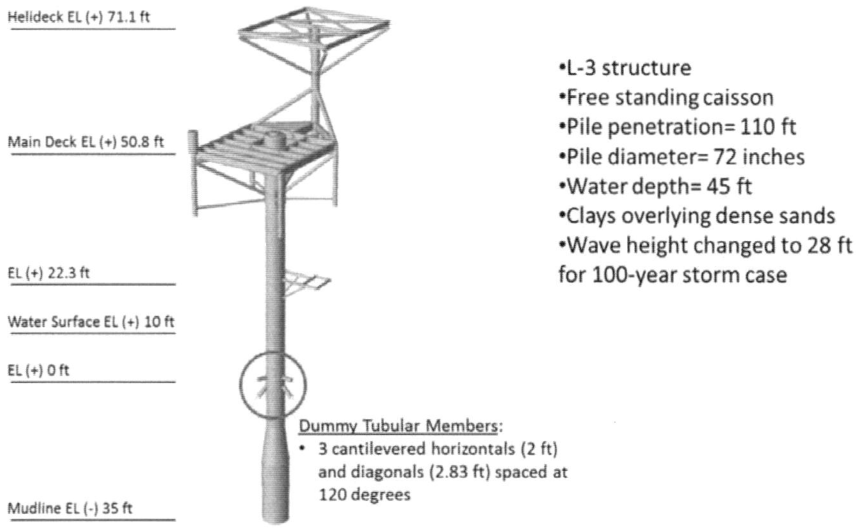

Figure 7 3-D Rendering and Key Characteristics of AKADEME Case Study Platform 3

Table 9 Unity Check Ratios of AKADEME Platform 3 Conical Transition Member Using API RP 2A-WSD 21st Edition

API RP 2A-WSD 21st edition Conical Transition Requirements	UC Ratio
Combined Stresses for Tubular Member	0.17

*** Formatting Issue - Best Available Paper/Graphic ***

$\frac{}{0.6} + \frac{\sqrt{\ ^2 + \ ^2}}{}$ $\frac{}{} + \frac{\sqrt{[\frac{}{1-\frac{}{,}}]^2 + [\frac{}{1-\frac{}{,}}]^2}}{}$	
Total Stress Limited to Tensile Strength $\frac{(+ + ')}{}$	0.20
Allowable Stress $(+) \leq 0.6$ ≤ 0.8	0.19
Hoop Stress $' = 0.45 \sqrt{-(+)} \quad \leq 0.6$ ≤ 0.5	0.06 0.08

Table 10 Unity Check Ratios of AKADEME Platform 3 Conical Transition Member Using ISO 19902:2007 1st Edition

ISO 19902:2007 1st edition Conical Transition Requirements	UC Ratios
Local Buckling $\frac{, ,}{}$	0.23
Junction Yielding (when hoop stress is tensile) $\frac{\sqrt{\ ^2 + ( )^2 -}}{,}$	Tubular side 0.25 Cone side 0.24

Proceedings of the 22nd Offshore Symposium, February 2017, Houston, Texas
Texas Section of the Society of Naval Architects and Marine Engineers

$\sqrt{\ ^2 \quad + (\quad)^2 + \quad \| \quad \|}$,	
Junction Buckling (when hoop stress is compressive) $^2 + \ ^2 + 2 \quad \times$ — — , — — ,	0.09 0.36 0.06

Table 11 Unity Check Ratios of AKADEME Platform 3 Conical Transition Member Using NORSOK Standard

NORSOK Conical Transition Requirements	UC Ratios
Local Buckling —	0.22
Junction Yielding (when hoop stress is tensile) $\sqrt{\ ^2 \quad + (\quad)^2 -}$ $\sqrt{\ ^2 \quad + (\quad)^2 + \quad \| \quad \|}$	Tubular side 0.17 Cone side 0.17
Junction Buckling (when hoop stress is compressive) $^2 + \ ^2 + 2 \quad \times$ — —	0.05 0.23 0.06

Proceedings of the 22nd Offshore Symposium, February 2017, Houston, Texas
Texas Section of the Society of Naval Architects and Marine Engineers

— —	

Nonlinear FEM Study of Conical Transition Members (Maersk and DNV GL)

A nonlinear FEM study on the capacities of conical transition members with various cross-sectional properties was conducted by Maersk and DNV GL. Table 12 summarizes the parameters of the conical transition members used in this study (Maersk and DNV GL). The definitions of these parameters are illustrated in Figure 8 (API RP 2A-WSD, 2007). A total of 100 conical transition members were modeled separately, and the corresponding capacities from FEM analyses were compared to the calculated capacities based on code equations (API RP 2A-WSD, ISO 19902 and NORSOK).

Figures 9, 10 and 11 present the comparisons of axial compression, axial tension, and bending moment capacities between FEM analysis results and calculated capacities using code equations (Maersk and DNV GL). The results were normalized to the plastic axial and bending moment capacities. Each case was numbered by changing Parameter 3, Parameter 2, and then Parameter 1 in Table 12. Note that the safety factors and resistance factors were excluded in the comparisons. The results show that ISO 19902:2007 1st edition equations produce the most conservative results. NORSOK standard produces less conservative results than API RP 2A-WSD 21st edition, except for the cases with small diameter-to-thickness (D/t) ratios and small transition angles (α). The results from NORSOK standard remain on conservative side as compared to the FEM analysis results, demonstrating adequacy of the NORSOK code check equations.

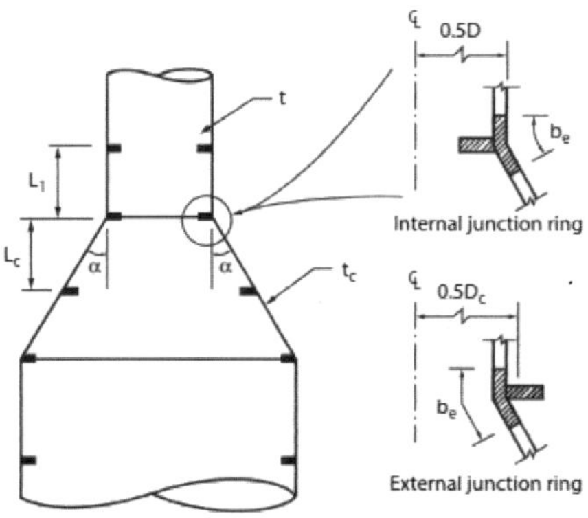

Figure 8 Definitions of Conical Transition Parameters (API RP 2A-WSD, 2007)

Table 12 Parameters in Nonlinear FEM Study of Capacities for Conical Transition Members (Maersk and DNV GL)

Nonlinear FEM Study Parameters		
Parameter 1	—	18, 33.3, 50, 66.7, and 100
Parameter 2		3°, 6°, 10°, and 15°
Parameter 3	—	3.0, 2.0, 1.5, 1.2, and 1.0

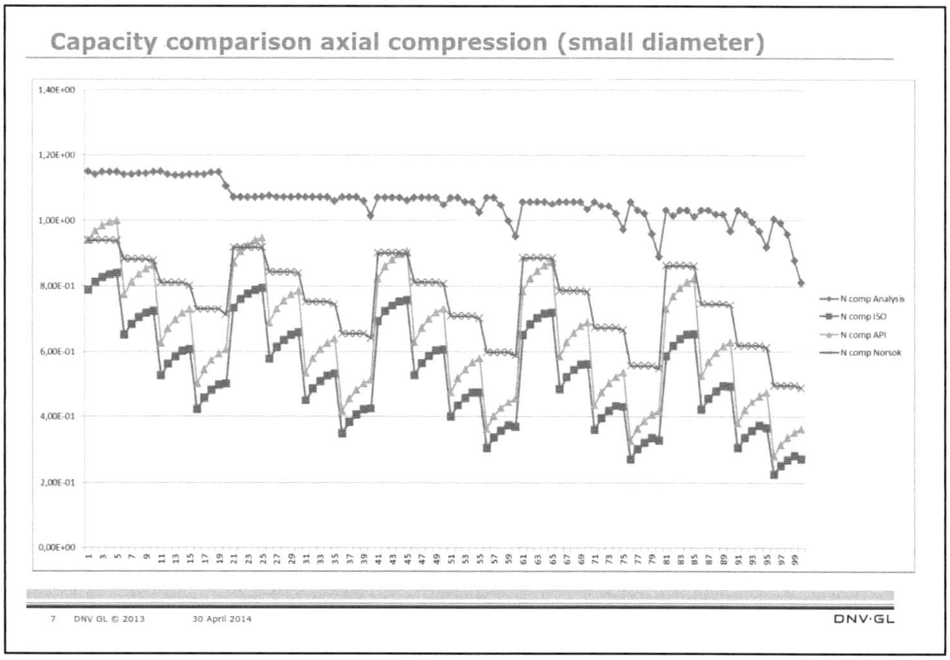

Figure 9 Comparison of Axial Compression Capacities between FEM Analyses and Design Codes (Maersk and DNV GL)

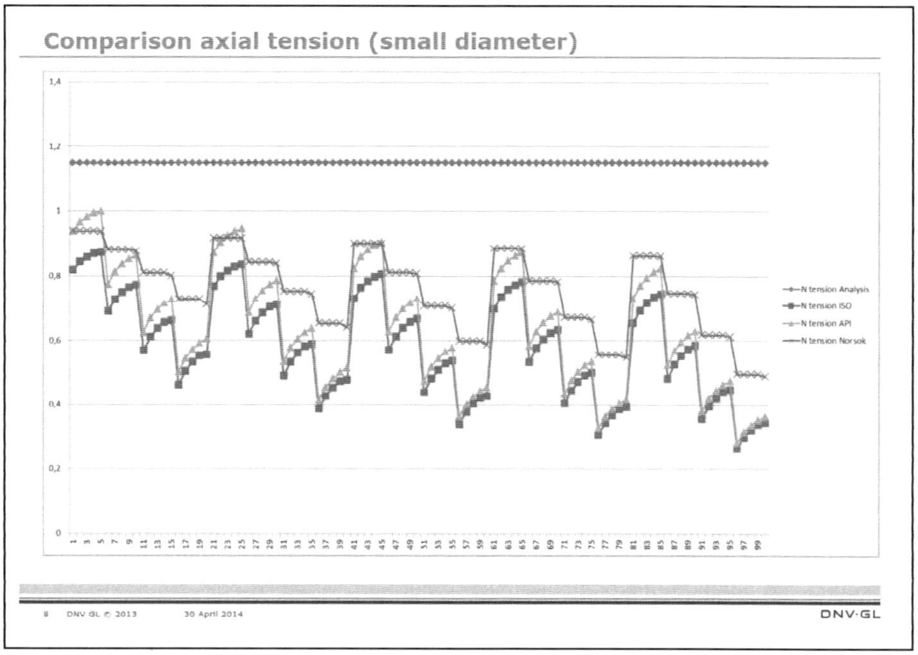

Figure 10 Comparison of Axial Tension Capacities between FEM Analyses and Design Codes (Maersk and DNV GL)

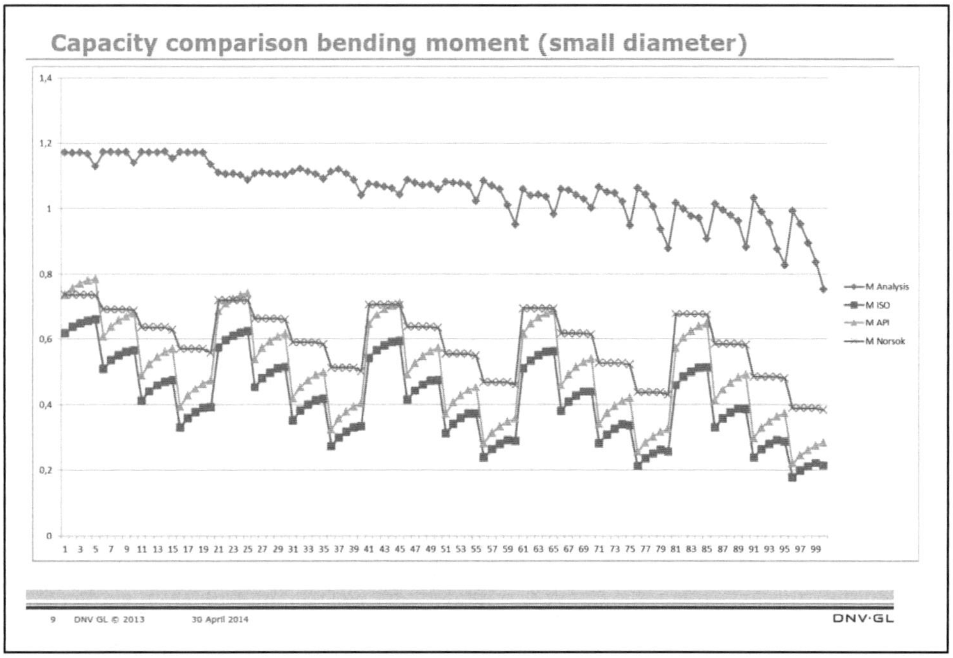

Figure 11 Comparison of Bending Moment Capacities between FEM Analyses and Design Codes (Maersk and DNV GL)

STUDY ON MINIMUM CAPACITY REQUIREMENTS OF TUBULAR JOINTS

The minimum capacity requirements for tubular joints in API RP 2A-WSD 21st edition and ISO 19902:2007 1st edition are different. API RP 2A-WSD 21st edition requires a joint capacity to be higher than 50% of the effective strengths of the adjoining members. ISO 19902:2007 1st edition requires the minimum joint strength to exceed the strength of all adjoining members and requires an additional partial resistance factor. Sensitivity analyses were performed in the AKADEME project to assess the effect of including the minimum joint capacity requirements from API RP 2A-WSD 21st edition and ISO 19902:2007 1st edition on the joint UC comparisons for the case study platforms. This section presents the results of AKADEME Case Study Platform 5 as an example. Figure 12 shows the 3-D rendering and key characteristics of Platform 5. Figure 13 shows the scatter diagrams of joint UC comparisons based on API RP 2A-WSD 21st edition and ISO 19902:2007 1st edition. Figure 13a includes the API 50% effective strength requirement and ISO 100% minimum strength requirement, while Figure 13b excludes these requirements and is only based on strength required by the design loads. When the API 50% effective strength requirement and ISO 100% minimum strength requirement are considered, the results show significantly more scatters than those without the minimum capacity requirements. Additionally, ISO UCs become a lot higher than API UCs, as compared to the case without the minimum strength requirements (Figure 13a). These results indicate that the minimum joint strength requirement in ISO 19902:2007 1st edition is overly conservative, which is consistent with the general perception in the offshore structural community.

Proceedings of the 22nd Offshore Symposium, February 2017, Houston, Texas
Texas Section of the Society of Naval Architects and Marine Engineers

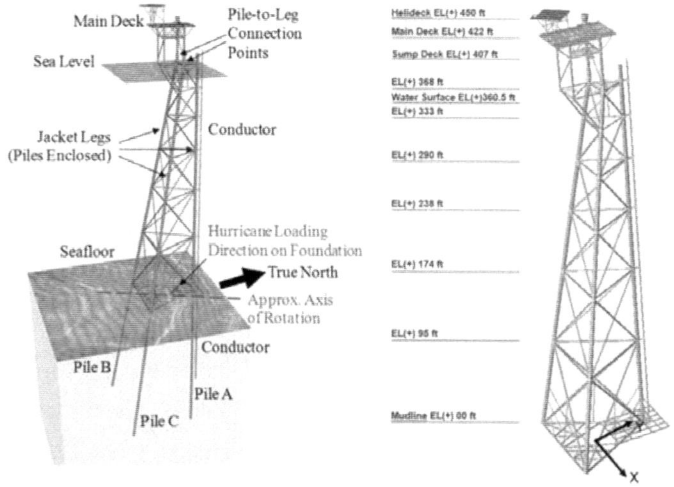

Figure 12 3-D Rendering and Key Characteristics of AKADEME Case Study Platform 5

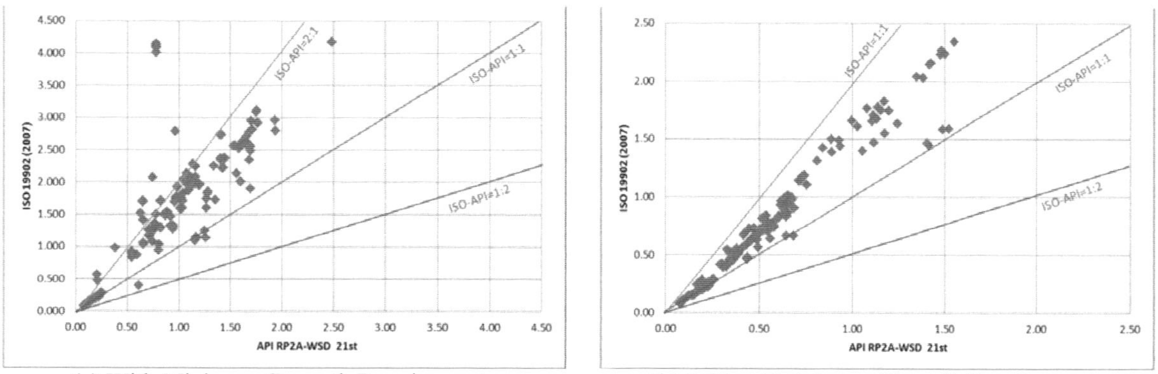

(a) With Minimum Strength Requirements (b) Without Minimum Strength Requirements

Figure 13 AKADEME Platform 5 Joint UC Comparisons with and without API and ISO Minimum Strength Requirements

CONCLUSIONS

Based on the study results from the AKADEME project presented herein, the following conclusions can be made:

- The study on hydrostatic pressure shows that API RP 2A-WSD 21st edition and ISO 19902:2007 1st edition have identical code check equations when members are in axial compression with hydrostatic pressure. For members in axial tension with hydrostatic pressure, the code check equations are different. API RP 2A-WSD 21st edition code check equations provide a reasonably conservative fit to existing test data. Hence, it is recommended to adopt API RP 2A-WSD 21st edition equations for hydrostatic code checks.
- The UC value for the conical transition member of Platform 3 based on ISO 19902:2007 1st edition is much higher than the UC value based on API RP 2A-WSD 21st edition (nearly twice as high), which indicates that ISO 19902:2007 1st edition code check equations for conical transitions are more conservative. A review of an

FEM study on the capacities of conical transitions shows that ISO 19902:2007 1st edition produces the most conservative results, while the NORSOK standard produces the least conservative results. It is recommended to consider adopting the NORSOK conical transition code check equations for the future revisions of API RP 2A-LRFD and ISO 19902.

- When the API 50% effective strength requirement and ISO 100% minimum strength requirement are considered in the joint code checks, a lot more scatter is found in the ISO 19902:2007 1st edition vs. API RP 2A-WSD 21st edition joint UC comparison than the case without considering these requirements. ISO 19902:2007 100% minimum joint strength requirement is considered to be overly conservative. As such, adopting the API 50% effective strength requirement is recommended for API RP 2A-LRFD and ISO 19902 2nd editions.

REFERENCES

American Petroleum Institute, "Recommended Practice for Planning, Designing and Constructing Fixed Offshore Platforms – Working Stress Design Supplement 3", API RP 2A-WSD, 21st Edition, 2007.

American Petroleum Institute, "Recommended Practice for Planning, Designing and Constructing Fixed Offshore Platforms – Load and Resistance Factor Design" API RP 2A-LRFD, 1st Edition, 1993.

Energo Engineering, "AKADEME Project – API RP 2A vs. ISO 19902 Member and Joint UC Comparisons", API AKADEME Project Report, 2016.

International Standard Organization, "Petroleum and Natural Gas Industries – Fixed Steel Offshore Structures", ISO 19902, 2007.

Maersk and DNV GL, "Plastic Hinge Model for Tubulars and Conical Transitions. Results of detailed FEM analyses condensed into a beam Model," Rambøll Ref. 862601/340_0001(1).

NORSOK Standard, "Design of Steel Structures", NORSOK N-004, Rev. 3, 2013.

SNAME
THE INTERNATIONAL COMMUNITY
FOR MARITIME AND OCEAN PROFESSIONALS

HOW CLASSIFICATION SOCIETIES CAN BE EFFICIENTLY INVOLVED IN THE CURRENT OFFSHORE INDUSTRY COST REDUCTION ERA

OLIVIER BENYESSAAD
OFFSHORE BUSINESS DEVELOPMENT MANAGER, BUREAU VERITAS

STEVEN BARRAS
OFFSHORE GENERAL MANAGER FOR US AND CANADA, BUREAU VERITAS

GUSTAVO ROCHA
REGIONAL DIRECTOR OF OPERATION, BUREAU VERITAS

Proceedings of the 22nd Offshore Symposium, February 2017, Houston, Texas
Texas Section of the Society of Naval Architects and Marine Engineers

ABSTRACT

Classification Societies (CS) have accompanied Offshore Oil & Gas industry for decades to guarantee and enhance the safety of their installations through continuous improvement of design codes, standards, rules and methodologies. In today's environment, where the offshore industry business model is threatened, CS have joined the industry efforts to reduce development costs by proposing solutions to fight against cost escalation without compromising safety.

CS main role is to ensure compliance of projects with pre-defined set of rules, standards or specifications. As a guarantor of the safety, they have an in-depth understanding of the technical and regulatory challenges related to an offshore development and are regularly consulted on technical & economic questions. To accompany the Oil & Gas industry further, CS have extended their classic role related to safety and are now looking to propose integrated and collaborative solutions to assist the full supply chain in their cost reduction efforts. Standardization, integrated QA/QC, AIM are some of the key solutions investigated in this paper.

For almost 200 years, CS have brought their expertise to service the Marine and then the Offshore industry with impartiality. By publishing rules and participating in projects from concept stage to decommissioning, they have contributed to the sustainability of the industry through the continuous integration of industry experience and lessons learnt in order to ensure continuous improvement in asset operational safety and integrity.

Cost efficiency is not only motivated by the current crisis, it is the new reality of the Oil & Gas industry. Future generations of Oil & Gas companies will have to develop their solutions on new bases and this paper aims at providing an outline of the key areas in which the Classification Societies can provide expert assistance in the cost reduction efforts.

Keywords: SNAME, Classification, Cost Reduction, Offshore, Symposium

INTRODUCTION

Classification Society (CS) concept appeared in the second half of the 18th century when marine insurers were developing a system to assess conditions of ships presented to them for insurance cover. While the concept emerged firstly in the UK under the name of Register Society, the formal classification societies appeared finally about 60 years later with Bureau Veritas in 1828, Lloyd's Register in 1834 and DNV in 1864.

Classification Societies' role, as defined by the International Association of Classification Societies (IACS), is "*to provide classification and statutory services and assistance to the marine industry and regulatory bodies as regards maritime safety and pollution prevention, based on the accumulation of maritime knowledge and technology*". In simple words, todays major objective of CS is to ensure safety & integrity of Marine & Offshore installations to avoid any major human or environmental disaster, which is often a synonym of economic disaster for the asset owner.

In many instances, Classification Societies possess more detailed knowledge about an installation than its owner or the flag state. For this reason, Marine & Offshore industry has always used Classification Societies as a source of return of experience to improve the safety of their installations, firstly, by proving guidance and rules that incorporate up to date operational experience, and secondly, by providing third party technical assessments & inspections to guarantee the safety of the asset from its early development phase to the end of operations.

Modern role of Classification is not only related to safety assurance but are more and more focus on overall project assurance. Main idea behind this notion of "assurance" is to provide the industry with the capability to cover any single phase of its development as well as any actor of the supply chain within a global assurance scheme that will bring confidence in the asset in term of safety and integrity, but that will also provide key inputs to improve projects economics. Providing assurance of safety, integrity but also availability performance for a project allows companies to manage efficiently risks related to their business. For these reasons, Classification Societies have developed unique competencies transverse to their core business in order to accompany the whole Marine & Offshore sector in its evolution and especially during crises.

The following section will thus develop the methodology that has been applied to identify cost reduction efforts in which Classification Societies can really add value as well as general observations related to industry cost structure that could help to identify other sources of cost reduction.

METHODOLOGY

Methodology of this paper is based on three main steps summarized as follows:

- Identification of offshore oil & gas projects phases and related cost constraints;
- Identification of offshore oil & gas projects economics key parameters & qualitative assessment of the economic impact of each parameter identified; and
- Identification of classification societies key competencies / involvement capable to impact positively key project economics parameters.

The first step consists in defining in a high level manner the main cost constraints of each phase of a project, from exploration to decommissioning, and to identify main sources of costs and related areas of improvement to increase the overall project value. Main purpose of this first step is to ensure cost reduction efforts will be oriented toward the most effective project phases.

The second step consists in defining economics parameters related to Net Present Value (NPV) calculation that can be summarized as the measure of the profitability of a project. Main goal is to understand underlying mechanisms and principles making a project profitable or not. Authors have then performed a qualitative assessment of the key parameters in high level manner to quantify impact of each of them on project profitability. Main purpose is to ensure cost reduction efforts will be oriented toward the most effective parameters.

During final step, the authors have used their knowledge related to the capabilities of the main IACS classification societies to identify the likely sources of cost reduction in direct link with the variables identified in step 1 and 2.

Finally, the paper will highlight the key actions a classification society can be involved with in order to reduce overall project cost and positively impact the profitability of a project.

1. Definitions and Terminology

Before detailing the three-step methodology and related results, it is important to first establish some important definitions and terms.

Capital Expenditure

Capital Expenditure (CAPEX) is commonly defined as money spent to acquire or refurbish installations. It can be further broken down into costs for pre-development, development, engineering, construction and include all infrastructures & personnel, temporary or not, used for the project until operation phase. It is also common to consider exploration, appraisal and decommissioning spending within the CAPEX.

Discount Rate

Discount Rate is the multiplier that converts anticipated returns from an investment project to their current market value (present value). It is always less than 1, and depends on the cost of capital (current compound interest rate) and the time interval between the investment date and the date when returns start to flow.

Internal Rate of Return

Internal rate of return (IRR) is a metric used in capital budgeting measuring the profitability of potential investments. Internal rate of return is a discount rate that makes the net present value (NPV) of all cash flows from a particular project equal to zero. IRR calculations rely on the same formula as NPV does.

Net Present Value

Net present value (NPV) is a measurement of the profitability of an undertaking that is calculated by subtracting the present values of cash outflows (including initial cost) from the present values of cash inflows over a period of time

Operating Expenditure

Operating Expenditure (OPEX) is commonly defined as the expenditure related to the operation and maintenance of an installation. It includes logistics, insurance, labor costs, taxes, maintenance contracts…etc.

2. Offshore Development Key Phases & Related Economic Constraints

Offshore Oil & Gas developments are generally long term investments. Figure 1 shows the typical development phases that are common to any Oil & Gas field development.

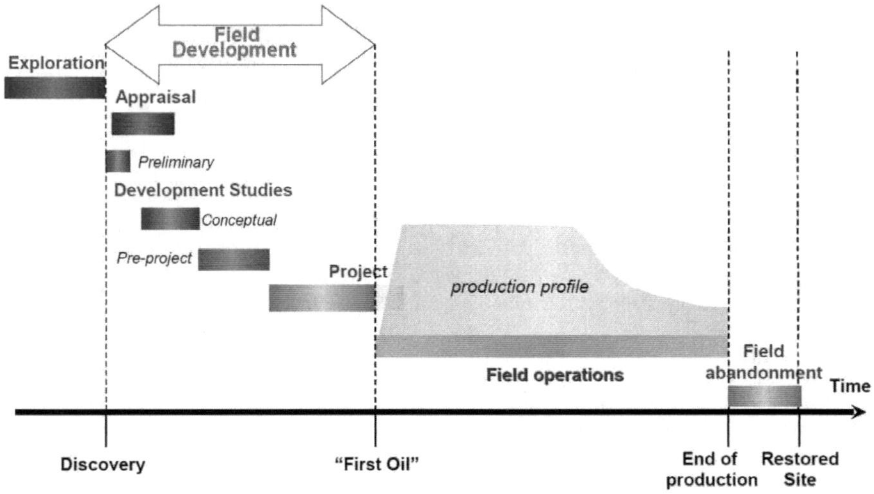

Figure 1 – Oil & Gas Field Development Phases

Level of investment for each phase and related economic impact on the overall project are key information that will be discussed in the following sections.

Exploration & Appraisal

As Oil & Gas are trapped within reservoir rocks in the ground, the first phase of a field development consists in locating likely hydrocarbon accumulation within a block. Once a prospect has been selected and acquired by the company, additional investigation will be carried out by performing additional seismic survey and appraisal well drilling.

Investments during the phases of exploration & appraisal are particularly important as installations & technologies used to obtain reliable data and images are relatively expensive (order of magnitude of hundreds million dollars for an offshore exploration). Prices of specialized vessels such as seismic vessels or drilling rigs are also a major cost contributor within the exploration & appraisal phase. In addition, when it comes to well testing, success rate is very low. With about 8 wells dry for 10 wells drilled, it is easily understandable that exploration investments are made with a relatively high risk profile.

Once a reservoir has been identified, uncertainties related to its characterization can also lead to huge variation in reserve estimation that could impact greatly project economics. Appraisal phase will focus on reducing considerably this uncertainty that will remain a major bottleneck for the project as development plan will be highly dependent on reservoir characterization.

Block acquisitions conditions (Exploration & Production licenses) are some other important economic constraints. In case all wells drilled are dry or not productive enough, block is abandoned and all previous investments are lost together with a part of the block acquisition fees. Similarly, as acquisition fees are generally paid for a limited duration, acquisition conditions could limit the return of investment in case production period exceeds the contractual period.

Once presence of oil & gas has been confirmed, it is then time to assess economic potential of the future development to ensure economic viability of the project.

Development Studies Phase & Front End Engineering Stage

During the development phase, some preliminary studies are performed based on data collected during the appraisal phase. Development studies will focus on selecting the best field development plan and related organization in order to match economic requirements.

While development studies represent generally about 5% of the overall CAPEX of a project, they represent the highest value for the project itself, meaning they have a very high impact on the profitability. During this early phase, it is possible to investigate alternate designs and architectures as well as innovative solutions at very low cost as it is only "paper work" at this stage. While the level of uncertainties related to the reservoir(s) and the field itself is decreasing, development studies are getting more and more precise to reach an acceptable level of details and allow operating companies to step in the Final Investment Decision with a solid development plan.

It is considered that potential of cost reduction at the very early stage of the development studies phase can reach about 40% between a fast track "standard" development plan and an optimized one. It becomes clear that the "copy / paste" solution from one field development plan to another is not the best solution as it will clearly reduce the innovative process. However, in a time constraint business, it is often the shorter way to an acceptable design that will rarely be the optimum one. A good recent example, still under development phase, is the Zinia Phase 2 project that is a tie back planned for the TOTAL FPSO Pazflor. The project has seen its CAPEX reduced by about 50% (30% just for the development phase) between first development plan proposition and the optimized one that has been proposed several years later. Main reasons for this cost reduction are split as follows:

- Optimization of the field architecture that have contributed to CAPEX reduction of about 20%
- Reduction in project specifications usually required by TOTAL led to a CAPEX reduction of about 10%
- Additionally, renegotiations of contracts with FEED, EPC, Drilling and OSV contractors as well as negotiation with States are responsible for another 20% of CAPEX reduction.

Contractual approach with the project supply chain / contractors is an important factor in the value generation process at FEED stage. When a FEED is made by a contractor under a Lump Sum contract, it is often a break to the optimization process. Alternate designs are not covered at their maximum, project definition is often limited to its minimum and largely inspired by some similar projects, deviations to specifications are considered as a loss of time specially if FEED contractor is unlikely to be the EPC contractor and finally, if the Lump Sum contract has been roughly negotiated, FEED contractor will provide with the minimum contractual work. On the other hands, if this contractual approach can limit the value of the project, it has the merit of being an efficient cost control solution as CAPEX will be delimited straight from the beginning.

Once the development plan including the likely final design has been validated by the client company, EPCs will be awarded with the various packages. EPCI phase will fix the best but also the worse of the solutions proposed during this pre-project phase.

Engineering, Procurement, Construction & Installation Phase

During EPCI phase, EPC contractors will have to endorse the work perform during the previous FEED phase. On this basis, they will carry out the detailed engineering design, procure all the equipment and materials and then construct and deliver the final facility on site. EPCI phase represents about 90% of the overall project CAPEX mainly due to equipment cost, raw material, labor cost and contractor costs & profits. During this last phase before the operating phase begins, adding value becomes almost impossible as every details have been fixed by contract during the FEED that has been endorsed by contractors. However, cost escalation is a common issue as any single mistake made during previous phases, and reported within the contract signed with the contractors, can induce high variation on the total CAPEX (delays, change orders, permits refused, quality issues, etc.).

As it is the case for the FEED stage, the contractual aspect of the EPCI phase is one very sensitive aspect when it comes to control the cost while keeping project schedule and quality on track. Indeed, several options can be proposed depending on operator motivation. One of the most common structures is the split of EPC work in packages, each of them being under a Lump Sum contract. It means that each discipline (i.e. Subsea Production System, Subsea Umbilical Riser & Flowlines, Topsides, Floating unit…etc) are handled by a separate contractor that is responsible for his own package under a fixed total price. This contractual structure is generally a non-negligible source of complexity as the difficulty is often directly linked to the number of project interfaces. More packages the project has, more complex the interface management will be. On the top of this, a Lump Sum contract will generally be fixed on a relatively low price basis that will limit considerably the EPC contractor capability to deviate from initial scope of work, even if it can lower the cost for everybody. As the contract by nature places the risk for schedule & budget on the EPC contractor, the EPC will then focus on delivering on time and within budget, and not always under the optimum conditions.

Finally, installation methods are highly dependent on the architecture selected. Weight, size, type of equipment and location are some of the key parameters that can influence largely the overall CAPEX. It is then important to think from the early stage of the development to the most cost effective way to install the various units /systems on the field and solving the tradeoff between the cost of the asset and the cost of its installation.

At the end of the EPCIC phase, the field is ready to produce its first oil / first gas and step in its production phase.

Operation phase including Enhanced Oil Recovery Phase

Operation phase will sanction the design selected by demonstrating its capability to produce safely and in an economic manner the reservoir fluids. While CAPEX teams have focused their work on developing an installation producing economically a field, OPEX teams will have to maintain best operating conditions by focusing on the integrity management as well as production maximization over the decades of production.

Roughly, OPEX spending per year can be considered as equal to about 7-8% of the overall CAPEX. It means that for a one billion dollar CAPEX project, about 80 Million dollars will be spent each year within the IMR tasks, labor costs, consumables, etc. In case export is made by pipeline, transportation costs can be very high and add another 7-8 % of the overall CAPEX per year related to pipeline tariff.

Another important source of NPV reduction is related to production downtime. While the production installation has been developed to produce a certain amount of gas or oil over 20 or 30 years, it becomes also of the utmost importance to ensure reliability and availability targets will be met. It is estimated that each percentage of loss of production for a large offshore oil development (200kb/d) is reducing NPV of up to USD 0.5 billion. Solutions to increase production availability are unfortunately very limited when the unit is already in operation as modification are costly and sometimes even impossible because of the lack of space onboard the installation. However, some projects have been planned with a likely extension in order to perform enhanced oil recovery for the latest operational phases. Additional processing system, subsea boosting systems or wells can be added to the initial development in order to maximize production of the field.

Field Abandonment

Strategy related to field abandonment needs to be thought and economically assessed during the CAPEX phase. Generally, strategy will be based on local requirements set up by the state itself or based on international standards. Reusing partially assets deployed on the field or dismantling the complete field to restore the offshore location will not have the same cost but generally they have the same environmental and human risks. As before 2013 it was estimated that UK spent related to decommissioning was of about USD 600 million, it has today nearly double. As environmental legislation is changing, investments related to field abandonment have also increased even if they still represent a marginal part an overall project expenditure.

The largest cost sensitivity is related to rig and Offshore Service Vessels (OSV) rates that are likely to impact massively overall field abandonment cost. Recent available data from UK North Sea allows a rough estimation related to facilities removal costs at $3000 per ton for topsides and $6000 per ton for subsea infrastructure.

3. Offshore projects economics key parameters

It has been the choice of the authors to perform project economics assessment based on the definition of the NPV. It is obvious that other "entry" points than NPV could have been chosen to perform such exercise but it seemed to the authors that starting the work based on the NPV would be the most time effective and straight forward approach. Indeed, NPV is a very convenient indicator that will take into account the tradeoff between CAPEX & OPEX and that will consider as well a large number of variables external to the technical components of the project itself (i.e. tax regime, inflation, oil price, etc.).

Thus a good way to illustrate in an intuitive manner the Net Present Value is by having a look to the cost structure hidden behind breakeven price of a barrel of oil (or oil equivalent) depending on its production location. As it is shown on Figure 2, minimum viable production cost of a barrel (i.e. breakeven price) is not the same everywhere in the world and its internal cost structure can vary considerably depending on the ease to extract the product, the operability costs as well as the taxation regime.

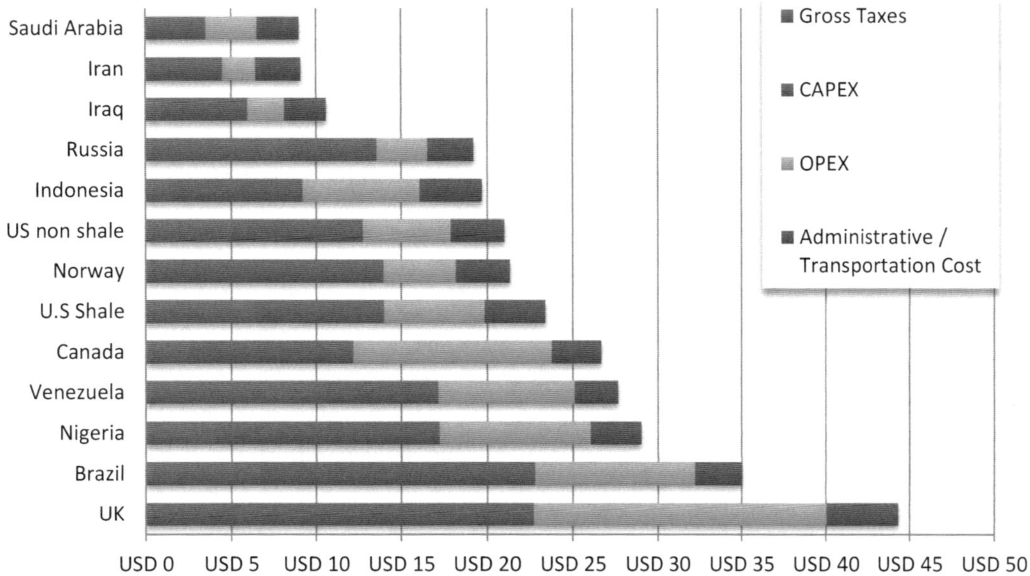

Figure 2 - Cost of producing a barrel of oil or gas equivalent in March 2016 *(source: Rystad Energy Ucube)*

The NPV can then be seen as the comparison between the barrel cost structure reached for a project and the breakeven price cost structure. In the case cost of producing a barrel is inferior to the breakeven price as presented on Figure 2, it basically means that the "empty" part to reach the breakeven price can be filled with an additional value representing profit of the company. This profit generated is then the NPV of the project. It is also important to understand that the different components of the barrel of oil cost structure are dependent on each other. Modifying CAPEX can impact directly OPEX as well as taxation regime for instance.

Thus, NPV calculation will depend on the four following key parameters:

- CAPEX;
- OPEX;
- Operating Revenues; and
- Tax Regime.

The calculation formula for NPV is then as follows:

$$\text{NPV} = \sum_{n=0}^{n} \frac{CF_n}{(1+i)^n} \qquad (1)$$

- **CF_n**: Constant Money Cash Flows that is the difference between the Operating Revenues and CAPEX, OPEX and Taxes
- **n**: number of years
- **i**: Discount Rate

In simple words, if NPV > 0, it can be considered that the revenues of the project will be sufficient to pay all types of expenses, repay the capital invested and create a profit equal to this NPV value (generally calculated in Million or Billion Dollars)

The following will now investigate the four main components of the NPV calculation to identify what are the main underlying mechanisms capable to impact project value.

Capital Expenditure

CAPEX takes into account all costs to develop the field including block acquisition, data acquisition, appraisal wells, engineering, construction and installation of the field facilities and generally, all the costs related to their abandonment as well. It also includes all the non-material costs such as the ones related to the development studies, project management, insurance, etc.

Accuracy of CAPEX estimates increases with the progression of the project. At Concept stage, CAPEX estimation can vary from -25% to +40% compare to the final one. At pre-project phase, it can vary from -15% to +25% and finally during project execution (i.e. EPCI phase), it is estimated at about -10% to +15%. Uncertainty on CAPEX estimation is reduced as additional data are collected on the field, number of field architectures is limited, layout is fixed, likely suppliers are identified and financial assumptions such as contract types and taxation regime are validated.

Key CAPEX contributors impacting positively or negatively the NPV are summarized in Table 1. "*CAPEX contribution*" represents the economic weight of the contributor within the CAPEX. Each parameter identified includes a qualitative estimation of its sensitivity that can be defined as its capability to increase or decrease rapidly its relative contribution to the profit of the project in case one of its sub-parameters is modified.

CAPEX CONTRIBUTORS TO NPV VARIATION	CAPEX CONTRIBUTION	IMPACT ON NPV	SENSITIVITY
Exploration & Appraisal Phase (incl. Vessel & Labor)	*MEDIUM*	*HIGH*	*HIGH*
Development Phases	*LOW*	*HIGH*	*HIGH*
EPCI Phase (incl. Raw Material, Equipment, Labor...)	*HIGH*	*MEDIUM*	*LOW*
Decommissioning	*LOW*	*LOW*	*LOW*
Contractors Profits	*LOW*	*LOW*	*LOW*
Taxes & Insurances	*MEDIUM*	*LOW*	*LOW*
Quality	*MEDIUM*	*HIGH*	*LOW*
Currency	*MEDIUM*	*MEDIUM*	*HIGH*

Table 1 – CAPEX Contributors to NPV Variation and related qualitative impact

Operating Expenditure

OPEX takes into account all costs to maintain field production including crews, spare parts, consumable, and fuel for energy production. It also includes all the non-material costs such as the ones related to the taxes, insurance, Inspection, Maintenance & Repair contracts and product transportation fees.

Key OPEX contributors impacting positively or negatively the NPV are summarized in Table 2. "*OPEX contribution*" represents the economic weight of the contributor within the OPEX. Each parameter identified includes a qualitative estimation of its sensitivity that can be defined as its capability to increase or decrease rapidly its relative contribution to the profit of the project in case one of its sub-parameters is modified.

OPEX CONTRIBUTORS TO NPV VARIATION	OPEX CONTRIBUTION	IMPACT ON NPV	SENSITIVITY
Labor / Crews to operate the field	*MEDIUM*	*MEDIUM*	*MEDIUM*
Material, Spare Equipment & Consumables	*HIGH*	*HIGH*	*MEDIUM*
IMR cost (incl. contracts, vessels, labor)	*HIGH*	*HIGH*	*HIGH*
Logistics / Servicing of the field	*LOW*	*LOW*	*HIGH*
Product transportation (ex: export pipeline tariff)	*HIGH*	*HIGH*	*MEDIUM*
Contractors Profits	*MEDIUM*	*MEDIUM*	*LOW*
Taxes & Insurances	*MEDIUM*	*HIGH*	*HIGH*
Quality	*LOW*	*HIGH*	*LOW*
Currency	*LOW*	*LOW*	*HIGH*

Table 2 – OPEX Contributors to NPV Variation and related qualitative impact

Operating Revenues

When crude-oil prices were more than $100 a barrel in 2014, the ensuing profits were huge, filling government taxes and maximizing company earnings. Now prices barely cover the average cost to get the oil produced in places like the U.K. Additional expenses, like taxes on profits, mean that the actual breakeven price for many projects is higher. New and / or more complex projects generally fall above the average cash cost of production if they are not carefully optimized. Thus, price of oil is the major driver when it comes to operating revenue.

As it can be seen on Figure 3, industry provided oil and gas at a relatively stable and low price for about 100 years (1880 to 1970). The various oil crises that happened successively in 1973 and 1979 as well as economic changes in 2008 and 2014 made oil as well as gas prices very volatile, bringing additional uncertainty in economic assessment and profitability of an Oil & Gas development. In other words, one of the key input data required for any field development economic assessment is a very unpredictable variable.

However, oil price is not the only parameter impacting operating revenue. Capability to produce at the planned rate, within specified quality range and during planned duration are also key concerns for project profitability.

For instance, volume of hydrocarbon in place and related recoverable reserves have been defined from first stages of development and evolve during the whole life of the field. In case these estimations are wrong, what is very likely, impact on revenue and project profitability can be huge. For instance, if plateau phase cannot be reached, quantity produced will be lower than what was expected and finally economic target will never be achieved.

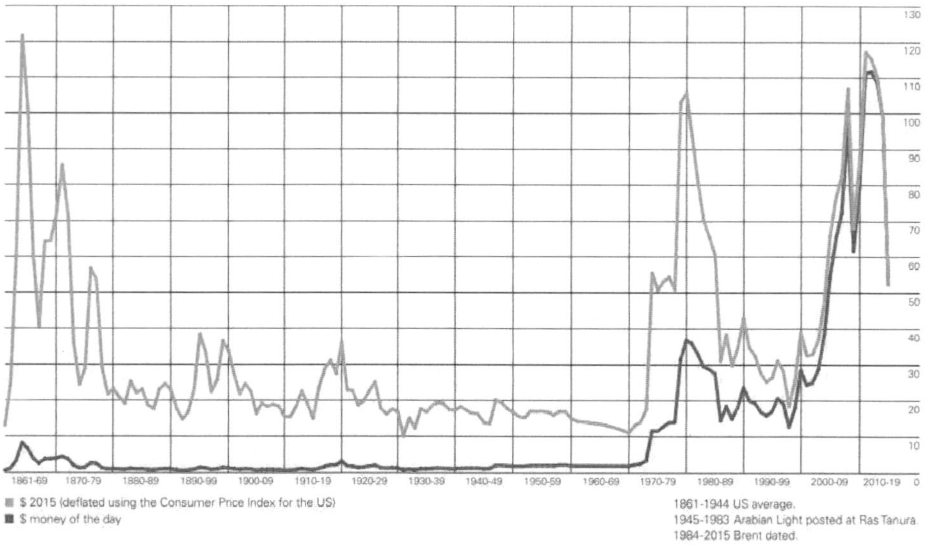

Figure 3 - Historical Price of Oil from BP Statistical Review 2016

Production profile is another key input that can play an important role within the Return on Investment process. Taxation regime imposed by States to an International Oil Company (IOC) is generally softer during first years of production. Similarly, CAPEX fiscal depreciation will be generally carried out during first years of production reducing considerably losses related to taxes. Maximizing production straight from the beginning to reach the plateau phase as early as possible becomes often the best strategy. However, for National Oil Company, a leaner supply of product will generally be more cost effective as they will not be exposed to the same tax regime and will generally use a part of its production for local consumption. Production profile will then define revenue profile.

Whereas installations are designed for a certain period of time, unplanned failures are always likely to happen. Production availability is then a vital input to successfully pass the Final Investment Decision as it will allow ensuring the overall unit can produce economically the field, considering the planned maintenance strategy but also assuring all the unplanned failures that can be statistically determined as well as related maintenance & operating strategy will not be a bottleneck for the project profitability.

Taxation Regime

When an operating company decides to invest in a field and starts a development, they need first to establish a formal contract with the State owning the reserves. Contracts are broadly split in 3 categories framing from the beginning the Taxation Regime to be applied to the International Oil Company:

- Concession, meaning the full delegation of mining rights in exchange of royalties & taxes;
- Production Sharing Contract (PSC), meaning delegation of operations in exchange of access to oil; and
- Service Contract, meaning delegation of operations in exchange of direct payment.

Depending on the type of contract, taxes will be dependent on the quantity of oil produced, quantity of oil directly send to owning State, shares within the field, percentage of local content, etc. Tax Regime is definitely a major contributor to project profitability. Even if Classification Societies have only a very marginal role to play with the agreement between states and companies, taxes impact on the NPV has been assessed in the Table 1 and Table 2.

4. Classification Societies key involvement to overall cost reduction strategy

As Classification Societies are often consulted only for matters related to safety, and more precisely, within their limited role of classification, certification or testing companies, this section will finally propose a different vision of CS role and involvement. The following will consider again the main NPV components (CAPEX, OPEX, Revenues & Taxes) and will define a more adapted and beneficial role of CS to add value to the cost reduction strategies.

Reducing Capital Expenditure and Maximizing NPV by involving CS at early phases

Development Phase is one of the key phases to maximize NPV and this is also the key phase a CS should be involved with. As it has been shown with the example of the Zinia 2 project and within Table 1, improving development phase can reduce considerably CAPEX, but will also ensure the field development plan will optimize production phase (i.e. revenue). CS have a large operational experience of the oil & gas installations and have experienced safety, reliability, integrity and other operational issues related to a large number of field developments. Even if all the field development plans are site specific, similarities can always be found and CS can influence strongly the selection of the architecture based on their experience. The earliest a CS will be involved, the easiest it will be easy to implement return of experience and good practices and finally fix the basis to guarantee overall project assurance.

Main role of CS consists in ensuring development plan will be in line with project QHSE targets. CS are able to provide operators with the most suitable applicable standards, reduce the conservatism of some costly and not always justified specifications and validate international regulation application with respect to local authorities requirements. CS can also go deeper in the technical assessment by advising on the best layout in term of safety, that is one of the classic role of CS, but also in term of production availability, reliability and integrity performance. CS have often collected a large number of performance related data and they have then a strong capacity to perform or review independently performance and integrity assessments. Risks related to a project are also very complicated to assess properly as it should cover a large number of threats and opportunities, and it often necessitates an extensive practical and operational experience from the design teams / field architects. One more time, CS can pretend to be one of the largest sources of experience transverse to all Oil & Gas supply chain, especially when it comes to safety and environmental risk. CS are able to assess risks within the full supply chain, whatever if it is technological, technical or organizational risks.

Operating companies use their return of experience to upgrade their internal specifications that can be considered as a good knowledge capitalization practice. However, Return of Experience integration often leads to an "inflation" of the volume of documentation to be produced by the supply chain as well as additional constraints for field developers. Surprisingly, return of experience becomes then, not a source of optimization, but a source of loss of time and source of economic show stoppers. As specifications and integration of return of experience are vital for an industry, their good application requires a lot experience as well as a critical opinion. Obviously, one of the core roles of CS, is to develop rules and standards by incorporating latest return of experience. They are then ideally positioned to provide any project with an adapted level of specification to be applied and then reduce considerably over-work and associated cost. In addition to the cost reduction induced by the simplification of the documentations to be produced, it will allow the full project supply chain to step in the design challenge with more flexibility, and sometime, to come back to the essential that is the ability to design a system maximizing profits and reducing risks as at an acceptable level.

A pretty recent role of CS is the one they are currently playing within the development of new technologies. As they were only involved as a simple "stamp" provider few years ago, they have now the will to accompany developers of new technologies all along the product development, from specification phase, through qualification test plan definition and until witnessing of technology readiness proven during final tests. By adapting their rules but also by accepting more and more alternate schemes and solutions, CS can support and can give more credits to the innovation process. Indeed, insurances and financial world does not like risks. Thus, in the case a CS validates or at least take part of an innovative process, it gives often more weight for companies to negotiate with banks and other financial institution the likely use of the novelty. Companies can win on two aspects; reducing financial exposure by having involved a classification society in the development of the new technology, and improving NPV by reducing CAPEX or OPEX or by improving revenues through the use of this new technology.

"Standardization" has come as a fancy word for the last 2 years as the economic crisis within the Oil & Gas Industry was allegedly due to the fact that each project was made so unique that the lack of standardization would have made the cost exploded. However, tailor made projects have nothing to do with such massive cost increase as each project is made on an in-depth economic review and no operating company would have sanction a project if they deem it not economically viable. Actually, it is now commonly admit that standardization is more related to the concept of being able to replicate all the steps and elements that make a project a success while removing efforts that are not contributing to maximize the NPV. Some good initiatives started with Joint Industry Projects led by CS that is one of the only most suitable ways to get a consensus within the industry. Standardization will then begin by fixing common understanding of the different concepts within the whole supply chain. All the requirements coming from operators down to the supplier needs are to be structured in a unified format in order to reduce quality issues but also in order to reduce "grey" areas related to a lack of understanding of the suppliers that is often due to a poor expression of the needs from the operator. Once communication channel will be standardized, it would be the time to step in the standardization of the interfaces that is one of the key challenges of any industry. Aeronautics, railway or automotive industries have defined the engineering of the interfaces within the System Engineering (SE). SE covers all the technical related interfaces as well as all the human and organizational risks hidden behind interfaces. Some international companies and CS have teamed up with aeronautic or automotive specialists to be able to deploy properly SE within their own industry. Most of the operators and EPC contractors focus their future development based on this approach and have required in some extend the assistance of CS to support them in this relatively new exercise. Standardization of equipment or systems, according to paper authors, will not necessary lead to a massive cost reduction as scale effect in Oil & Gas industry is not strong enough to be a leverage and tailor made product will often be one of the most effective solution when it comes to NPV maximization.

If CS have been involved early enough, EPCI phase should then focus on overall project quality & performance control. Overall assurance plan established previously for all the future phases and for the whole project supply chain needs to be careful followed up to ensure the engineering phase has been performed with respect of all the key indicators and assumptions defined during FEED phase, equipment & material provided have been manufacturer according the quality, reliability and availability targets, and that construction & installation are performed on time within the best safety conditions. It is then classic role of CS to perform an in-depth review of the designs including construction, installation and test / commissioning plan. However, the real added value of CS in this EPC phase would be able to perform an integrated QA/QC work bringing together all project parties teams in one unique team. As some of the highest risks of non-conformities and source of delays are often hidden within the interfaces of the various inspections / surveys scope of work, it would be of the utmost importance to align all the QA/QC objectives of all the parties in order to improve interface coverage as well as reducing cost by avoiding extra layers of inspections on the same items. CS would be able to play the role of overall coordinator and take a larger share in project risks to maximize chance of project success.

Reducing Operating Expenditure and Maximizing revenues with the long term assistance of CS

CS have been used for decades as a living source of knowledge for the marine & offshore industry. However, sharing operational knowledge takes time and is rarely effective if it is not directly apply on a project. Thus, since the beginning of the crisis in 2014 several Classification Societies have massively invested in the development of new Asset Integrity Management tools capable to digest and restitute all the valuable information than can be extracted of a given installation. Most of the existing concepts are based on 3D models, called digital twins, and use data collected from inspections or sensors installed on physical asset to represent their near real-time status, working condition or position. Investments in this direction are driven by the capabilities for a CS to have access to a large number of installations by their status of classification body, from which they could eventually extract crucial information and trends to be organized by locations, type of equipment, type of assets, type of maintenance schemes, type of operational conditions, etc. This digital solution, that is not only the future of the Oil & Gas industry but the future of all type of industries, is generally combined with a large panel of calculation and optimization tools that can, in near real time, assess different scenarios and become then an unavoidable decision making tool for any field in operation. Reducing response time, performing intelligent data collection to predict failures, assessing a large number of likely scenarios , being able to have access to an infinite number of data everywhere in the world and finally being in capacity of solving problems before they appear is one of the key challenges the CS are now focusing on with the whole industry. As stated previously, digital solutions are the matter of all the industries, and

CS having the largest footprint within the industrial world could be the most successful in this digital rush.

Another workable area is Oil & Gas industry capability to perform "cross-fertilization", i.e. to adapt solutions or tools developed and used in other industries and to apply it efficiently in the Oil & Gas industry. Military or railway sectors have for instance excellent capabilities to perform condition based maintenance, to manage their spare parts in a cost effective manner and optimize their stock to maximize production availability while not exploding their OPEX. Many CS are now focusing on this proven tools and solutions issued from other industries that could really benefits easily to Oil & Gas industries. CS have understood that it is the opportunity for them to demonstrate their ability to solve operational problems with new solutions that have been proven in other industries.

As each downtime can lead to hundred million dollars of loss of revenue, it is obvious that main operational concern is not only limited to operating spending only, but also to the maximization of production availability. Production availability is dependent on two major parameters that are the failure rates of installations components and time to restore failed equipment. Reducing failure rate means basically developing more reliable products but direct consequence would be a massive increase in cost. Keeping same failure rate and making product easy and quick to repair could be a convenient solution but equipment complexity and spare part strategy limits drastically repair time optimization. Thus, even if CS have developed their own Production Availability (or RAM) assessment tools aiming at solving the tradeoff between CAPEX & OPEX and identifying sources of non-productivity, data related to random failures rate and repair time are often of poor quality or based on a very limited number of items making statistics presented in some literature not applicable. One of the most suitable solutions that could benefit the whole industry would be to involve CS in systematic data collection campaign in order to set up a powerful database that could be used by the whole industry.

Finally, one of the paramount of any industry would be related to the capability to capitalize the knowledge within database and trends encapsulated in software. It is already partially the role of CS that often use their key experience in naval architecture to develop their own calculation tools that are generally used by the widely in the industry. Reorienting R&D capabilities under the lead of the industry to focus on the most critical aspect of the cost structure of Oil & Gas project could be an area of common work between CS and the industry supply chain.

RESULTS

By involving Classification Societies at the earliest phases of a project, the oil & gas industry could benefit of the large experience accumulated by the CS that goes far beyond their classic role. Novelty of the approach discuss in this paper focus on the strong capability for CS to provide the whole project with a QHSE and performance assurance plan covering all actors and all phases of development based on their extensive experience. At early phases, CS can share return of experience with the various project parties, provide independent technical, risk, QHSE and performance assessment of the project, encourage utilization of new technologies and be an easy facilitator between the operating companies, its contractors as well as the various State authorities that will be involved in the future project. During operation, CS have already a major role to play by performing regular inspection on units they class and provide operators with strong support related to many types of failures or degradation mechanisms they can face with. To go beyond this classic role during operation, many CS have developed strong Asset Integrity Management systems, based on digital twins, and capable to collect a wide range of information to maximize operation availability, reduce operational risks and size properly OPEX at the earliest phase of development based on a solid database fed with return of experience of the whole supply chain.

With no direct interest in the projects but with a major responsibility role within the industry, CS can be largely consulted without adding any additional competition within the supply chain. By structuring its knowledge under the impulsion of the whole industry, CS would be able to participate to the current re-shape of the Oil & Gas business by providing key solutions aiming at improving NPV while keeping development on the safe side.

CONCLUSIONS

Oil & Gas industry cost structure is a living entity evolving day by day that depends on market conditions and global geopolitical stability. While Drillings Rigs and OSV markets were some of the major sources of cost in 2013, they have now dropped so low than contractors cost & margins are not anymore a likely source of cost optimization. It is now largely admitted that sustainable sources of cost reduction will be conditioned by in-depth culture changes. Segmentation of supply chain roles needs to be modified in a more collaborative way of working. Paper work developed on digitized supports needs to be enhanced and step in a digital workspace. New solutions can come from our own industry but getting inspiration from others can bring considerable added value. Classification Societies have been created about two centuries ago to respond to an insurance need, their role has evolved to cover all safety related issues and they have finally adapted their role again to provide valuable return of experience in order to optimize project overall QHSE but also to maximize return of investment of the whole supply chain. New role of CS is now reshaped because of the recent crisis and new services and approaches encouraged by CS could be a new axe of work to improve cost efficiency within the overall industry.

REFERENCES

International Association of Classification Societies (IACS), "*Classification Societies – What, Why and How?*"

Wall Street Journal, *"Barrel Breakdown"*, April 2016

Petroleum, petrochemical and natural gas industries *"Collection and exchange of reliability and maintenance data for equipment"*, ISO 14224:2006

SINTEF Industrial Management *"Offshore Reliability Data Handbook (OREDA)"*, 5th Ed, 2009

http://www.businessdictionary.com, December 2016

http://www.investopedia.com, December 2016

Decommissioning Insight 2014, Oil & Gas UK

Decom North Sea, *"The real Costs of Decommissioning"*, http://decomnorthsea.com/news/the-real-costs-of-decommissioning, July 2015

SNAME
THE INTERNATIONAL COMMUNITY
FOR MARITIME AND OCEAN PROFESSIONALS

CONCEPTUAL STUDY OF A PAIRED-COLUMN SEMI-SUBMERSIBLE PLATFORM FOR A 1.5 MTPA FLNG

JUN ZOU

DIRECTOR OF NAVAL ARCHITEURE
HOUSTON OFFSHORE ENGINEERING, LLC
AN ATKINS COMPANY

Proceedings of the 22nd Offshore Symposium, February 2017, Houston, Texas
Texas Section of the Society of Naval Architects and Marine Engineers

ABSTRACT

The hull forms of the sanctioned FLNG projects are all ship-shape with either internal or external turret systems. The cylindrical hull for FLNG application without turret system has been studied for years. Recently, Sansom and Ankarsward (2016) presented a six-column rectangular shape Semi for medium FLNG. In this paper, a conceptual study of an alternative, a Paired-Column Semi-Submersible (PC Semi) platform with spread moorings, is presented for a 1.5 MTPA FLNG application in Western Australia.

This study is a generic one to design an economical small FLNG which can be achieved through modular topsides layout, robust hull configuration and turret-less mooring arrangement without scarifying safety and operability. The aim of this study was to explore an alternative option within current industry capability to reduce the overall investment, complexity and risk as much as possible.

A PC semi has low motion characteristics with eight columns suitable for large storage requirements and can efficiently support large deck area for production facility with gas processing, treatment and liquefaction facilities, offloading facilities and Living Quarter.

This conceptual study presents a generic exercise of efficient topsides and hull configuration and layout with considerations of its constructability, transportability, installability and operability. Advantages of a PC Semi vs. a ship-shaped hull are addressed and highlighted. Standardized and phased approach with multiple small FLNG units vs. a single mega FLNG unit is also discussed.

Keywords: Conceptual Study, PC Semi, FLNG, Alternative Hull, Constructability, Transportability, Installability, Operability, Standardized and Phased Approach

INTRODUCTION

Even oil and gas prices are still low and may last for years, FLNG (floating liquefied natural gas) market is predicted to grow steadily in Asia, Pacific region, Latin America and Africa since there is a long-term predicted growth in LNG demand. FLNG is complementary to onshore LNG with advantage close to gas resource and eliminating long distance piping which is able to monetize a larger amount of gas resources in a more cost effective way. In addition, LNG business is known as a long term business with fixed contracts for a significant volume.

Production capability is a key driver for a FLNG to be cost competitive. The small size, midsize and large size

Proceedings of the 22nd Offshore Symposium, February 2017, Houston, Texas
Texas Section of the Society of Naval Architects and Marine Engineers

capabilities are normally defined as 0.5~1.5 MTPA (million tonnes per annum); 2,0~3.0 MTPA; 3.5~6.0 MTPA respectively. Pek and Velde (2013) investigated economics of a midsize FLNG and a large size FLNG by the method called normalized UTC (unit technical cost). They found a midsize FLNG has about 30% higher UTC than that of a large size FLNG and a small FLNG did not look attractive from the normalized UTC perspective. However, recent industry trends and feedbacks from sanctioned projects led us to revisit this conclusion since more essential factors, such as reservoir risk, financial risk and delivering risk, shall be taken into account for field development evaluation.

Shell Prelude with equivalent 5.3 MTPA capability was sanctioned in May 2011 and its budget was estimated around $US12.6 billion at that time. However, it is expected to significantly increase, even double the original budget to reach production. It was reported per news of the West Australia that its sail away date was originally planned at April, 2016 might be delayed to July to October, 2017. On the contrast, for a small process capability, such as Petronas PFLNG-1 with equivalent 1.5 MTPA capability has a shorter construction cycle. PFLNG-1was the second sanctioned FLNG in June 2012. On May, 2016, it finished construction and headed to installation site. It took about 4 years for PFLNG-1 while Prelude might take as long as more than 6 years.

Large CAPEX and long deliver period of large FLNG is leading to difficulties in making investment decisions for future projects. However, a standardized and phased approach with two or more small units (e.g. 1.5 MTPA) in a fast-track and staged fashion to allow the execution, scope and CAPEX requirements to be more manageable. Early production and early cash flow income yields heathy business. Thus, small CAPEX with short deliver period for small capability FLNG (e.g. 1.5 MTPA) might be preferred option for more future projects.

As pointed by Terashima et al (2011), "some motion sensitive process facilities, such as cryogenic heat exchangers and acid gas absorbers, whose performances degrades with the given motion criteria, and the degree of degradation varies with concurrent motion period." Among 6-DOF motions and accelerations, they are more sensitive to roll and pitch motions and accelerations. A typical complimented plan of motion/acceleration degradation is to add additional margins to these facilities which results in significant cost increase. Based Australia regulation, the platform or vessel shall be designed for "manned" even in 10,000-yrp events. Survivability and people comfort levels are extreme challenges.

The hull forms of the sanctioned FLNG projects are all ship-shape with either internal or external turret systems. In addition, thruster system is employed for dynamic positions and safe tanker berthing. The cylindrical hull for FLNG application without turret system has been studied for years (Odeskaug, 2015). Recently, Sansom and Ankarsward (2016) presented a six-column rectangular shape semi-submersible for medium capability FLNG. Since PC Semi has superior low motion characteristics, a conceptual study of an alternative hull, a PC Semi platform with spread moorings, was employed.

In current low commodity market, oil and gas prices will stay in new norm $35 ~ $55 per barrel for many years which calls for more economical solution with innovative low cost hull form and efficient field development strategy. There are extremely strong pressures on offshore industry to improve field economics and keep floating liquefied natural gas project afloat. To achieve major capital cost reductions, it calls for entire industry collaborations to accumulate savings on drilling, topsides (equipment and facilities), hull, moorings, risers, foundations and construction, transportation and installation. This paper presents one example of PC Semi for small FLNG application in Western Australia through innovation hull arrangement to achieve savings on deck, hull, mooring and foundation.

The outline of the paper is as follows. First, the PC Semi development chronicle is reviewed. Second, descriptions of sizing basis are highlighted. Third, the topsides layouts and hull configuration are presented. Fourth, the study cases including a generic ship-shape FLNG are given; Fifth, numerical results of characteristic responses in operating sea states are summarized. Sixth, constructability, transportability, installability and operability have been addressed. Finally, discussions are given and main conclusions are drawn.

PC SEMI DEVELOPMENT CHRONICLE

PC-Semi concept was first introduced by Zou (2008). This hull form was selected by RPSEA for dry tree development in ultra-deep water in 2009 and study results have been summarized in study report, Ref [8]. Extensive wind tunnel tests and wave basin model tests funded by RPSEA were completed in 2009 and 2010 respectively. A series of additional developments and examinations have been carried out, such as VIM response characteristics (Zou et al, 2013 and Zou et al, 2014) and characteristic responses of PC Semi hull features including extreme shear forces and pry/squeeze loads (Das and Zou, 2015).

A brief summary of PC semi technology development and reviews of PC Semi technical readiness with government authorities, class societies and offshore industries is presented as follows:

- In September 2009, technology review meeting with BOEM (then MMS) and USCG was held;
- In March 2010, Technology Readiness Review Workshop by RPSEA with industries including major oil companies and engineering firms was held;
- In March 2011, DNV GL (then DNV) Technology Assessment and two-day Workshop with industries including major oil companies and engineering firms was held.
- In 2014, DNV GL Technology Qualification and two-day Workshop with industries including major oil companies and engineering firms was held.
- In 2014, Approval in Principle (AiP) of this development was awarded by DNV GL

PC Semi technology development for GoM has been matured from "potential idea" to "technically feasible" to "major capital project ready" after more than 8-year persistent efforts.

PC Semi developments and applications have been extended from GoM to Offshore Western Australia (Zou and Chianis, 2011 and Zou, 2012) and Offshore Brazil (Zou, 2012).

SIZING BASIS

Major basis of design data, such as process design (liquefaction process), LNG and condensate storage requirements, topsides modules weights, electrical power requirement were similar to those presented in Jeanneau et al (2013) with additional margins for this generic study. It is not main goal to achieve an optimized design but to have a generically standardized design with robustness built-in.

Process Basis of Design

- 1.2 MTPA of LNG
- No LPG's production
- 0.3 MTPA of Condensate

Total equivalent production capability is 1.5 MTPA for this study.

Hull LNG and Condensate Storage

For this generic study, LNG and Condensate storage capabilities were selected and summarized in Table 1.

Table 1 LNG and Condensate Storage

Items	Current Study (m^3)	Jeanneau et al (2013) (m^3)
LNG	92,500	75,000
Condensate	25,000	NA

Topsides Module Weights and Required Deck Area

For this generic study, topsides module weights and required deck area were summarized in Table 2.

*** Formatting Issue - Best Available Paper/Graphic ***

Table 2 Topsides Module Dry Weights and Required Deck Area

Items	Current Study	Jeanneau et al (2013)
Topsides module dry weights (mt)	31,500	28,000
Required deck area (m^2)	15,000	14,800

The required deck area in Jeannean et al (2013) was estimated per author by considering total area 300m (length) x 61 m (width) subtracting turret area.

In this generic study, almost all topsides modules are located on the upper deck (13,500 m^2) and only very minor facilities are located on lower deck (1,500 m^2). There is plenty of space on lower deck and can be easily expanded. More details are described and displayed in **Topsides Layout**. In addition, 2,000mt SCRs and Umbilicals vertical loads was considered for sizing.

Western Australia Metocean Criteria

In this study, it is our intention to select the most severe metocean criteria for sizing as presented in Zou and Chianis (2011) and Zou (2012). A generic extreme Hs-Tp contour plot of 10,000-yrp, 1,000-yrp and 100-yrp wave is displayed in Figure1. In order for easy discussion, the table and plots are reproduced in Table 3 and Figures 2 and 3 respectively.

Table 3 Generic Western Australia Extreme Sea States

Parameters	Units	10,000-yr RP	1,000-yr RP	100-yr RP
Significant wave height	m	22.5	20.5	16.8
Wave spectral peak period	sec	18.1	17.1	15.3
Gamma		2.5	2.5	2.5

For a floater design, its dynamic response is not only a function of significant wave height Hs but also is a function of wave peak period Tp. To seek the governing dynamic response, it is essential to run many Hs-Tp pairs along contour to ensure all critical cases having been captured.

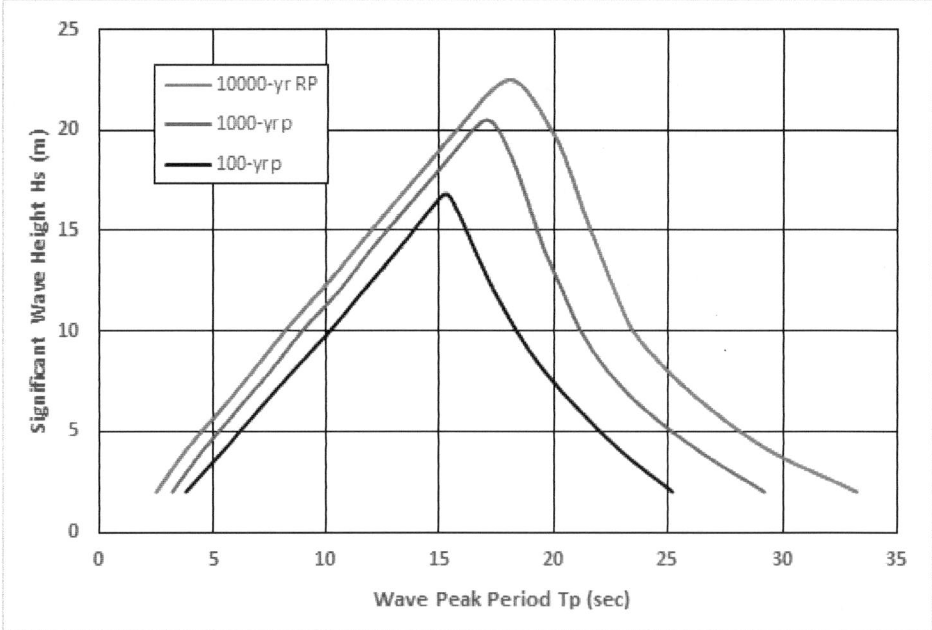

***** Formatting Issue - Best Available Paper/Graphic *****

Figure 1 Generic Hs-Tp Contour Plot of 10,000-yrp, 1,000-yrp and 100-yrp in WA

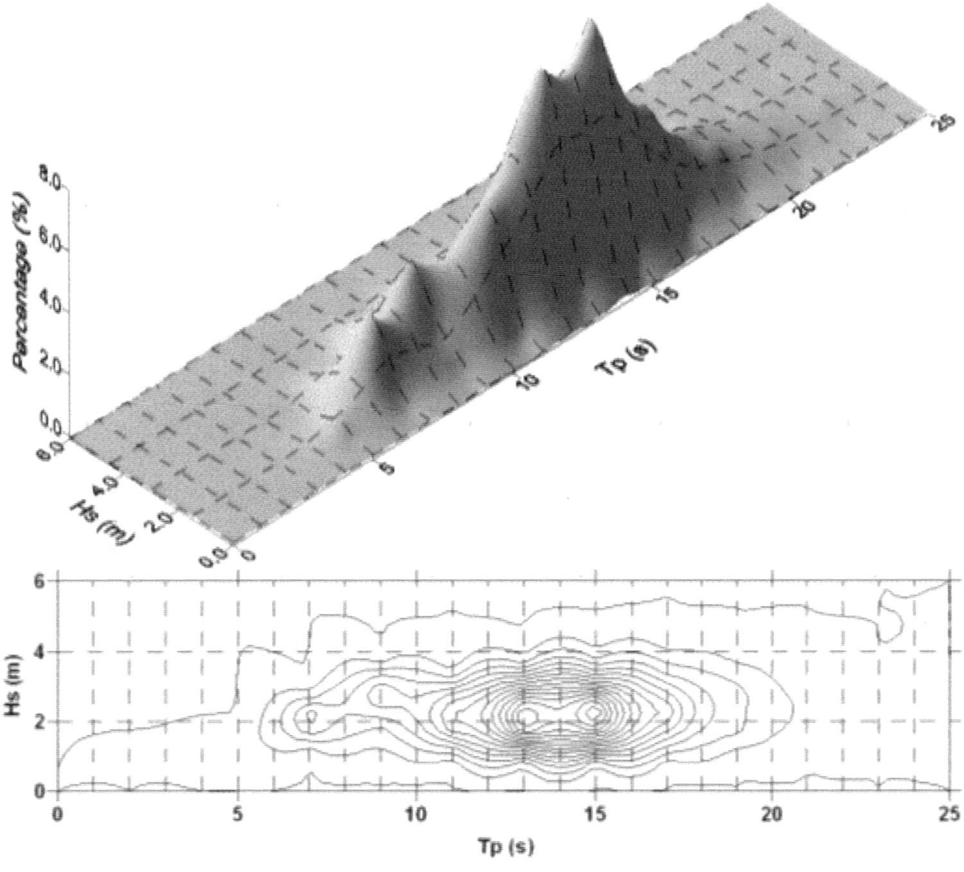

Figure 2 Generic Western Australia Wave Scatter Diagram

*** Formatting Issue - Best Available Paper/Graphic ***

Proceedings of the 22nd Offshore Symposium, February 2017, Houston, Texas
Texas Section of the Society of Naval Architects and Marine Engineers

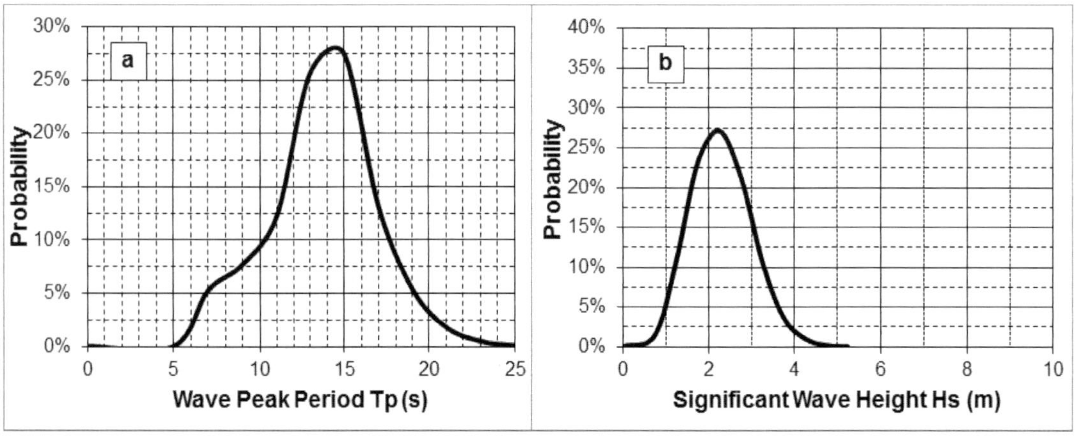

Figure 3 Generic Western Australia Wave Scatter Diagram as a Function of Tp (a) and Hs (b)

For the generic Western Australia wave scatter diagram in Figure 1, there exists a large percentage of waves in the approximate range of 1 to 4 m Hs and 8 to 17 sec Tp. There are almost no waves with Tp less than 5 sec. The percentages of occurrence of Western Australia wave scatter diagram presented in Figure 3 as a function of both wave peak period Tp (left) and significant wave height Hs (right).

The characteristics of the Western Australia (WA) wave scatter diagrams are displayed in Figures 2 and 3 respectively. The key aspects which affect hull configuration are as follows:

- The dominant wave peak period occurrence is approximately 14 sec. There are near zero occurrences of wave peak periods below 5 sec and above 25 sec. It should be pointed out for the sanctioned ship-shape FLNG in WA,
    - Heave and pitch natural periods are around 12-14 sec,
    - Roll natural period is about 20 sec
- The dominant occurrence of significant wave height is approximately 2.3 m. The occurrence of significant wave height drops dramatically to approximately 4 m and essentially goes to nearly zero beyond 5 m.

WA extreme waves given in Table 3 have been considered in hull and mooring sizing. WA characteristic waves as illustrated in Figures 2 and 3 on topsides and hull structure fatigue as well as operability of FLNG and offloading operation shall be carefully investigated. A more in-depth discussion follows in a later section.

TOPSIDES LAYOUT

Topsides Arrangement

The general philosophy of topsides arrangement is to ensure process and operation safety and efficiency and operation friendly and easy maintenance in service phase and easy construction (modular size and weight) and commissioning in pre-service stage.

Topsides Modular Design

Topsides modular design is a popular choice for current FLNG projects, Pek and Velde (2013), Jeanneau et al (2013) and Talib and Price (2011). This approach is adopted for this study.

Product Offloading Systems

There are two typical offloading systems, one is side-by-side offloading and the other is tandem (stern to bow) offloading. The selection of offloading systems is depended on uptime requirement and relative motions between a LNG carrier (LNGC) and FLNG offloading arms. To achieve a sufficient uptime, the side-by-side offloading option requires relatively benign environmental condition while the tandem offloading system is suitable for relatively rough metocean condition. The details on offloading systems and technical readiness are given in Le Devehat (2011), Le Devehat et al (2012) and Duhamel (2016).

Plan views of main deck and cellar deck are illustrated in Figures 4 and 5 respectively. Elevation views of looking West/East and North/South are displayed in Figures 6 and 7 respectively. It has been seen that there are very minor facilities/utilities located on cellar deck and can be easily expanded in future.

Proceedings of the 22nd Offshore Symposium, February 2017, Houston, Texas
Texas Section of the Society of Naval Architects and Marine Engineers

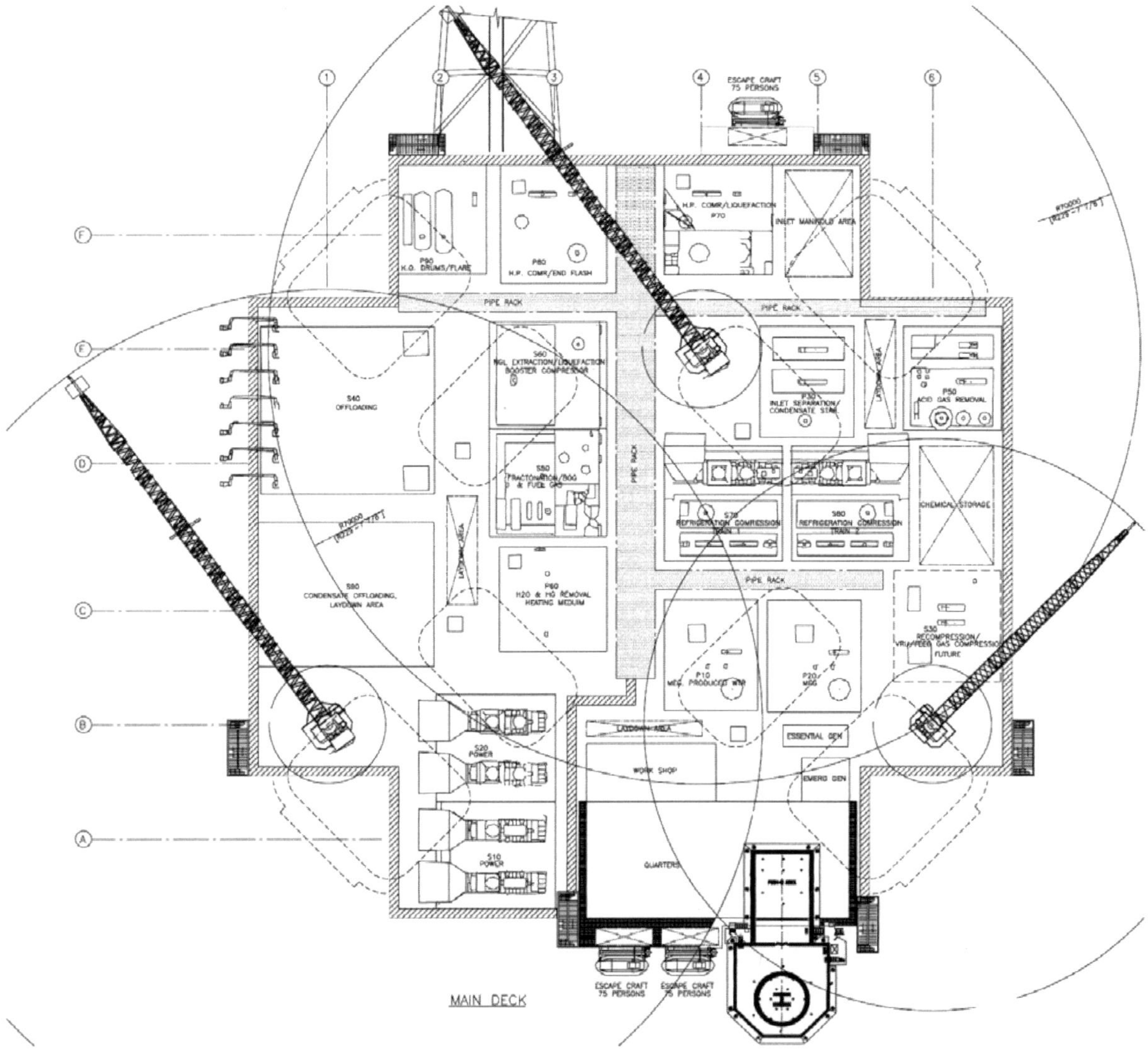

Figure 4 Generic Topsides Layout – Main Deck

Proceedings of the 22nd Offshore Symposium, February 2017, Houston, Texas
Texas Section of the Society of Naval Architects and Marine Engineers

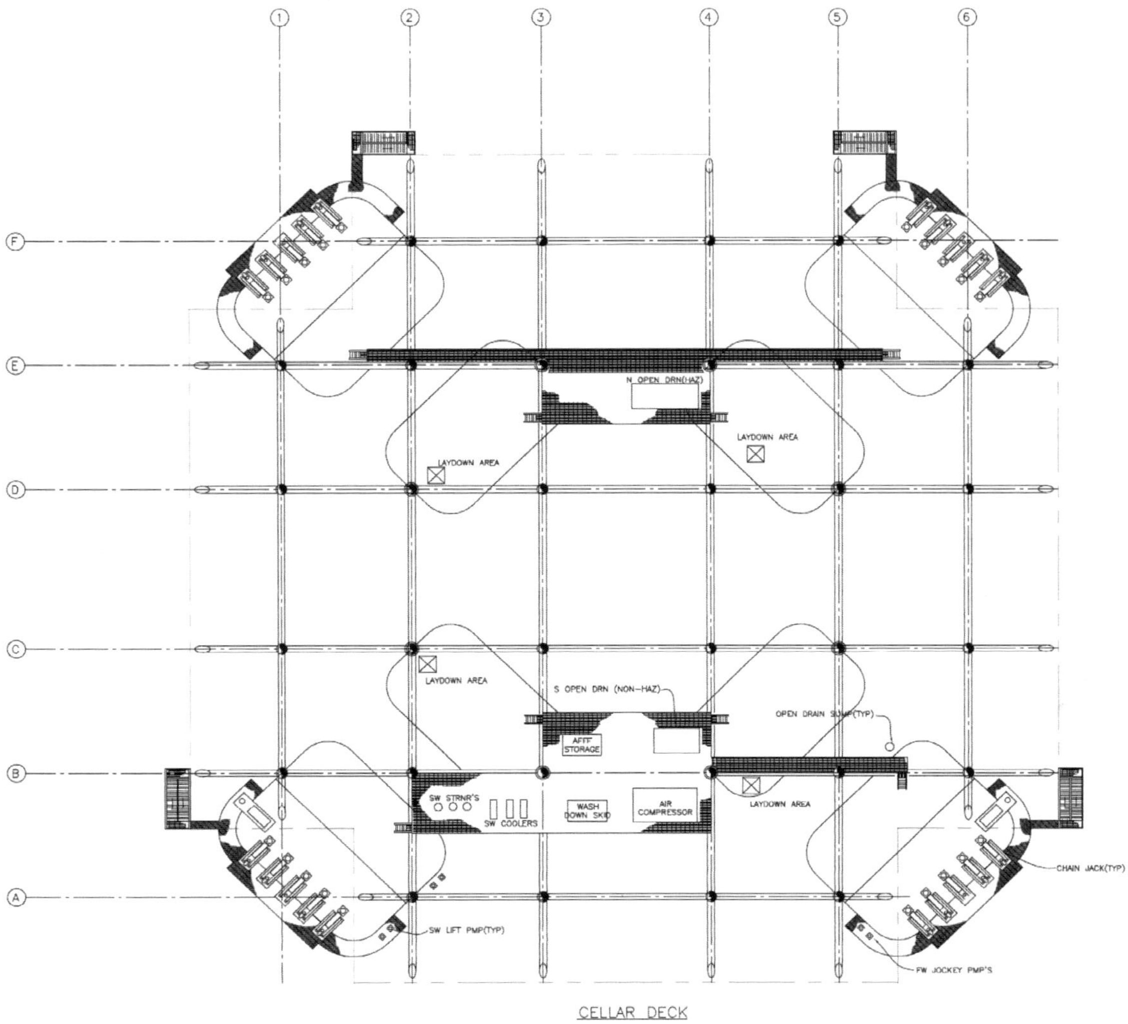

Figure 5 Generic Topsides Layout – Cellar Deck

*** Formatting Issue - Best Available Paper/Graphic ***

*** Formatting Issue - Best Available Paper/Graphic ***

Proceedings of the 22nd Offshore Symposium, February 2017, Houston, Texas
Texas Section of the Society of Naval Architects and Marine Engineers

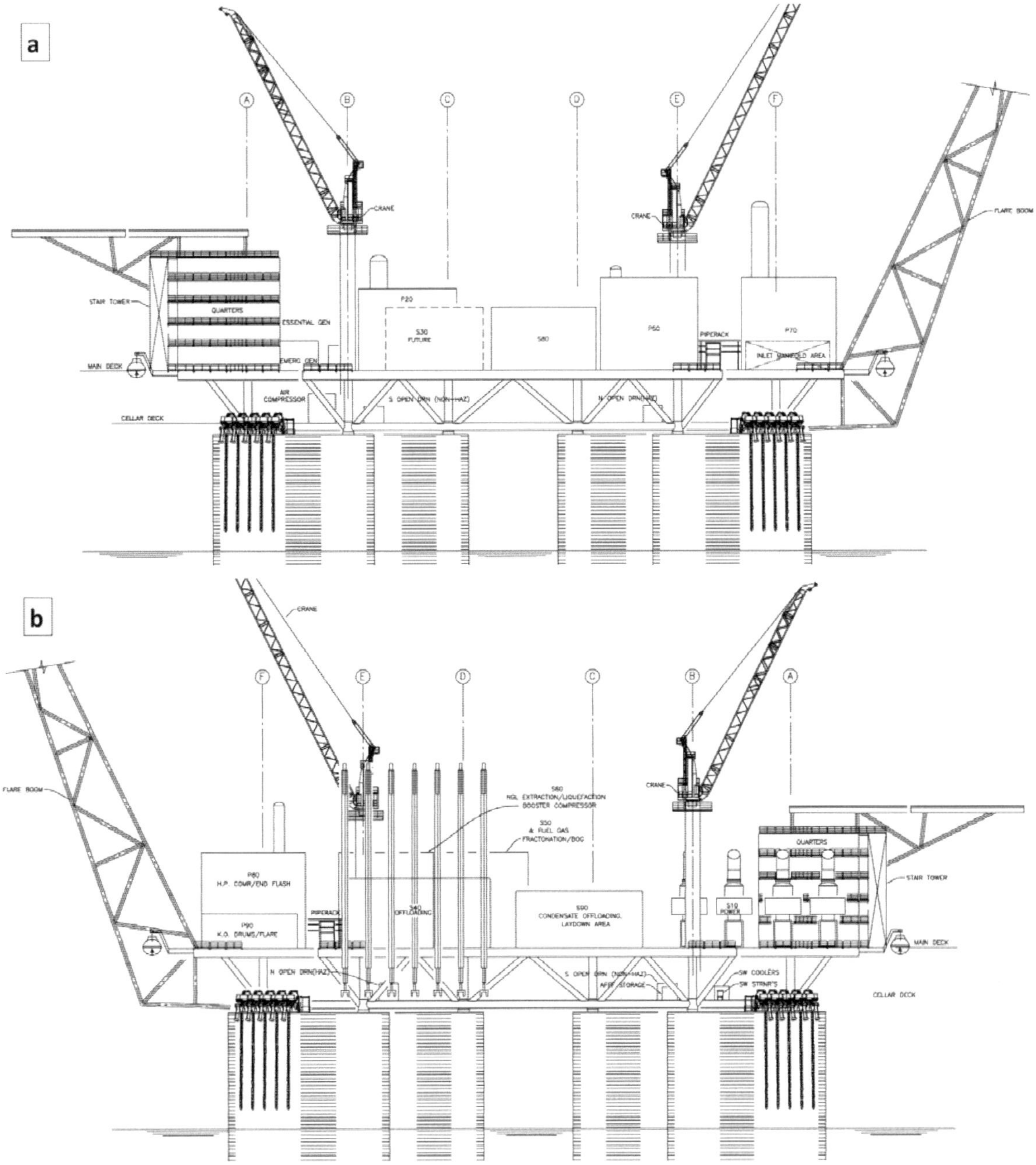

Figure 6 Generic Topsides Layout – Elevation Views, a) Looking West; b) Looking East

*** Formatting Issue - Best Available Paper/Graphic ***

Proceedings of the 22nd Offshore Symposium, February 2017, Houston, Texas
Texas Section of the Society of Naval Architects and Marine Engineers

*** Formatting Issue - Best Available Paper/Graphic ***

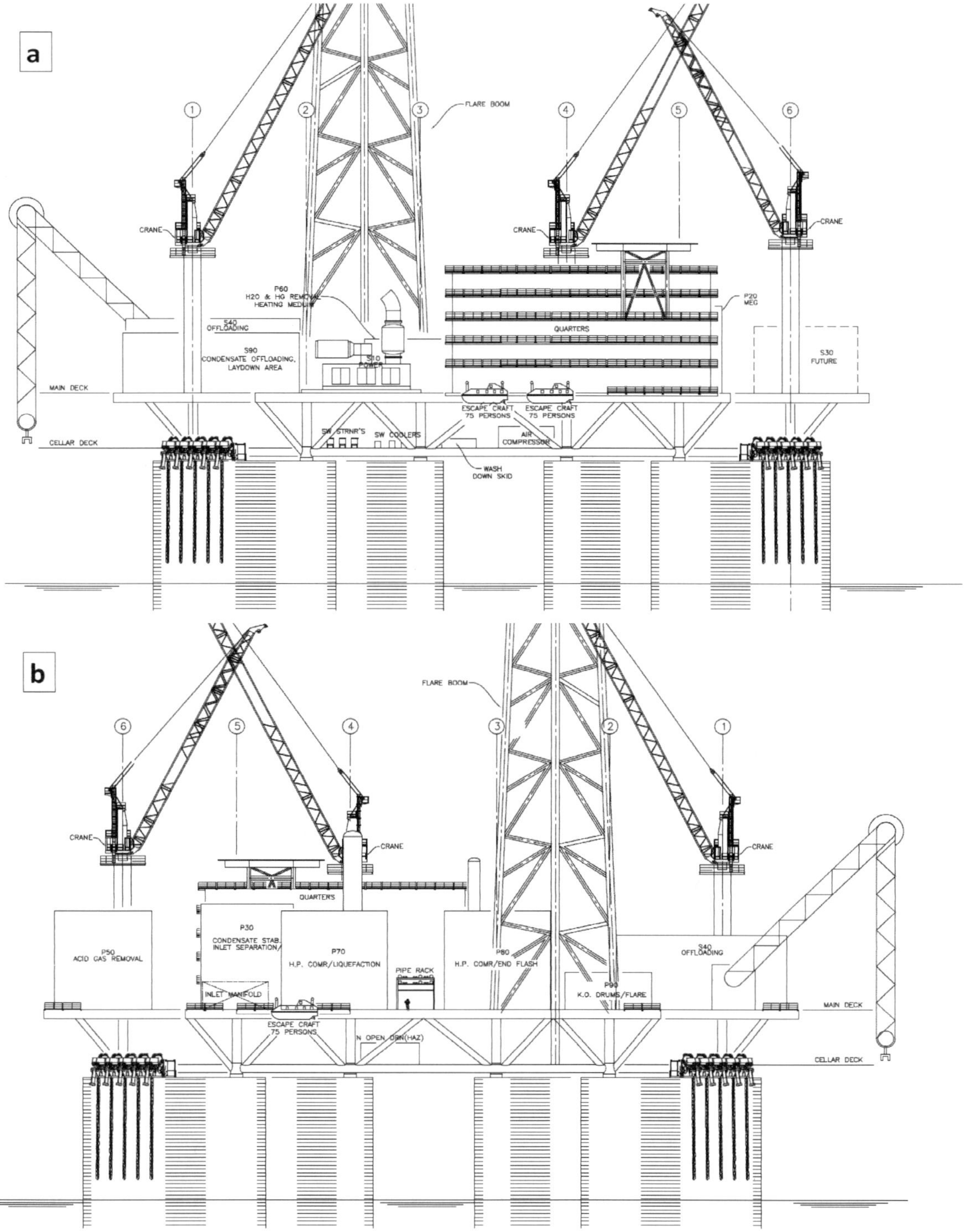

*** Formatting Issue - Best Available Paper/Graphic ***

Figure 7 Generic Topsides Layout – Elevation Views, a) Looking North; b) Looking South

HULL CONFIGURATION

Hull Key Figures

The hull configuration is illustrated in Figure 8 and key particulars are summarized in Table 4. Since the general philosophy of topsides arrangement is to adopt modular approach, the deck steel is treated as part of hull steel for this study.

Table 4 Hull and Deck Particulars

Parameters	Units	Hull
Fully loaded platform draft	m	32
Fully loaded platform displacement	mt	180,000
Column freeboard	m	21
Total column height	m	53
Pontoon height	m	12
Maximum hull edge to edge width	m	110
		Deck
Main deck top elevation	m	65
Cellar deck top elevation	m	55
Main deck area	m^2	13,500
Cellar deck area	m^2	1,500
Total deck area	m^2	15,000

Cargo Tanks

Cargo tank capabilities of LNG and Condensate are summarized in Table 5 and permeability 98% was considered. Since SPB type B tank has all shapes as displayed in Figure 9 (Watanabe, 2012) and self-supporting and independent, therefore SPB type B tank was considered for this study. Detailed features of SPB tanks can be found in Watanabe (2012). However, membrane type tank is also suitable and can be employed in future.

The LNG and Condensate tanks are separated and protected by outside cofferdams and double bottoms in columns.

Table 5 Cargo Tank Capabilities

Items	# of Tanks	Volume per tank (m^3)	Total Volume (m^3)	Tank Locations
LNG Large	4	16,875	67,500	Outer Columns
LNG Small	2	12,500	25,000	NE & SW Inner Columns
Total LNG	6		92,500	
Condensate	2	12,500	25,000	NW & SE Inner Columns

Mooring System

So far, the sanctioned FLNG projects are all ship-shape with either internal or external turret systems. For this generic study, a typical spreading mooring system is employed to eliminate complicated and expensive turret system and dynamic positioning (no thrusters) to reduce complexity of hull arrangement and cost of operations.

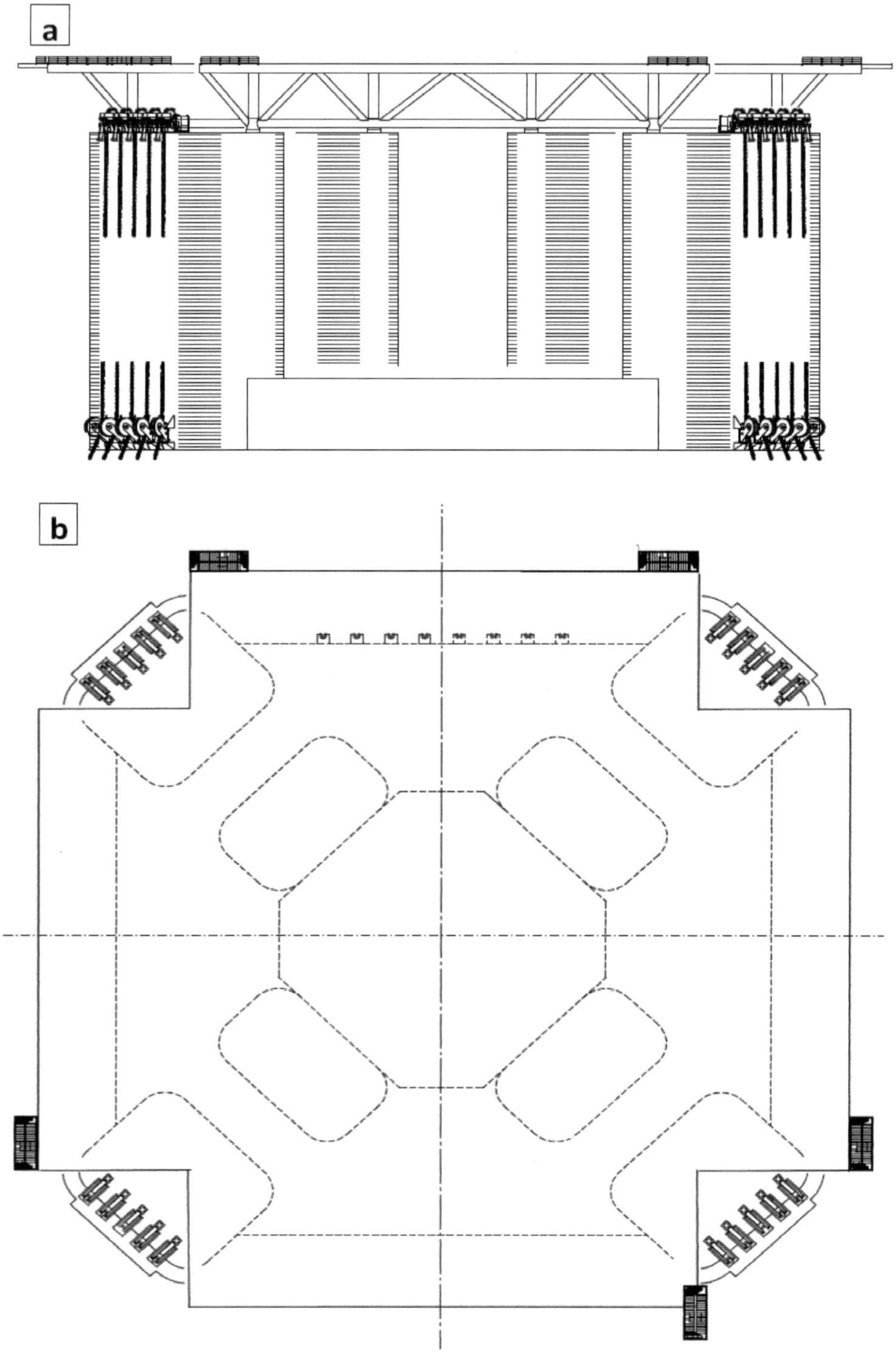
a
b

*** Formatting Issue - Best Available Paper/Graphic ***

Figure 8 Hull Configuration – a) Elevation View; b) Top View

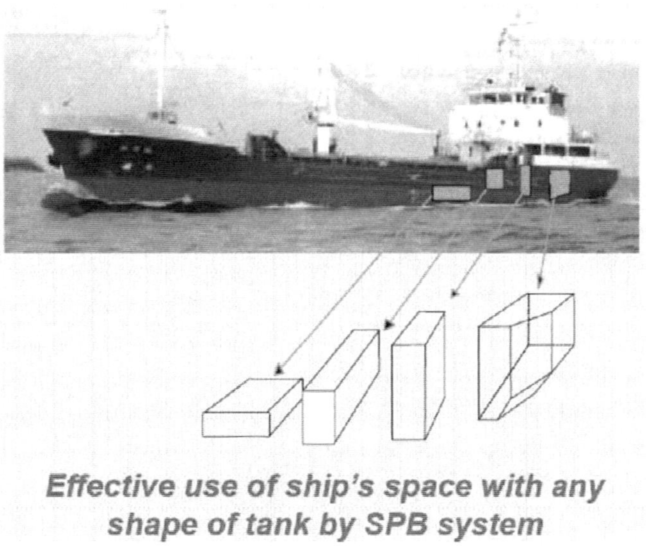

Figure 9 Shapes of SPB Tank System (Watanabe, 2012)

STUDY CASES

In order to understand the characteristic responses of ship-shape FLNG in WA, a simplified and generic ship-shape FLNG (G-FLNG) with internal turret mooring system has 5.0 MTPA process capability (3.5 MTPA LNG, 1.0 MTPA LPG and 0.5 MTPA Condensate). The main particulars are summarized in Table 6 and illustrated in Figure 10 respectively.

Table 6 Main Particulars of Ship-Shape G-FLNG

Parameters	Units	Hull
Fully loaded draft	m	20
Fully loaded platform displacement	mt	659,000
Fully loaded depth	m	42
Length between perpendiculars (LPP)	m	468
Breadth	m	74
Longitudinal metacenter (GMl)	m	940
Transverse metacenter (GMt)	m	34

In this study, study cases are focused on investigating operability by running sea states in wave scatter diagram. Dynamic responses of PC Semi and G-FLNG in sea states given in Table 3 are excluded for this paper and will be addressed in a separated paper.

*** Formatting Issue - Best Available Paper/Graphic ***

Figure 10 A Generic Ship-Shape FLNG Hull

NUMERCIAL RESULTS

Diffraction Model

The diffraction models of PC Semi FLNG and G-FLNG are displayed in Figure 11.

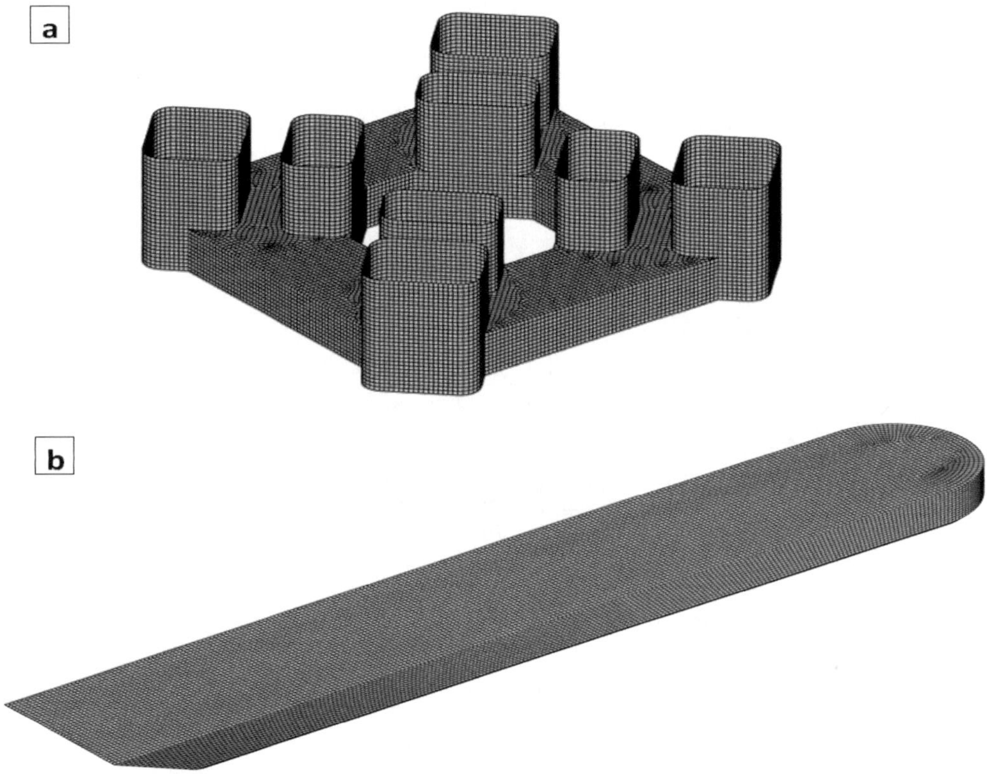

Proceedings of the 22nd Offshore Symposium, February 2017, Houston, Texas
Texas Section of the Society of Naval Architects and Marine Engineers

Figure 11 Diffraction Models of PC Semi FLNG (a) and G-FLNG (b)

Natural Periods

Heave, roll and pitch natural periods are given in Table 7.

Table 7 Heave, Roll and Pitch Natural Periods

Items	Units	PCS-FLNG	G-FLNG
Heave	Sec	21.4	13.3
Roll	Sec	39.8	19.9
Pitch	Sec	39.1	12.6

In Table 7, "PCS-FLNG" stands for PC Semi FLNG and "G-FLNG" represents generic ship-shape FLNG for this study.

From Figures 2 and 3, there are large percentages of waves with periods of 12-14 seconds in WA metocean criteria. Even around 20 sec, there are some percentages of waves which are not negligible. Although extreme responses in 100-yr, 1,000-yr and 10,000-yr sea states presented in Table 3 and Hs-Tp contours shown in Figure 1 are excluded in this paper, it is anticipated that G-FLNG will experience considerably larger dynamic responses and its survivability will be challenged tremendously and suitability of ship-shape FLNG hull form in WA metocean will be questionable.

Locations of the Interested Points

For comparison purpose, there are four interested points which were selected, coordinates are defined in Figures 12 and 13. Locations are summarized in Table 8 and illustrated in Figures 12 and 13 respectively.

Four interested points picked for this study are highlighted as follows,

- P1, central point of offloading zone (red dot)
- P2, central of upper deck (blue dot)
- P3, middle point of living quarter (green dot)
- P4, middle point of cryogenic heat exchanger and fractionation towers (purple dot)

Table 8 Heave, Roll and Pitch Natural Periods

Items	Units	PCS-FLNG	G-FLNG
P1 (x, y, z)	m	(-66.0, 21.0, 64.5)	(-52.0, -37.0, 52.0)
P2 (x, y, z)	m	(0.0, 0.0, 64.5)	(0.0,0.0, 52.0)
P3 (x, y, z)	m	(14.5, -52.0, 75.0)	(-233.0, 0.0, 62.0)
P4 (x, y, z)	m	(11.5, 45.5, 76.0)	(-127.0, 22.0, 72.0)

Coordinates definitions are summarized as follows:

- Positive x: from platform center toward platform East or from vessel center toward vessel Bow
- Positive y: from platform center toward platform North or from vessel center toward vessel Port side
- Positive z: from platform or vessel keel upward
- Positive roll, pitch and yaw: by right-hand-side rule.

Heading definitions are given as follows:

- 0 degree is from platform West toward platform East or from vessel Stern toward vessel Bow
- 90 degrees is from platform South toward platform North or from vessel Starboard side toward vessel Port side.
- Other headings are defined in same fashion and rotated in counter-clockwise direction.

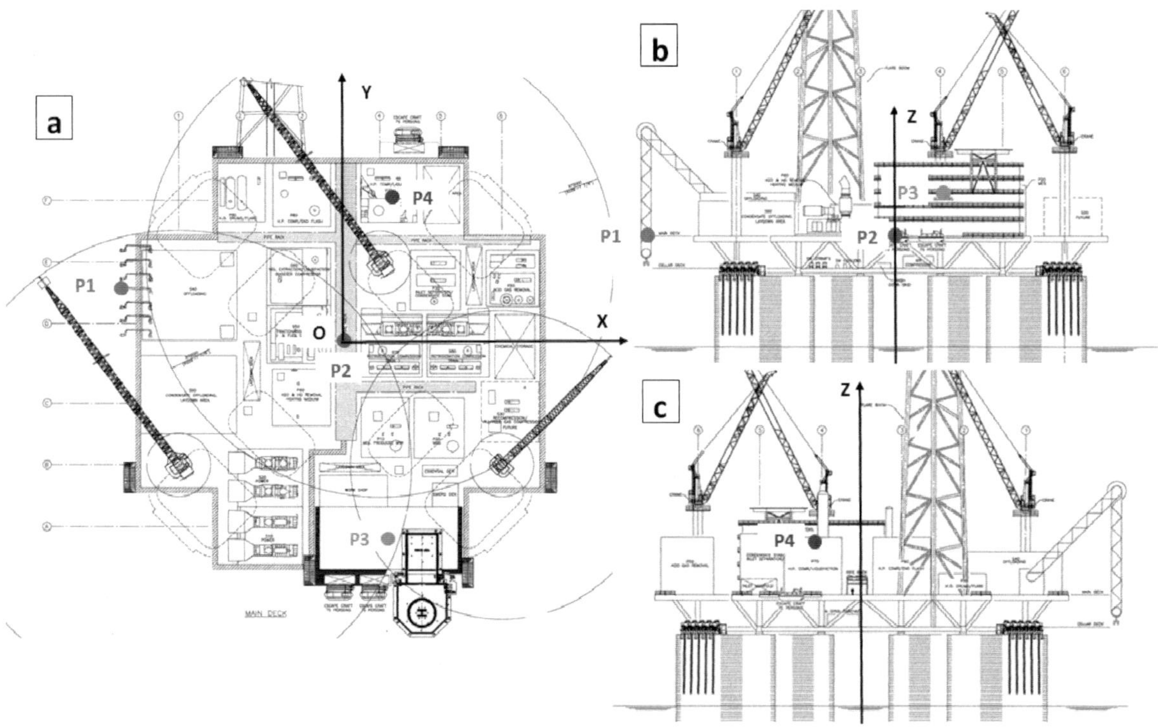

Proceedings of the 22nd Offshore Symposium, February 2017, Houston, Texas
Texas Section of the Society of Naval Architects and Marine Engineers

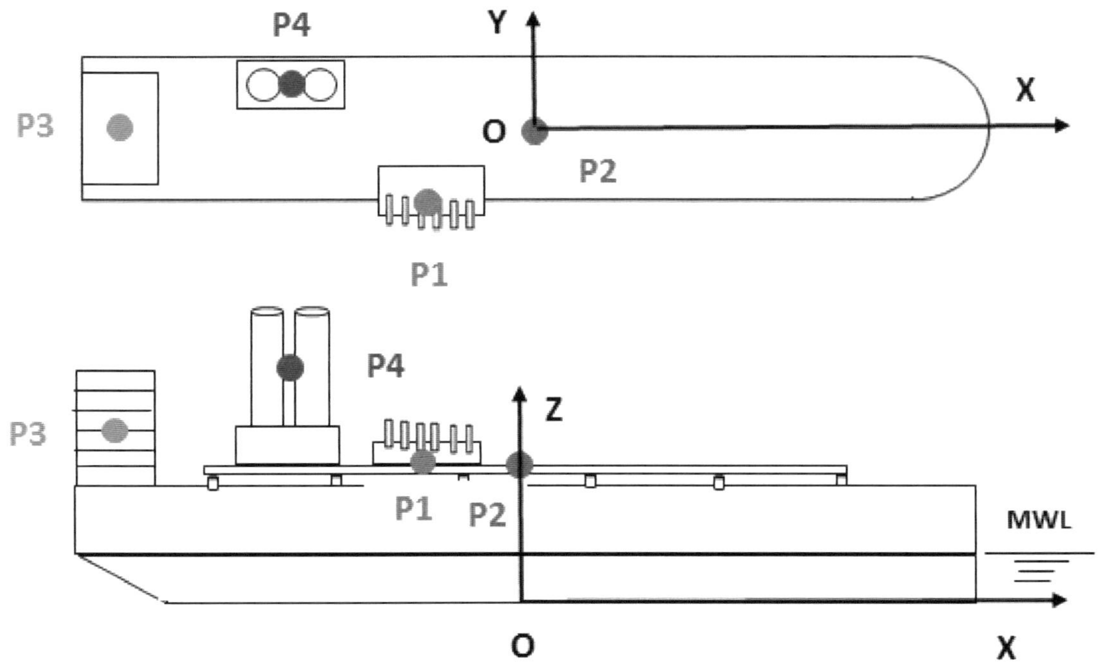

Figure 12 Four Selected Points on PCS-FLNG, a) Top View; b) Looking North; c) Looking South

Figure 13 Four Selected Points on G-FLNG

Vertical Motions

By applying generic wave scatter diagrams as presented in Figures 2 and 3, single amplitude significant vertical motions of points P1 to P4 of PCS-FLNG and G-FLNG as a function of non-exceeding probabilities are plotted in Figures 14a to 14d respectively.

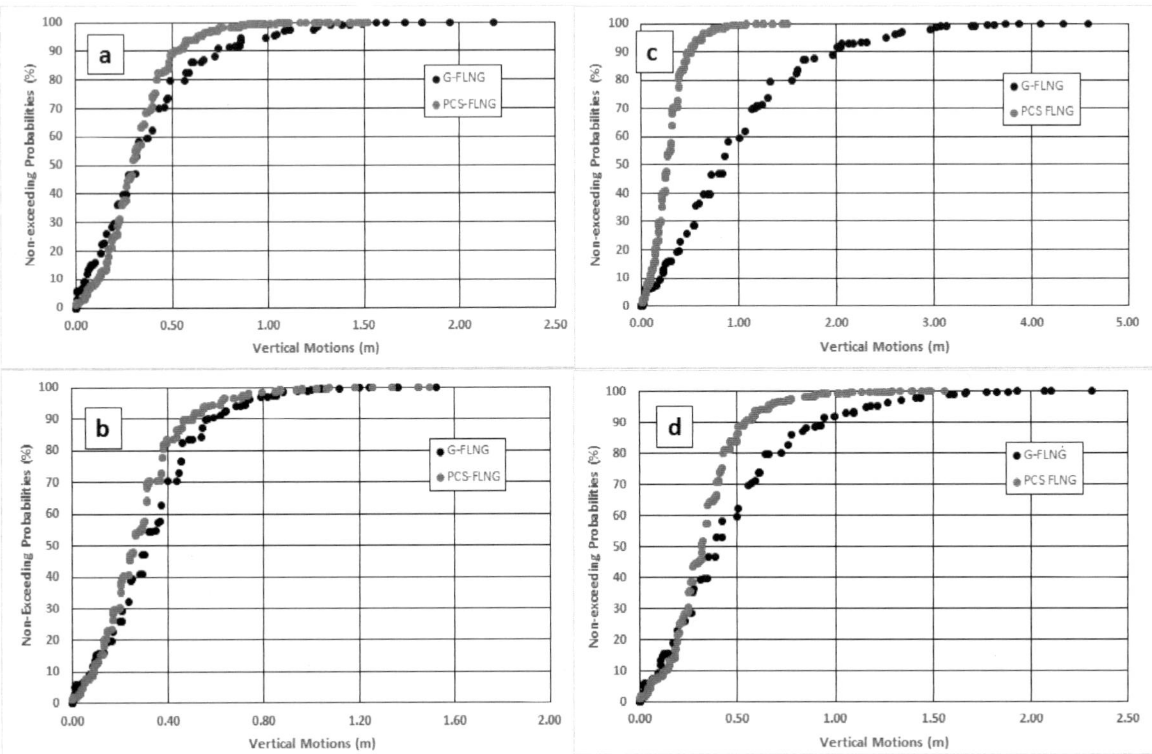

Figure 14 Single Amplitude Vertical Motions, a) at P1; b) at P2; c) at P3; d) at P4

From Figure 14, a few observations are highlighted as follows,

- As anticipated, the worst vertical motions for G-FLNG occurred at point 3 (living quarter) since its location is far away from middle ship and least vertical motions for G-FLNG was at the point 2 (middle ship). For PCS-FLNG, the differences among four points are small.
- At point 1 (offloading) of PCS-FLNG in Figure 13a, its single amplitude vertical motion is less than 0.5 m and 1.0m in more than 90% and 99% of times respectively.
- At point 4 (location of motion-sensitive facilities) of PCS-FLNG in Figure 13d, its single amplitude vertical motion is less than 0.5 m and 1.0m in more than 88% and 98% of times respectively while corresponding probabilities of G-FLNG is 60% and 91% respectively.
- The most distinct single amplitude vertical motion differences between G-FLNG and PCS-FLNG are at point 3 (living quarter).
    - For less than 1.0m vertical motion: 99% for PCS-FLNG; 60% for G-FLNG
    - For less than 2.0m vertical motion: 90% for G-FLNG
    - For less than 3.0m vertical motion: 97% for G-FLNG
    - For less than 4.0m vertical motion: 99% for G-FLNG
    - About 4.6m single amplitude vertical motion in a relatively small wave with Hs=4.75m and Tp=21.0s.

Proceedings of the 22nd Offshore Symposium, February 2017, Houston, Texas
Texas Section of the Society of Naval Architects and Marine Engineers

It shall point out the all floaters shall be designed as "manned" in 100-yrp, 1,000-yrp and 10,000-yr events per Australia regulation. The generic extreme sea states are given in Table 3 and Hs-Tp contours are illustrated in Figure 1. It is anticipated that tremendous vertical motions at Living Quarter will occur and motion sickness of people on board will experience.

Roll and Pitch Motions

Assume either G-FLNG or PCS-FLNG is rigid body, roll and pitch motions at four points are same. By employing generic wave scatter diagrams as presented in Figures 2 and 3, single amplitude significant roll and pitch motions of points P1 to P4 of PCS-FLNG and G-FLNG as a function of non-exceeding probabilities are plotted in Figures 15a and 15b respectively. For G-FLNG, since it has turret system, wave headings were assumed to be nearly head sea within 180 deg +/- 15 deg. Only waves induced roll and pitch motions were considered. No wind and current was employed for this exercise.

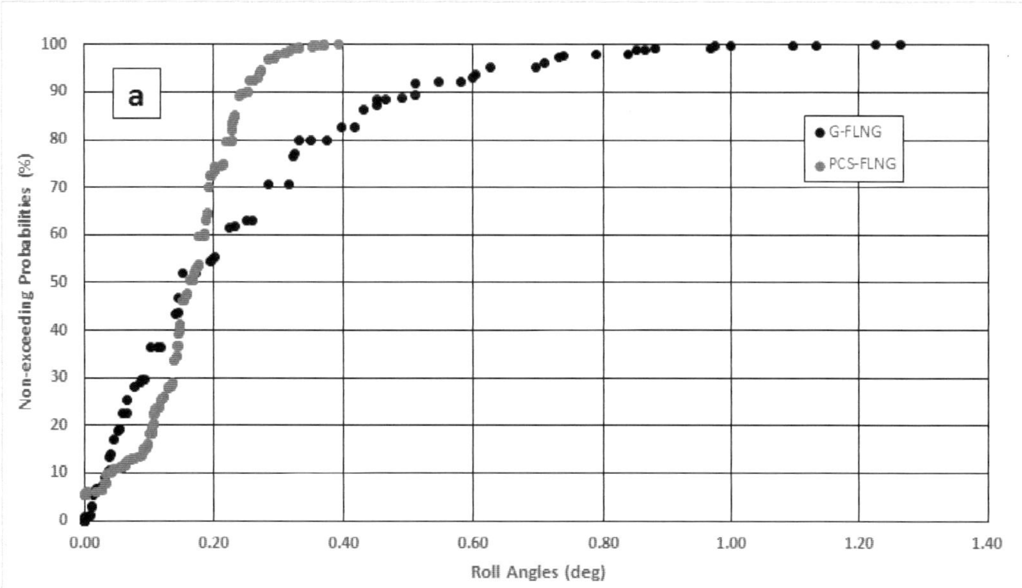

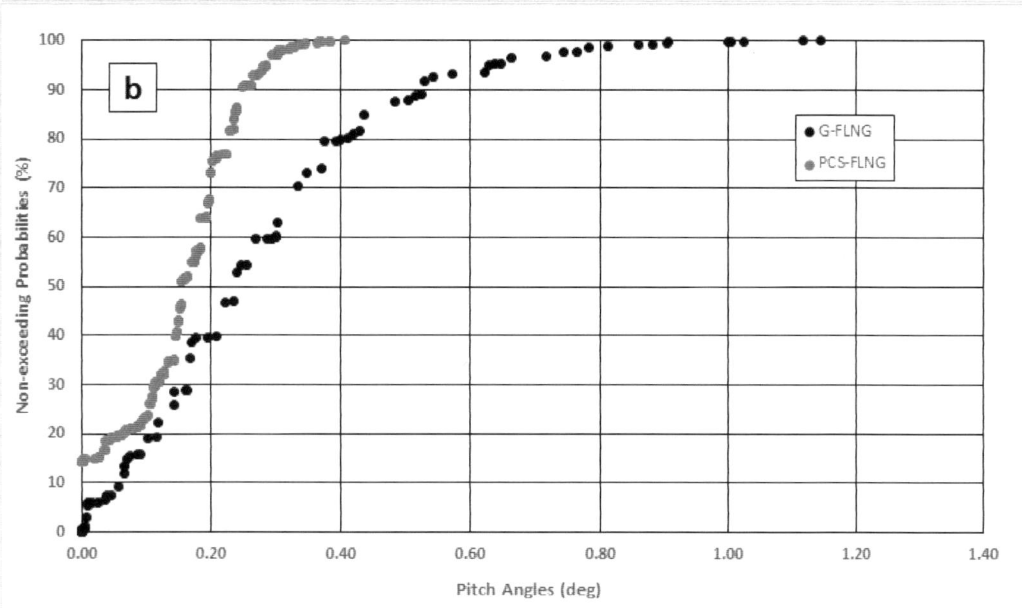

***** Formatting Issue - Best Available Paper/Graphic *****

Figure 15 Single Amplitude Roll (a) and Pitch (b) Motions

Terashima et al (2011) investigated motion criteria for performances of FLNG process facilities and pointed out performances of some process facilities, such as cryogenic heat exchangers and acid gas absorbers, degrade with the motion criteria, especially sensitive to roll and pitch motions and accelerations.

From Figure 15, a few findings are identified as follows,

- The distinct single amplitude roll and pitch motion differences between G-FLNG and PCS-FLNG are noticed.
- For Roll angular motions,
    - For less than 0.2 deg roll: 72% for PCS-FLNG; 55% for G-FLNG
    - For less than 0.4 deg roll: nearly 100% for PCS-FLNG; 82% for G-FLNG
    - For less than 0.6 deg roll: 92% for G-FLNG
    - For less than 1.0 deg roll, 99% for G-FLNG
    - About 1.3 deg single amplitude roll motion in a relatively small wave with Hs=4.75m and Tp=21.0s.
- For Pitch angular motions,
    - For less than 0.2 deg pitch: 75% for PCS-FLNG; 40% for G-FLNG
    - For less than 0.4 deg pitch: nearly 100% for PCS-FLNG; 80% for G-FLNG
    - For less than 0.6 deg pitch: 93% for G-FLNG
    - For less than 1.0 deg pitch, 99% for G-FLNG
    - About 1.15 deg single amplitude pitch motion in a relatively small wave with Hs=4.75m and Tp=21.0s.

As shown in Table 7, roll natural period is about 20 sec. As illustrated in Figure 1, there is one Hs-Tp pair that is Hs=20.0m and Tp=20.0 sec. It is anticipated that considerably large roll motions will occur which will induce board impacts of G-FLNG responses. Similarly, governing extreme pitch responses in 100-yrp, 1,000-yrp and 10,000-yrp sea states shall be screened through Hs-Tp contours as displayed in Figure 1. These extreme sea states are not the scope of work of this study and thus extreme responses are excluded from this paper.

Vertical Accelerations

Similarly, single amplitude significant vertical accelerations of points P2 to P4 of PCS-FLNG and G-FLNG as a function of non-exceeding probabilities are plotted in Figures 16a to 16c respectively.

From Figure 16, a few observations are highlighted as follows,

- Vertical acceleration at a point consists of three contributions, 1) heave acceleration; 2) roll induced and 3) pitch induced. At point 2, it is center of the platform or vessel, vertical acceleration just comes from heave acceleration. At points 3 and 4, vertical accelerations include roll and pitch induced vertical accelerations.
- From 16a, single amplitude vertical acceleration of G-FLNG is slightly outperformed PCS-FLNG for amplitude less than 0.006g. For amplitude larger than 0.006g, the trend is reversed.
- Form 16b, single amplitude vertical acceleration of PCS-FLNG at Living Quarter is considerably outperformed G-FLNG.
    - For less than 0.01g: 97% for PCS-FLNG; 28% for G-FLNG
    - For less than 0.015g: 100% for PCS-FLNG; 49% for G-FLNG
    - For less than 0.02g: 75% for G-FLNG

*** Formatting Issue - Best Available Paper/Graphic ***

- For less than 0.03g: 94% for G-FLNG
- For less than 0.04g: 99% for G-FLNG
- The largest vertical acceleration at P3 is 0.0495g with relatively small wave Hs=4.75m and Tp=19.0s.

Similarly, it is anticipated that tremendous vertical acceleration at Living Quarter will occur and motion sickness of people on board will experience in 100-yrp, 1,000-yrp and 10,000-yrp sea states as given in Table 3 and illustrated in Figure 1.

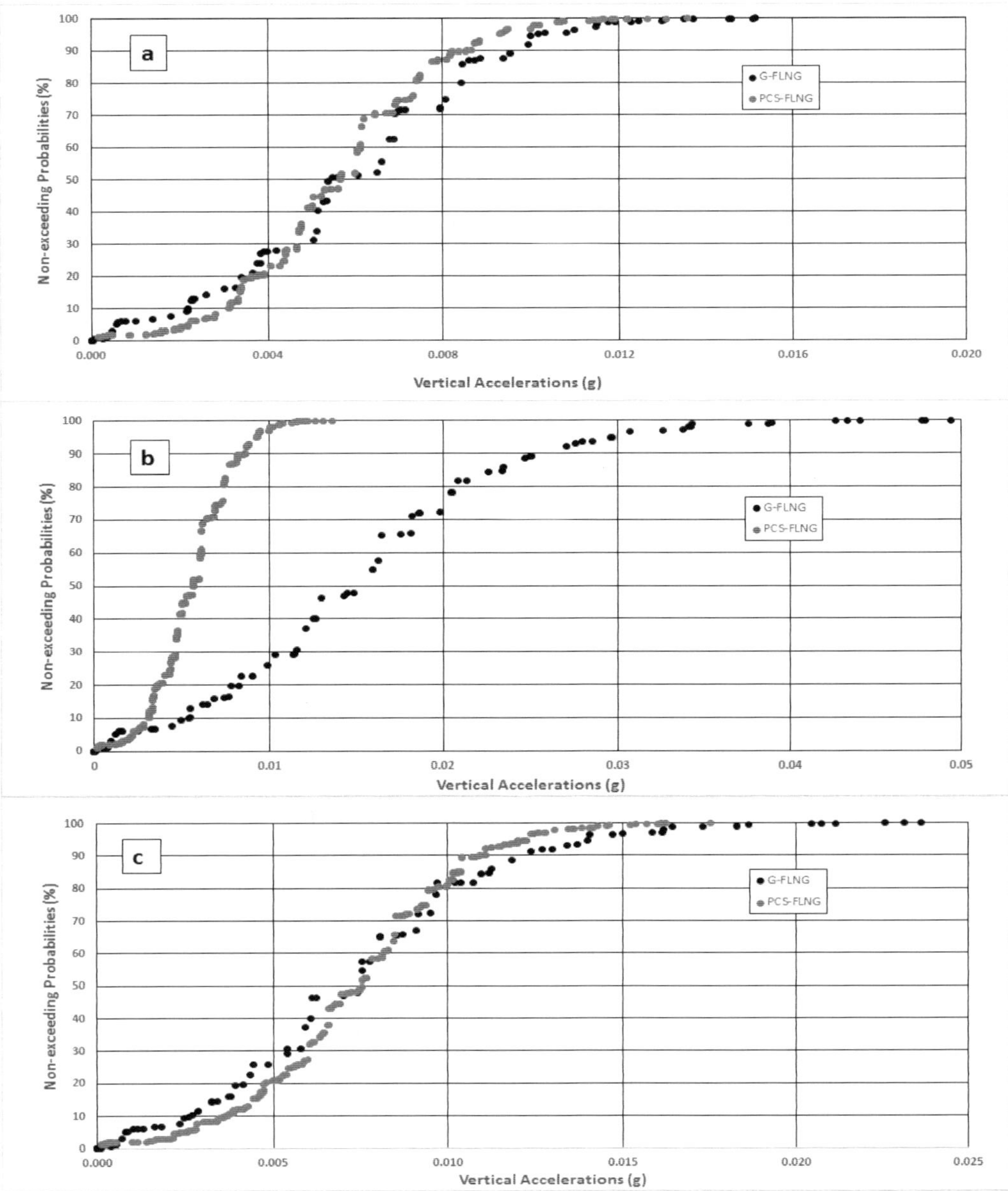

*** Formatting Issue - Best Available Paper/Graphic ***

Figure 16 Single Amplitude Vertical Accelerations, a) at P2; b) at P3; c) at P4

Roll and Pitch Accelerations

From Figures 17a and 17b, it has been noticed that roll and pitch accelerations of G-FLNG and PCS-FLNG are nearly identical.

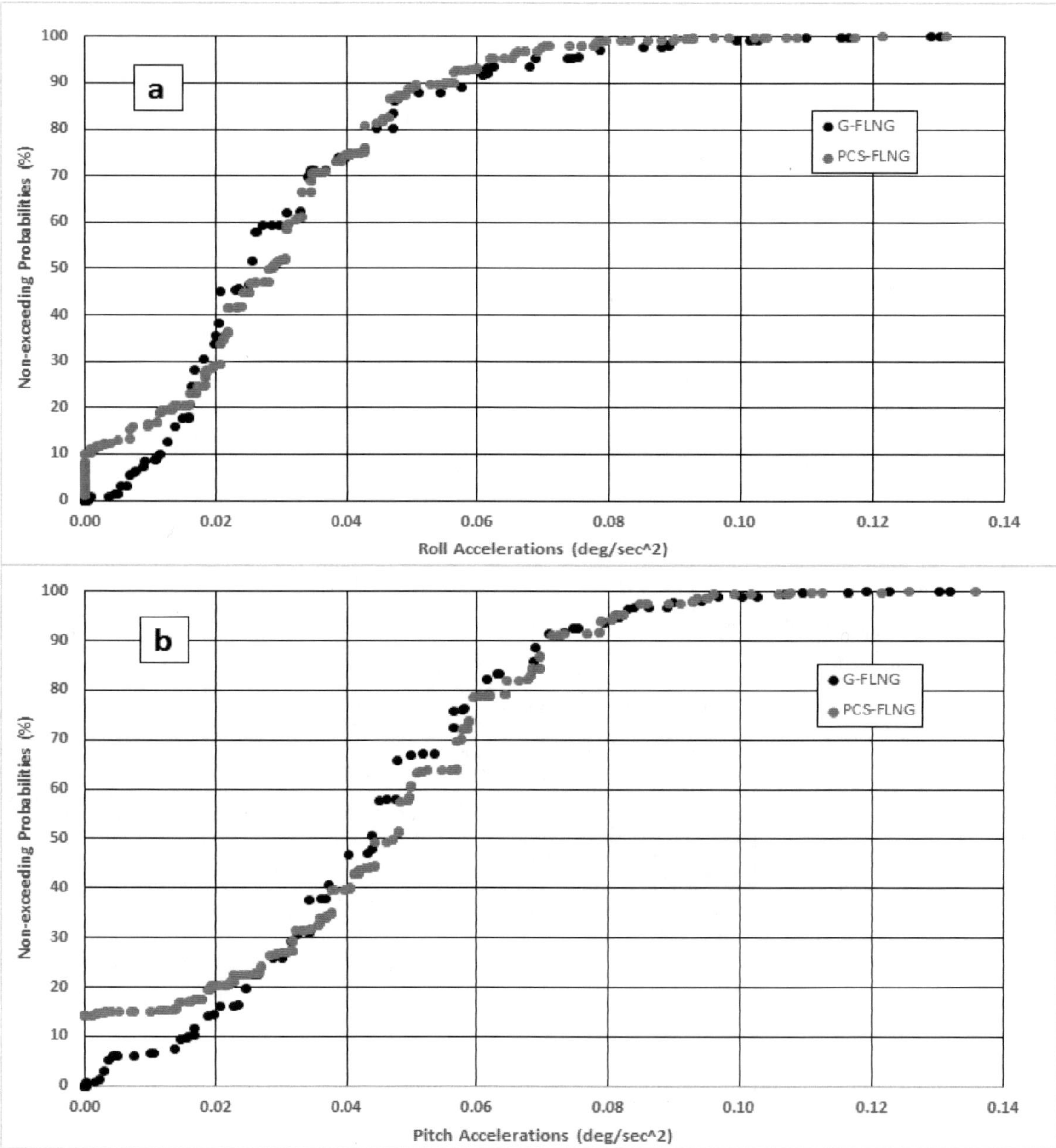

Figure 17 Single Amplitude Roll (a) and Pitch (b) Accelerations

CONSTRUCTABILITY

Hull Constructability

The hull structure as illustrated in Figure 8 is comprised of traditional stiffened flat plate panels for simplified fabrication. The flat panels are especially beneficial for fabrication yards with existing panel-line assembly equipment that can efficiently assemble stiffeners to plate panels prior to subsequent assembly steps. The hull plate thickness, stiffener sizes and girder sizes were conventionally standard hull materials. Stiffener sizes are also selected from commonly available shapes. Girders are fabricated "T" sections, and the associated spans are suitable for the arrangement of the structural panels.

The typical construction sequence for the hull and deck is displayed in Figure 18 and highlighted as follows,

1) One quarter of base subassembly
2) Lift bottom half of inner column block on base subassembly
3) Lift bottom half of outer column block on base subassembly
4) Lift upper halves of inner and outer column blocks
5) Repeat steps 1-4 to assemble one quarter hull
6) Mate two one quarter to assemble half of hull
7) Repeat steps 1-6 to assemble another half of hull and assemble them into the complete hull either in a dry dock or on land nearby a skid way for offloading the hull onto a transport vessel
8) The deck structure, which is considered part of the hull in this study, is assembled at the top of column elevation

In general, the block sizes, material usage, plate and shape sizes and overall structural arrangements are well suited to fabrication at many worldwide locations. Any fabrication yards with experience assembling medium to large hulls for floating production facilities would be capable of building the proposed hull configuration. A typical candidate list (not complete) is presented in Table 9.

Table 9 Potential Construction Yards and Facility Dimensions

Potential Construction Yards	Facility Dimensions
SHI (Korea)	Floating Dock (G5) 157m x 131m
DSME (Korea)	Dry Dock 530m x 131m x 14.5m
Yanti CIMC Raffles (China)	Dry Dock 380m x 120m x 14m
HHI (Korea) Offshore yard	Dry Dock 490m x 115m x 13.5m

Topsides Constructability

As described in previous section, general philosophy of topsides layout is to have modular design which maintains competitions between different yards. The weight and dimension of each modular was limited which may offer opportunity to local fabrication yards to meet local content requirement. Project execution plan may vary, however, quayside integration is the preferred option for this study. As an example, Figure 19 shows lifting the largest modular Living Quarter with Helideck at Kiewit yard. Since helideck extends widely in horizontal direction which prevents the reach of crane, therefore helideck will be pre-cut half (showing in dash-line) and install after lifting. This alternative method has been proven and employed in previous projects. It is obvious from Figure 19 there are adequate vertical clearance.

Proceedings of the 22nd Offshore Symposium, February 2017, Houston, Texas
Texas Section of the Society of Naval Architects and Marine Engineers

*** Formatting Issue - Best Available Paper/Graphic ***

Figure 18 One Hull Construction Assembly Sequence

Proceedings of the 22nd Offshore Symposium, February 2017, Houston, Texas
Texas Section of the Society of Naval Architects and Marine Engineers

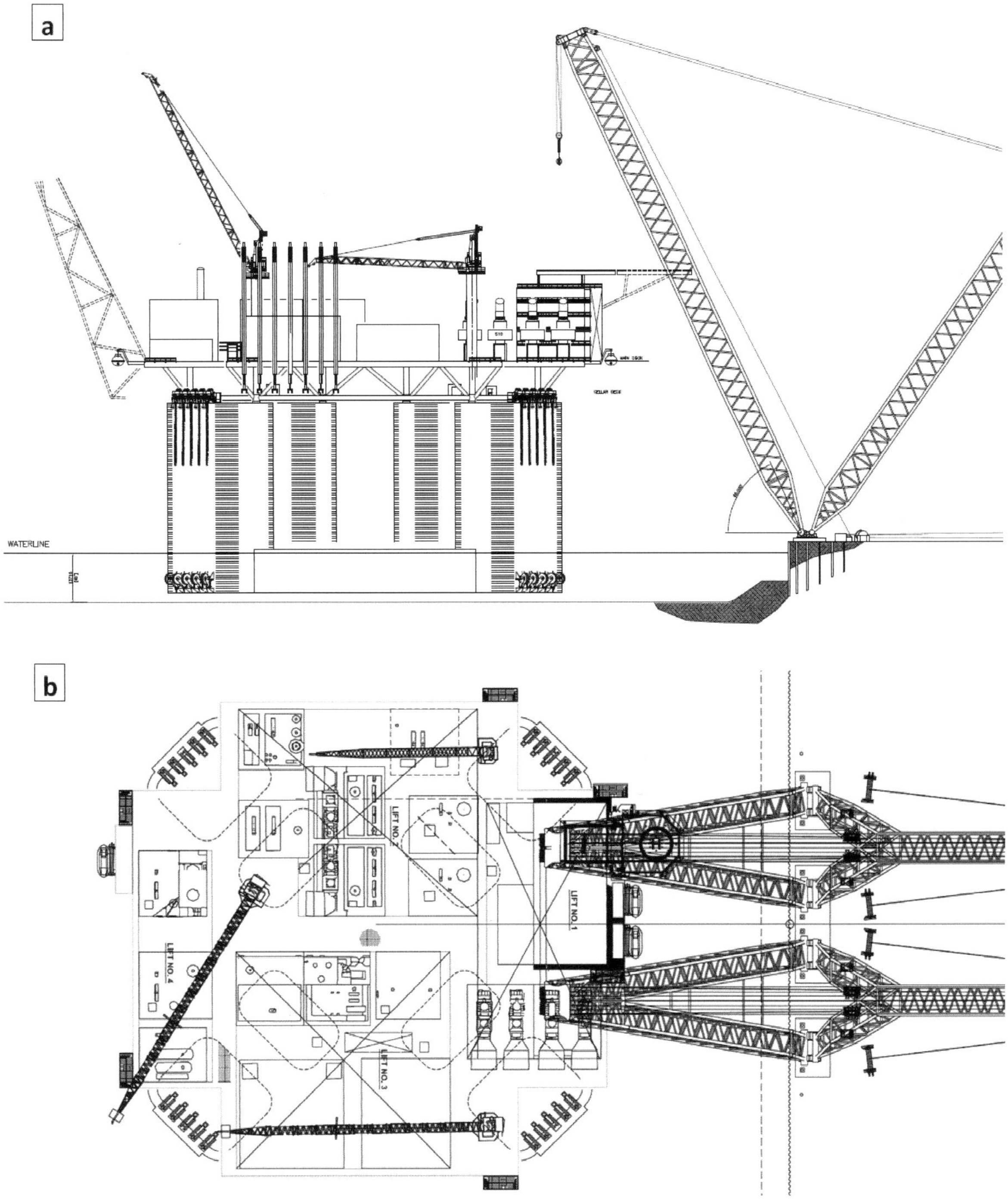

*** Formatting Issue - Best Available Paper/Graphic ***

Figure 19 Quayside Integration – One Example at Kiewit Yard, a) Elevation View; b) Top View

TRANSPORTABILITY

Project execution plan may vary, such as hull and topsides are constructed at same fabrication yard or hull and topsides are constructed at different locations. The current hull sizing has adequate feasibility and is suitable for either execution plan.

Hull with Deck Only

As illustrated in Figure 20, hull with deck only (no topsides facilities/utilities) can be dry transported on a heavy lift vessel. In this example, Dockwise Vanguard is considered. Main characteristic data can be found in Dockwise website and a few selected parameters are reproduced in Table 10 for convenient discussion purpose.

Table 10 Main Characteristic Data of Dockwise Vanguard

Parameters	Dimensions or Weights
Overall length	275m
Breadth molded/max	70m/78.75m
Depth	15.5m
Max submerged draft at FPP/App	31.5m
Max draft	10.99m
Dead weight	117,000 mt

Weights and required platform draft of hull with deck only are summarized in Table 11.

Table 11 Weights of Hull with Deck only

Parameters	Dimensions or Weights
Weights	56,660 mt
Platform draft	7.2 m
Maximum hull edge to edge width	110 m
Hull overhang on Vanguard	20 m

From Tables 10 and 11, it is obvious that weights of hull with deck only is well within the limit of Vanguard and overhang on both sides of HLV is acceptable.

Hull with Integrated Topsides

As presented in Figure 21, hull with integrated topsides can be dry transported on a heavy lift vessel, Dockwise Vanguard. Weights and required platform draft of hull with integrated topsides are summarized in Table 12.

Table 12 Weights of Hull with Integrated Deck

Parameters	Dimensions or Weights
Weights	78,360 mt
Platform draft	10 m

Maximum hull edge to edge width	110 m
Hull overhang on Vanguard	20 m

From Tables 10 and 12, it is obvious that weights of hull with integrated topsides is well within the limit of Vanguard and overhang on both sides of HLV is acceptable.

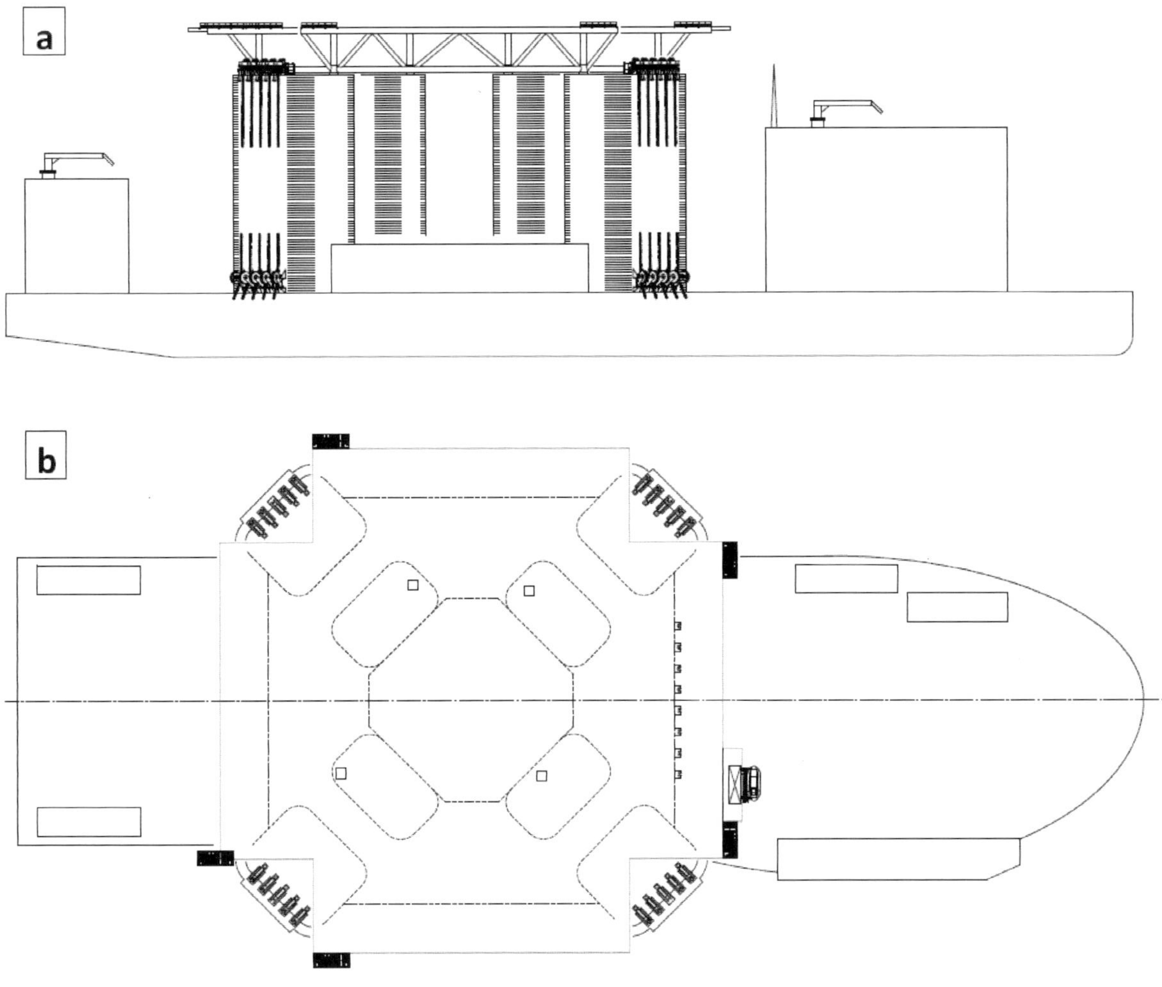

Figure 20 Hull with Deck only – On Dockwise Vanguard, a) Elevation View; b) Top View

***** Formatting Issue - Best Available Paper/Graphic *****

Proceedings of the 22nd Offshore Symposium, February 2017, Houston, Texas
Texas Section of the Society of Naval Architects and Marine Engineers

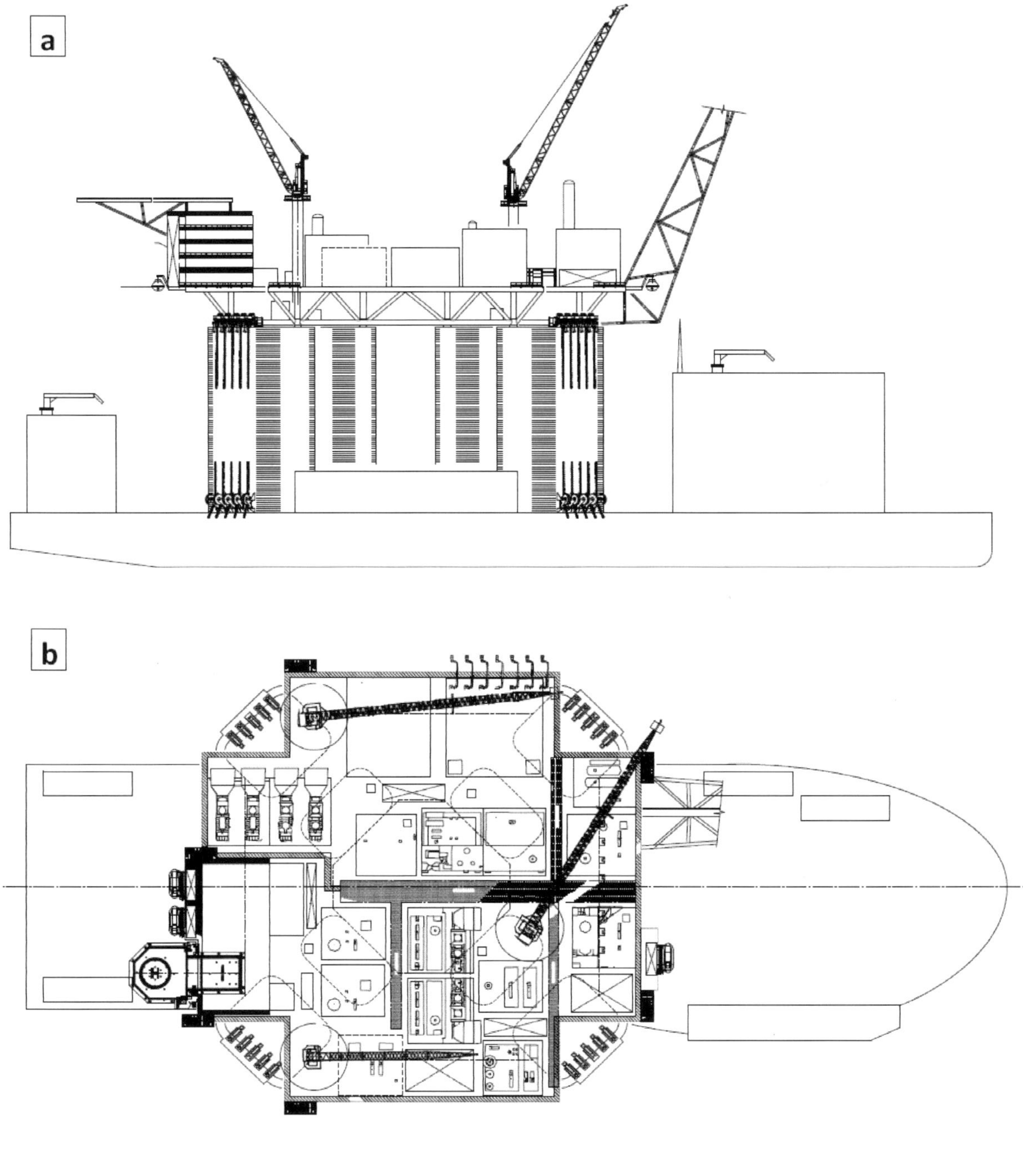

Figure 21 Hull with Integrated Topsides – On Dockwise Vanguard, a) Elevation View; b) Top View

INSTALLABILITY

The integrated platform is dry transported to the nearby sheltered area and offloaded. Then it is wet towed to the installation site. The mooring system is traditional and can be pre-laid and hull is a typical Semi-submersible with adequate stability. Obviously, there is no "show stopper" to hoop-up hull and mooring. Similarly, it is believed there is "no concern" to pull-in risers.

OPERABILITY

As described in previous sections, dynamic responses of four representative points picked for this study in operating sea states have been analyzed and main findings are highlighted as follows,

- Better performances of PCS-FLNG at Point P1 (central point of offloading zone) in terms of vertical motions (Figure 14a) and roll (Figure 15a) and pitch (Figure 15b) motions
- Out performances of PCS-FLNG at Point P2 (center of upper deck) in terms of vertical motions (Figure 14b); roll (Figure 15a) and pitch (Figure 15b); and vertical accelerations (Figure 16a)
- Significantly better performances of PCG-FLN at Point P3 (middle point of Living Quarter) in terms of vertical motions (Figure 14c); roll (Figure 15a) and pitch (Figure 15b) and vertical accelerations (Figure 16b)
- Out performances of PCS-FLNG at Point P4 (middle point of cryogenic heat exchanger and fractionation towers) in terms of vertical motions (Figure 14d); roll (Figure 15a) and pitch (Figure 15b); and similar vertical accelerations (Figure 16c).
- Almost identical roll (Figure 17a) and pitch (Figure 17b) accelerations at all four points.

Overall, it is concluded that operability of PCS-FLNG is better than that of G-FLNG.

DISCUSSIONS

Current commodity market calls for innovative and economical solutions from all possible perspectives keep floating liquefied natural gas project afloat at low gas prices for a prolonged period. In this study, traditional field development strategy and rethinking suitability of ship-shape hull for WA metocean are addressed below.

Field Development Strategy

As Pek and Velde (2013), the normalized UTC (unit technical cost) was presented to compare economics of a midsize FLNG and large size FLNG and concluded that midsize FLNG has about 30% higher UTC than that of a large size FLNG and a small FLNG does not look attractive from the normalized UTC perspective. However, there are some key uncertain aspects which should be factored into decision making and final conclusion might be altered. Therefore, it is worth revisiting field development strategy and criteria of selection of most economic and least risk solution.

- Reservoir Risk

  It is a challenging task to estimate reasonably accurate size and characteristics of reservoir which affects selection of process facilities of LNG, LPG and condensate. It will take more exploration time to yield a more accurate estimate of reservoir. However, for a standardized and phased approach, it can start a relatively smaller but certain production capability FLNG first and keep exploring more wells in parallel. If it is found out larger field size later, it can consider second FLNG or even third one. By this way,

- It can save 1-2 years or more and result in early production and early cash flow for the project
- It can avoid overestimating the filed size and thus oversizing FLNG and wasting investment
- It can have better definition of process capabilities of LNG, LPG and Condensate for second one to mitigate reservoir size overestimate risk.

- Financial Risk
  Large FLNG requires large CAPEX which is leading to difficulties in making investment decisions for low commodity market. It is foreseen that oil and gas prices will be low and may last for many years. The financial risk might be too high to be manageable for large FLNG. However, for a standardized and phased approach, multiple small standardized units in a phase-wise fashion to allow the execution, scope and CAPEX requirements to be more manageable. Early production and early cash flow income yields heathy business. Thus, small CAPEX for small capability FLNG (about 1.5 Mtpa) might be preferred option for future projects.
- Delivering Risk
  Large FLNG, such as Prelude, has
  - 488m in length and 74m in breadth
  - ~660,000 mt in displacement
  - 260,00 mt steel
  - World largest turret, weighing more than 11,000 mt
  - Three largest Rolls-Royce electrically powered azimuth thrusters will ensure it can keep its heading against the wind, current and waves to allow safe tanker berthing

With many world records in hull size, turret system and thrusters, delivering project on time is facing tremendous challenges during design, construction, transportation and installation. It has been reported Prelude FLNG took longer time than it was planned due to all kinds of realities. Again, for a standardized and phased approach, smaller and standardized unit is to allow the execution, scope and delivering time to be well manageable and delivering risk will be greatly mitigated.

Suitability of Ship-Shape Hull for WA Field Development

Per Australia regulation, the FLNG shall be designed in accordance with the safety criteria and it is "manned" in 10,000-yr event. The tolerable limit for the frequency of catastrophic failure is less than 1 x 10-4 per annum for each major accident event consequence category.

For a large ship-shape FLNG, such as Prelude, the ratio of vessel length over vessel breadth is somewhere 6 to 7 which means many equipment/facilities/utilities are far away from middle ship, especially Living Quarter. The harshest WA metocean criteria as presented in Table 3 and displayed in Figure 1 impose extreme challenges for a ship-shape hull even with a turret system in extreme sea states. For example, maximum significant single amplitude vertical motions in 100-yrp and 10,000-yrp will reach about 10m and 20m respectively. The suitability of ship-shape hull for WA metocean to meet Australia regulation "manned" requirements is questionable. This study presents a better alternative hull form which is more suitable and cost effective solution.

In Gulf of Mexico with similar metocean criteria as those in WA, two ship-shape FPSOs will disconnect if the forecasted sea states are sever than 10-year hurricanes. So far, no ship-shape FPSO is reported to stay "manned" in 100-yr and 1,000-yr hurricanes in GoM.

In addition, complicated and expensive turret system and costly thrusters are also attributed to the inherited shortcomings of ship-shape hull form.

In this study, equivalent process capability 1.5 MTPA was assumed and can be expanded to 2.0 MTPA within current industry capability of available construction dry dock and dry transportation vessel.

*** Formatting Issue - Best Available Paper/Graphic ***

CONCLUSIONS

An alternative FLNG hull, PCS-FLNG, is proposed and its feasibility has been demonstrated through

- A robust sizing basis for 1.5 MTPA
- Topside layout
- Hull configuration
- Numerical simulations
- Topsides and hull constructability
- Transportability of hull with integrated topsides
- Installability
- Operability

Advantages of a PCS-FLNG vs. a ship-shaped FLNG are highlighted as follows,

- Less complexity and capital expenditure with spread-mooring vs. internal or external turret system
- Better heave motion characteristics which leads to better offloading operations and offers large offloading window and less requirement on offloading arms.
- Better roll motion which results in better performances of cryogenic heat exchangers and acid gas absorbers and less sloshing in LNG tanks.
- No thrusters reduces complexity of hull arrangement and cost of operations

If the standardized and phased approach is acceptable, then there is no need to consider large ship-shape FLNG for large field development in WA. For a mega field, two or three 1.5 MTPA units might be considered with the standardized, fast-track and phrased approach. Thus, it is believed that the proposed PCS-FLNG alternative is a safe and cost competitive candidate for the project in future.

ACKNOWLEDGEMENTS

The author would like to acknowledge the management of Houston Offshore Engineering, LLC, an Atkins Company for their supports of the research on application of PC Semi for small standardized FLNG and for allowing the publication of the study results. The author is also grateful for valuable comments and assistances from many colleagues, especial thanks go to Mr. Chuck Brown for topsides layouts, Mr. Joey Lopez for construction drawings, Dr. Vimal Vinayan, Mr. Brajesh Kumar and Mr. Dongxing Jia on numerical results.

REFERENCES

1. Das, S and Zou, J., "Characteristic Responses of a Dry Tree Paired-Column and Deep Draft Semisubmersible in Central Gulf of Mexico", SNAME2015, Feb. 2015.

2. Duhamel, N., "LNG Tandem Offloading Solution: Ready for Development", Offshore Technology Conference 2016, Houston, Texas, USA. OTC 27166.

3. Jeanneau, E., Chretien, D. and Fuchs, V., "How to Develop An Economical Small Capability Floating LNG Taking Advantage of A Design for a Larger Capability Unit", 17th International Conference and Exhibition on Liquefied Natural Gas, 2013.

4. Le Devehat, R., "The Offshore Vital Link for Tandem Transfer of LNG", Offshore Technology Conference 2011, Houston, Texas, USA. OTC 21709.

5. Le Devehat, Poidevin, L., Pek, B., Thomson, D. and van Eekeren, J., "Offshore LNG Transfer: the Hard

Link", World Gas Conference, Kuala Lumpur, 2012.

6. Odeskaug, L., "The Cylindrical Hull Concept for FLNG Application", Offshore Technology Conference 2015, Houston, Texas, USA. OTC 25703.
7. Pek, B., and van der Velde, H., "A High Capacity Floating LNG Design", 17th International Conference and Exhibition on Liquefied Natural Gas, 2013.
8. Research Partnership to Secure Energy for America; RPSEA (2009-2010). "Ultra Deepwater Dry Tree System for Drilling and Production", RPSEA UDW 1402, Stage 1 Final Report & Stage 2 Summary Report, H08130-G-RPT-GN-15002 & H08130-G-RPT-GN-15003.
9. Sansom, P. and Ankarsward, P., "Semi Solution for FLNG", Upstream, the International Oil and Gas Newspaper, April 15, 2016.
10. Talib, J., and Price, B.C., "Development of Floating LNG Production Units with Modular/Scalable SMR Processes", Offshore Technology Conference 2011, Houston, Texas, USA. OTC 21976.
11. Terashima, T., Shaimada, K., Ichimaru, Y., and Orimo, Y., "Methodology to Define Design Motion Criteria for Performance of Floating LNG Process Facilities", Offshore Technology Conference 2011, Houston, Texas, USA. OTC 21292.
12. Watanabe, K., "IHI-SPB Tank System for FLNG", IHI Marine United Inc., Nov, 2012.
13. Zou, J., "Dynamic Responses of a Dry Tree Semi-submersible Platform with Ram Style Tensioners in the Post-Katrina Irregular Seas." International Society of Offshore and Polar Engineering Conference 2008, Vancouver, Canada.
14. Zou, J., and Chianis, J., "Paired-Column Semisubmersible Platform for Wet Tree with Steel Catenary Riser Application in Offshore Western Australia." 1st International Conference on Ocean System Engineering, 2011, Seoul, Korean.
15. Zou, J., "Dry Tree Paired-Column Semisubmersible Platform for Ultra-Deepwater Offshore Brazil." 17th SNAME Texas Section Offshore Symposium 2012, Houston, Texas, USA.
16. Zou, J., "Semisubmersible platforms with Steel Catenary Risers for Western Australia and Gulf of Mexico", International Journal on Ocean System Engineering, June, 2012, vol. (2), pp 99-113.
17. Zou, J., Poll, P., Antony, A., Das, S., Padmanabhan, R., Vinayan, V., and Parambath, A., "VIM Model Testing and VIM Induced Mooring Fatigue of a Dry Tree Paired-Column Semisubmersible Platform", Offshore Technology Conference 2014, Houston, Texas, USA. OTC 25427.
18. Zou, J., Poll, P., Roddier, D., Tom, N., and Peiffer, A., "VIM Testing of A Paired Column Semi-submersible", Proceedings of International Conference on Ocean, Offshore and Arctic Engineering, June 9 - 14, 2013, Nantes, France. OMAE2013-10001.

SNAME
THE INTERNATIONAL COMMUNITY
FOR MARITIME AND OCEAN PROFESSIONALS

REVISITING GLOBAL RESPONSE OF FPSOS IN SHALLOW WATER AND THE RISER ANALYSIS REQUIREMENTS

AMIR H. IZADPARAST
SENIOR RESEARCH ENGINEER, HYDRODYNAMICS AND MOORING TECHNOLOGY, SOFEC

JIAXING CHEN
RESEARCH ENGINEER, HYDRODYNAMICS AND MOORING TECHNOLOGY, SOFEC

ARUN DUGGAL
CHIEF OF TECHNOLOGY, SOFEC

Proceedings of the 22nd Offshore Symposium, February 2017, Houston, Texas
Texas Section of the Society of Naval Architects and Marine Engineers

ABSTRACT

Analysis and design of riser systems for permanently moored FPSOs/FSOs in shallow water are known to be extremely challenging. The complex nature of shallow water hydrodynamics and dynamic responses of the FPSO in extreme environmental conditions, limitations caused by water depth, and the close proximity of risers to each other and other structures, define significant technical and practical challenges for the riser designer. These challenges are typically magnified by conservative estimates of the global response of the FPSO system.

The new riser design codes tend to be vague about global riser analysis requirements and do not clearly define the global mooring analysis outputs, i.e. vessel offset and correlation between low-frequency offset and maximum wave height, to be used in riser detailed analysis. The codes commonly refer to the fully coupled mooring-riser system analysis for more realistic representation of the behavior of the system but do not provide guidance on the application of various offsets, associated directions, and environmental conditions. However, due to the large analysis matrix and level of numerical modeling required for detailed riser analysis, the fully coupled mooring-riser system analysis is not typically used by riser designers in the actual project phase. It is common to perform the riser analysis by assuming the vessel is at an extreme offset and performing dynamic analysis about that location with the maximum wave by using a time history of regular waves, or segments of random wave elevation time histories that contain the maximum wave. While this approach may not cause much difficulty for deepwater riser systems, it has a profound effect on the performance of riser systems in shallow water, leading to challenges in both the design and costs associated with such a system as it exaggerates the dynamic response of the floater, and the loading on the riser system.

In this paper, model test data and numerical time-domain simulations have been utilized to revisit the response of typical shallow-water FPSOs in extreme conditions. The main purpose here is to discuss the natural response of the system and highlight the correlation between the low-frequency and wave-frequency responses.

Keywords: FPSO Response, Shallow Water, Riser System Design, Offset Requirement

INTRODUCTION

In most offshore projects, the mooring and riser designs are typically performed by separate teams and the main interface between these groups is the definition of the global response of the floating system in different environmental conditions. There is, however, a disconnect between the way the vessel offsets are developed and reported by the mooring designer and how they are then applied in the riser system design. As a solution, the new riser design codes and standards rely on advancements in numerical calculations and commonly refer to the fully coupled mooring-riser system analysis for more realistic representation of the response of the system. However, due to the large analysis matrix utilized in the riser system design and the level of details needed in numerical modeling of risers, the fully coupled mooring-riser system analysis is not typically used in detailed analysis and design of the riser system in the actual project phase. Typically a simplified uncoupled approach is utilized where extreme quasi-static offsets defined by the mooring analyst are combined with a simplified representation of extreme wave-frequency responses.

In general, the total global response of the mooring system to environmental loadings can be divided into three main components: mean, low-frequency (LF), and wave-frequency (WF). The mean component is the static response of the system to the mean environmental loadings, LF response refers to the response of the system to the slow drift forces and occurs at the system natural period, and the WF component refers to the dynamic response of the system to the wave action. As an example, the components of the turret offset for a shallow water system are presented in Figure 1. In frequency domain calculation, the contribution of each term is calculated separately while in the time domain approach the total response is typically calculated directly. The term quasi-static offset, called offset for simplicity, refers to the combination of mean and LF components of the total horizontal motion at the riser top connection, e.g. turret center. The extreme statistics of the offset can be easily obtained from the summation of the mean offset and the extreme LF offset in frequency domain approach. In time domain approach, the quasi-static offset timeseries can be obtained from low-pass filtering of the total horizontal motion signal and its extreme statistics can be estimated by performing extreme analysis on the sample data from multiple realizations.

For uncoupled riser analysis, the vessel is placed at a pre-defined quasi-static offset and the dynamic response of the riser is captured by using regular waves or segments from a random wave time history and allowing the vessel to only respond to the WF loads. The new design codes and industry guidelines tend to be vague about the vessel offset requirements for uncoupled riser analysis which adds to the confusion and results in conservative and in some cases unrealistic assumptions about the global response of the system. In some cases, the general recommendations about offset requirements for riser analysis purposes in the earlier revisions of the design codes have been removed from the latest revisions, e.g. API RP 17-B fourth and fifth editions. One main reason for these modifications can be related to the industry's general tendency to remove prescriptive requirements from the new codes and standards to give the users more flexibility to define appropriate criteria. This approach has provided room for the development of project specific requirements and criteria which in general is positive. On the other hand, the offshore industry has been mainly focused on deepwater developments in the past few years and the common practices have been updated to mainly consider the challenges associated with the deepwater related concerns. In some cases, these new common practices and requirements do not distinguish between deepwater and shallow water applications and miss the fact that the response of the deep water and shallow water systems could be significantly different.

As indicated in previous studies (Duggal et al. 2004), the complex nature of shallow water hydrodynamics in extreme environmental condition, non-linearity of mooring system response, limitations caused by water depth, and the close proximity of risers to each other and other structures define significant technical and practical challenges for the riser designer. These challenges are typically magnified by conservative estimates of the global response of the system resulting in significant technical challenges and cost impact on the riser system design. Therefore, it is prudent to evaluate the global response requirements and the common practices used in riser system design.

FPSOs in shallow water are characterized by stiff mooring systems and very low system damping, i.e. typically less than 10% of the critical damping, leading to high dynamic response primarily around the natural surge period, i.e. typically less than 150 seconds. This results in small mean offset to the environment load, coupled with large LF dynamic response which could be significantly larger than the mean offset in some cases. This can be seen in the example shown in Figure 1 where the main contributor to the total horizontal offset is the LF component and both mean and WF contributions are significantly smaller. In contrast deepwater systems are quite heavily damped, i.e. around 30-60% of critical and thus the low frequency dynamic offsets are less pronounced.

In recent years, there have been efforts to better incorporate the global response of the system into riser system design with the goal of eliminating the unnecessary conservative assumptions. For instance, an integrated mooring-riser system design approach is being used (Martens et al. 2011) more frequently in which the estimate of vessel offset for the specified seastates are directly used by the riser designer instead of using a pre-defined maximum offset for all environmental combinations.

Following these efforts, in this paper the correlation between the vessel LF offset and the maximum wave height is studied. The focus of this study is on the shallow water systems as the correlation between LF and WF responses tends to play a more important role in shallow water systems as compared to the deepwater systems. The main motive here is to shed light on some of the known basics and revisit the assumptions made in riser analysis and design with respect to the mooring system global response characteristics.

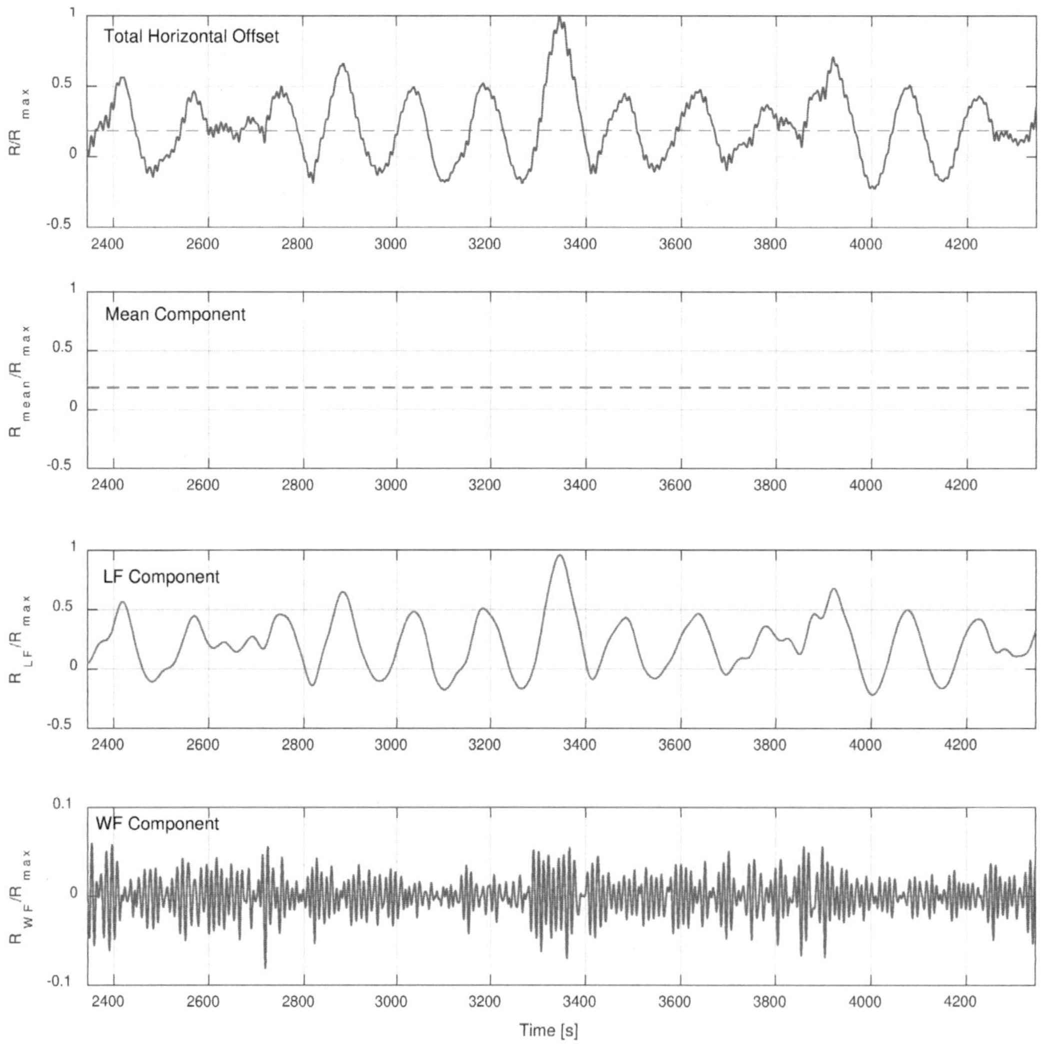

Figure 1 Example of turret offset of a shallow water system and the contributing components

UNCOUPLED RISER ANALYSIS REQUIREMENTS

The term uncoupled analysis has different applications in the field of ocean engineering so it is worth defining the term for riser analysis application. The uncoupled riser analysis is typically referred to the type of analysis in which the riser detailed analysis and design is done separately from the global mooring system response analysis. For this purpose, the vessel is placed at a representative quasi-static offset and the vessel WF responses and loads on the riser are captured by vessel motion RAOs and applied waves. These waves are typically applied as a short time history of regular waves with height equal to the maximum wave height or short segments of random wave elevations that include the maximum wave. The vessel offset itself may have been calculated from a coupled global mooring response analysis in which the effect of risers on the global system response, i.e. stiffness and damping effects, is captured. The main question to investigate here is how the representative quasi-static offset and the parameters of applied wave condition are defined and how the correlation between LF and WF responses are captured in the uncoupled riser analysis.

The mooring design codes, e.g. API RP 2SK, DNVGL-OS-E01, and ISO 19901-7, consistently define the total design offset, i.e. Most Probable Maximum (MPM) offset in a 3hr storm, using the following expression:

Design Offset = max [Mean Offset + MPM (LF Offset) + significant (WF Offset),
Mean Offset + significant (LF Offset) + MPM (WF Offset)]

The expression is recommended for frequency domain calculation where the LF and WF responses are calculated separately. In time domain approach, the total offset is calculated directly and the quasi-static offset is obtained by filtering the WF contribution. Multiple realizations are required to estimate the statistics of the extreme quasi-static offset in time domain approach.

The mooring codes, however, do not provide a clear guideline on the way the global analysis information should be used for the uncoupled riser analysis purposes. The riser design codes, e.g. API RP 17-B and DNV-OS-F201, are not very specific about how the global analysis inputs need to be considered for riser analysis and design either. Consequently, it is common for riser analysis and design to assume that the extreme LF offset happens at the time of the extreme wave height. Specifically, for riser analysis purposes the vessel is located at the extreme LF offset, i.e. MPM or Expected maximum LF offset in a 3-hour storm, and WF extreme responses are captured by applying either a regular wave with the MPM wave height and associated period or a short window of irregular wave timeseries containing the MPM wave height. The underlying assumption of this approach is that WF and LF responses are fully correlated and their peaks coincide. While this approach may not cause much difficulty for deepwater systems, it could have significant impact on the design of the shallow water riser systems. This approach is assumed to be conservative and if the riser system is not affected by the assumption is an efficient approach when it comes to riser analysis and design. However, for shallow water systems in relatively harsh environments, this conservative assumption can have a very large impact on the riser analysis and design, impacting the riser structure design, the riser configuration, and even the design of the FPSO and its mooring system.

This discussion can be extended to the ways the phasing between the vessel WF motions, e.g. turret vertical motion, and the wave loads on the riser are captured in uncoupled riser analysis. Furthermore, defining the parameters of the applied regular wave is debatable and attention has to be given to the response of the system and governing factors while defining the parameters of the applied regular wave. This paper briefly discusses some of these topics but the main focus remains on considerations made in combining LF response and the wave height.

CASE STUDY

As a case study the response of an external turret-moored system in extreme condition is evaluated. The studied case is an FSO with a typical external turret designed for a water depth of 60 meters in relatively benign conditions with a 100-year return period Hs of about 7.0 m. The mooring system consists of 9 catenary anchor legs arranged in 3 groups of 3 identical anchor legs. Each anchor leg is made of all chain components. The vessel is an Aframax size converted tanker with a relatively small topside area meaning that the vessel quasi-static response is dominated by wave loading. Schematic views of the mooring and riser system of the studied case are shown in Figure 2. As shown in this figure, the risers have a lazy S configuration supported by a midwater arch.

In this study, the scaled model test data for the most critical loading condition and the most critical environmental combination is used. The environmental condition represents a design 100-year return period condition. The data from 5 independent model test realizations have been utilized. Moreover, the results of time domain calculations of this FSO system have been used to obtain better estimates of the statistics. The surge natural period for this condition is about 130 seconds.

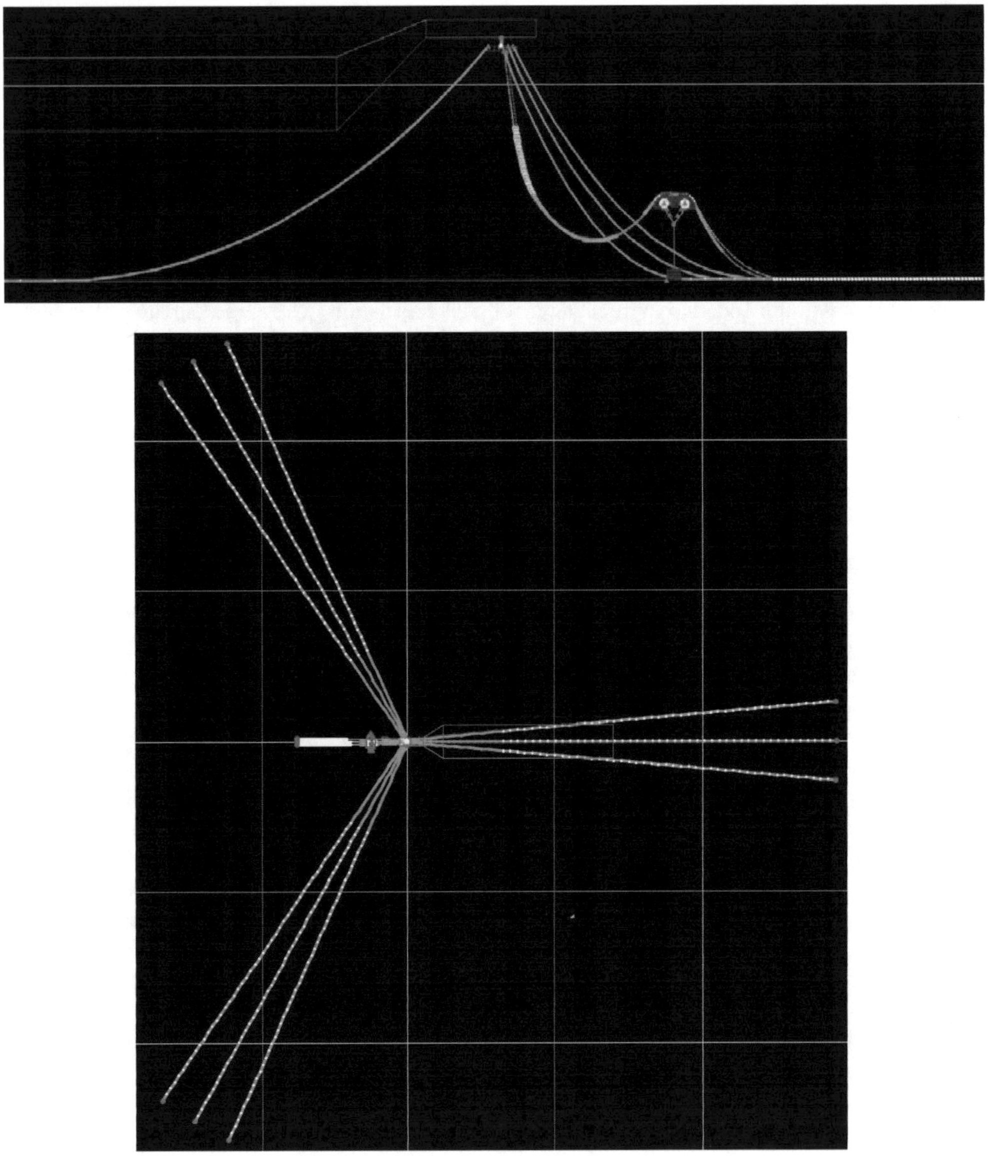

Figure 2 Schematic views of the studied mooring system

DISCUSSION ON GLOBAL RESPONSE OF THE MOORING SYSTEM

The vessel quasi-static offset is a combination of the static mean offset and the dynamic LF offset. Physically, large LF offsets for lightly damped systems can be seen to occur after a large group of waves within the realization interact with the vessel. Generally speaking, the vessel LF offset is a function of the intensity of the wave group and the vessel initial condition. The maximum individual wave height, however, may or may not be part of the wave group with the highest intensity. This is illustrated in Figure 3 which shows the measured model test timeseries of wave elevation and turret offset for the Case Study. In this figure, the results of 5 different realizations are presented and the location of maximum observed wave height and maximum observed vessel offset in the 3-hour test are highlighted. The offset estimates in Figure 3 are transposed into the wave direction and therefore a positive offset indicates a downstream turret offset and a negative offset refers to an upstream turret offset. For the ease of comparison, the timeseries in this figure are normalized by the average of the observed maxima of the five seeds, i.e. average maximum downstream offset and average maximum wave height. Due to the nonlinearity of waves, the crests and troughs are not symmetrical around the zero mean position but the wave heights seem to be almost linear and they follow the Rayleigh distribution relatively closely.

As shown In Figure 3, the absolute maximum observed offset and maximum observed wave height could be far apart indicating that the wave group with the highest intensity may not contain the largest wave. The five 3-hour simulations show that time from a few hundred to several thousand seconds can occur between the two maxima. This indicates that the maximum offset and maximum individual wave height are uncorrelated. However, the maximum offset happens after a large wave group that contains a relatively large individual wave. To investigate the correlation between the extreme offset and the extreme wave height, a few examples of extreme incidents from the 5 seeds are selected and the timeseries for a shorter period are presented in Figure 4.

The first example (Figure 4 (a)) is a typical response of the mooring system to a large wave group. As shown in this figure, the extreme offset happens with a time lag after the end of the large wave group. The large wave group in this case contains a relatively large individual wave height, i.e. smaller than the maximum observed wave height in the timeseries. The important note here is that even though an extreme wave height has resulted in the maximum offset, the extreme wave height and the maximum offset do not coincide. Like any spring and mass system, the time lag between the maximum loading and maximum response depends on the system characteristics, i.e. mass properties, damping, and initial condition, etc.

The first part of the second example (time <6250sec of Figure 4 (b), is similar to what was seen in the first example where a large wave group results in an extreme offset. It is however seen in the second example (time > 6250sec) that the wave group dies out and wave elevation remains low for quite some time. This allows the restored force in the mooring system to cause a significant upstream offset. On the cycle after the maximum upstream offset, the system experiences another large wave group that results in the second large offset. As can be seen here the wave group ends with a large individual wave and consequently a large offset that is slightly less than the maximum offset coincides with a large individual wave height that is smaller than the maximum observed wave height. This example highlights the importance of considering the randomness of wave elevation and vessel response in analysis and design process.

The third example (Figure 4 (c)) shows the maximum observed wave height that was not accompanied with a large wave group. As shown in this figure, the maximum observed wave height did not result in a significant offset.

These examples highlight that extreme wave height and extreme vessel offset have a cause and effect relationship but their peaks do not typically coincide and therefore assuming that MPM wave height occurs with the MPM offset for riser design is an extremely conservative approach. This is why the mooring design codes combine extreme quasi-static offsets with significant WF responses and vice-versa.

As a side note, this discussion is extended to the correlation between wave height and turret vertical motion. As indicated earlier in this paper, riser analysis is typically done using regular wave analysis with MPM wave height and associated wave period or irregular wave analysis of a short time span containing the MPM wave height. Similar to Figure 3, the timeseries of normalized wave elevation and turret vertical motion from the five 3hr seeds are presented in Figure 5. The turret Z motion is normalized with the average of maximum positive Z motion of the five seeds. The turret Z has a linear behavior with peaks following Rayleigh distribution. The turret vertical motion is a result of vessel pitch motion and vessel heave motion while the pitch motion is the dominant contributor. It is worth mentioning that the surface elevation is measured at turret original offset so a few seconds of time shift is expected between surface elevation and turret vertical motion. The location of maximum observed wave height and maximum and minimum observed turret vertical motions is indicated in Figure 5. As shown here, even though a strong correlation between vertical motion and wave elevation exists, the maximum wave height may not necessarily result in the maximum turret vertical motion. As expected the wave period associated with the extreme wave height plays an important role on the vessel WF response.

Similar to Figure 4, a few incidents of extreme wave height and extreme turret vertical motion are highlighted in Figure 6. The main remark here is that in case the riser design is sensitive to turret vertical motion, which is typically the case in shallow water designs, attention has to be given to the wave period associated with the maximum wave height. This is more relevant when regular wave analysis is being used for riser analysis purposes.

STATISTICAL ANALYSIS OF EXTREME OFFSET AND EXTREME WAVE HEIGHT

To perform a statistical analysis on the extreme offset and extreme wave height, the results of numerical time domain simulations have been used. The numerical model is based on the same system that was model tested and its performance has been verified using the model test data. The same load case and vessel loading condition as used in the model test campaign has been used in this example. For this purpose, the results from 50 independent 3-hour simulations have been utilized. To increase the sample size the peaks-over-threshold method has been used and threshold is defined as 90% of average observed maxima from 50 realizations.

In Figure 7, the correlation between extreme offsets and associated wave heights, and extreme wave height and associated offset is presented. The data points in this figure are normalized with the appropriate average observed maxima from 50 realizations. Figure 7 (a) shows the wave height at the time of the maximum observed offset. As can be seen in Figure 7 (a), the maximum observed offset from the 50 realization is equal to 1.15 times of the expected maxima. These maximum observed offsets occur with an individual wave height that is less than 60% of the expected maximum wave height. Similarly in Figure 7 (b) the offset at the time of the maximum observed wave heights are reported. As shown in this figure, the maximum observed wave height in 50 seeds is more than 1.15 time the expected wave height and this wave height is associated with an offset that is less than 65% of the expected maximum offset. It is also important to note that the extreme offset and extreme wave height do not necessarily happen at the same time and as can be seen in Figure 7 (a) and Figure 7 (b), there is no clear correlation between the extreme observations of the two random variables.

In Table 1, the statistics associated with the extreme events are presented. In this table, Hs and Rs refer to the significant wave height and significant vessel offset, respectively. As shown in this table, the average wave height associated with maximum offset is about 65% of Hs and 36% of expected maximum wave height. Similarly, the average offset associated with maximum wave height is about 62% of significant quasi-static offset and 48% of expected maximum offset. To compare, the mooring design codes combine the MPM quasi-static offset with significant WF response, i.e. E (H) / Hs = 1.0, and combine the MPM WF response with significant quasi-static offset, i.e. E (R) / Rs = 1.0, which is more conservative than the results obtained here. Therefore, the recommendation made by mooring design codes to combine the MPM, or expected maximum, quasi-static offset with maximum WF response and vice-versa is better suited for riser analysis purposes.

Finally in Figure 7 (c), the correlation between extreme wave height and extreme turret vertical motion has been shown. The results show some level of correlation between the two variables but as discussed before some variability mainly due to the variation of wave period associated with maximum wave height can be seen.

CONCLUDING REMARKS

This paper aims at reviewing the characteristics of global response of permanently moored FPSOs/FSOs in shallow water and revisiting the inputs to the riser analysis. The main conclusions are listed below:

- The comments made here are based on one example. However, based on authors' experience with the analysis and design of mooring system for several shallow water FPSOs, the main observations are dictated by the basic response of the system and can be extended to other systems.

- As shown here, the assumption used in the uncoupled riser analysis that maximum wave height and maximum vessel quasi-static offset coincide is very conservative. In the author's experience using this approach can lead to requirements for very heavy riser structures, more complex riser configurations, and have a large impact on the FSO hull, turret and mooring system, all leading to increases in CAPEX.

- The vessel offset is a function of wave group intensity and the large wave group may not contain the individual wave with the maximum wave height. Moreover, a singular large wave does not cause a significant offset. Therefore, the maximum offset is not typically caused by the maximum wave height.

- The natural response of the mooring system typically causes a lag between the extreme offset and the largest wave height in the wave group forcing the extreme offset. As a result, the peaks of offset and wave height do not necessarily coincide.

- The statistical analysis performed here using 50 numerically generated realizations confirmed the qualitative observations made from the model test data. Specifically, it has been confirmed that the peaks of wave height and vessel offset do not regularly coincide and the extreme values of the two variables are not correlated.

- Based on the results shown here, it is recommended to combine the MPM (or expected maximum) quasi-static offset with the significant wave height and the MPM wave height with the significant offset for uncoupled riser analysis purposes. This is in-line with the way the total response is calculated from LF and WF contributions for the global mooring response calculation purposes.

- As expected, the wave period plays an important role in vessel WF response. Therefore, attention has to be paid to the wave period associated with the wave height, specifically in regular wave riser analysis.

- This paper illustrates the importance of understanding the physics associated with the system dynamics before embarking on riser analysis and design. As mooring design (and FSO global analysis) and the riser analysis and design are typically performed by different entities, the definition of the offsets and the proper translation of them from global analysis output to riser analysis input is a critical interfaces during project execution that needs to be recognized by the end user when defining mooring and riser requirements, and ensuring consistency between them. This is also important when defining interfaces between mooring and riser teams in projects. It is quite common to restrict mooring offsets to a pre-determined value to serve as an extreme interface offset for the riser system design. As demonstrated in this paper this offset would be defined by the MPM quasi-static offset (mean + LF) and should be combined with the significant wave, rather than the maximum wave for shallow water systems.

REFERENCES

Duggal, A. S., Liu, Y. H., Heyl, C. N., "Global Analysis of Shallow Water FPSOs", 2004, Offshore Technology Conference, OTC 16720.

Martens, M., Whelan, J. R., Drobyshevski, Y., "Integrated Analysis of Mooring and Riser Systems for FPSO's in Harsh Shallow Water Environments", 2011, International Conference on Ocean, Offshore and Arctic Engineering, OMAE2011-50007.

API RP 17B, "Recommended Practice for Flexible Pipe", 2008.

API RP 17B, "Recommended Practice for Flexible Pipe", 2014.

API RP 2SK, "Design and Analysis of Stationkeeping Systems for Floating Structures", 2005.

DNVGL-OS-E01, "Position Mooring", 2015.

ISO 19901-7, "Petroleum and natural gas industries – Specific Requirements f or Offshore Structures", 2013.

DNV-OS-F201, "Dynamic Risers", 2010.

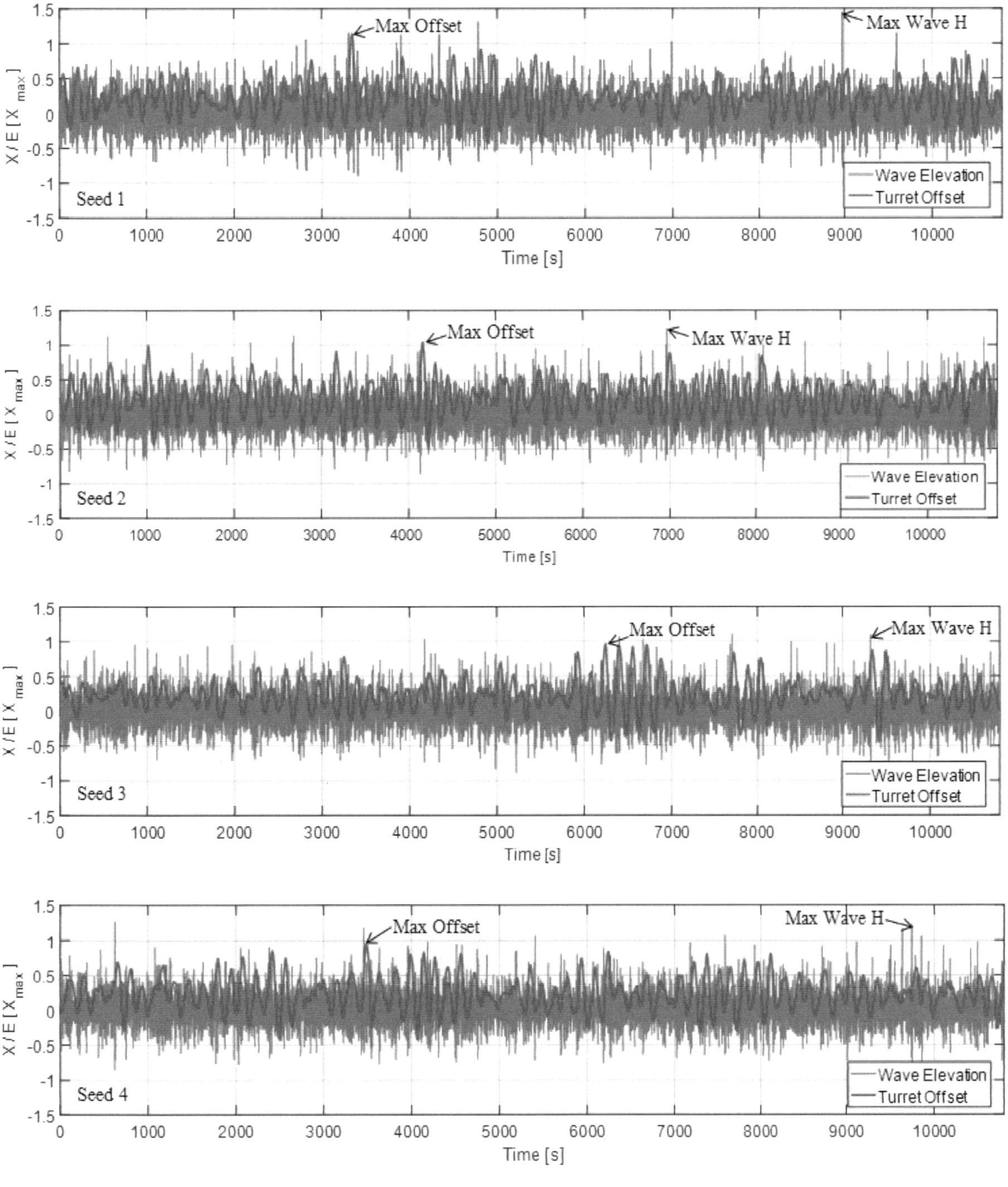

Figure 3 Normalized vessel offset and wave elevation

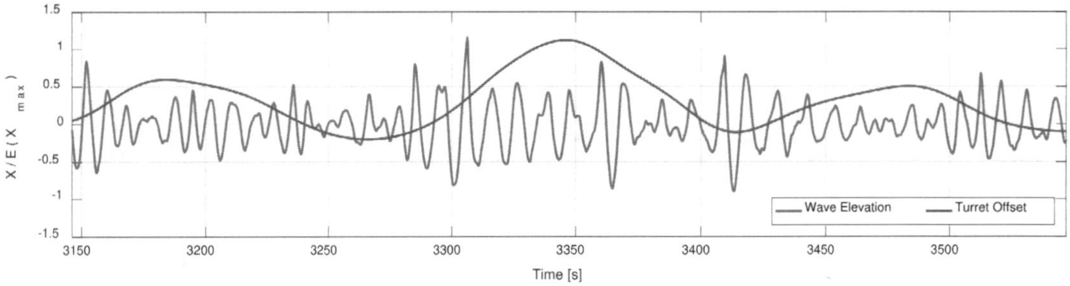

Seed 1 – Max Offset

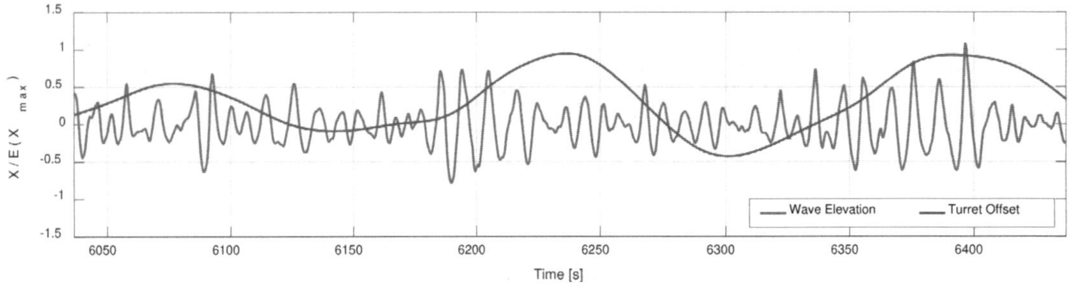

Seed 3 – Max Offset

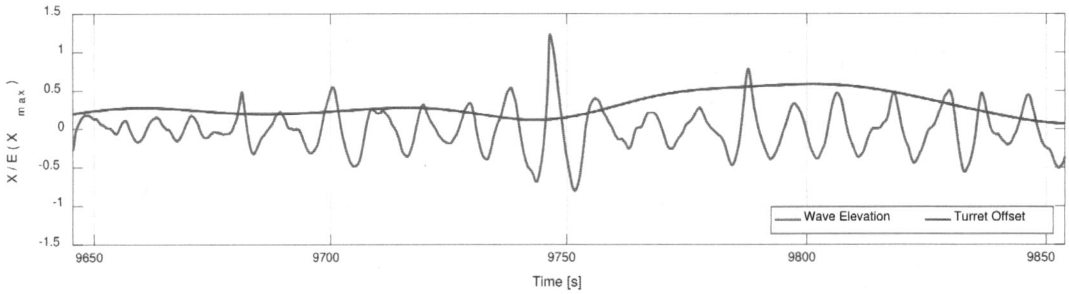

Seed 4 – Max Wave Height

Figure 4 Examples of maximum offset and maximum wave height incidents

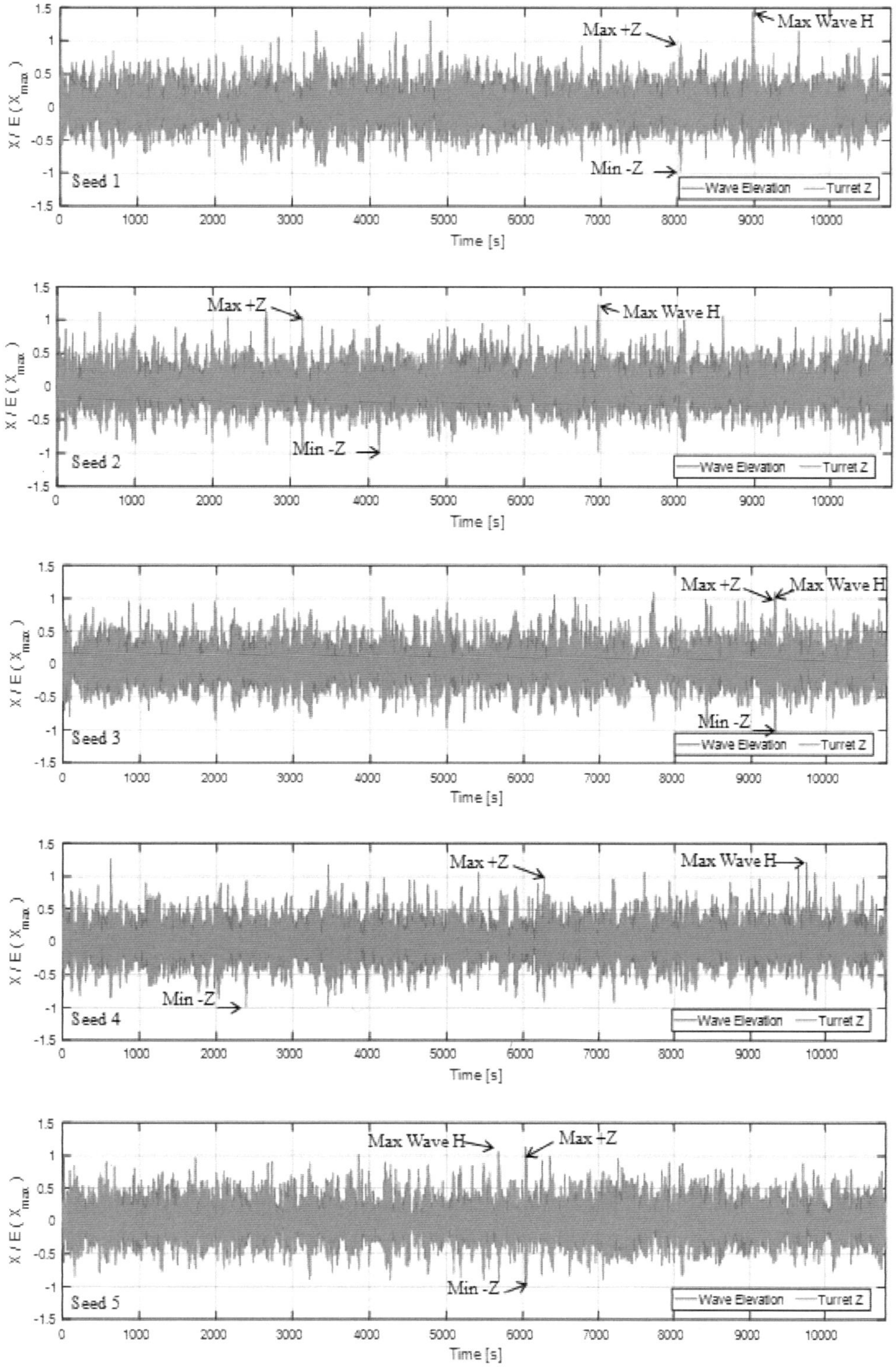

Figure 5 Normalized turret vertical motion and wave elevation

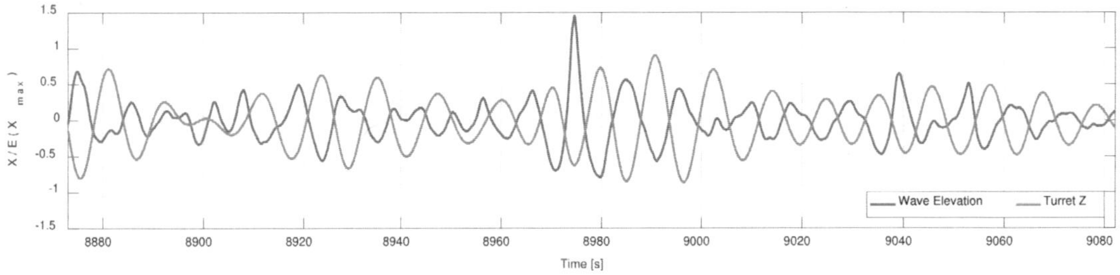

Seed 1 – Max Wave Height

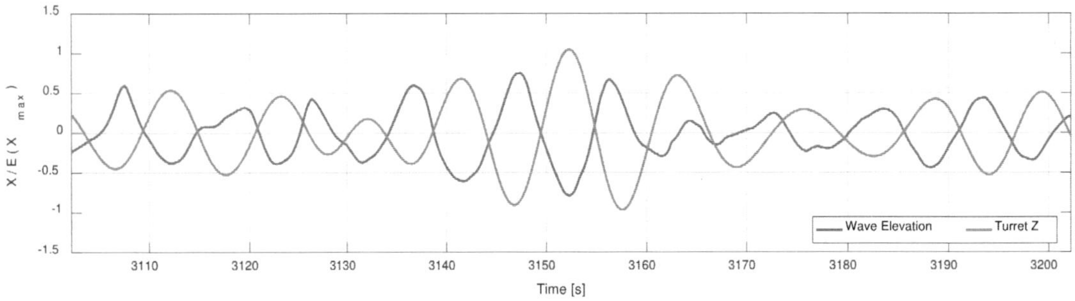

Seed 2 – Max Turret Z

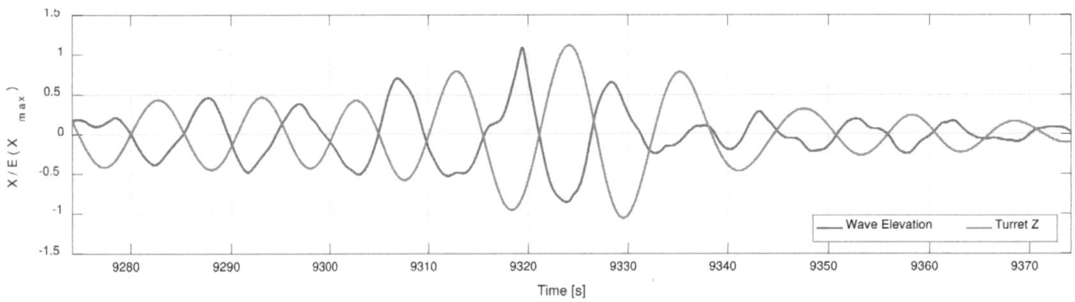

Seed 3 – Max Turret Z and Max Wave Height

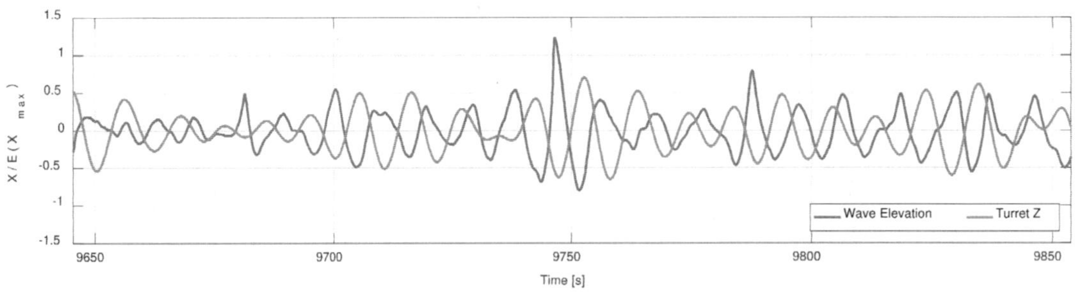

Seed 5 – Max Wave Height

Figure 6 Examples of maximum turret vertical motion and maximum wave height incidents

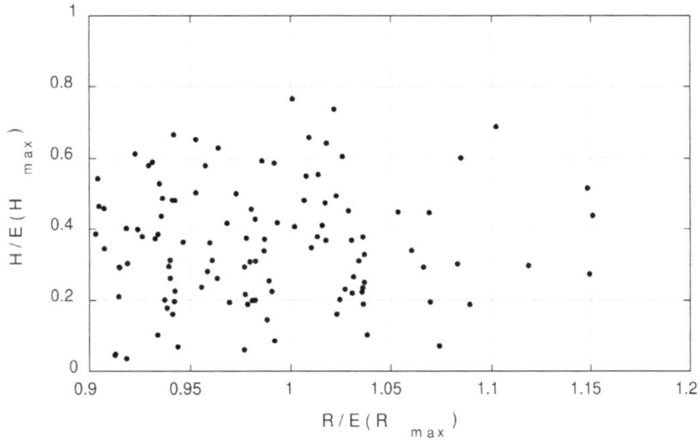

Maximum offset and associated wave height

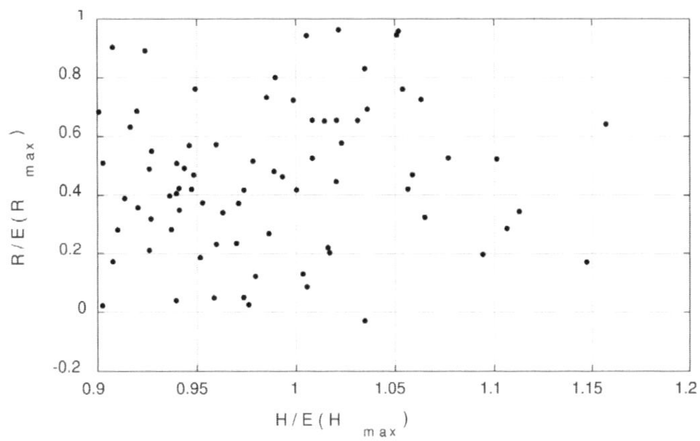

Maximum wave height and associated offset

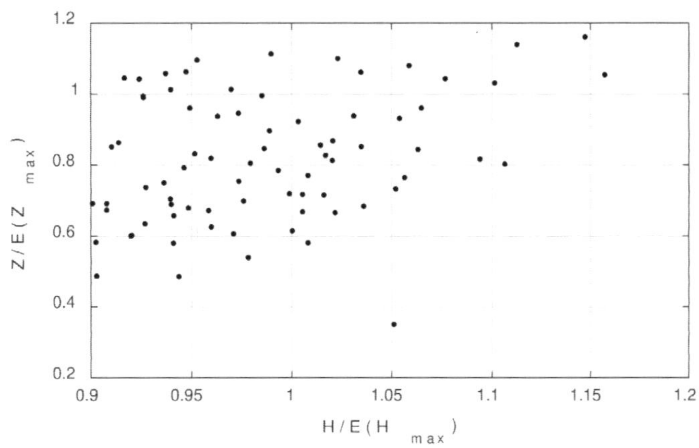

Maximum wave height and associated turret vertical motion

Figure 7 Extreme events and associated values

Table 1 Extreme events and associated values

	Average of Associated Wave Height	
At Maximum Offset	E (H) / E (Hmax)	E (H) / Hs
	0.36	0.65

	Average of Associated Offset	
At Maximum Wave Height	E (R) / E (Rmax)	E (R) / Rs
	0.48	0.62

EDP BUOY

ANIL SABLOK
CHIEF ENGINEER, TECHNIPFMC
JIM O'SULLIVAN
CHIEF TECHNOLOGY OFFICER, TECHNIPFMC
JOHYUN KYOUNG
PRINCIPAL TECHNICAL ADVISOR, TECHNIPFMC

Proceedings of the 22nd Offshore Symposium, February 2017, Houston, Texas
Texas Section of the Society of Naval Architects and Marine Engineers

ABSTRACT

A normally un-manned semisubmersible has been developed, consisting of three columns spanning between a deck box structure and a very deep submerged pontoon. The platform, called the EDP (Extended Draft Platform) Buoy, can be used for various applications such supporting a low cost subsea development of a marginal field with controls and power and other surface processes, or serve as a science platform for gathering long term oceanographic, meteorological and geotechnical data. The EDP Buoy is designed to be towed to site in a stable collapsed configuration like a Jack-up MODU by offshore tug supply vessel and/or Anchor Handling vessel, and self-erect with minimal human intervention. The columns and pontoon of the EDP Buoy are raised for construction and installation. When lowered, the structure behaves like a very stable deep draft semisubmersible. It is suitable for very harsh, but ice free, environments and no water depth limit with use of synthetic mooring lines.

The deck box and the pontoon structures are Y-shaped. Within and atop the spacious EDP deck box there is sufficient area to locate processing, power, communications, instrumentation and material handling equipment, and provide sufficient bulk storage space for fuel and other fluids. Multiple risers and umbilicals can be accommodated on the platform. The facilities on the platform can be expanded in the future including adding more equipment, risers and umbilicals.

The configuration and size of the platform lends itself to simple low cost fabrication with rolled steel plate columns, and box girder deck and pontoon. The three leg design also makes for easy self erection.

The paper describes the configuration, installation procedure and in-place performance of the EDP Buoy.

Keywords: Extended Draft Platform, Marginal Field, Unmanned System, Low Cost, Semisubmersible

INTRODUCTION

In terms of cost savings, hull efficiency and safety, an un-manned platform becomes attractive as offshore operators seek alternatives for developing economically challenging offshore fields. A recent study to evaluate different unmanned wellhead platforms installed in the North Sea found the unmanned platform to be a cost and production effective development concept having the same functionality and robustness as subsea developments [1]. The main issue for the un-manned platform was found to be the platform design which should minimize manned operation & maintenance onboard. Also it is imperative that the initial cost should be minimized by simplifying the fabrication and installation of the platform to enable developing the field economically feasible. One such application of the low cost unmanned platforms would be to support subsea production systems near the subsea wells in deep water with very long tie backs to the host that can provide power, communication, glycol regeneration, and processing produced water.

In a previous study of floating unmanned science platforms, the often used Spar-type buoy has evolved into the Extended Draft Platform (EDP) [2]. In order to reduce the CAPEX of the platforms, similar platforms can be configured to support subsea production systems with different equipment on board and support risers and umbilicals in deep water offshore even with very harsh environments. The configuration also lends itself to add future equipment or replace equipment, move to another field, if required. The paper describes the configuration, installation procedure and in-place performance of a EDP Buoy configured for supporting subsea production system in deep water with harsh environments.

CONFIGURATION

The proposed EDP Buoy consists of an upper deck box, three columns and a lower pontoon as shown in **Figure 1**. The columns and pontoon will be raised for construction and installation. When the EDP Buoy is installed by lowering columns and pontoon, the response is comparable to those of a deep draft semi. The mooring system is a combination chain/steel wire or chain/polyester lines to hold the EDP on station. Use of the near neutrally buoyant polyester makes the buoy essentially water depth independent. After completing commissioning at quayside, the EDP Buoy designed for minimal heavy lift transport will be wet towed to site.

Application of the EDP Buoy as an un-manned platform has the following benefits compared to other concepts:

- The EDP Buoy is self-installing without upending or offshore lifts.
- All equipment and systems can be pre-commissioned.
- The EDP Buoy can be wet towed to final site using one single vessel.
- The EDP Buoy draft can be adjusted by site environment. More severe environment requires deeper draft.
- The EDP column diameter is a function of mission payload. Maximum diameter is kept less than 25 ft for use of double steel sheet for rolling with simple ring stiffeners. However, large payloads can be accommodated with larger column diameter and deck box which would require different column fabrication method than rolling.
- The EDP Buoy overall width is function of surface towing stability, which is primarily function of column height and diameter.
- The equipment located on the deck box is ventilated, accessible and easier to maintain and replace, if required
- Can provide multiple risers and umbilicals supports

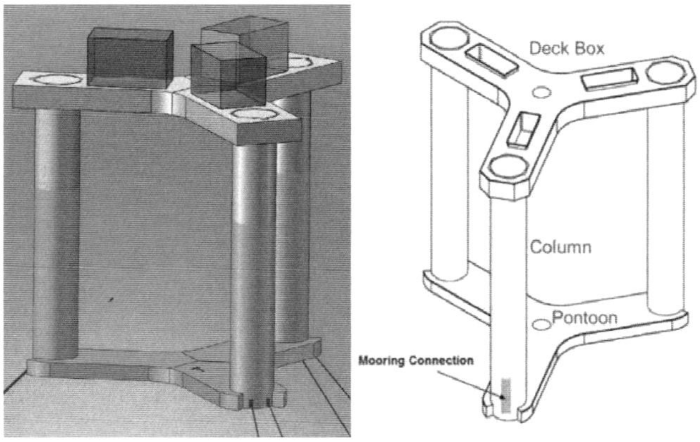

Figure 1 Extended Draft Platform

HULL MISSION AND CONFIGURATION

The EDP Buoy is intended for unmanned missions. The traditional mission would be a Power Buoy to support subsea production systems. The EDP Buoy has been configured to support subsea production systems with power, communication, glycol regeneration and residual hydrocarbon disposal. Power was provided from gas turbines utilizing a small tap of the dehydrated gas from the seafloor via a flexible riser.

Presently, the EDP hull is designed to operate in the world's harshest ice free regions such as Mid North Atlantic, North Sea or offshore North West Australia. No attempt has been made to design the light and robust hull for sea ice of any kind. **Figure 2** illustrates the dimension of an EDP Buoy configured for an application in very harsh environment in Western Australia with a water depth of 3,609 ft (1,100 m). It has a deck area of 8,300 sq. ft.

Table 1 Weight Distribution for the Production EDP Buoy

Weights	
Payload (Topside deck, Facility, Riser load) (kips)	3,207
Hull Weight (incl. hull steel and outfitting) (kips)	4,063
Ballast (kips)	7,249
Mooring Vertical Load (kips)	471
Total Displacement (kips)	14,990
Metacentric Height, GM (ft), In-place	5.9

Both the deck box and the pontoon have "Y" shapes. However, the deck box and the pontoon can be of other shapes like a triangle with three sides between the columns or one complete triangular shape filling all the area between the columns or other shapes. Both structures internally are compartmentalized. For the production EDP Buoy, the deck box includes space for topside facility such as power generator, HPU, water treatment, MEG storage tank and piping. The applicable payload for present design is about 3,200 kips. The generators are housed in a separate container on the top deck. Fuel oil is stored in tanks that are built into the deck structure. Pontoon compartments are utilized to control flooding during installation and re-deployment operations.

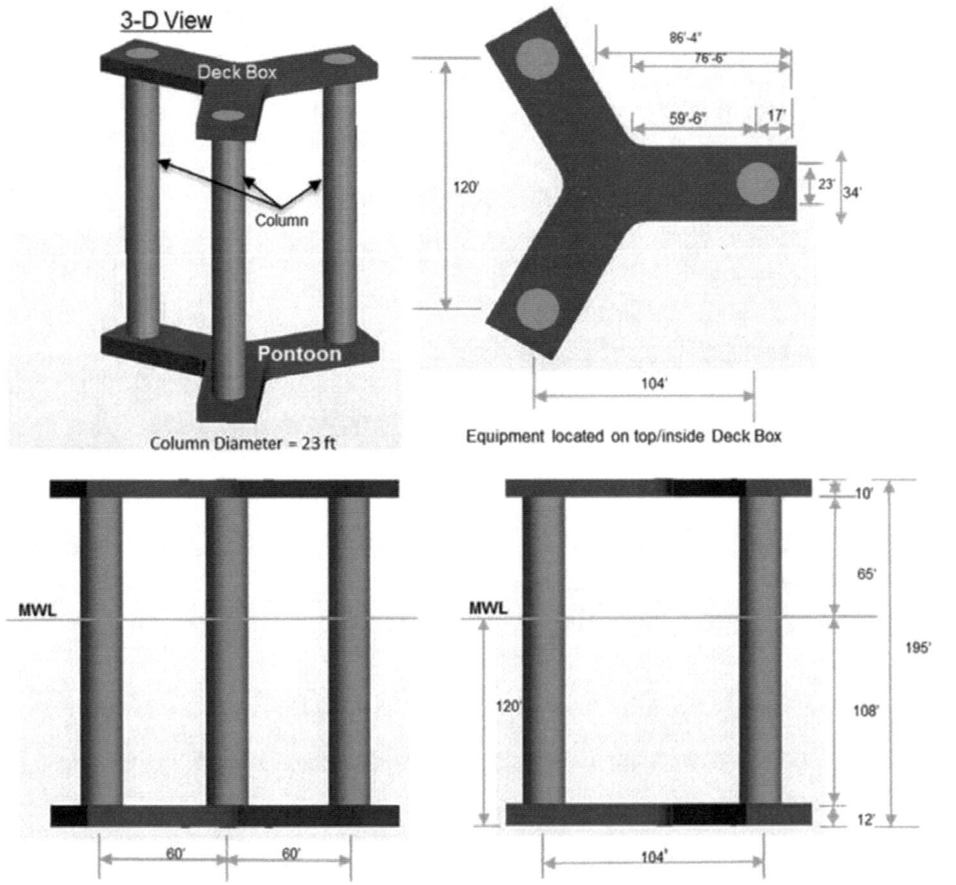

Figure 2 Buoy Hull Dimensions

The mooring system for this example consists of six lines, two connected to each Column. The configuration of each mooring line is summarized in **Table 2**. Each mooring line consists of a Chain/Polyester/Chain system. It is proposed to locate fairleads with chain stoppers near the bottom of the Column and use underwater tensioning system to adjust the chain lengths [3, 4]. Only one tensioner is required to adjust all the lines. For this option, no chain lockers are required in the Columns or Deck Box. Alternatively, one chain jack can be used on top of each Column to adjust the line. It is anticipated that line adjustments will not be required during the platform life.

Table 2 Mooring Line Configuration

6 lines, 3 groups of 2	Type	Length	Diameter	MBL
		(ft)	(in.)	(kips)
Platform Chain below Fairlead	R4S	150	3.4	1,891
Polyester Rope		4,963	6.625	1,800
Anchor Chain	R4S	300	3.4	1,891

As shown in **Figure 3**, the characteristic behavior of the EDP hull is similar to that of a deep draft semi. This should help to use low costs SCRs in deep water or to provide a less demanding and lower cost specification for dynamic flexible risers.

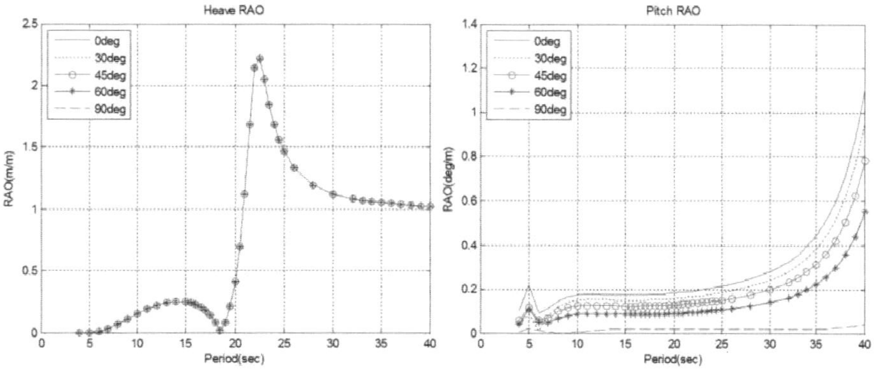

Figure 3 Hull Motion RAO

FABRICATION AND INSTALLATION EFFICIENCIES

The EDP hull is sectionalized into a limited number of components. The EDP has structurally identical column sections. As shown in **Figure 4**, it has two types of deck box sections and two types of pontoon sections. The column sections will be outfitted with air piping for ballasting and HP air accumulators in upper section for ballast air. Top view of the top deck box and pontoon geometry, shown in **Figure 4**, shows identical extension sections and the center connection section. Each section can be fabricated and transported to assembly site by barge or truck. In the fabrication yard, the assembly will be done on a launch skid way. The assembly sequence includes:

- Assembly of the pontoon
- Install spacer bracket on the top of pontoon
- Assembly of the top deck box
- Install the lower column sections to pontoon
- Lift/swing the upper column assembly using crawler crane
- Install the upper column assembly

The unique feature of the EDP is that the columns penetrate the top deck, and the pontoon can be positioned on the ground just below the top deck for construction of the topside facilities at near ground level. After final structural assembly including installation of marine and locking system, the platform is launched into the sea by the skid way, or skidded onto a small heavy lift transit vessel for mobilization to location of final equipment installation and integration, if different from the hull fabrication yard.

After final commissioning, the platform is transported to an unload site nearest the final installation site, from where it is wet towed to site by one vessel while the top deck and pontoon are closely connected as shown in **Figure 5**. The top deck box will provide sufficient buoyancy for offshore tow.

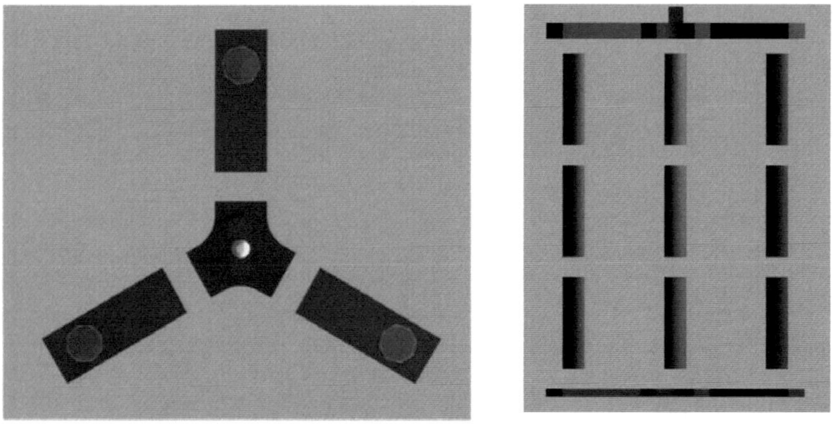

Figure 4 Sectioned EDP Hull

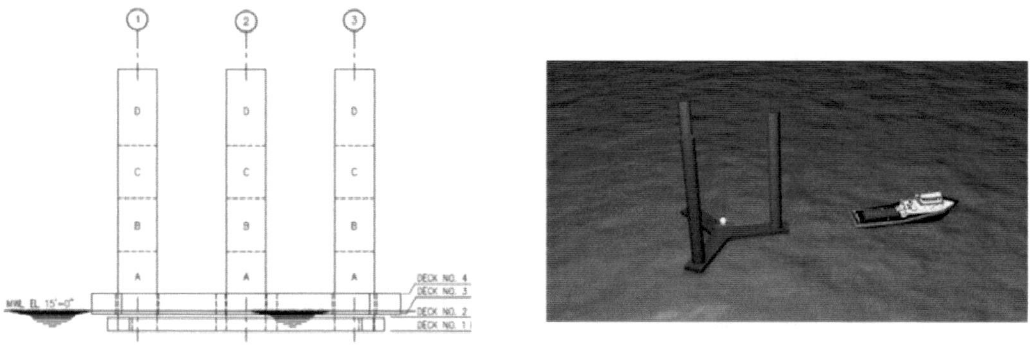

Figure 5 EDP Hull Configuration for Wet Towing and Installation

At location, the boat will keep the platform on location during pontoon lowering and mooring connection. In the lowering operation, the mechanical transit lock will be released. The pontoon will be lowered by synchronous jacks and ballasting of the pontoons. Flooding of the pontoon extends 30% of final draft. Controlled flooding of columns achieves full draft extension as shown in **Figure 6**. A Pawl system, shown in **Figure 7**, will be extended to secure against dynamic column/deck box motions. HP air from the upper columns serving as accumulators displaces column ballast water to lock the pawls with the deck box. Alternatively, the Column lowering operation, connecting Deck Box to Column and engaging the pawls can be done in a deep water sheltered area, if available and then wet tow the platform to the final site.

Following the lowering connection of the Column to the Deck Box, the mooring lines are attached. For the option with Chain Jacks on the top of the columns, pre-installed messenger wires retrieve permanent mooring lines and mooring lines pulled in, pre-tensioned and locked into chain stopper. Similar operations would be performed for the option with integrated fairleads/chain stoppers and using underwater tensioners. Once moored on station with the pontoon and legs lowered, more compressed air released into the lower columns adjusts buoyancy to final operational draft.

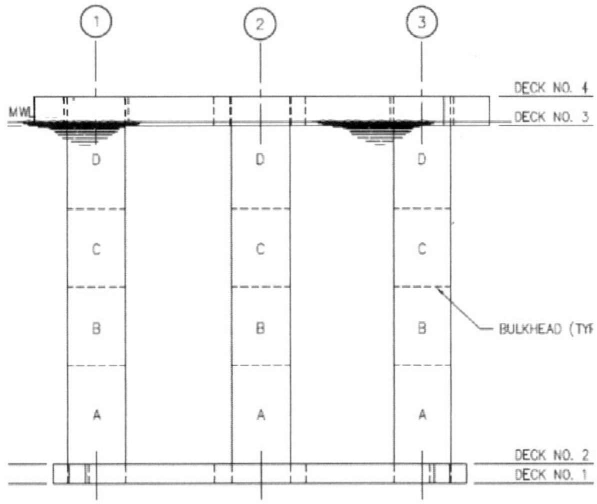

Figure 6 Columns Lowered, Buoy Floating at the Deck Box

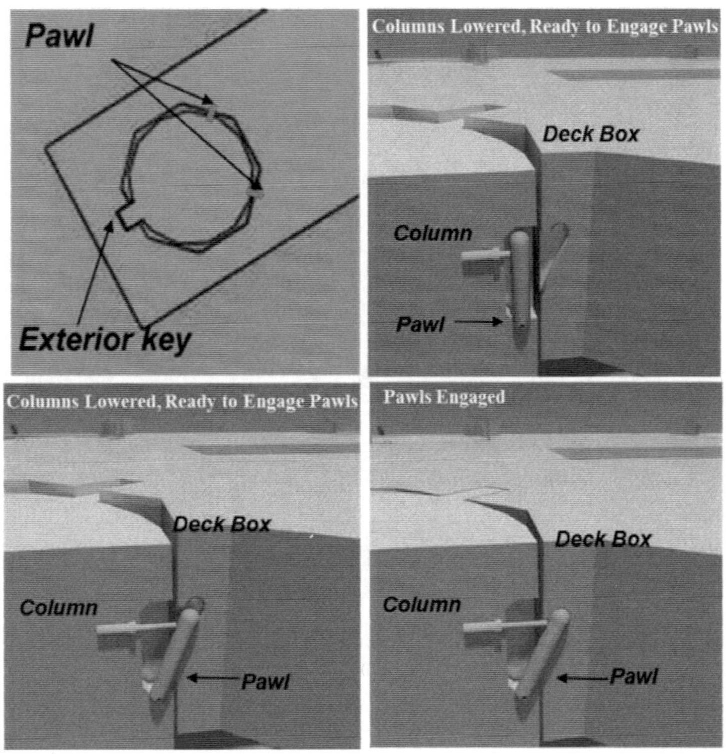

Figure 7 Pawl Configuration and Engagement Procedure

Marine vessels are required to set anchors and mooring lines, deploy subsea instrumentation and cables, install the buoy, hook up the mooring and instrumentation lines, hook-up risers and umbilicals and to change out or maintain the equipment. Most over-the-side operations are limited to sea state 4 with significant wave heights of about 2.5m. Some of the more critical operations, e.g. personnel transfer, typically require less than sea state 3 with about 1.5 m seas. Accordingly, buoy installation and maintenance will be a primary challenge where mild seas are rare. The availability and cost of vessels required for buoy installation may be a driver for the buoy design. At present, North Sea rated AHT (anchor handling tugs) will be needed for sites having severe environment.

OPERATIONS AND MAINTENANCE

The EDP Buoy will operate unmanned and would require occasional inspection and maintenance of the equipment on board. To do this, a supply vessel will approach the EDP stern with lines attached to the adjacent columns. A davit positioned in the center of the EDP will be capable of reaching any of the service vessel for replacement of equipment if needed. Flexible hoses will be used to replenish fuel and lube oil. The EDP hull is designed so that a supply vessel may back down adjacent to the deck box where an air-actuated davit permanently mounted on the EDP hull may move equipment between the EDP hull and the supply vessel. People can access the top of the Deck box either with the help of davit onboard the buoy or by walk-to-work bridge on a support vessel.

CONCLUSIONS

The EDP hull is designed for an un-manned platform supporting cost savings, hull efficiency and most of all safety. The EDP minimizes manned operation for installation & maintenance onboard by providing the following features.

- High reliability, able to operate without manned service.
- Applicable for stand-alone operation, installable as production platform far from existing infrastructure in the harsh environment site.
- Supporting spacious deck space for topside facility required to support process and power to the operation of subsea facility on the sea floor.

ACKNOWLEDGEMENTS

The authors would like to thank TechnipFMC for allowing and providing support to publish this paper

REFERENCES

[1] Asmus Nielsen, "Unmanned Wellhead Platforms – UWHP Summary report", 1140000528/ROGC-Z-RA-000027, 2016.

[2] Atle Steen, John Halkyard, Jon Berger, "New Buoy Designs for Ocean Observatories", Proc. of 26th International Conference on Offshore Mechanics and Arctic Engineering, San Diego, California, USA, OMAE2007-29609, 2007.

[3] Martyn Conroy, Marco Teixeira, "Tether Top Connector & Subsea Tensioning System for Buoyancy Supported Risers", Deep Offshore Technology International Conference, Houston, 2015.

[4] Anil Sablok, Andrew Blundon, Trond Landbo, Johyun Kyoung, Chad Fowlow, Svein Stromme, "Disconnectable Concrete Spar FPSO", Arctic Technology Conference 2016, St. John's Canada, OTC 27467, 2016.

IMPROVING OFFSHORE INSPECTION EFFICIENCY THROUGH UAV INSPECTIONS

STEVE MOIR
ENGINEERING MANAGER, SKY-FUTURES PARTNERS LIMITED

Proceedings of the 22nd Offshore Symposium, February 2017, Houston, Texas
Texas Section of the Society of Naval Architects and Marine Engineers

ABSTRACT

In particularly hostile and challenging deep water environments offshore, it is crucial to ensure asset management is effective, efficient and safe. Knowing the condition of assets is a pre-requisite for successful structural integrity assessments and to support risk based inspection strategies. UAV (Unmanned Aerial Vehicle) inspections gather HD imagery, video and thermal data for both General Visual Inspections (GVI) and Close Visual Inspections (CVI) akin to traditional rope-access inspection techniques. The fundamental benefit of UAV inspections are the vast savings in time and cost to perform in comparison to rope access and the significant reduction in consequent risk to inspectors. These savings will be analysed and compared with traditional techniques using a typical UAV inspection case-study where an oil and gas client saved 90%+ compared to traditional inspection techniques including rope-access inspection work. This representative case study will be used to investigate how a structural inspection of an offshore rig saved the equivalent of 392 days of rope access inspection work, 40 days of support vessel hire for supporting rope access inspection work and 28 days of work that would have been required during a shutdown by comparing and contrasting to traditional inspection techniques. Themes explored will include data capture, speed, safety and cost. More data can be gathered by the UAV in optical (video and stills) and thermal formats, from vantage points otherwise inaccessible to rope-access inspectors physically constrained to the platform. However, the data must be of high quality to add more value to decision-making. UAV pilots can be deployed offshore within a matter of hours to do emergency inspection work as and when required, freeing up bed space offshore. UAVs can typically inspect five structures (e.g. flare boom/tip, telecoms tower, helideck, bridge, under deck) within one week and all inspection reports and imagery can be delivered two weeks later. Rope-access inspections cost more than UAV inspections due to taking longer to complete and being man-power intensive. In the traditional model a 6 week shutdown would be planned periodically for major maintenance work. UAV inspections establish whether shutdowns are needed before they occur. Inspections of live flares before shutdowns can also allow replacement parts for flares to be ordered before shutdowns, minimising downtime and allowing the correct flare parts to be ordered. The key to better maintenance across the lifecycle of a rig is better flexibility, availability and speed of inspections. This paper will assess the best means of achieving this.

INTRODUCTION

The inspection of oil and gas installations is a long established requirement in order to meet international, national and local legislation and regulation and plays a key role in Integrity and Maintenance Management Strategy. The purpose of legislation is to ensure that; the asset is operated safely to prevent injury to people, there is no damage to plant and equipment and impacts to the environment are minimized.

Against this, oil and gas Operators face huge challenges associated with working in mature basins. As plant and equipment age, this typically leads to an increase in the frequency of inspections. This is set against further competition between resources as maintenance, repairs, modifications and alteration scopes also increase.

Despite the additional resource requirement, there is also a need to make permanent and structural changes across the supply chain to counteract oil price volatilities and reduce production costs, yet at the same time increase operational performance and efficiency.

At present, UAV inspections are not an embedded tool of an Operator's Integrity and Maintenance Management Strategy, though they have routinely been used as an aid to the inspection, Non-Destructive Testing (NDT) and maintenance scopes of work.

UAV inspections can form an important part of the Strategy through vertical integration, the benefits of which are:

a) Reduced risks to the safety of people and plant.
b) Increased efficiency of the Asset Plan by focusing inspection and maintenance scope of works to those critical areas which require inspection.
c) Cost savings (of up to 90% on some scopes of work).

The purpose of this paper is twofold. Firstly, to discuss the reasons why UAV inspections are not presently a recognised tool for the Integrity and Maintenance Strategy and how the UAV Industry can overcome this. Secondly, the paper will discuss the benefits of UAV inspections in further detail using a case study whereby a workscope which could have taken 700 man-days offshore, 28 of which would need to be during a shutdown and 40 days of associated standby vessel cover, was reduced to take just 24 offshore man-days with no requirement for a standby vessel and no work needing to be performed during a shutdown. This represents a saving of 90% over traditional methods of visual inspection.

METHODOLOGY

2.1. Perception of the UAV Industry

Although drone technology has been around and used in the military and commercial arenas for 15 years, and specifically used in the oil and gas industry for 5 years there is a perception that this is new and untested technology, particularly in commercial use.

This is a challenge for the UAV industry to overcome because whether the perception is correct or not, the effect is that engineers think that a lot of their time will be taken up with ensuring technical feasibility and competency of the UAV company and personnel.

Certainly some high profile incidents involving UAVs, and subsequent media coverage have not helped and have possibly led to the perception that UAVs are used in a commercial capacity by almost anyone, with minimal training required to fly them.

2.2. How does the commercial UAV industry overcome this?

a) Standard Operating Procedures (SOPs)

It could be argued that given the degree of control that the human team have over the small UAV, that the word unmanned cannot strictly be justified. Each UAV company should have their own SOPs, which are tailored to their own flying requirements but given the manned interaction it is recommended that they are based on established SOPS for manned operations.

As Nick Rogers, Chief Regulatory and Training Officer**and Co-founder at Sky-Futures who is also a qualified commercial airline pilot states:**

"...it is certainly recommended to develop a personal and deep understanding of what you want to achieve using your UAS (UAV) and focus almost entirely on doing this as safely as possible. The important words in your operations manual will help you define this and hone your operational professionalism, instill

confidence from your clients and keep you safer."[1]

The development and continual review of SOPs help to ensure that the pilots flying operation is as consistent as possible.

b) Pilot Competency

In the UK, the Civil Aviation Authority (CAA) requirements state that UAV pilots must undergo a 1 ½ day competency course before they can be allowed to fly in a commercial capacity.

However, the prudent approach taken by the leading commercial UAV operators is to define this as a minimum requirement. Theoretical and practical training courses should be documented and reviewed in line with updated SOPs, whilst flying hours on vehicle type should be logged and records kept.

c) Experience

To date, of the two largest commercial UAV operators in the UK, one has flown over 10,000 flights whilst the other has amassed over 5000 hours of flight time.

Summary

Taken together, the evolution of SOPs, enhanced training, certification and experience might help dispel some myths regarding safe operation of UAVs, such as, that anybody can fly commercially and that it is extremely new.

2.3. UAV Inspection Limitations and Applications

Before discussing the application of UAV inspections within the oil and gas industry, it is worthwhile considering the limitations of the survey and what can be done.

a) Limitations

Safety is the main limiting factor restricting UAV inspections at present, as they cannot fly in between obstructions or underneath decks. The UAV team must have visual line of sight (VLOS) with the vehicle and maintain communication with the GPS network.

Depending on the type of UAV being flown there are environmental weather limitations as well as localised temperature limitations associated with flares and exhausts.

Current technological limitations mean that UAVs do not have the capability of being in contact with the substrate. NDT cannot be performed and cleaning of the component cannot be undertaken for further analysis of indications.

For purposes of clarity, one oil and gas UAV operator's SOPs limit the UAV to fly up to 5 metres of a structure or obstruction and up to 20 metres of a live flare, whilst flying in wind speeds up to 29 knots.

b) What Can Be Done?

The quality of the images on the cameras is one factor in explaining how UAV inspections can add more value than is currently derived from Helicopter fly-by inspections and rope-access inspections This field of technology is rapidly evolving and lens, zoom, imaging software and gyrostabiliser advances mean that detailed surveys can be performed even when the UAV is over 20 metres from the area to be inspected. Figure 1 was captured approximately 15 metres from the live exhaust stack.

---

[1] Rogers, N. 2015. LinkedIn Blog. *"Me, My Drone and My Ops Manual"*, 4th Feb.

Figure 1. Image captured 15 metres from the exhaust by UAV.

A combination of developing detailed flight paths and pilot skill, whilst maintaining safety, mean that as much of the component can be inspected as possible.

Evolving SOPs, UAV technology, training and safety procedures mean that flying can now be done in weather conditions they previously could not.

The industry has recognized the need for external accreditation of UAV inspections and advanced stages of collaboration are on-going with external accreditation bodies in order to provide the third party certification.

c) Summary

It must be understood that UAV inspections cannot replace rope access teams in order to perform NDT and there will be anomalous areas which require close contact with the substrate material in order to gauge fitness for purpose. What it can do is allow the rope access team's workpack to focus on these areas.

As important, UAV inspection reports detail those areas not inspected, so that a campaign rope access workpack instructs the team to visit these areas. With this in mind, considerable resource savings can be achieved.

2.4. UAV inspection within the Integrity and Maintenance Strategy

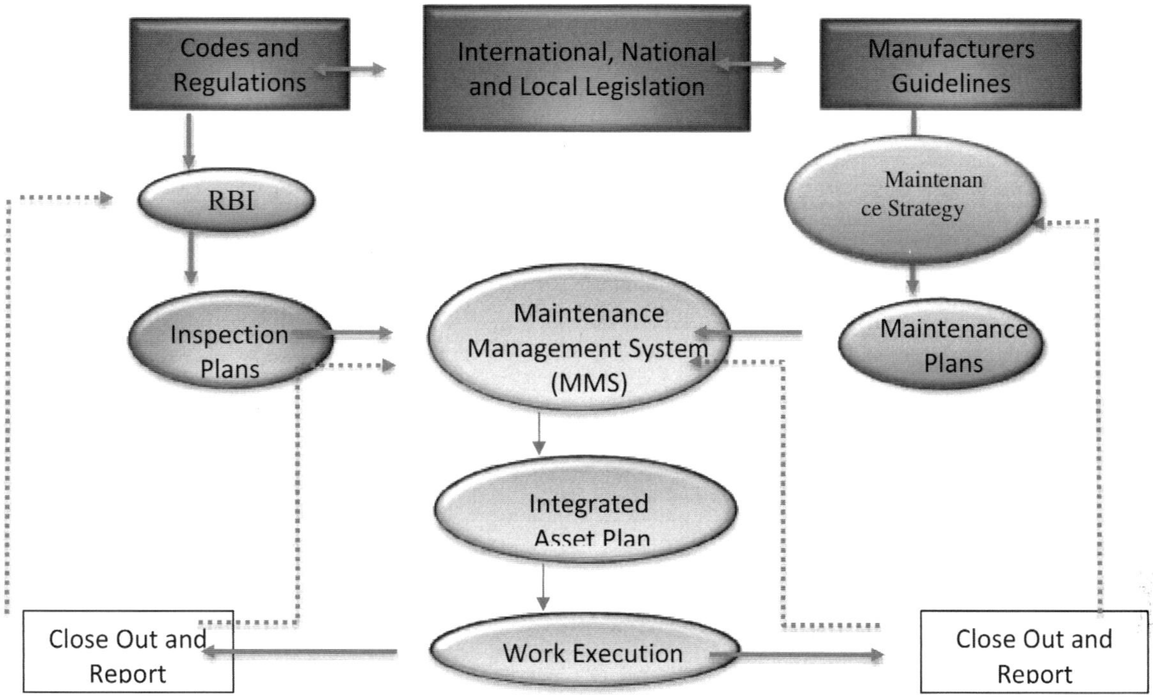

Figure 1

At present, UAV inspections are not routinely included as part of an Operator's Integrity and Maintenance Management Strategy. All inspections done to date have been deployed to aid the inspection, NDT and maintenance processes. Within the diagram above (Figure 2), this is at the Integrated Asset Plan or even at the Work Execution phase. It is believed that UAV inspections can and should be incorporated higher up the chain and form a key element of the RBI (or Written Scheme of Examinations) and Maintenance Strategy.

a) Typical UAV Inspections Done to Date

To date, the vast majority of UAV inspections have been on those components which cannot be reached by people during its operation, typically live flares, exhaust stacks and communication towers utilising microwaves.

UAVs can have a significant impact on the operational efficiency of an asset in the following ways;

i) Reduce immediate unplanned shutdowns. Known anomalies on hard to access areas can be further inspected whilst online, allowing for an engineering decision to be made on the risk of failure up to a planned shutdown.
ii) Help keep planned shutdown plans on track by reducing "hidden surprises." By inspecting flares and exhausts stacks up front of a planned shutdown, anomalous areas are known prior to the shutdown. The TAR team are able to keep their focus on the planned tasks.
iii) Minimise future unplanned shutdowns. UAV inspections upfront of a shutdown allow the time to plan and execute permanent repairs, modifications or alterations to the Operator's procedures, reducing follow up shutdown-dependent work to remove temporary solutions at a future date.
iv) Reduce overdue inspections and aid the deferral process. Safety critical inspections often go overdue as access can be a problem. UAV inspections might be able to help in the risk assessment of the deferral.

However, UAV inspections can play a fundamental role in the integrity and maintenance management strategy, leading to cost savings and driving limited resources to those areas that merit further inspection, so long as the limitations are understood and known.

b) Vertical Integration to the Integrity and Maintenance Management Strategy

By embracing UAV inspections as an integral component of the RBI process, major reductions in man hours can be achieved in the Inspection Plan and in the MMS, creating a streamlined Integrated Asset Plan which targets crucial resources to the critical areas that require them. Crucial to the success of this strategy, is that the UAV inspection report must highlight the areas of the component that were not inspected as well as those areas which warrant further inspection and target inspection workpacks.

Significant efficiencies and cost savings can be realised with the maintenance strategy by using planned and routine UAV inspections, especially prior to a shutdown.

2.3. Case Study

A recent UAV inspection scope was undertaken on a multi-jacketed offshore asset, this included 14 separate scopes of work on the three bridge connected structures.

The scope included items such as the wellheads jacket, twin derricks, cranes, bridges, riser platform jackets, helidecks, flare towers and tips. These scopes of work were completed ahead of schedule, in 12 days by the UAV team offshore.

Each of the individual structural scopes of work has been entered in to the MMS, through a Planned Maintenance Routine (PMR) and each scope resourced as taking 3-men a total of 14 days to complete the scope. An estimate therefore for the completion of this work comprising 14 scopes using rope access techniques is that the work would require a 3-man rope access team to be on the platform for a total of 196 days. This works out at 588 man-days to complete the entire scope of work.

A rope access team would still be required to mobilise to target those areas which require further inspection and to inspect the areas not captured by the UAV. Even a conservative estimate of 4 days per scope to complete the inspection in its entirety, would reduce the entire rope access scope to 56 days for a 3-man rope access team or 168 man days.

Taking in to account the UAV man days gives:

a) 168 + 28 = 196 man days.
b) Original scope = 588 man days.
c) Time saved = 392 man days or 67%.

As well as the significant time savings in completion of the scopes, there are also savings in the use of resources and costs:

a) On top of this the rope access team would require up to 10 more crew changes, flights and mobilisations.
b) The figures above do not include additional costs associated with stand-by vessel cover, possible 4th man cover for stand-by and additional permitry.

When all of the above is taken in to consideration, it is believed that UAV inspections can save up to 90% in terms of total costs.

CONCLUSIONS

Reduced safety risks, significant cost savings and operational efficiencies can be achieved by the ability of UAV inspections to capture large amounts of inspection and maintenance data in short periods of time.

With current technology, UAV inspections cannot replace rope access inspections. However, it has been demonstrated that significant cost savings and resource efficiencies can be achieved by targeting the rope access teams to those areas of a component which require the physical contact of a close visual inspection and NDT and to those areas not inspected with the UAV.

At present, risk based inspection, written scheme of examinations and structural modelling of load paths determine the areas of focus for topside inspection workpacks. Whilst these methods will take into account previous inspection histories, UAV inspections can go even further in targeting limited inspection resources to the most critical areas of inspection.

As well as being a valuable component to the integrity management system, UAV inspections can make further contributions to an Operator's Engineering, Maintenance and TAR teams. Those areas which cannot be inspected under normal operating conditions can be inspected by the UAV system, allowing for planned repairs, modifications and alterations to be done ahead of a planned shutdown. This results in there being no surprises to the TAR team, allowing their efforts to be focused on the planned work and minimising shutdown durations.

USING CEL TO ACCOUNT FOR SEABED DEFORMATION EFFECTS FOR JACK-UPS GOING ON LOCATION

JOSE H. VAZQUEZ, PH.D., BARTON D. GRASSO, P.E. AND MARCUS A. GAMINO
3DENT TECHNOLOGY

JACK S. TEMPLETON, III, D. ENG., P.E.
SAGE USA

Proceedings of the 22nd Offshore Symposium, February 2017, Houston, Texas
Texas Section of the Society of Naval Architects and Marine Engineers

ABSTRACT

There have been a number recent improvements for the analysis of self-elevating units during the transition phase commonly referred to as the "going on location" or "emplacement" phase (i.e., the transition from the afloat condition to the pinning of the legs). Unlike the well-established guidelines from Class or ISO for units in the elevated condition, the guidelines for going on location analyses have yet to be fully developed, let alone become standard. State-of-the-art analysis methods to establish adequate limits for going on location include nonlinear time-domain response analysis of the units, accounting for jacking speed and spud can shape as well as wave directionality, water depth and soil conditions. The majority of analyses represent the seabed as a nonlinear spring, attempting to match the penetration curves which account for spud can shape and soil properties.

There are two significant effects that have not been addressed by these methodologies: the fact the seabed profile changes after each leg impact, and the energy dissipation properties of the soil. In this paper, the Combined Eulerian-Lagrangian (CEL) capabilities of Abaqus are used to determine both vertical and lateral soil resistance to leg motions of a representative jack-up while going on location. The CEL results are compared with the forces from idealized representations of the soil-structure interaction in the nonlinear response simulation.

A single wave period is considered. Two general soil conditions are considered: a stiff soil and an intermediate soil. The forces from the CEL analysis are then used to replace the elastic seabed on the nonlinear response analysis to see if the jack-up motions are affected.

The results of this study show that the impact loads for jack-ups going on location based on elastic representations of the seabed are over-predicted, and as such they can be considered to be on the conservative side when used in the determination of permissible wave heights.

Keywords: ***Jack-ups going on location, spud can impact loads, CEL analysis of spud cans in clay***

INTRODUCTION AND BACKGROUND

Jack-ups are mobile offshore units made up of a hull and legs and having two main modes of operation: afloat and elevated. Jack-up legs are fitted with footings, known as spud cans, intended to reduce the bearing pressure and minimize leg penetration. In the elevated mode, the hull is completely out of the water and the hull is supported by the legs, which rest firmly in the seabed, thereby providing a stable platform. In the afloat mode, the jack-up is supported by buoyancy on the hull. In the presence of waves, a jack-up in the afloat mode is also referred to as being in free-floating mode (i.e., the wave loads acting on the hull and legs are balanced mostly by the inertia response of the unit), since there is no mooring.

Transitioning between the afloat mode to the elevated mode involves jacking of the legs and preloading of the soil. This is known as "going on location" (GoL). In the absence of waves, the going on location process is of no concern, as the jacking speed of the legs is relatively low and there is minimal impact load when the spud cans make initial contact with the seabed. In the presence of waves, the potential exists for large impact loads between the spud cans and the seabed.

Until recently, GoL limits provided in a unit's marine operating manual typically listed a single wave height or inclination angle for all conditions, regardless of direction and only occasionally depending on water depth, wave period, or broad "stiff/soft" soil conditions. While these limits have proved to be safe, the expectation is that they are extremely conservative, causing unnecessary waiting on weather delays.

There have been a number recent improvements for the analysis of jack-ups during the going on location phase. However, unlike the well-established guidelines from Class or ISO for units in the elevated condition, the guidelines for going on location analyses have yet to be fully developed, let alone become standard. State-of-the-art analysis methods to establish adequate limits for going on location include nonlinear time-domain response analysis of the units, accounting for jacking speed and spud can shape as well as wave directionality, water depth and soil conditions. The majority of analyses represent the seabed as a nonlinear spring, attempting to match the industry-standard penetration curves.

An advanced going on location analysis methodology was developed by three of the authors of this paper (Ref 1). When this analysis methodology is used to check the induced loads on the leg members and jacking system of a given jack-up, permissible wave height curves (wave height vs period) for going on location can be produced for a given water depth, wave direction and soil condition.

The going on location analysis methodology (described in the next section) utilizes the commercial software OrcaFlex (Ref 2) to carry out the nonlinear response analysis. OrcaFlex offers various options for representing the seabed. At its most basic, the seabed can be represented using an elastic block. OrcaFlex also allows for the seabed to be modelled so as to have the capacity to introduce "suction" when the penetrating object is being retrieved. Neither of these seabed representations in OrcaFlex allow for the seabed to see permanent deformation after each impact between the spud can and the seabed.

This paper presents the results of a study which uses the spud can motions from a linearized GoL analysis as input to a Combined Eulerian-Lagrangian (CEL) model to compare the resulting forces between the soil and the spud can. The CEL analyses are carried out using Abaqus (Ref 4), a commercial software package commonly used for soil-structure interaction analysis.

SIMPLIFIED GOING ON LOCATION ANALYSIS METHODOLOGY

In a 2016 paper by three of the authors of this paper, entitled *Jackups Going on Location – Understanding Energy Principles on Leg Impact Loads* (Ref 1), a simplified GoL analysis methodology was introduced. The simplified GoL analysis methodology in Ref 1 was subsequently expanded (Ref 3) as outlined below.

The GoL analysis methodology is made up of three types of analyses, as outlined below.

- *Frequency-domain diffraction analysis* to calculate the wave load as well as the added mass/damping coefficients of the jack-up at the target depth with its legs lowered near the seabed.
- *Structural analysis* to establish the structural capacity envelope of the legs (combinations of vertical force and moment reactions that cause either the pinions, chords or braces to reach their structural capacity).
- *Time-domain response analysis* to calculate the motions and spud can loads during the jacking operations as the legs make contact with the seabed.

The time-domain response simulation is considered to be divided into four segments, as described in Table 1 and illustrated in Figure 1 (depicting regular waves, but also applicable to random waves).

Table 1 Description of GoL Time-Domain Simulation Segments

Segment	Description
Ramping	Time prior to the actual start of the simulation, taken to be 100 seconds, during which all wave heights are gradually increased from 0 to the target value.
Free-Floating	Time starting at t=0 sec, to the first time there is contact between the seabed and any of the members representing the spud cans.
Transition	Time from the first time there is contact between the seabed and any of the members representing the spud cans, until all legs have made contact and the hull starts to come out of the water.
Elevating	Time after all legs have made and remain in contact with the seabed, resulting in minimal motions except for the elevation of the hull.

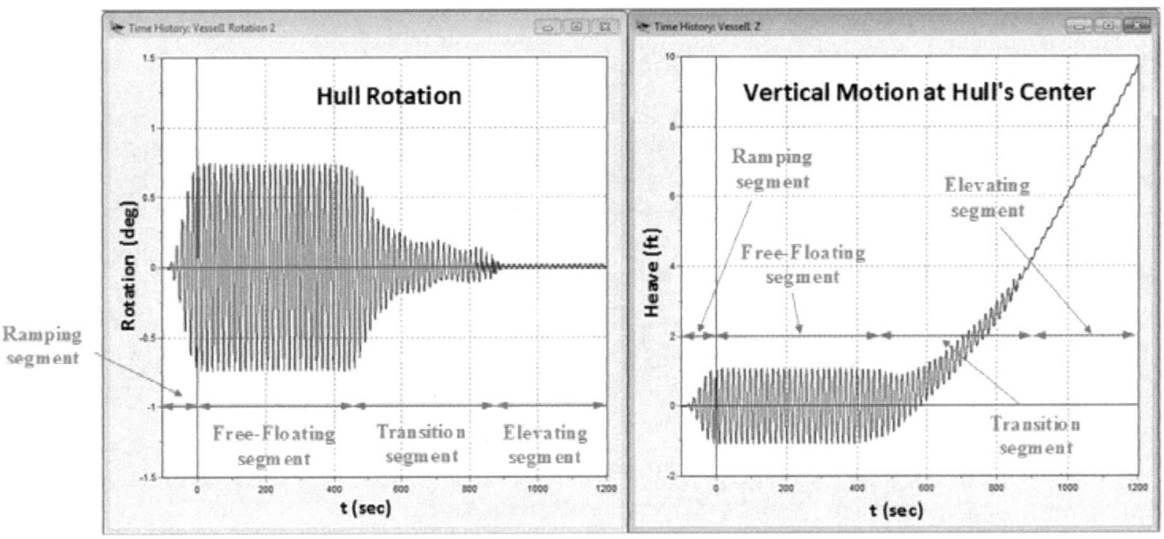

Figure 1 Illustration of GoL Time-Domain Simulation Segments

The methodology for establishing permissible wave height curves (for a given water depth and wave direction relative to the orientation of the hull and soil condition) has a number of simplifying assumptions, as listed below:

1. A structural capacity curve (combinations of leg axial load and bending moment) is established for the leg connected to the hull at a position corresponding to the target water depth.

2. Wind and current effects are ignored.

3. The diffraction analyses include the hull and spud cans (i.e., drag and inertial loads on the truss legs are ignored).

4. Wave excitation loads and added mass/damping coefficients are determined for a single tip-of-can position (typically 4 ft from the seabed).

5. Linear soil stiffness is used.

6. Lateral loads on the leg when it makes impact with the seabed are assumed to be proportional to the vertical impact load, as well as the relative magnitude of the lateral velocity of the spud can.

7. A moving element with a calibrated contact stiffness is used to represent the relative movement of the legs and seabed (thereby simulating jacking of the legs).

8. For regular waves, wave heights are iterated in 0.25ft increments to identify the highest wave height for which all loads are within the structural capacity curve.
9. For the random waves, iterative runs of the significant wave height value are made until a calculated percent of passing (PoP) value within a range of 95%-97% is reached.

The approach used for the time-domain analysis is to treat the hull and legs as a single body with the stiffness characteristics of the jacking system, legs and soil combined into an equivalent stiffness value. Ten (10) levels of contact elements outlining the shape and associated bearing area of the spud can (see Figure 2). The contact elements are assigned appropriate stiffness values to reproduce the target Force vs Penetration plus the stiffness effects of the legs and jacking system.

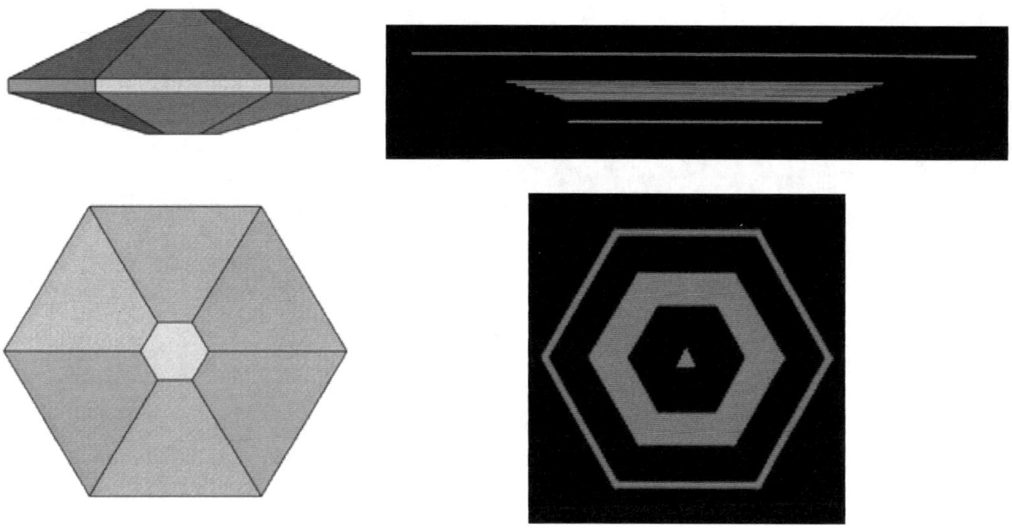

Figure 2 Representation of Spud can in OrcaFlex using Contact Elements

COMBINED EULERIAN-LAGRANGIAN MODEL

Analyses with the Abaqus program were used to provide numerical solutions for the soil reactions of the Forward spud can while the jack-up is going on location. Abaqus (Ref 4) is a large, nonlinear, general purpose, commercial finite element analysis program with substantial soil mechanics capabilities. It has a long history of successful use in offshore geotechnical analysis (Ref 5). The analyses reported here were performed using the Combined Eulerian-Lagrangian capabilities available in Abaqus.

Traditionally, finite element analysis has been accomplished using what is now called a Lagrangian approach. In this approach, nodes and element boundaries are tied to material, and a fixed body of material is contained in each element. The mesh deforms with the material, and the amount of material deformation that can be achieved is ultimately limited by errors resulting from element deformation. With an Eulerian approach the nodes and element boundaries are fixed in space and material flows across element boundaries as material deformations occur without any element deformation. In this way arbitrarily large material deformations can be achieved without element performance problems. In Abaqus Combined Eulerian-Lagrangian (CEL) analysis, Lagrangian meshing for bodies with limited deformation can be combined with Eulerian meshing for bodies with large deformations, and the two kinds of bodies can interact.

In the analyses reported here, the spud can was modeled as a Lagrangian body, and the seafloor soil was modeled as an Eulerian body. The geometry of the model used is shown in Figure 3. Note that this is a half-space model taking advantage of the symmetry of the spud can and the fact that the motion studied is confined to a plane (the x-z plane in Figure 3). The spud can was made rigid. The soil behavior was modeled using Abaqus' kinematic hardening model. This model can produce cyclic and dynamic behavior quite representative of the known characteristics of soils, and this material representation has been used successfully on a number of offshore geotechnical analyses. (See, e.g., Refs 6 through 9.) Specifically, the normalized stress-strain behavior shown as medium in Ref 7, Figure 11, was used, with the stiff case being used for sensitivity.

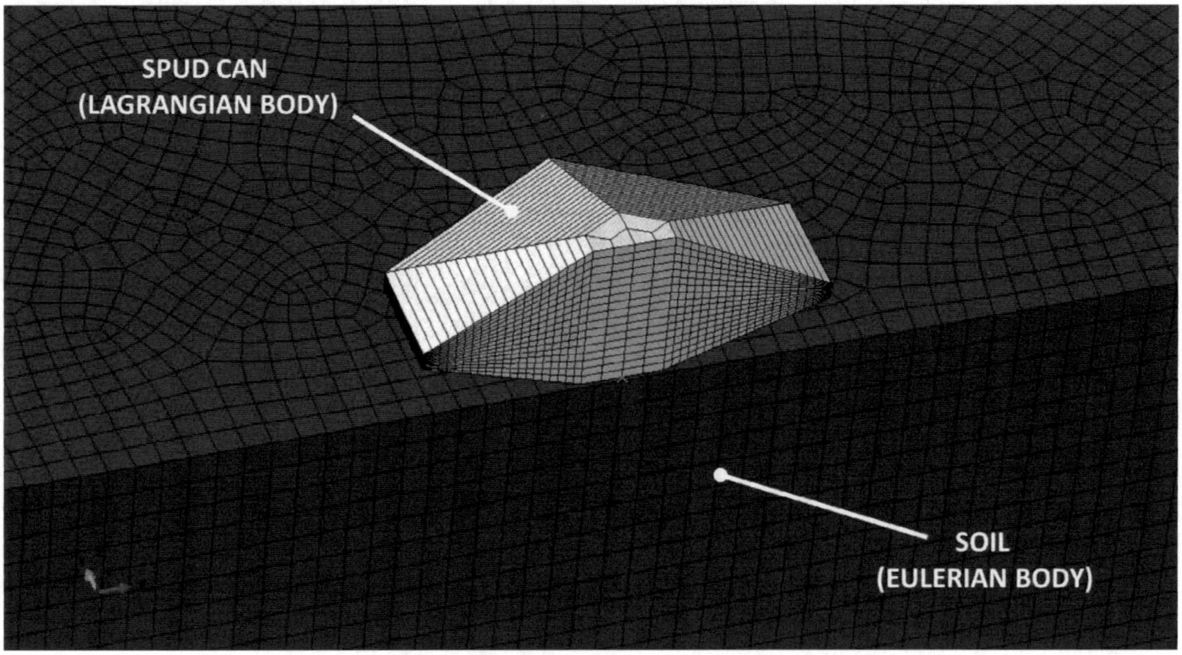

Figure 3 Half-Space Spud can Model in Abaqus

METHODOLOGY AND GOL CONDITIONS USED IN THIS STUDY

The methodology to compare loads between elastic soils represented in OrcaFlex and the nonlinear soils represented in Abaqus is explained below.

1. Select a water depth and wave (wave height, wave period and direction) to carry out a simplified GoL analysis for the above-described jack-up.
2. Select two representative soil conditions, corresponding to different penetrations at full preload.
3. Obtain motions. Due to symmetry, only heave, pitch and surge are tracked).
4. Impose spud can motions from the GoL simulation in Orcaflex to the spud can in the Abaqus CEL model.
5. Compare the vertical and horizontal reactions obtained from the two models.

Two clay soils were selected for the analyses, representing stiff and intermediate soils, producing leg penetrations of 2 and 10ft (see Table 2). The specifics of the GoL analyses are listed in Table 3.

Table 2 Properties of Clay Soils for the Study

ISO Penetration Estimate (ft)	Su (psf)
2	7,000
14	1,300

Table 3 Specifics of GoL Analysis

Water Depth	300 ft
Wave Height	6 ft
Wave Period	13 sec
Wave Direction	Head seas
Jacking Speed	1.5 ft/min

Figure 4 shows a comparison of leg penetration curves for the intermediate soil, as predicted from a) ISO 19905-1, b) OrcaFlex model with linear soil and 10 contact layers representing the spud can, and c) the CEL model of the spud can being pushed into the soil slowly and continuously (i.e., minimal dynamic effects).

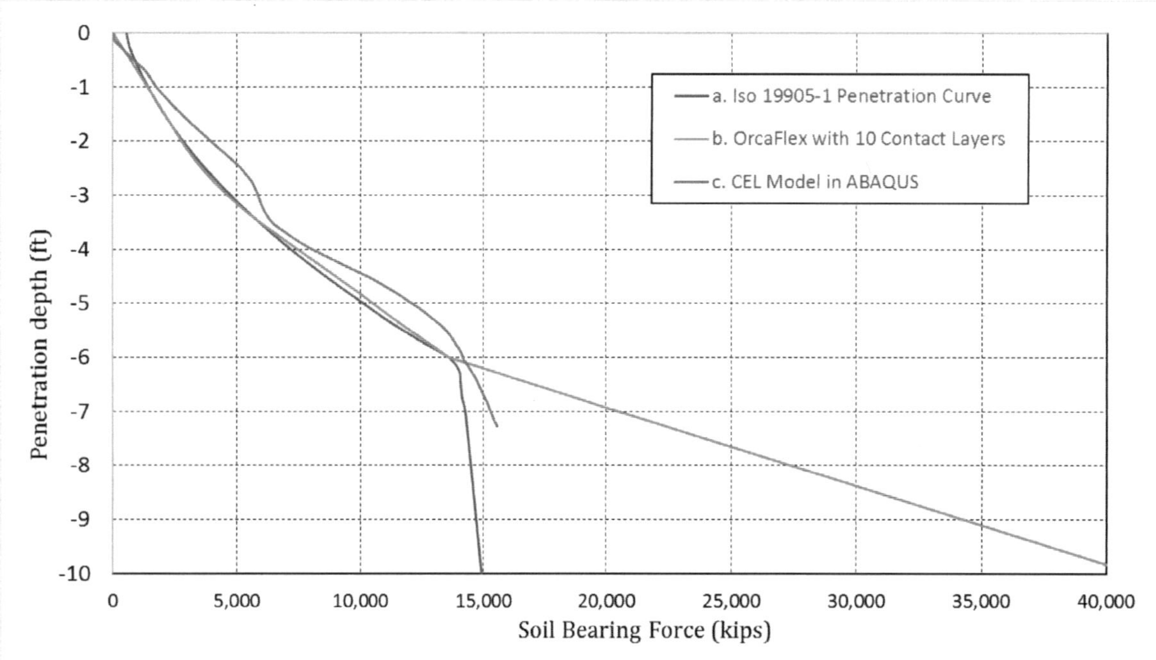

Figure 4 Leg Penetration Curves for Intermediate Clay

The results of the GoL simulation (forward spud can motions), which serve as input to the CEL simulation, are shown in Figures 5 and 6, for the stiff and intermediate soils, respectively.

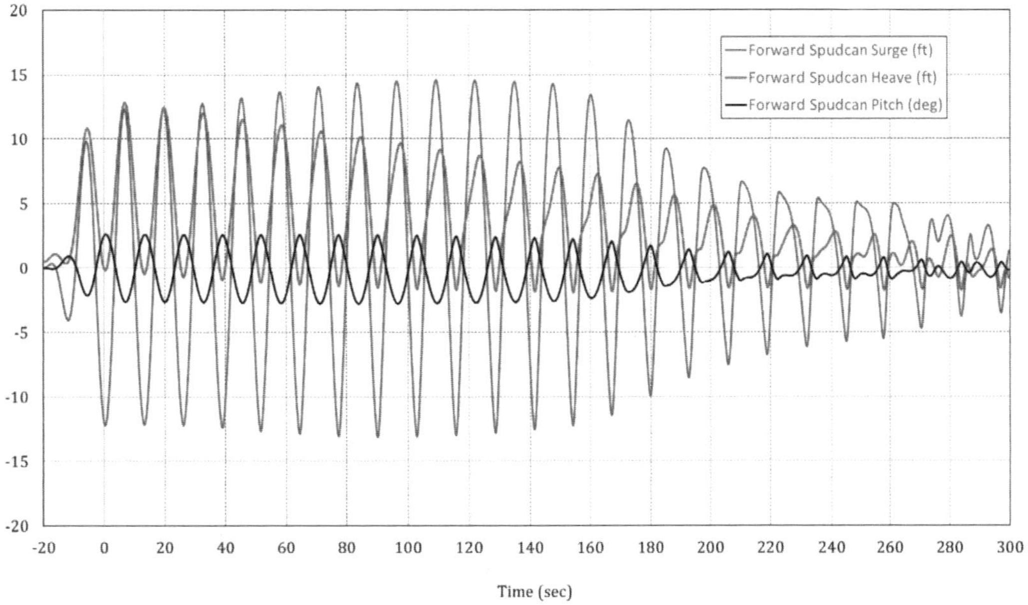

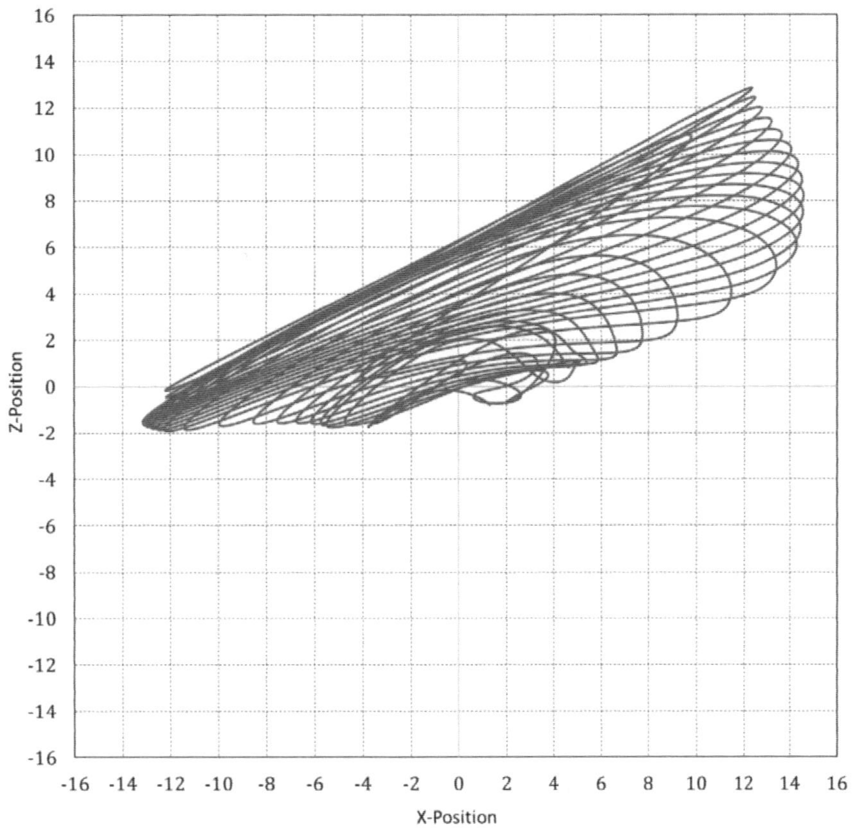

Figure 5 Forward Spud can Motions from GoL Simulation – Stiff Soil

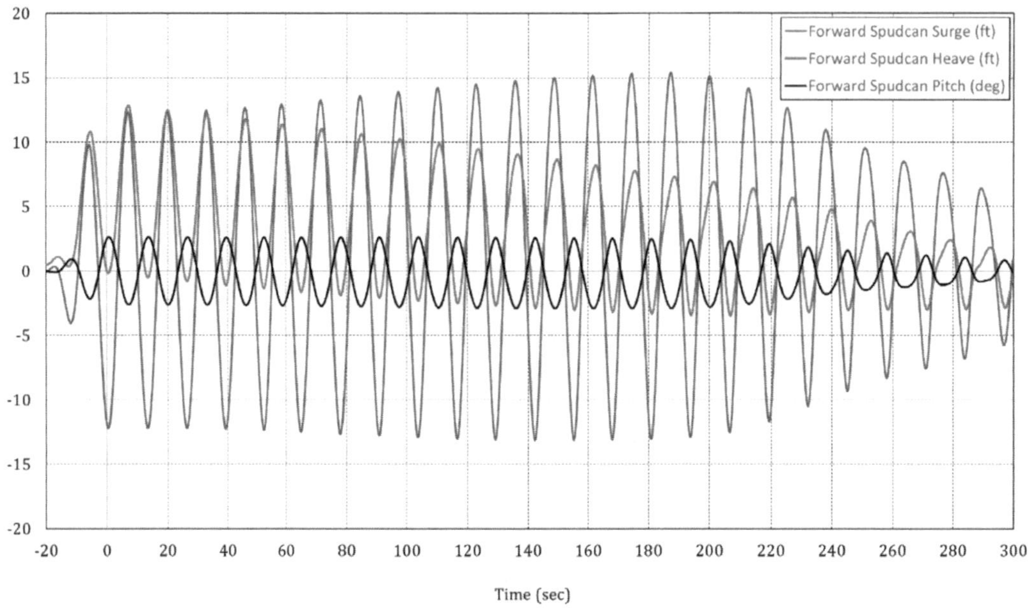

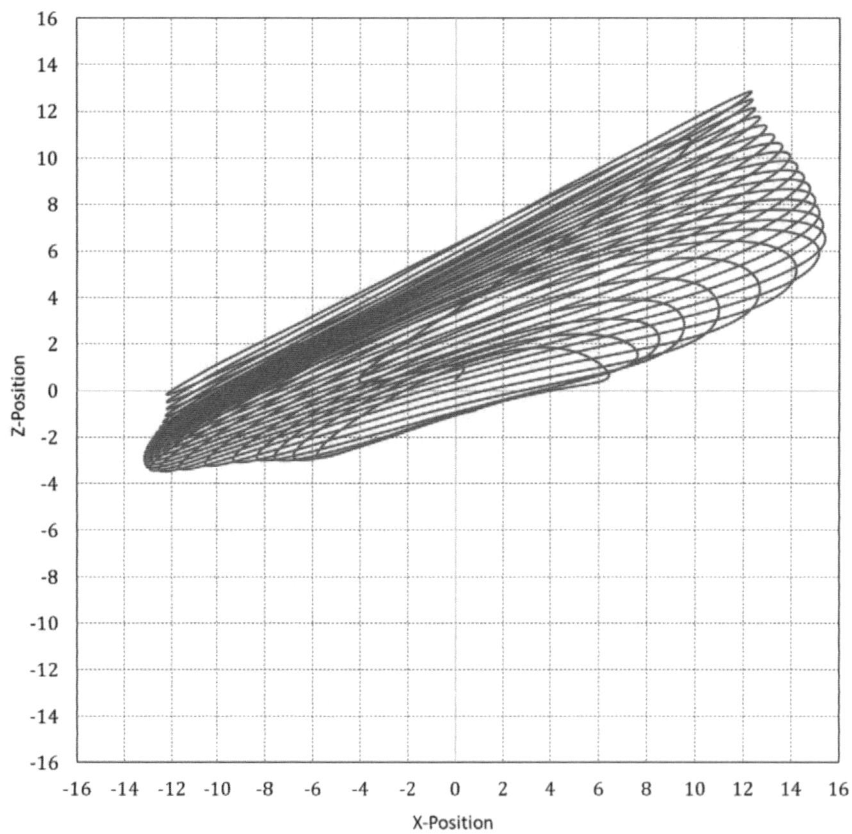

Figure 6 Forward Spud can Motions from GoL Simulation – Intermediate Soil

FORWARD LEG REACTIONS (STIFF SOIL)

Two soil conditions were used in the comparison of the loads obtained from OrcaFlex (with a linear seabed model) and Abaqus (with a nonlinear explicitly dynamic model accounting for plastic deformation of the seabed). The results of the OrcaFlex and Abaqus studies using stiff soil are presented here. The simulations were carried out for 300 seconds (i.e., about 7.5ft of leg jacking).

Figure 7 shows a comparison of the vertical reactions for the H=6ft, T=13sec, intermediate soil GoL simulation, as obtained in OrcaFlex from a linear seabed with 10-buoy layers to simulate the spud can and from the CEL Abaqus simulation. Figure 7 has three parts. The top part has a direct comparison of the vertical reactions from the two simulations, over the 300-sec simulation. The middle and bottom parts of Figure 7 focus on the first ten leg impacts. Figure 8 shows a comparison of the corresponding horizontal reactions. The horizontal reactions provide both positive and negative values. For the results from both OrcaFlex and Abaqus (CEL), the horizontal reaction at the forward spud can corresponds to the direction of motion of the spud can during the simulation. With the positive horizontal direction pointing from the center of the legs to the bow, forward spud can motion toward the direction of the aft of the hull provides negative horizontal reactions when in contact with the seabed. Forward spud can motion in the same direction as the positive horizontal direction provides positive horizontal reactions when in contact with the seabed.

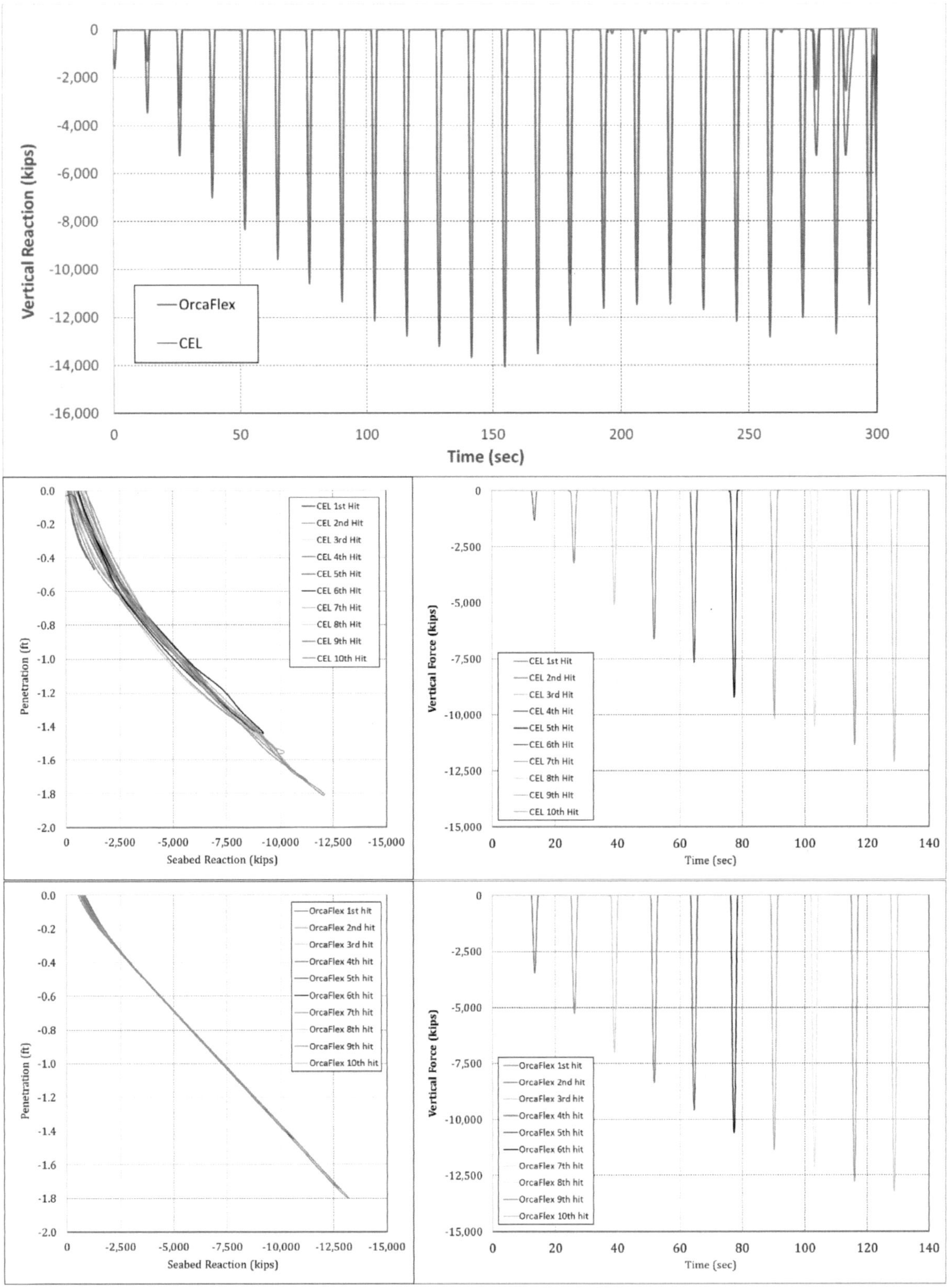

Figure 7 Vertical Reactions for Stiff Soil

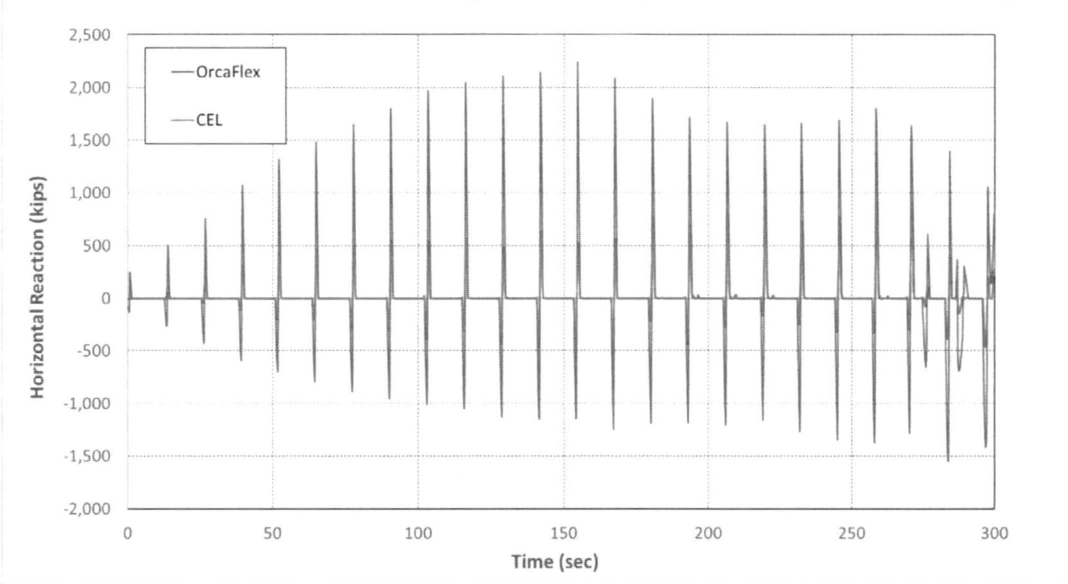

Figure 8 Horizontal Reactions for Stiff Soil

FORWARD LEG REACTIONS (INTERMEDIATE SOIL)

Two soil conditions were used in the comparison of the loads obtained from OrcaFlex (with a linear seabed model) and Abaqus (with a nonlinear explicitly dynamic model accounting for plastic deformation of the seabed). The results of the OrcaFlex and Abaqus studies using intermediate soil are presented here. The simulations were carried out for 300 seconds (i.e., about 7.5ft of leg jacking).

Figure 9 shows a comparison of the vertical reactions for the H=6ft, T=13sec, intermediate soil GoL simulation, as obtained in OrcaFlex from a linear seabed with 10-buoy layers to simulate the spud can and from the CEL Abaqus simulation. Figure 9 has three parts. The top part has a direct comparison of the vertical reactions from the two simulations, over the 300-sec simulation. The middle and bottom parts of Figure 9 focus on the first ten leg impacts. Figure 10 shows a comparison of the corresponding horizontal reactions. As was the case in the stiff soil results, the horizontal reactions provide both positive and negative values.

Table 4 shows a summary of the peak vertical and horizontal reactions. As expected, the ration of peak horizontal to vertical reaction is ~15% for both the stiff and intermediate soil OrcaFlex runs. For the CEL runs, the ratios are 13% and 7% for the intermediate and stiff soils, respectively.

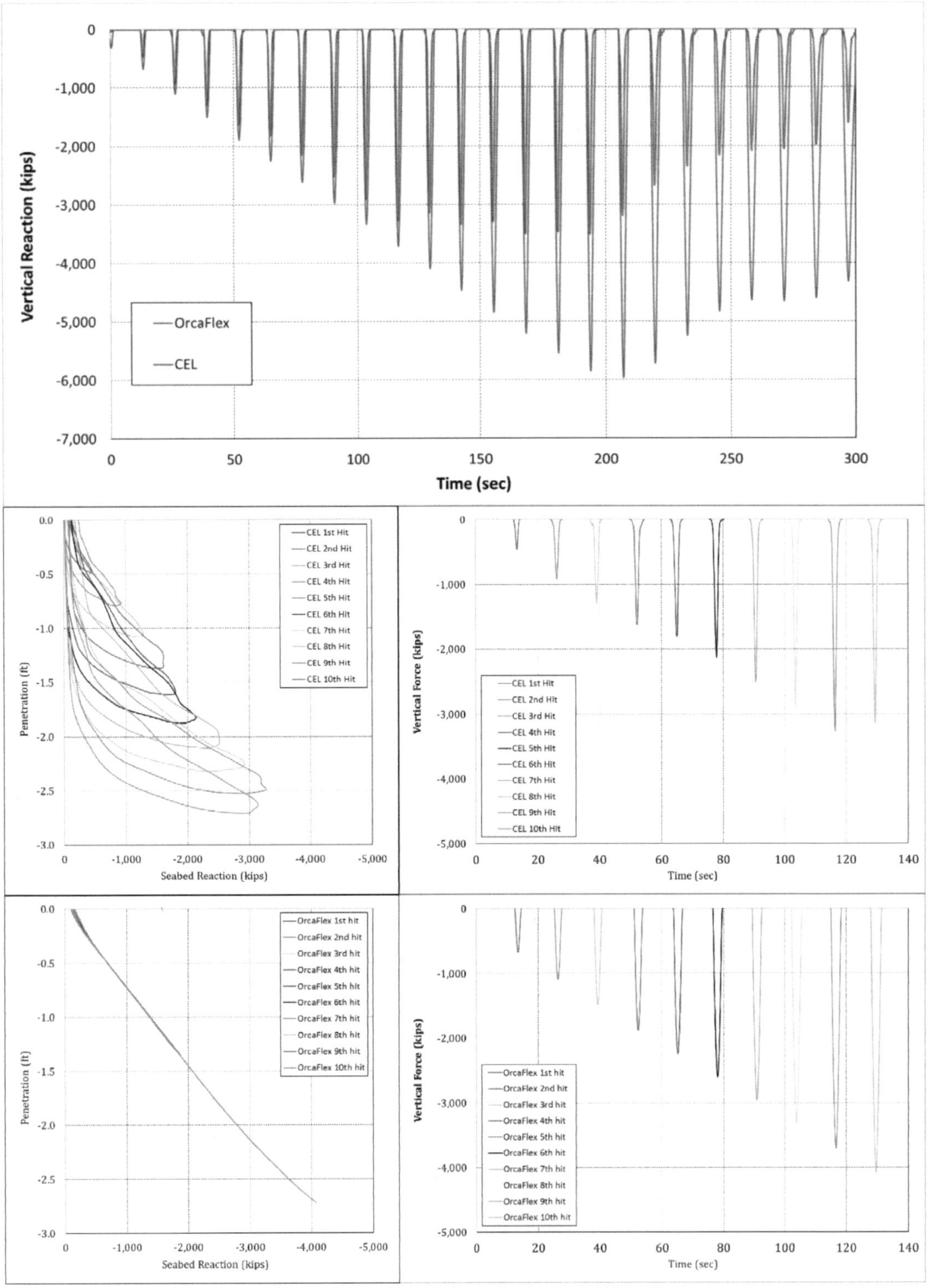

Figure 9 Vertical Reactions for Intermediate Soil

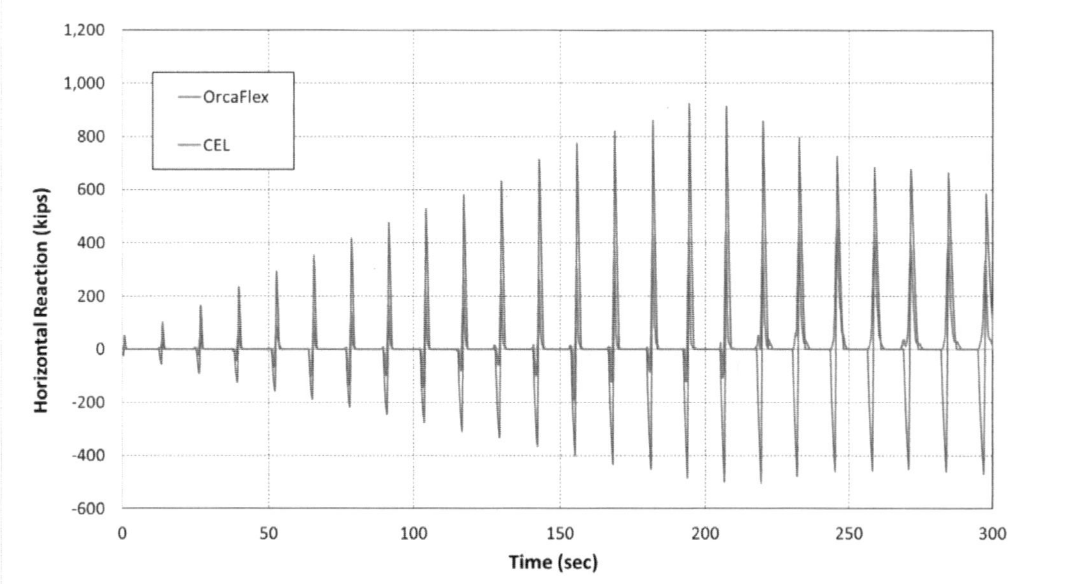

Figure 10 Horizontal Reactions for Intermediate Soil

A comparison of the resulting peak vertical and horizontal reactions from Orcaflex and the CEL analyses are given in the following table for the two soil stiffness values analyzed.

Table 4 Comparison of Orcaflex and CEL Peak Loads

Soil Stiffness	Reaction	Orcaflex	CEL	Percentage Difference
Intermediate	Vertical	5,968k	3,505k	170%
	Horizontal	921k	467k	197%
Stiff	Vertical	14,052k	12,811k	110%
	Horizontal	2,227k	905k	246%

For visualization, Figure 11 illustrates the spud can-soil interaction in CEL within Abaqus, showing the stress in the soil as the spud can penetrates into the seabed.

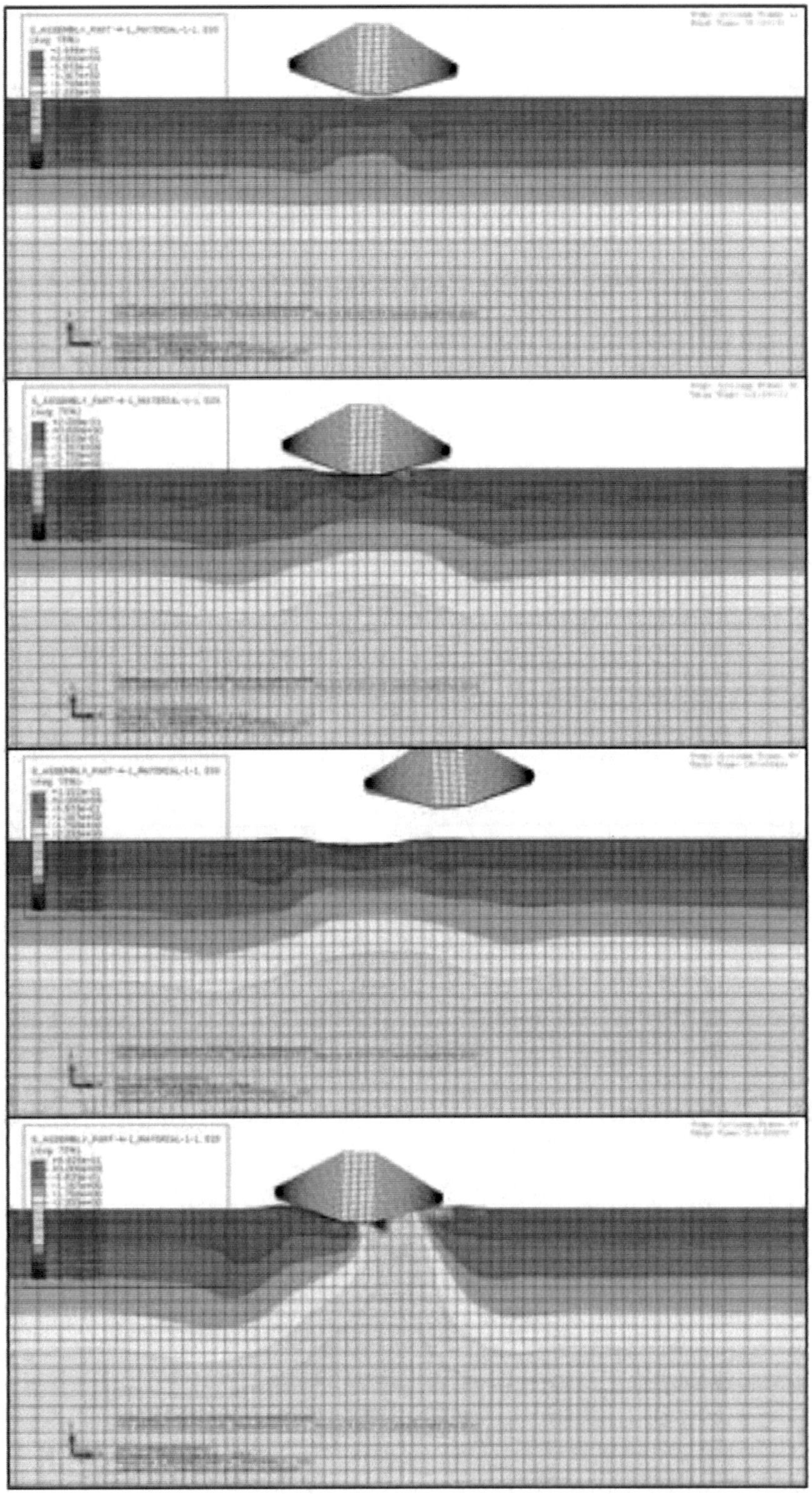

Figure 11 Spud can-Soil Interaction with Abaqus CEL Model

SUMMARY AND CONCLUSIONS

A comparison of two different methods to model the seabed during a GoL analysis was performed using OrcaFlex and Abaqus. In *Jackups Going on Location – Understanding Energy Principles on Leg Impact Loads* (Ref 1), a simplified GoL analysis methodology was introduced, and this methodology uses OrcaFlex as its time domain solver to calculate rig motions and leg reactions during the GoL phase. However, the seabed in OrcaFlex does not plastically deform after a spud can impacts its surface; therefore, the time domain solver has the spud cans impacting an elastic undeformed seabed (i.e., linear spring) at each collision with the seabed. To model the plastic deformation and energy dissipation of the seabed during the GoL phase, the motions obtained for the forward spud can during an OrcaFlex run was input into a CEL Abaqus model. Figure 12 shows the two different models impacting with their respective seabed.

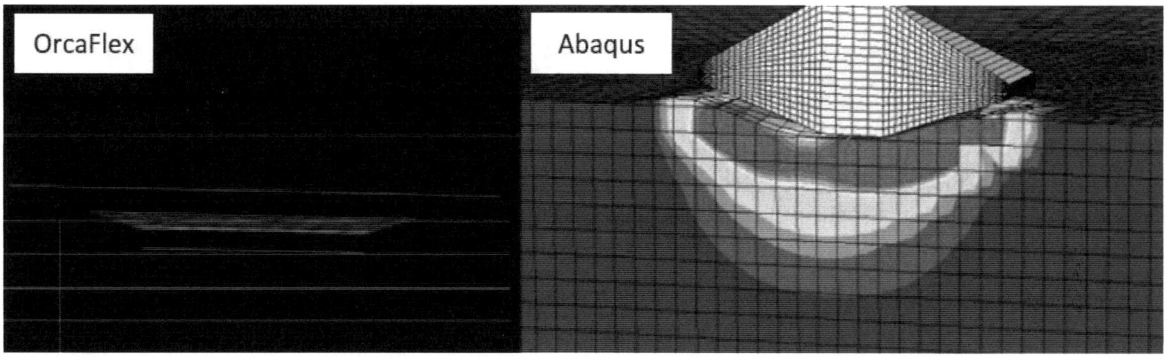

Figure 12 Spud can-Soil Interaction with OrcaFlex and Abaqus

In conclusion, the simplified GoL methodology produces conservative, yet reasonable, results in comparison with loads obtained from a highly nonlinear CEL model from Abaqus. With continuous improvements to computing power and software, conservatism may continue to be removed from current GoL analyses by taking into account the fact the seabed profile changes after each leg impact and the energy dissipation properties of the soil.

ACKNOWLEDGEMENTS

The authors would like to thank Wei Wang of 3Dent Technology and Frances Biegler of SAGE USA for their contributions to the work contained in this paper. Gratitude is also expressed to Bryan Lane of JBL Technologies for his assistance with certain of the Abaqus solutions.

REFERENCES,

1. Vazquez J, Grasso B, Gamino M, Wang, W "Jackups Going on Location - Understanding Energy Principles on Leg Impact Loads." Proceedings of the 21st Offshore Symposium, February 2016, Houston, Texas
2. ORCAFLEX User Manual, by ORCINA.
3. 100-AP-001 Rev 01 Analysis Procedure for Simplified GoL Analysis by 3DENT, dated October 5, 2015.
4. SIMULIA, Abaqus User's Guide, v.6.14 Dassault Systemes, Providence RI, 2014
5. Templeton, J. S. III. Finite Element Analysis in Offshore Geotechnics – A Thirty-Year Retrospective, Proceedings, SIMULIA Community Conference, 2012
6. Templeton, J. S. III. Time Domain FE Seismic Analysis of Mat-Supported Jack-Up Structure on Soft Clay, Paper No. OTC 19645. In Proceedings, Fortieth Offshore Technology Conference, Houston, Texas, 2008.
7. Templeton, J. S. III. Finite Element Analysis of Conductor/Seafloor Interaction, Paper No. OTC 20197 in Proceedings, Forty-First Offshore Technology Conference, Houston, TX, 2009
8. Templeton, J. S. III, , Continuum Foundation Analysis of Jack-up Rigs for Earthquake Conditions., Proceedings, Fourteenth International Conference, The Jack-up Platform, City University, 2013 London,
9. Templeton, J. S. III, M. J. Barrett and E. C. Clukey, Finite element analysis of soil damping for pipeline slugging, Proceedings, Int. Symp. on Frontiers in Offshore Geotech., ISFOG, Oslo, 2015

SNAME
THE INTERNATIONAL COMMUNITY
FOR MARITIME AND OCEAN PROFESSIONALS

SUBCHAPTER -M :AN OVERVIEW

RAMESH S. RAGHAVAN
PRINCIPAL ENGINEER, V.L. OFSHORE

STEFFEN A. SHELLEY
VICE-PRESIDENT, V.L. OFSHORE

Proceedings of the 22nd Offshore Symposium, February 2017, Houston, Texas
Texas Section of the Society of Naval Architects and Marine Engineers

ABSTRACT

The final version of Subchapter M (Sub M) is released and is effective July 2016. This subchapter adds new rules for a comprehensive safety system, that includes company compliance, vessel compliance, vessel standards and oversight in a new Code of Federal Regulations (CFR) subchapter dedicated to the towing vessels. The objective of this regulatory action is to enhance safe operations of the towing vessels in US waterways. The requirements of final rules are designed to encourage companies to engage at every level to improve safe operations, maintenance and design and adhere to prescribed safety standards.

Owners and managing operators of inspected towing vessels would be required to either develop and maintain documentation for their safety management system and arrange periodic audits and surveys through third party organizations, or to demonstrate compliance with the Subchapter – M to Coast Guard inspectors. This rule also includes provisions for crew and vessel operational safety standards (life-saving and fire-protection), vessel equipment and safety standards, procedures and schedules for routine tests and inspections, crew trainings and audit, electrical and machinery requirements which are applicable to new and existing vessels.

This paper presents an overview of Subchapter- M which includes Safety Management System to be adopted, procedures for safe vessel operations, the inclusion of additional requirements in electrical, mechanical and construction and other requirements for the rule compliance and timeframe for their implementation.

Keywords: Subchapter-M, Coast Guard, Towing Safety Management System, Third Party Organization.

INTRODUCTION

The Coast Guard and Maritime Transportation Act of 2004 added towing vessels as a class of inspected vessels and authorized the Secretary of the Department of Homeland Security to establish requirements for a safety management system for towing vessels. Towing vessel means a commercial vessel engaged in or intending to engage in the service of pulling, pushing, or hauling alongside.

US Coast Guard has established towing vessel regulations establishing new requirements for the design, construction, onboard equipment and operation of towing vessels with the input from Towing Safety Advisory Committee (TSAC) and the towing vessel industry. This new regulation will be incorporated into Title 46 of the

Code of Federal Regulations as a new subchapter called Subchapter M or Sub M. These regulations became effective July 20, 2016

This subchapter introduces safety regulations that govern characteristics, methods of operations, and nature of service of towing vessels. Inspection of towing vessels for compliance of these new safety regulations will be performed by the Coast Guard or by third party organizations (TPO's) such as classification societies. These regulations include provisions covering specific electrical and machinery requirements for new and existing towing vessels, the use and approval of third party organizations, and procedures for obtaining Certificates of Inspection.

WHY SUBCHAPTER -M ?

According to the Coast Guard and government research:

"towing vessels account for slightly more than two-thirds (67.5%) of the five-year average number of all vessels involved in collisions, allisions and groundings, almost one-third (29.8%) of the five-year average number of all commercial mariner deaths and injuries, and almost one-third (30.2%) of the five-year average number of all chemical discharge incidents and oil spills greater than 100 gallons into navigable waters for calendar years 2010 through 2014."

Based on Coast Guard Marine Information for Safety and Law Enforcement (MISLE), each year there is an average of 18 fatalities, 35 injuries, $66 million in property damages, and 446,000 gallons of oil spilled by incidents involving towing vessels. Additional damages occur after towing vessel causalities in the form of delays from lock and waterway closures. The primary objective of Sub M is to reduce fatalities, injuries, property damage and oil spills. [Ref -1]

The top ten common towing vessel deficiencies compiled by Towing Vessel National Center of Expertise (TVNCOE), Puducah, KY, are:

(1) Improper location of remote fuel shutoff valve
(2) Improper installation of general alarms
(3) Improper navigation lights
(4) Inadequate drug and alcohol testing and improper record of test.
(5) Fire detection control panels aren't fully functioning
(6) Improper log book entries or failure to record information
(7) Improper or non-functioning vessel compasses
(8) Malfunctioning marine sanitation devices.
(9) Improper official number markings on vessels
(10) Improper official number markings.

PREPARING VESSELS FOR SUBCHAPTER -M

In general, this regulation would apply to towing vessels of length >= 26 feet, as well as any towing vessels pushing, pulling or hauling a barge carrying dangerous or hazardous materials. The process is:
(1) Pull your vessels out of the water and get them inspected.

(2) Begin developing safety management system.
(3) Develop internal audit expertise
(4) Install new required equipment's like automatic external defibrillators
(5) Analyze training gaps
(6) Learn and prepare for new underway procedure

This rule creates a comprehensive safety system that includes company compliance, vessel compliance, vessel standards, and oversight. Subchapter-M compliance flowchart is shown in Figure -1

There are two inspection options:

(1) Coast Guard Option
(2) Towing Safety Management System (TSMS)

If the first option (Coast Guard) is selected, the vessel shall meet the safety management system requirements by using an operations manual, towing vessel record (TVR), or logbook and these should be approved by the Coast Guard.

TSMS will give flexibility to operators to tailor their safety management system to their own needs, while still ensuring an overall level of safety acceptable to the Coast Guard.

Regardless of the option chosen, the Coast Guard is responsible for issuing a towing vessel Certificate of Inspection (COI), and may board a vessel at any time to verify the compliance and take appropriate action. A COI is mandatory for operating the vessel. The COI phase-in period depends on the number of vessels under and operator. Operators / owners with only one towing vessel is required to have obtained COI by July 20,2020.

CERTIFICATE OF INSPECTION (COI)

Certificate of Inspection (COI), which is issued by Coast Guard, furnishes information about the vessel particulars, class, routes that it may travel, personnel onboard, lifesaving appliances (LSA) and firefighting appliances onboard, vessel power and other information which are deemed necessary for the safe operation of the vessel as deemed necessary by the Officer in Charge, Marine Inspection (OCMI). The original COI must be framed and posted in a conspicuous place or if this is impracticable, the COI must be kept onboard in a weathertight container and must be readily available. COI is valid for five years.

Route of operation of the vessel will be recorded in the COI and operation on a route of lesser severity is permitted, unless the route is expressly prohibited on the COI. If the towing vessel has to go to another port for repairs , permission may be required.

COAST GUARD OPTION

Annual inspection is required by the Coast Guard for this option. Annual inspection is to be done by the Coast Guard personnel and do not require participation of a surveyor from a TPO. Owners or managing operators of the towing vessel must contact the cognizant OCMI to schedule an inspection and for recordkeeping management system, vessel operations manual, towing vessel record or log book should be used.

Dry dock and Internal structural surveys shall be performed at regular intervals depending on the exposure to salt water irrespective of the option chosen to obtain a COI. Proper documentation is required for these surveys as described in Sub M to the satisfaction of the OCMI.

A drydock and internal structure examination performed for a towing vessel to maintain class by a recognized classification society, would count as an examination. Underwater Survey In Lieu of Dry-docking (UWILD) is acceptable depending on various criteria .

TSMS OPTION

TSMS is a documented and audited safety management system for meeting the owners established goals and also for ensuring regulatory (Sub M) compliance. Operators choosing the TSMS option must implement an external or internal survey program for vessel compliance. External surveys should be done by an approved Third Party Organization (TPO). Internal survey program can be done using internal resources or contracted surveyors. The internal survey program would be conducted with the oversight of a TPO. United States Coast Guard (USCG) approves TPOs.

The TSMS option requires a report detailing each internal survey of a towing vessel , including vessel information, name of surveyors, descriptive list of all non-conformities etc.

An external survey is to be conducted annually by a surveyor from TPO. Also, if a company has a TSMS and chooses an internal survey program, the surveys can be conducted by a qualified member of the company and would not require a TPO. The Coast Guard OCMI may request for an audit or survey in his/her presence if required and he/she can remove the owner or operator from using the TSMS option in case of deficient surveys. The scope of responsibility in selecting the TSMS option includes verification of:

(a) Policies and procedures, audits, and recordkeeping
(b) Compliance of hull structure and appliances
(c) Machinery, fuel and piping
(d) Steering systems
(e) Pressure vessels and boilers
(f) Electrical
(g) Lifesaving
(h) Fire protection
(i) Towing gear which includes deck machinery, hawsers, wires, bridles etc.
(j) Navigation systems and equipment
(k) Sanitary examination
(l) Unsafe practices
(m) Personnel

TSMS must:

(a) Demonstrate management responsibility
(b) Document Management Procedures
(c) Ensure document and data control
(d) Provide a process and criteria for selection of third parties

(e) Establish a system of recordkeeping
(f) Identify and meet training needs
(g) Ensure adequate resources.

This document should include policies and procedures to provide direction for the safe operation of towing vessels and protection of the environment which also includes defined levels of authority, procedures to prepare for and respond to emergency situations, shore side and vessel personnel, procedures for external and internal audits, designated person, master authority etc. TSMS internal audit should be done annually.

A safety management system (SMS) which is fully compliant with the International Safety Management (ISM) Code requirements will be deemed in compliance with the TSMS related requirements. The Safety Management System is inherently scalable. TSMS Compliance certificate shall be issued by the TPO at least 6 months before COI. Sub-M requirements for conducting internal and external audits shall be adhered to. The towing company is responsible for maintaining vessels in compliance with regulations and TSMS.

WHAT IS TPO ?

Third Party Organization (TPO) conducts audits and surveys for towing vessels as described in the TSMS option. Recognized Classification Societies can function as TPOs with the approval of Coast Guard. Other organizations, such as design and consulting engineering firms, can also seek approval to function as a TPO. The TPO issues TSMS certificates after verifying compliance.

SUBCHAPTER-M REQUIREMENTS

The following sections / parts are included in this subchapter:

(1) Part 136: Certification
(2) Part 137: Vessel Compliance
(3) Part 138: TSMS
(4) Part 139: TPO
(5) Part 140: Operations
(6) Part 141: Lifesaving
(7) Part 142: Fire Protection
(8) Part 143: Machinery and Electrical Systems and Equipment
(9) Part 144: Construction and Arrangement

The timeline for Subchapter M compliance is below

TimeLine	Activity
June 20, 2016	•Subchapter M Published
July 20, 2016	•Subchapter M regulations effective, but most of Parts 140-144 delayed for 2 years or after vessel is issued a COI. •All vessels must comply with watertight and weathertight requirements of 46 CFR 144.320
July 20, 2018	•Parts 140-144 implemented, with exception of Health and Safety Plan, Pilothouse Alerter System, Towing Machinery requirements.
July 20, 2019	•25% of a company's fleet must have a valid COI except for companies with only 1 existing towing vessel. •Towing Vessel owner or managing operator must implement a Health & Safety Plan. 46 CFR 140.500
July 20, 2020	•50% of a company's fleet must have a valid COI. •Companies with only 1 existing towing vessel must obtain a COI
July 19, 2021	•75% of a company's fleet must have a valid COI.
July 19, 2022	•100% of a company's fleet must have a valid COI.
Year 7 to Year 11	•5 years after issuance of initial COI, existing vessels comply with 46 CFR 143.450 (Pilothouse Alerter System) and 46 CFR 143.460 (Towing Machinery requirements).
Year 11	•Subchapter M Implementation is Complete.

RESPONSIBILITY OF MASTER AND CREW

Crew safety requirements is topmost in the list. Vessel operational safety is the responsibility of the master and crew which includes the vessel stability, examination and tests of steering gear, navigation lights, communication systems etc. and examination of navigational safety equipment's such as radar, gyrocompass etc.

Navigation and Communication Equipment. This includes carrying charts and nautical publications, presence of radar and communication equipments, proper navigation lights etc.

Towing Safety. – Ensuring the proper functioning of towing gear, awareness of the environmental conditions expected during the intended service, emergency procedures etc.

Vessel Records includes logbooks and Towing Vessel Records (TVR)

LIFE-SAVING EQUIPEMNT REQUIREMENTS

Existing towing vessel much comply with the below requirements no later than July 20, 2018 or the date the vessel obtains a COI, whichever date is earlier.

a) Survival craft requirements of towing vessels.
    - Carriage based on:

        - Area of Operation
        - Warm / Cold Water Operation
    - Stowage, marking, equipment requirements
    - Certain vessels may use skiff as a substitute

b) Lifejackets
c) Immersion suits
d) Lifebuoys.
e) Visual distress signals
f) Emergency position indicating radio beacon (EPIRB)
g) Line Throwing Appliance

Coast Guard will approve novel lifesaving appliance or arrangement if it demonstrates the compliance with the requirements.

FIRE PROTECTION

Fire hazards are to be minimized by ensuring no accumulation of combustible and flammable liquids at bilges and void spaces and also storage area areas as far as achievable. This includes fire extinguishing and detection equipment requirements

The requirements in this section include:

(a) Storage of flammable or combustible products should be done safely at storage rooms/ designated places and shall meet the standards specified.
(b) Firefighter's outfit
(c) Fire axe (at least one). This is mandatory for all vessels
(d) Portable and semi-portable fire extinguishers
(e) Training of crews to respond to fires. (Monthly fire-fighting drills and instruction and new crew member safety orientation)
(f) Requirements for fire pumps, fire mains, and fire hoses.
(g) Fire-detection system requirements. Engine room fire detection system has to be present.
(h) Smoke detection in lounges and berthing spaces.

Existing towing vessel much comply with the below requirements no later than July 20, 2018 or the date the vessel obtains a COI, whichever date is earlier

MACHINERY AND ELECTRICAL SYSTEMS AND EQUIPMENT

Existing towing vessel much comply with the below requirements no later than July 20, 2018 or the date the vessel obtains a COI, whichever date is earlier.Existing vessels should meet the pilothouse alerter and towing machinery requirements no later than 5 years after the issuance of the first COI for the vessel.

Requirements include:

(a) Machinery space fire prevention requirements
(b) Control and Monitoring requirements for thrust and rudder angle.
(c) Alarms and monitoring which includes alarm for main engines, auxiliary engines and generators and general alarms.

(d) Communication requirements like two-way voice communication and calling between the engine room and pilothouse.
(e) System isolation and markings. Isolation devices and markings shall be provided for electrical equipment, piping for flammable or combustible liquid etc.
(f) Fuel shutoff requirements

Portable fuel systems shall be avoided unless deemed necessary for outboard engines or portable bilge or fire pumps. Additional requirements include fuel pipe and vent pipe requirements. Electrical systems and equipment requirements shall be met which includes requirements for shipboard lighting, navigation lights, pilothouse alerter system to detect when its master or mate(pilot) becomes incapacitated,

REQUIREMENTS FOR NEW TOWING VESSELS

There are additional requirements for Machinery and Electrical systems and Equipment requirements for new towing vessels. New towing vessels means:

(a) Had its keel laid or was at similar stage of construction on or after July 20, 2017 or
(b) Underwent a major conversion that was initiated on or after July 20,2017

Pumps, pipes, valves, pressure vessels, steering systems etc. must meet the requirements of ABS Rules for Building and Classing Steel Vessels under 90 meters. Requirements apply to electrical distribution panels and switchboards, overcurrent protection devices, electrical grounding, and ground detection etc.

Regarding propulsion, there must be an alternate means to control the propulsion and steering system which much be independent of the primary control and shall be located at or near the propulsion and be readily accessible and suitable for prolonged operation.

Design verification of new vessels must be done by a Registered Professional Engineer (P.E.), Authorized classification society, or by Coast Guard.

CONSTRUCTION AND ARRANGEMENT

Existing vessels must have watertight or weathertight integrity starting July 20, 2016. Vessels with bulwarks must have sufficient freeing ports or scuppers to allow water to run off the deck quickly without adversely affecting the stability of the vessel. Deckhouse and hull penetrations shall have closure devices before it's voyage.

The below requirements shall be met no later than July 20, 2018 or the date the vessel obtains a COI, whichever date is earlier. A vessel which is currently classed by a recognized classification society meets the Structure, Stability and Watertight integrity requirements.

Structural standards for a new vessel shall comply with the standards established by American Bureau of Shipping (ABS) or another recognized classification society provided they are accepted by the Coast Guard as providing an equivalent level of safety.

<u>Stability and Watertight Integrity</u>
Existing vessels shall have a stability document or should provide proof for vessels operational history showing

adequate stability and handling characteristics.

Fire Protection
Machinery and fuel tank spaces must be separated from accommodation spaces by bulkheads. Engine exhaust, galley exhaust ducts shall be insulated.

Emergency Escape
Each space where crew may be quartered or normally employed must have at least two means of widely separated escape.

Ventilation
Accommodation space on a vessel must be properly ventilated. Mechanical ventilation system may be required for vessels of more than 65 feet in length

Crew Spaces:
Deck above a crew accommodation space must be located above the deepest load waterline. Overnight accommodation must be provided for crew members if it is operated for more than 12 hours in a 24 hour period unless the crew is put ashore and vessel is provided with a new crew.

Rails and Guards
Strom rails, Guards in dangerous places, protection against hot piping shall be provided

Visibility
Adequate visibility shall be ensured. Mechanical means such as windshield wipers, defoggers etc. may be provided.

REFERENCES

[1] Subchapter – M Rule: Federal Register/Vol. 81, No 118/Monday, June 20, 2016/Rules and Regulations

[2] ABS Rules for Building and Classing Steel Vessels Under 90 meters in Length

[3] US Coast Guard, Towing Vessel National Center of Expertise (TVNCOE) Webinar about Subchapter-M

[4] US Coast Guard, Towing Vessel National Center of Expertise (TVNCOE) website. 'https://www.uscg.mil/hq/cg5/TVNCOE/'

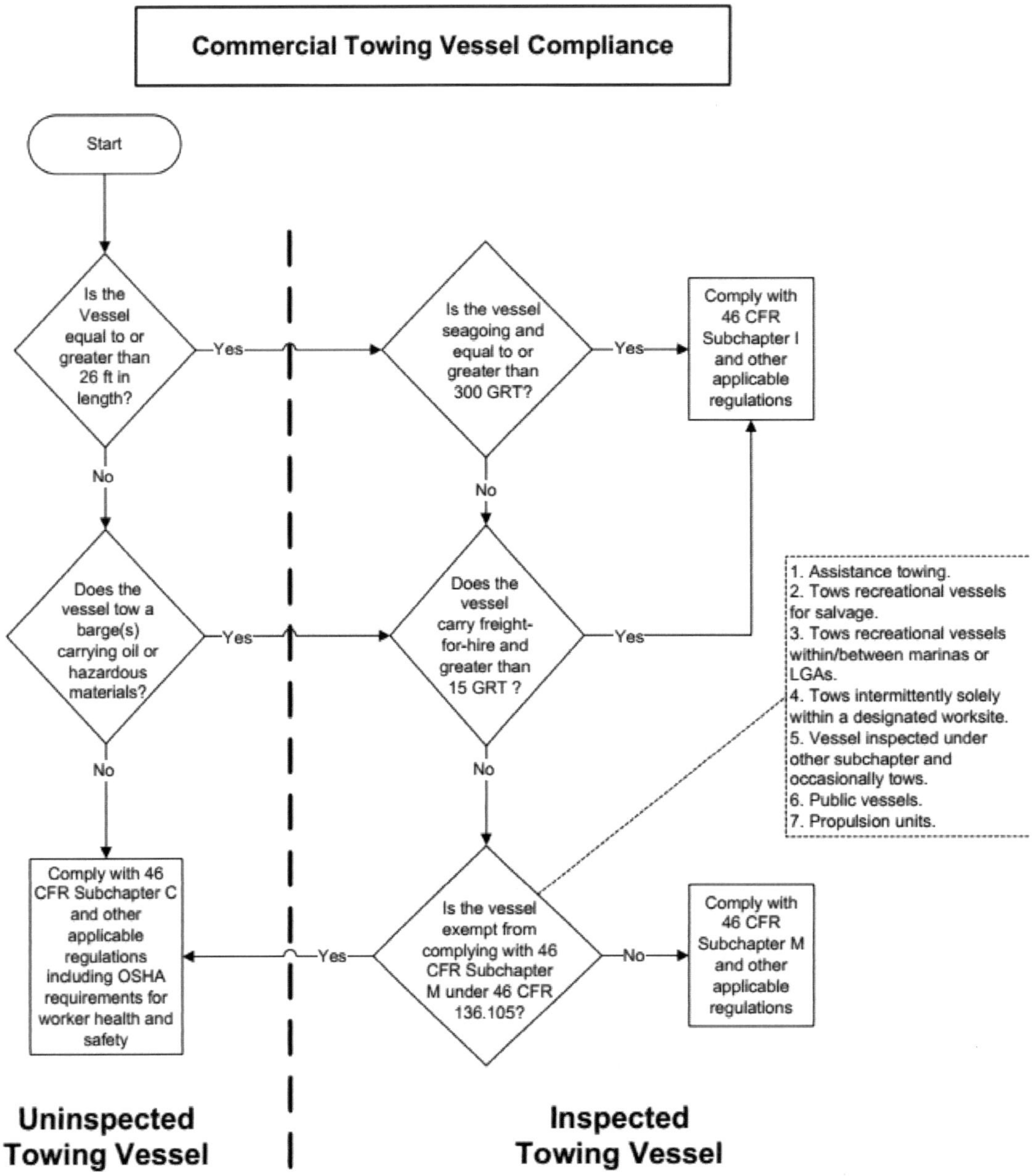

Figure -1: Subchapter – M Compliance Flow Chart [Ref -4]

THE INTERNATIONAL COMMUNITY
FOR MARITIME AND OCEAN PROFESSIONALS

APPLICATION OF A PANEL METHOD TO THE PREDICTION OF PERFORMANCE OF DUCTED PROPELLERS AT DESIGN AND LOW ADVANCE RATIOS

SEUNGNAM KIM
MASTER'S STUDENT, OCEAN ENGINEERING GROUP, THE UNIVERSITY OF TEXAS AT AUSTIN

SPYROS A. KINNAS
PROFESSOR, OCEAN ENGINEERING GROUP, THE UNIVERSITY OF TEXAS AT AUSTIN

Proceedings of the 22nd Offshore Symposium, February 2017, Houston, Texas
Texas Section of the Society of Naval Architects and Marine Engineers

ABSTRACT

Ducted propellers have been used as an effective method of propulsion due to its higher efficiency especially when high thrust is required at low speeds. In particular, the ducted propellers are highly useful in station keeping of floating offshore structures. For example, both the floating production, storage and offloading unit (FPSO) and liquid natural gas (LNG) hulls keep their positions with the help of the dynamic positioning by ducted propellers with a rotating pod. Predicting the performance of the ducted propellers in low advance ratios is very important since most of the floating structures are designed to keep their position at a fixed location.

In the 21st Offshore Symposium of Texas SNAME, Ocean Engineering Group (OEG) of UT Austin presented a full wake alignment scheme in which the trailing vortex wake sheets of the blades are aligned with the local flow velocity considering the effect of duct and duct wake. The repaneling process applied to the existing panel method was also investigated to show the improved accuracy of the predicted forces on the blade and duct.

In this paper, improvements in the predicted pressure distribution on the blade by using a panel method are described. A square-tip propeller inside a sharp-trailing edge duct is considered. The two repaneling processes on the duct and duct wake are also applied to improve the predicted forces and pressure distributions on the blade and duct. In order to improve the predicted results at lower advance ratios special care must be given to the arrangment of control points on the duct, as well to avoid penetration of the blade wake into the duct surface.

The results from the panel method with the two repaneling processes are compared with those from a three-dimensional full-blown RANS simulation as well as the experimental data. The predicted pressure distributions on the blades show very good agreements with the results from the full-blown RANS simulation. Also, the forces on the blade and duct are in very good agreement with the experimental data, at design and at lower advance ratios.

Keywords: Ducted Propeller, Thruster, Panel Method

INTRODUCTION

As viable alternatives of propulsion, ducted propellers can produce higher efficiency than open propellers especially at the high loading conditions, or low advance ratios. In the offshore industry, in particular, the ducted propellers are highly important to keep the positions of the floating offshore facilities such as the floating production, storage and offloading unit (FPSO) and liquid natural gas (LNG) hulls. With the help of the dynamic positioning system (DPS) and the rotating pod, it became more practical and comfortable for those offshore structures to keep their positions at specific locations. Considering that those mechanical structures use the ducted propeller as their primary propulsion sources to be stationary at that specific locations, predicting the performance of ducted propellers in the low advance ratios is essential. The sturdy structure of the ducted propeller from the harsh circumstance of the ocean makes it even suitable as a primary propulsion system of the offshore facilities.

As introduced in the previous Offshore Symposium of Texas SNAME by Ocean Engineering Group (OEG) at UT Austin, Tian et al. (2012) already showed that the wake alignment model could improve the predicted propeller performance by introducing a pseudo-unsteady wake alignment scheme (full wake alignment, or FWA). Kinnas et al. (2015) extended the scheme for the ducted propellers by including the effects of the duct and duct wake for the evaluation of the potentials on the discretized geometries. The same effects are also considered when the FWA scheme evaluates the induced velocities on the blade wake with the careful attention with the wake panels, which are very close to the duct inner surface. Considering that the outer edge of the blade wakes gets extremely closer to both the duct inner side and duct trailing wake, the repaneling process, which aligns the panels on the duct and duct wake according to the blade wake, has been introduced to avoid singular behaviors coming from the panel mismatching by Fan (2015). He introduced a variation of the repaneling option 1, as defined in the following chapter, on the duct and duct wake to improve the accuracy of the predicted force performances of the ducted propeller.

Kinnas et al. (2016) presented the further improved wake alignment scheme with an emphasis on the duct and duct wake paneling. In that work, proper paneling along the duct inner surface enhanced not only the predicted circulation distribution toward the blade tip but also the pressure distribution on the blade sections ranging from near the hub to the blade tip. Additionally, the two different options for the repaneling processes on the duct and duct wake based on the blade wake were developed. Throughout this paper, the same definitions defined in Kinnas et al. (2016) will also be used to be consistent with and avoid any confusions from the previous work. Brief descriptions for each repaneling option are following:

(a) Option 1: The duct and duct wake panels are kept the same as those in the start of the FWA scheme, based on the helical blade wake. Once the duct panels are formed at the initial iteration, those are not modified throughout the rest of the iterations. The repaneling scheme introduced in Fan (2015) is a variation of this option.

(b) Option 2: The duct and duct wake panels are adapted to the blade wake through all the iterations in FWA scheme. Therefore, duct and duct wake panels are keep changing with the blade wake.

Although both options produced the results which are very close to each other, especially at the higher advance ratios, the option 2 was not stable, and the predicted results were not accurate either at the lower advance ratios. Although the other choice, option 1, produced the accurate final results over the most advance ratios ranging from Js=0.3 to Js=0.7, when it comes to the lower advance rations, the option 1 also was not stable enough in terms of the convergence history. To compensate this drawback of the option 1, it was crucial to repanel the duct and duct wake repeatedly, as option 2 to avoid any singular behaviors coming from the close distance between the control points on the duct inner surface and the outer edge of the blade wake. As will be seen shortly in the figures of the convergence history of the predicted force and torque, whether the duct panels are repaneled repeatedly (option 2) or not (option 1) will significantly affect the final results, especially at the lower advance ratios.

In the present work, many efforts have been given to improve the option 2 so that it can produce not only the stable but also the accurate results at the lower advance ratios. To achieve this, the control points on the duct panels need to be carefully located such that the panel method can avoid very high peaks of the induced velocity on the wake panels near the duct inner surface. Otherwise, very close distance between the control points on the duct and the outer edge of the wake panels causes the singular behaviors resulting in the distortion of the blade wake followed by the crash of the duct and duct wake panels accordingly. This paper will address the reason for the wrong positions of the control points along with the treatments to resolve this singular problem in detail. To produce the accurate results, the source disk will be introduced to replace the transitional wake with the distributed source potentials. Finally, the predicted propeller performances not only at the design advance ratio but also over the most advance ratios down to *Js*=0.20 will be investigated. The experimental data and the full blown RANS simulations are adopted for the sake of the correlation with the current panel method. To this end, the two different repaneling options are applied to the ducted propeller with the square tip blade for the validation of the repaneling options over the most advance ratios.

METHODOLOGY

1. Full Wake Alignment Scheme with Repaneling Options

A detailed description of the FWA scheme with its formulations are well presented by Tian and Kinnas (2012), so this chapter will only focus on how the repaneling processes on the duct panels are applied within the iterative steps in the FWA scheme. For the first iteration, the panel method generates a helical wake based on the advance ratio, and then the panels on the duct and duct wake are adapted according to the helical wake. After this initial adaptation, the duct and duct wake panels do not change during the rest of the iterations in the case of the option 1. With these initial geometries, we solve the potential equations to evaluate the potentials over the whole discretized geometries for the first outer iteration. Then, the panel method starts the inner iterations to determine the blade wakes, which are going to be aligned based on both the inflow and the induced velocities on the wake panels. During these inner iterations, the potentials on the geometries (blade, hub, and duct) do not change. However, the total velocities, i.e. inflow plus the induced velocities evaluated at the four corners of the wake panels keep changing since the wake panels are repeatedly modified based on the potentials and the total velocities until the inner iteration reaches its maximum number or the convergence criteria are satisfied in terms of the wake geometry. Due to the movements of the wake panels during the inner iterations, the induced velocities on the blade wake by itself are also reevaluated. Once the inner iteration terminates with the newly aligned wake for the next outer iteration, repaneling process in the option 2 enters at this point, i.e. just before the beginning of the next outer iteration. The panels on the duct and duct wake are adapted again according to the new blade wake from the previous outer iteration. Then, the panel method restarts to solve the potential equations to evaluate the potential over the discretized geometries. The outer iterations will be repeated until the predicted force performance converges or the maximum iterations numbers are reached. Detailed descriptions of the iterative steps are presented in the Figure 1.1.

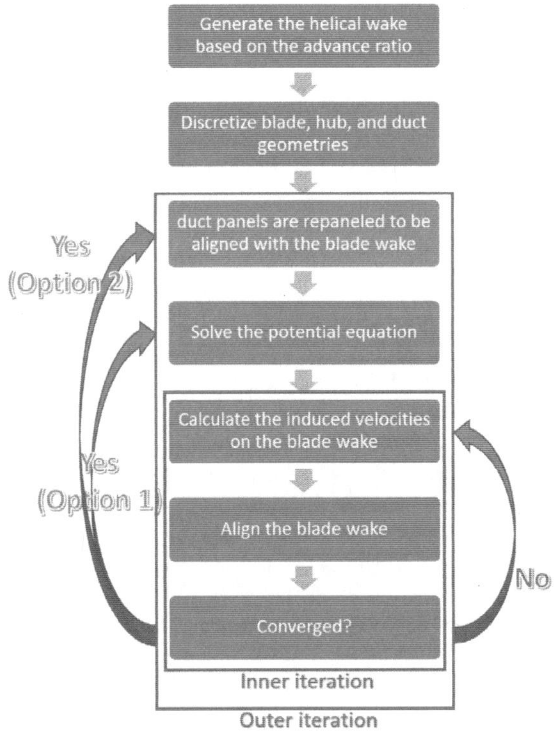

Figure 1.1 Flowchart of the iteration algorithm in the FWA scheme with the two repaneling options

2. Panel Penetration of the Blade Wake on the Duct and Duct Wake

Just before repaneling the panels on the duct and duct wake, an additional treatment is required on the blade wake in the case when the outer edge of the blade panels penetrates on the duct inner side or the duct wake. In the present panel method, we regard the duct wake as the cylindrical geometry as shown in the Figure 2.1. Considering that the wake from the trailing edge of the duct is the material surface, which has to be aligned with the local flow velocity as in the case of the blade wake, the uniform shape of the duct wake is somewhat artificial, even though it does not significantly affect the final results. Due to these panels, which are uniformly distributed along the stream-wise direction as well as the radial direction, the blade wake needs to be post processed such that the blade wake is cut off by the duct inner surface or by the cylindrical duct wake to prevent the penetrations between the blade/duct wakes, as in the Figure 2.2 and 2.3.

This penetration is due to the fact that, for more physical representation of the wake geometry, the radial component of the local velocity of the blade wake is not constrained such that the wake is located under the duct inner surface in the radial direction. Although the stiffened form of the duct wake does not have much effects on the predicted results due to its far distance from the blade. The blade wake still needs to be post processed to avoid any singularities coming from the penetrations between the panels. To this end, whenever the edge of the blade wake penetrates the duct inner side or even the suction side of the duct, the wake will be drawn down and cut off by the duct inner surface through cubic spline interpolation along the duct inner surface. After that, the locations of the points on the wake panels are reallocated based on the cubic spline interpolation along the radial direction of the wake. The same procedures are also applied when the blade wake penetrates on the duct wake, while the blade wake is cut off by the cylindrical duct wake in this case.

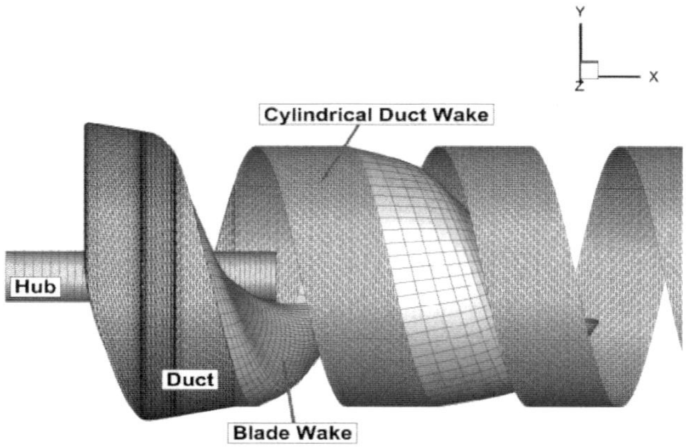

Figure 2.1 Ducted propeller with the cylindrical geometry of the duct wake. Only half and a quarter of the whole body are presented for each of the duct and the wake geometry.

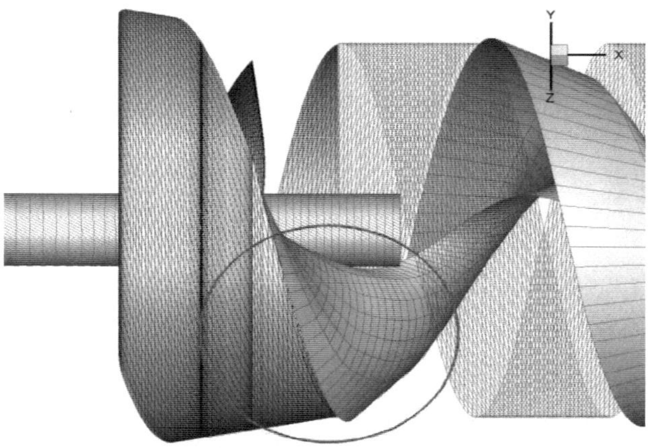

Figure 2.2 Penetration of the blade wake on the duct and duct wake.

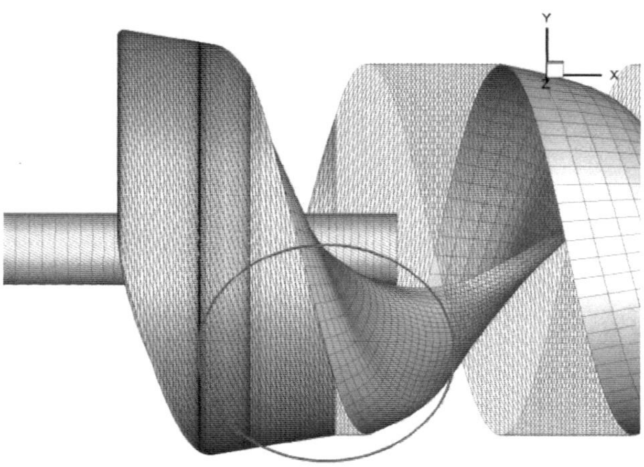

Figure 2.3 The paneling of the blade wake after resolving the penetration problem. Wake panels are cut off by the duct inner surface and the duct wake

3. Treatment of the Control Points on the Duct Panels

To predict the reasonable results using the repaneling option 2, especially at the lower advance ratios, careful attention needs to be given to the control points on the non-planar panels. The locations of the control points on the planar panels, which have its four corners on the same plane, are located at the same plane as corner points (Figure 3.1). However, in the case of the non-planar panels, as can be seen from the Figure 3.2 below, control points are no longer located on the same plane.

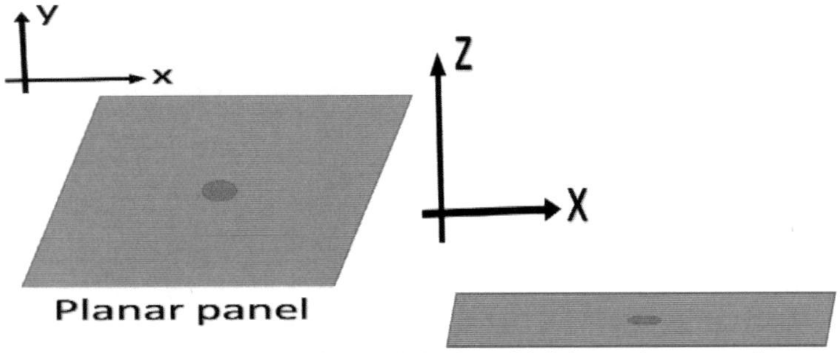

Figure 3.1 Planar panel and the its control point (red circle) which is on the same plane as the panel corners

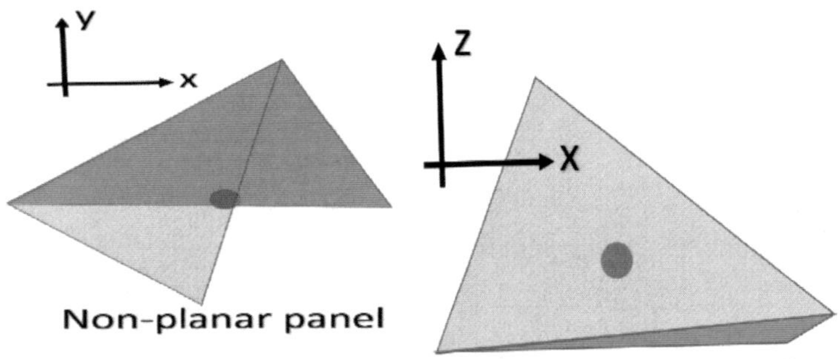

Figure 3.2 Non-planar panel and its control point (red circle) which is not on the same plane as the panel corners

In the case of the ducted propeller, if the locations of the control points on the duct panels are wrongly evaluated due to the non-planar panels, the control points can get closer to the wake panels in the radial direction. Then, it causes the abnormal peaks of the induced velocity on the wake panels near the duct inner surface. The Figures from 3.3 through 3.5 show the ducted propeller with its wake at several advance ratios (Js=0.10, 0.30, and 0.50). Only half and a quarter of the whole geometry for each of the duct and wake are presented in the figures. As it can be seen from the figures, in the case of the lower advance ratios, the blade wakes do not advance farther to the downstream from the blade than the cases with the higher advance ratios.

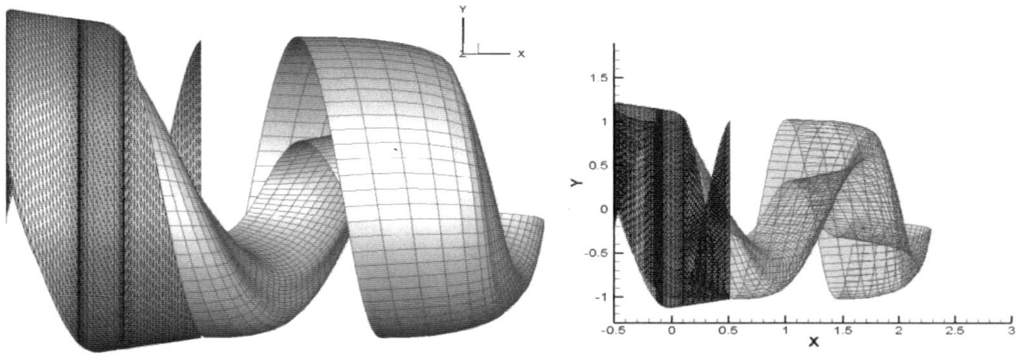

Figure 3.3 Wake and duct geometry in 3D (left) and 2D (right) at *Js*=0.10

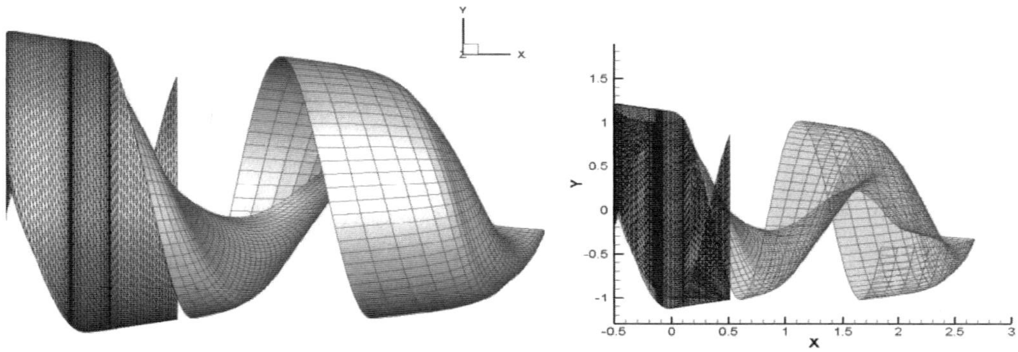

Figure 3.4 Wake and duct geometry in 3D (left) and 2D (right) at *Js*=0.30

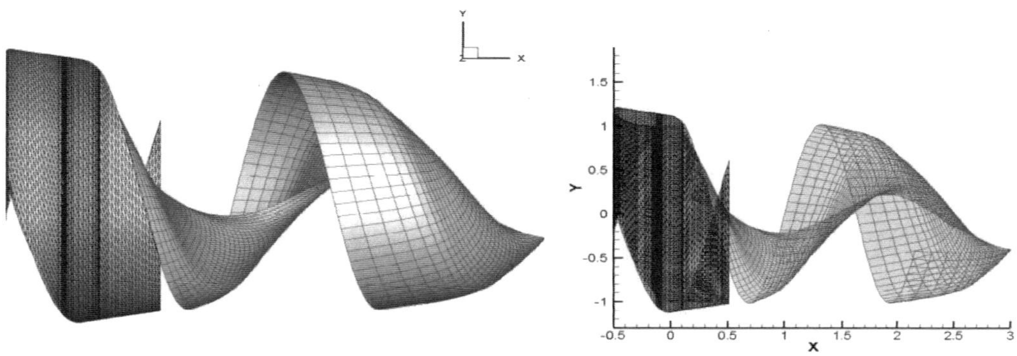

Figure 3.5 Wake and duct geometry in 3D (left) and 2D (right) at *Js*=0.50

Therefore, to be adapted to the wake panels which are compact in the axial direction, but have the fixed angle between the consecutive panel points in the circumferential direction at the lower advance ratio, duct panels become twisted resulting in the non-planar panels, as shown in the Figure 3.6. The twisted duct panels in the circumferential direction produce the wrong locations of the control points near the outer edge of the blade wake in the radial direction. As results, the abnormal velocities are induced on the wake panels followed by the crashes of the wake panels. The Figure 3.6 describes this procedure.

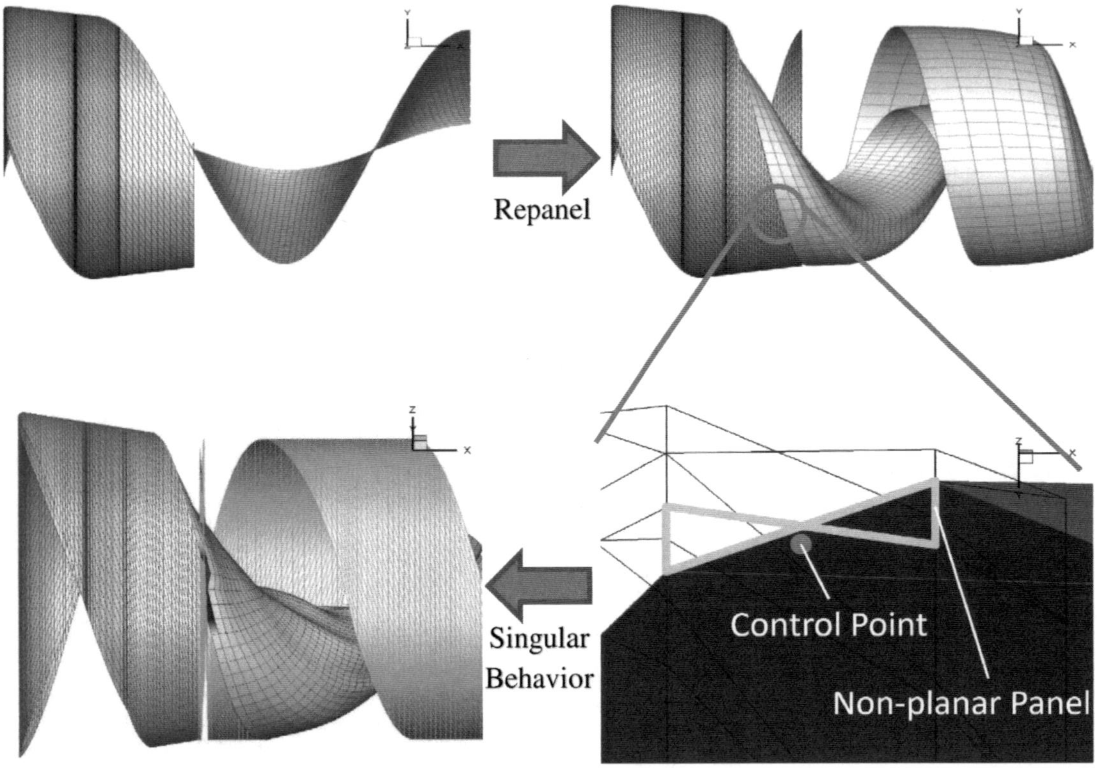

Figure 3.6 Abnormally induced velocity on the wake panels due to the non-planar panels at the lower advance ratios.

To resolve this problem, the non-planar panels are forced to become planar panels just before evaluating the centroid of the panels for the locations of control points. By setting the radial distance of the control points on the non-planar panels same as the ones on the planar panels, the singular behaviors on the blade wake can be avoided, particularly at the lower advance ratios. As results, this improved the convergence of the repaneling option 2, which now produces more stable convergence histories of the propeller performance at the lower advance ratios, compared to the option 1. It will be more addressed with the final results in the following part. For the validation of the panel method with the repaneling options, the experimental data, as well as the results from the full blown RANS simulations, are adopted for the correlations. The comparisons of the predicted propeller performances using the two repaneling options will be investigated for a range of the advance ratio from 0.2 to 0.7.

4. Results

The present panel method is applied to the case of the ducted propeller with the square tip blade and a sharp trailing edge duct. For the square tip blade case, we adopted the experimental data conducted by Bosschers (2009) and performed the full blown RANS simulations using ANSYS Fluent for the validation of the predicted results. Detailed running conditions of the full blown RANS simulations, i.e. boundary conditions and the turbulent model are presented in detail in Fan (2015). This paper will use the same geometry as presented in the 20th Offshore Symposium of Texas SNAME for the sake of the comparison of the improved results. The thrust and torque distributions for a range of *Js* from 0.2 to 0.7 will be investigated with the convergence histories.

4.1 Ducted Propeller with The Square Blade Tip and Sharp Trailing Edge Duct

The geometry of the ducted propeller consists of a 4-bladed propeller with a square blade tip and a sharp trailing edge duct 19Am. The duct 19Am is modified by MARIN from duct 19A which has a blunt trailing edge that the current panel method cannot well resolve due to the recirculation region around the duct trailing edge. The design advance ration of this propeller is around *Js*=0.50. Zero gap is assumed between the duct inner side and the blade tip for this geometry.

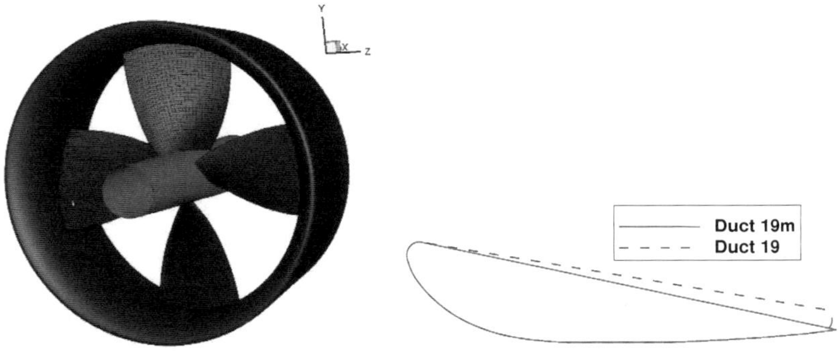

Figure 4.1 Geometry of the ducted propeller with square blade tip, and the section of the sharp trailing edge duct. The section of the original duct is shown in the dashed line.

4.2 Panel Method

For the panel method, 80 × 30 (chordwise × spanwise) and 200 × 80 (chordwise × curcumferential) panels are used to discretize the panels on the blade and duct. Previously, the FWA scheme models blade wakes of one revolution, while the current scheme models 100 panels to resolve the wake panels for both the option 1 and option 2 in the stream-wise direction. As described earlier, in the current panel method, transitional wakes from the blade are forced to be located under the duct wake to avoid the penetrations, which is somewhat violation of the natural behaviors of the wakes as the material surfaces. To avoid the errors coming from the excessively modified wakes, the blade wakes down below the specific axial location in downstream will be ignored in terms of the influence coefficients, as described in the Figure 4.2. Instead, the source disk is assumed at that specific location to replace the ultimate wake for both repaneling options, even though the duct wake is still aligned according to the blade wake in option 2. The blade wake and the duct geometries using the repaneling option 1 and 2 at the advance ratios from *Js*=0.20 to *Js*=0.50 are presented in the Figures 4.3 through 4.6.

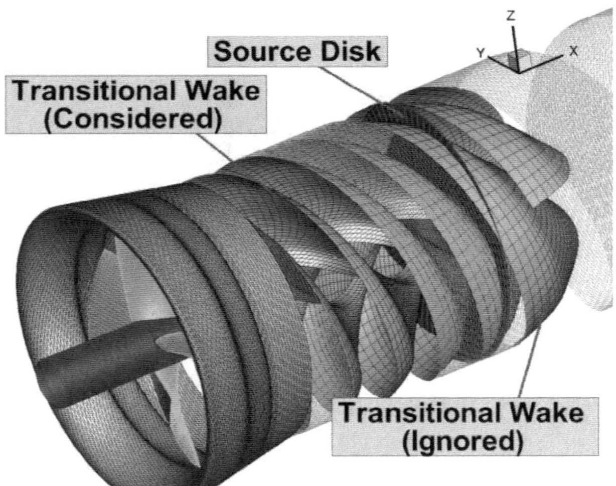

Figure 4.2 Transitional blade wake and the source disk replacing the ultimate wake in downstream

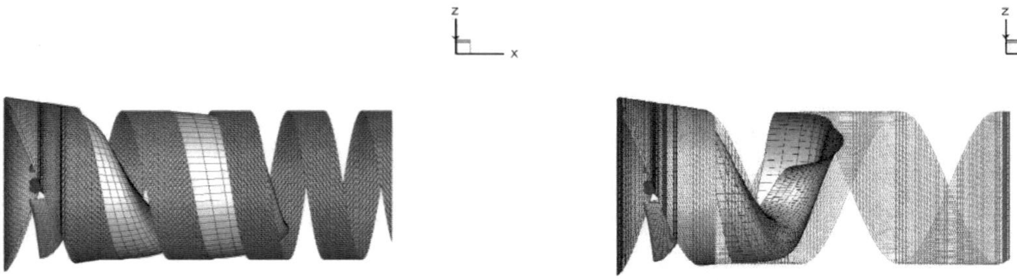

Figure 4.3 Wake and duct geometry using the repaneling option 2 (left) and the option 1 (right) at *Js*=0.20

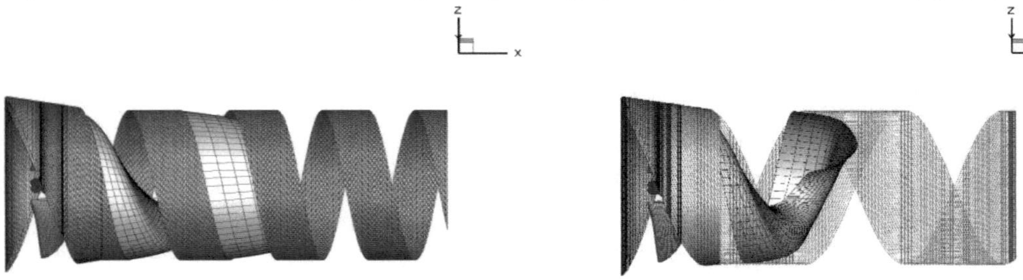

Figure 4.4 Wake and duct geometry using the repaneling option 2 (left) and the option 1 (right) at *Js*=0.30

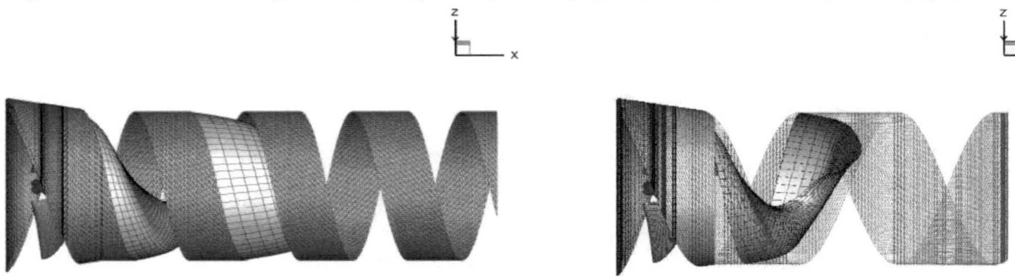

Figure 4.5 Wake and duct geometry using the repaneling option 2 (left) and the option 1 (right) at *Js*=0.40

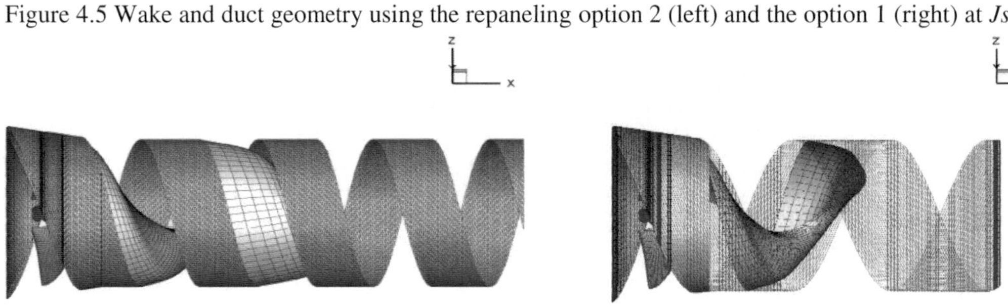

Figure 4.6 Wake and duct geometry using the repaneling option 2 (left) and the option 1 (right) at *Js*=0.50

4.3 Full Blown RANS Simulation

The full blown RANS simulations are performed with ANSYS/Fluent in which periodic interfaces have been applied such that for the present propeller only a quarter of the whole fluid domain is necessary for further simulations. To better resolve the boundary layer structures on the blades and duct, structured meshing model has been adopted. The same gridding model has been used for the trailing wake of the blades and duct in order to reduce the possible artificial diffusivity.

The detailed description of the RANS calculation is presented in Fan (2015), so summary of the calculating parameters will be presented in this part. As listed in the Table 4.1, k-ω SST turbulent model is adopted. QUICK scheme is used for the spatial discretization and SIMPLEC scheme for the pressure correction. Over 6 million polyhedral cells are used to simulate a quarter of the domain with periodic boundary condition. It took over 32 hours on 32 Intel Xeon 2.54 GHz CPUs for the residuals converged to 1.0E-6.

Cell Number	Over 6 million
Reynolds Number	1.0E+6
Turbulent Model	k-ω SST
Pressure Correction Scheme	*SIMPLEC*
Spatial Discretization	*QUICK*
CPU Type	Intel Xeon 2.54 GHz CPU
Total Time for Calculation (32 CPUs)	Over 32 Hours

Table 4.1 Parameter setting in the RANS simulation and the computational time, adjusted from Table 4.1 of Fan (2015)

4.4 Propeller Performance and Convergence History of Results

The overall force performances, i.e. the thrust and torque on the blade and duct are compared with the experimental data over the advance ratios from Js=0.20 to Js=0.70, as shown in the Figure 4.7. The Figure shows that the predicted forces using the FWA scheme with the option 1 predict the thrust and torque which are in very good agreements with the experimental values over the most advance ratios down to Js=0.20, while option 2 somewhat under predicts the blade torque even though not considerably. Thrust on the duct is overestimated at the high advance ratios. This needs further investigation in the near future. Compared to the predicted thrust and torque using the old panel method, PROPCAV 3.0 presented in the 21st Offshore Symposium as in the Figure 4.8, the current method shows much improved correlations with the experimental data especially at the lower advance ratios.

Given that the current FWA scheme is basically based on the iterative method as described in the previous chapter, the convergence histories which show not only the accurate but also the stable results are essential to be considered as a reliable wake alignment scheme. Figure 4.9 and 4.10 show the convergence history of the predicted thrust and torque by using the two repaneling options. As it can be seen from the figures, both options show stable convergence histories not only at high advance ratios but also at low advance ratios, even though only the option 2 keeps stable at very high loading condition, i.e. Js=0.2. This is due to the fact that the wake panels, which do not advance farther to the downstream at the lower advance ratios become compact near to the blade and duct inner side. The unmatched panels between the duct inner side and the outer edge of the blade wake in the option 1, thus, can cause the unstably-induced velocities on the wake panels, while the matched panels in the option 2 produce relatively stable convergence histories even at the very high loading condition. Due to the relatively unstable results from the option 1, we used the averaged values from the last 10 iterations for the values of the thrust and torque plotted in the Figure 4.7.

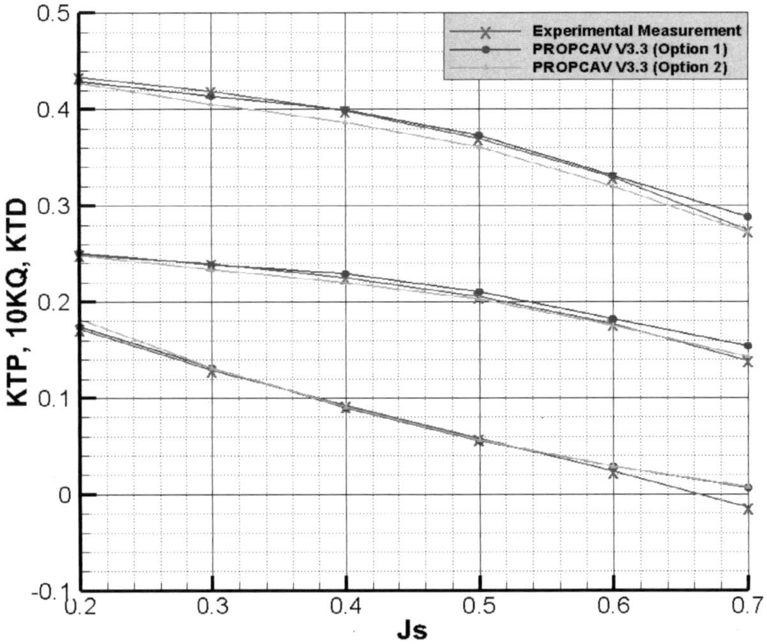

Figure 4.7 Predicted force performance of the KA4-70 ducted propeller from the experiment and the panel method with FWA scheme using different repaneling options

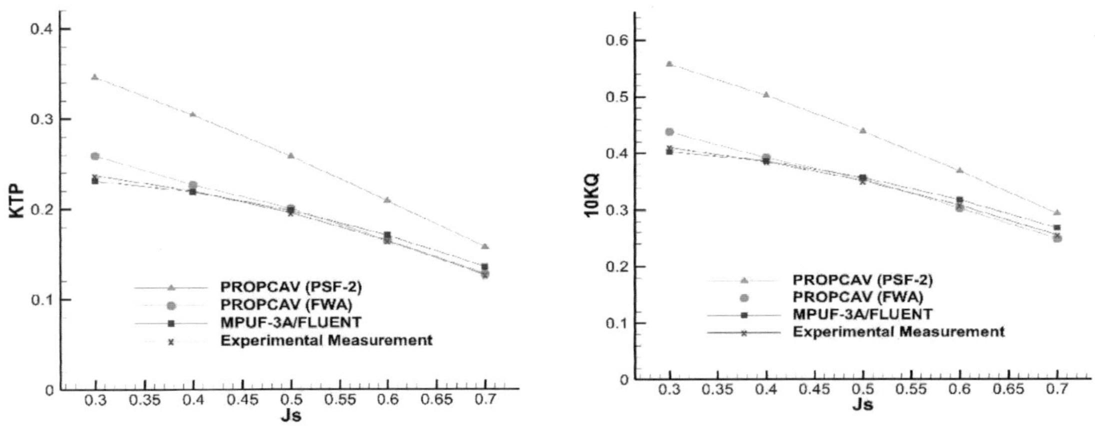

Figure 4.8 Predicted propeller performances of the KA4-70 ducted propeller from the old panel method, PROPCAV 3.0, and other methods developed in Ocean Engineering Group at UT Austin

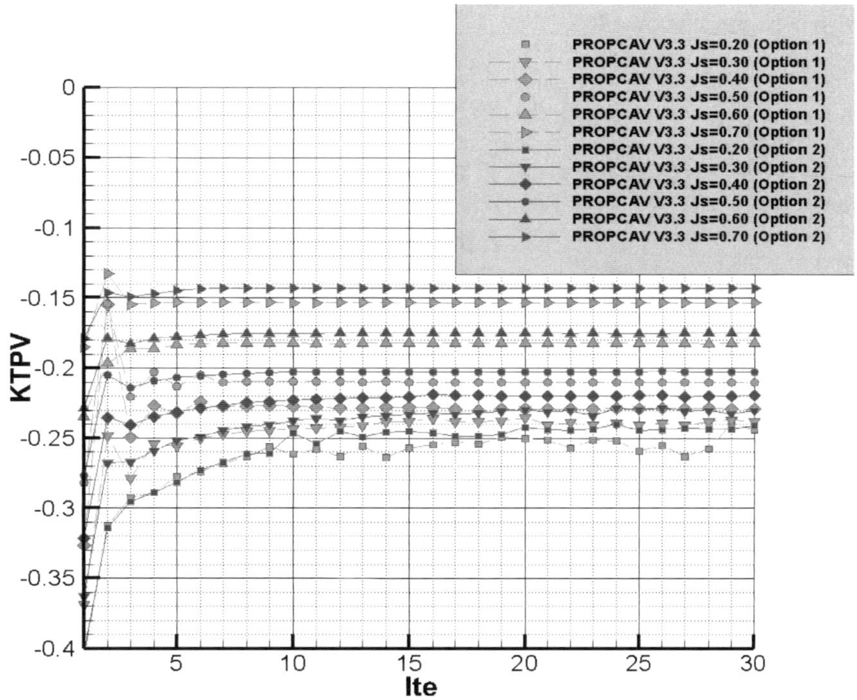

Figure 4.9 Convergence histories of the predicted thrust on the blade using the FWA scheme with the two repaneling options. The horizontal axis corresponds to the number of iterations in the FWA scheme.

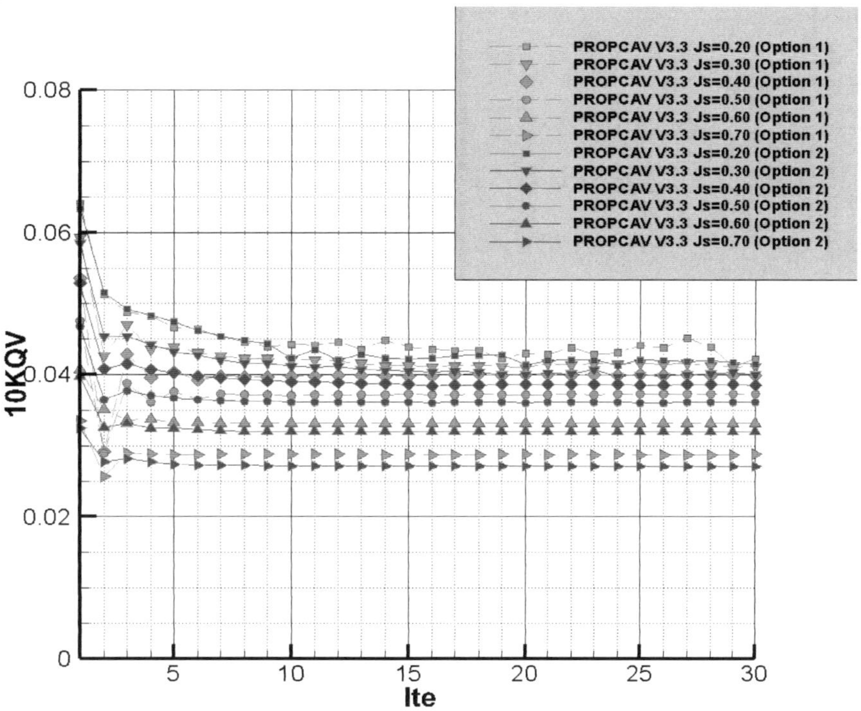

Figure 4.10 Convergence histories of the predicted torque on the blade using the FWA scheme with the two repaneling options. The horizontal axis corresponds to the number of iterations in the FWA scheme

4.5 Pressure Distributions on the Blade

Figure 4.11 through 4.14 present the detailed correlations of the pressure distribution along the several blade sections at the design advance ratio, i.e. Js=0.50. For the correlation of the results from the panel methods and the full blown RANS simulation, four blade sections in the radial direction are adopted: r/R=0.236, r/R=0.541, r/R=0.819, and r/R=0.902. The predicted results in the figures show that the FWA scheme with the repaneling option 1 and 2 are in very good agreements with the full blown RANS simulations over the most blade sections. The pressure predictions might be further improved by coupling the panel method on the blade with an integral boundary layer solver, as described in Kinnas et al (2012).

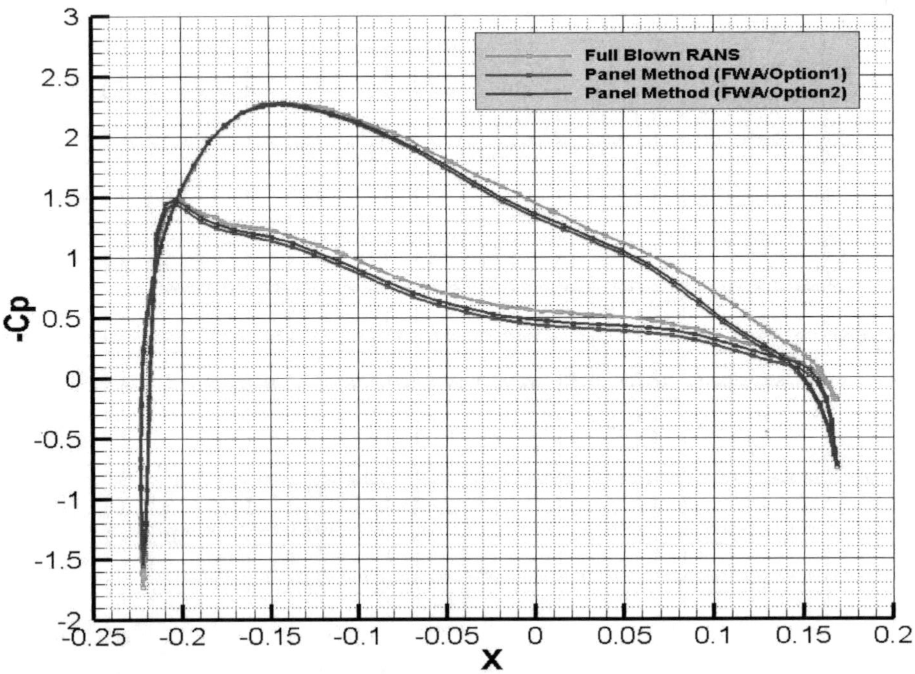

Figure 4.11 Correlation of the pressure distribution at radial section r/R=0.236

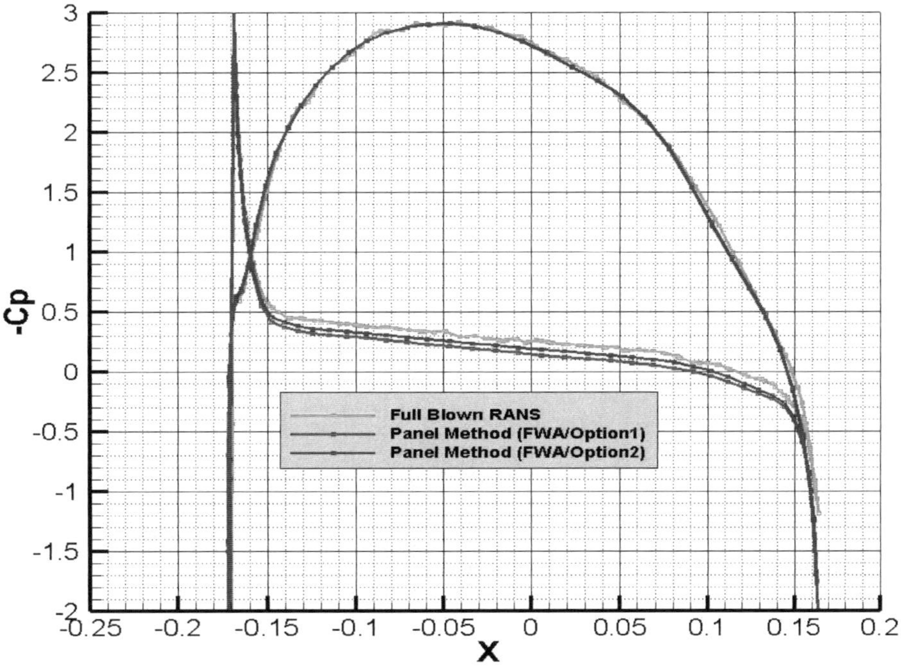

Figure 4.12 Correlation of the pressure distribution at radial section r/R=0.541

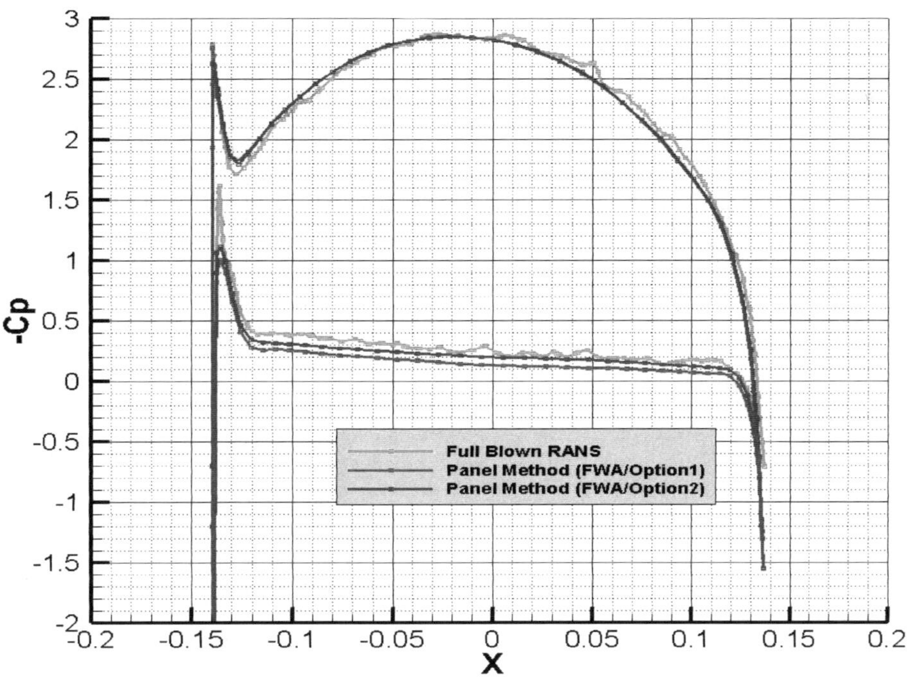

Figure 4.13 Correlation of the pressure distribution at radial section r/R=0.819

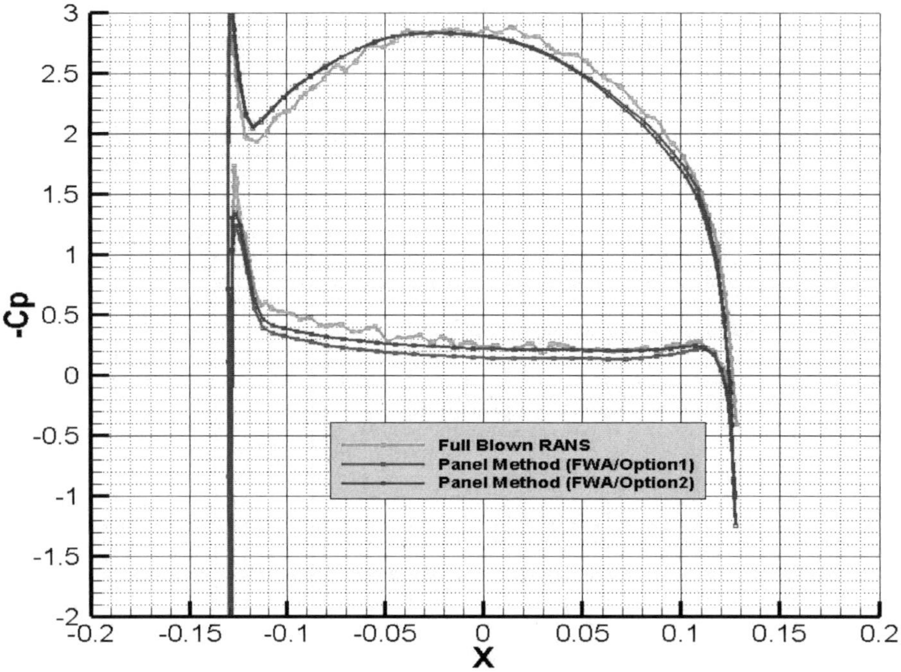

Figure 4.14 Correlation of the pressure distribution at radial section r/R=0.902

5. Conclusions and Future Work

In this work, the improved FWA scheme with the different repaneling options has been applied on a ducted propeller with the square tip blade and a sharp trailing edge duct. This paper has concisely introduced the two repaneling options with their definitions, i.e. option 1 and option 2, as well as the iterative algorithm built in the FWA scheme. By using the two repaneling options, this paper has investigated the predicted results of the current panel method at several advance ratios up to very high loading condition. To this end, this paper has also gone through the treatments on the penetration problems on the duct geometry as well as the issue at the ultimate wake, which was somewhat artificially contracted. For option 2, which repeatedly adapts the duct panels based on the blade wake, the control points on the duct panels should be readjusted so that the control points can have the same radial locations as the control points evaluated on the planar panels. This helps option 2 avoid the singular behavior of the induced velocity on the blade wake, and therefore improves the stability of the method. Especially, the matched panels between the duct inner surface and blade wake in the option 2 produced more stable convergence history down to the very low advance ratio than the case with the unmatched panels in the option 1. Along with the repaneling options, the source disk replacing the ultimate wake beyond the specific axial location downstream improved the final results by substituting the contracted wake panels with the distributed source potentials. As a result, not only the force performances but also the pressure distributions along the blade sections are in very good agreements with experimental data as well as those predicted from full blown RANS.

In the future the model will be applied to other ducted propeller geometries, including those with round tip blade, as well in the case of cavitating flow. Also, a systematic way of evaluating the axial location of the source disk needs to be developed such that the source disk does not affect the final results when applied to general propeller geometries. The best way of avoiding penetrations of the blade wake on the duct wake, which might be due the fixed cylindrical shape of the wake in the current panel method, is to also align the duct wake based on the local fluid velocity, as in the case of the blade wake. Additionally, as a final goal of the panel method, the effects of viscosity on the duct must be examined extensively, as described in the paper of Kinnas et al (2016).

6. Acknowledgments

Support for this research was provided by the U.S. Office of Naval Research (Grant No. N00014-14-1-0303; Dr. Ki-Han Kim) and Phase VII of the "Consortium on Cavitation Performance of High Speed Propulsors" with the following members: Kawasaki Heavy Industry Ltd., Rolls-Royce Marine AB, Rolls-Royce Marine AS, SSPA Sweden AB, Andritz Hydro GmbH, Wärtsilä Netherlands B.V., Wärtsilä Norway AS, Wärtsilä CME Zhenjiang Propeller Co. Ltd., and Steerprop Ltd.

References

Bosschers, J., and R. van der Veeken. "Open water tests for propeller Ka4-70 and duct 19A with a sharp trailing edge." MARIN Report 224457-2-VT (2008).

Fan H. "An Improved Panel Method for the Prediction of Performance of Ducted Propellers." MS Thesis, Ocean Engineering Group, CAEE Department, UT Austin, August 2015 (also UT-OE Report 15-02)

Tian, Y., and Kinnas, S.A. "A wake model for the prediction of propeller performance at low advance ratios." International Journal of Rotating Machinery (2012).

Kinnas, S.A., Y. Su, W. Du, and S. Kim. "A Viscous/Inviscid Interactive Method Applied to Ducted Propellers with Ducts of Sharp or Blunt Trailing Edge." In 31st Symposium on Naval Hydrodynamics, California, September 2016

Kinnas, S.A., H. Fan, and T. Ye. "A Panel Method Applied to The Prediction of The Performance of Ducted Propellers and Thrusters." Proceedings of the 21st Offshore Symposium, Houston, Texas, Feb 2016

Kinnas, S.A., X. Yu, and T. Ye. "Prediction of Propeller Performance under High Loading Conditions with Viscous/Inviscid Interaction and a New Wake Alignment Model." In 29th Symposium on Naval Hydrodynamics, August 2012